I0055521

Curves and Surfaces with Applications in CAGD

Curves and Surfaces with Applications in CAGD

EDITED BY

Alain Le Méhauté
University of Nantes
Nantes, France

Christophe Rabut
Institut National des Sciences Appliquées
University of Toulouse
Toulouse, France

Larry L. Schumaker
Department of Mathematics
Vanderbilt University
Nashville, Tennessee

(Volume I)

VANDERBILT UNIVERSITY PRESS
Nashville & London

Copyright © 1997 by Vanderbilt University Press
All Rights Reserved
No part of this publication may be reproduced or transmitted in any form or by any
means, electronic or mechanical, including photocopy, recording, or any information
storage and retrieval system, without permission in writing from the publisher.

First Edition 1997
97 98 99 00 01 5 4 3 2 1

Library of Congress Cataloging-in-Publication Data

Curves and surfaces with applications in CAGD / edited by Alain Le Méhauté,
  Christophe Rabut, and Larry L. Schumaker. -- 1st ed.
       p.   cm.
  Papers from the Third International Conference on Curves and Surfaces,
  held June 27–July 3, 1996, in Chamonix–Mont-Blanc, France.
  "Volume I."
  Includes bibliographical references.
  ISBN 0-8265-1293-3 (cloth : alk. paper)
  1. Approximation theory--Congresses. 2. Spline theory--Congresses.
3. Surfaces—Computer simulation—Congresses. 4. Curves—Computer simula-
tion—Congresses. I. Le Méhauté, Alain. II. Rabut, Christophe, 1951– .
III. Schumaker, Larry L., 1939–   . IV. International Conference on Curves and
Surfaces (3rd : 1996 : Chamonix–Mont-Blanc, France)
QA221.C876   1997
511'.4—dc21   97-9342
                                                                  CIP

CONTENTS

Preface ix

Contributors x

Parallelization Strategies for the B-Spline Curve Interpolation Problem
   M. D'Apuzzo and L. Maddalena 1

Connection-Matrix Splines by Divided Differences
   R. H. Bartels and J. C. Beatty 9

Massic-Vector-Based Conics Modelling
   J. P. Bécar and J. C. Fiorot 17

Globally G^1 Free Form Surfaces Using Kirchhoff-Love Plate
Energy Methods
   M. Bercovier, O. Volpin and T. Matskewich 25

A New Approach to Tchebycheffian B-Splines
   D. Bister and H. Prautzsch 35

Flaw Removal on Surfaces
   J. Bousquet and M. Daniel 43

On Some Polynomial Curves Derived from Trigonometric Kernels
   J. Cao, H. H. Gonska and D. P. Kacsó 53

Spline Curves in Polar and Cartesian Coordinates
   G. Casciola and S. Morigi 61

La Tolérance d'usinage chez Citroën dans les années (19)60
   P. de Faget de Casteljau 69

Avoiding Local Minima for Deformable Curves in Image Analysis
   L. D. Cohen 77

Variable Degree Polynomial Splines
   P. Costantini 85

Approximation of Monotone Functions: A Counter Example
   R. A. DeVore, D. Leviatan and I. A. Shevchuk 95

Uniform Point Distribution on a Circle
   J-C. Fiorot and I. Cattiaux-Huillard 103

BR-form of an Entire Rational Curve with a Preassigned Point
   J-C. Fiorot and P. Jeannin 111

Optimal Convexity Preserving Bases
   M. García-Esnaola and J. M. Peña 119

Rational Interpolation on the Unit Circle
Th. Gensane 127

Designing Nonlinear Models for Flexible Curves
S. Girard, B. Chalmond and J. M. Dinten 135

SK–spline Interpolation on the Torus using Number Theoretic Knots
S. Gomes, A. K. Kushpel, J. Levelsey and D. L. Ragozin 143

Shape Preserving Interpolation by G^2 Curves in Three Dimensions
T. N. T. Goodman and B. H. Ong 151

Fairing Bicubic B-Spline Surfaces using Simulated Annealing
S. Hahmann and S. Konz 159

Positivity and Convexity Criteria for Bernstein-Bézier Polynomials
Over Simplices
T. X. He 169

Fitting Uncertain Data with NURBS
W. Heidrich, R. Bartels and G. Labahn 177

Interpolation and Approximation with Developable Surfaces
J. Hoschek and M. Schneider 185

Sectional Curvature–Preserving Interpolation of Contour Lines
B. Jüttler 203

Analysis of Curvature Related Surface Shape Properties
J. Kaasa and G. Westgaard 211

On an Almost–convex–hull Property
D. Kacsó and H.-J. Wenz 217

Developable Surfaces with Creases
Y. L. Kergosien 223

Universal Parameterizations of Some Rational Surfaces
R. Krasauskas 231

Curves on Surfaces for Computer Graphics: Theoretical Results
Y. Kuzmin and M. Daniel 239

Generalized Tension B-splines
B. I. Kvasov and P. Sattayatham 247

Shape Effects with Polynomial Chebyshev Splines
P. J. Laurent, M. L. Mazure and G. Morin 255

Interpolation with Triangulations and Curvature Minimization
M. Léger 263

On Convexity and Subharmonicity of some Functions on Triangles
J. Lorente-Pardo, P. Sablonnière and M. C. Serrano-Pérez 271

Marching Methods in Surface-Surface Intersection
  E. Malgras 279

Algorithms from Blossoms
  S. Mann 287

Generalized Parameter Representations of Tori, Dupin Cyclides
and Supercyclides
  C. Mäurer 295

Korovkin Type Results for Shape Preserving Operators
  F.-J. Muñoz-Delgado and D. Cárdenas-Morales 303

Stable Progressive Smoothing
  A. Nigro 311

Riemannian Quadratics
  L. Noakes 319

Approximation of Offset Curves and Surfaces by Discrete
Smoothing D^m-splines
  M. Pasadas, J. J. Torrens and M. C. López de Silanes 329

Algorithm for Computing the Product of two B-splines
  L. Piegl and W. Tiller 337

Smoothing Spatial Cubic B-splines under Shape Constraints
  K. G. Pigounakis and P. D. Kaklis 345

Some Error Estimates for Periodic Interpolation on Full and
Sparse Grids
  G. Pöplau and F. Sprengel 355

Curved Surfaces Reconstruction Based on Parallels
  W. Puech, J.-M. Chassery and I. Pitas 363

Zonal Kernels, Approximations and Positive Definiteness on Spheres
and Compact Homogeneous Spaces
  D. L. Ragozin and J. Levesley 371

Interpolation by Pieces of Euler's Elastica
  K.-D. Reinsch 379

A Knot Insertion Algorithm for Weighted Cubic Splines
  Mladen Rogina 387

Rational Speed Pseudo-Quadratic B-Splines
  M. A. Sabin 395

A Parameterization Technique for the Control Point Form Method
  A. Sestini and R. Morandi 403

Planar Shape Enhancement and Exaggeration
  A. Steiner, R. Kimmel and A. M. Bruckstein 411

On Geometric Continuity of Isophotes
  H. Theisel 419

A Sequence of Bézier Curves Generated by Successive Pedal-Point
Constructions
  K. Ueda 427

From Degenerate Patches to Triangular and Trimmed Patches
  M. Vigo, N. Pla and P. Brunet 435

Spline Orbifolds
  J. Wallner and H. Pottmann 445

Numerically Stable Conversion Between The Bézier and B-Spline
Forms of a Curve
  J. R. Winkler 465

G^2 Continuous G-Splines: An Interpolation Property
  R. Zeifang 473

PREFACE

During the week of June 27–July 3, 1996, the Third International Conference on *Curves and Surfaces* was held in Chamonix-Mont-Blanc (France). It was organized by the *Association Française d'Approximation* (A. F. A.).

The Conference was attended by 275 mathematicians from 26 different countries, and the program included 10 invited one-hour lectures and 161 research talks. The survey lectures dealt with several particularly active subareas of *Approximation Theory*. A number of the research talks were presented in minisymposia on *Scientific Visualization, Multivariate Approximation and Radial Basis Functions, Nonlinear and Adaptive Wavelet Approximation, Multiresolution Methods in Computer Graphics*, and *Approximation Under Constraints*, organized by Greg Nielson, Robert Schaback, Albert Cohen, Hans-Peter Seidel, and Dany Leviatan, respectively.

The proceedings of this conference consist of a total of 88 papers in two volumes. This volume contains 56 papers on curves and surfaces related to computer-aided geometric design. The volume *Surface Fitting and Multiresolution Methods* also published by Vanderbilt University Press, contains 32 papers.

We would like to thank the following institutions for their financial or technical support: European Commission, DG XII, Brussels; Ministère de la Défense (D.G.A. - D.R.E.T.); US Air Force European Office of Aerospace Research and Development; US Navy, Office of Naval Research, European Office; US Army, Research Development and Standardization Group, UK; Ministère de l'Education Nationale de l'Enseignement Supérieur et de la Recherche; Conseil Régional de la Région Rhône-Alpes; Conseil Général de Haute Savoie; ENS Télécommunications Bretagne; ENS Arts et Métiers de Lille; Université Joseph Fourier (Grenoble); Université de Rennes I; INSA de Toulouse; and Vanderbilt University.

Finally, we want to express our special thanks to Yvon Lafranche for his help before, after, and especially during the meeting.

Nashville, Tennessee   February 28, 1997

CONTRIBUTORS

Numbers in parentheses indicate pages on which authors' contributions begin.

MARCO D'APUZZO (1), *Center for Research on Parallel Computing and Supercomputers (CPS), and University of Naples "Federico II", Via Cintia, 80126 Napoli, ITALY* [dapuzzo@matna2.dma.unina.it]

RICHARD BARTELS (9, 177), *Computer Science Department, University of Waterloo, Waterloo, Ontario N2L 3G1, CANADA* [rhbartel@uwaterloo.ca]

JOHN BEATTY (9), *Computer Science Department, University of Waterloo, Waterloo, Ontario N2L 3G1, CANADA* [jcbeatty@uwaterloo.ca]

JEAN-PAUL BÉCAR (17), *Université de Valenciennes et du Hainaut Cambrésis, IUT- GEII, Laboratoire LIMAV, B.P. 311, 59304 Valenciennes Cedex, FRANCE* [becar@ univ-valenciennes.fr]

MICHEL BERCOVIER (25), *Institute of Computer Science, Hebrew University, Jerusalem, 91904, ISRAEL* [berco@cs.huji.ac.il]

DANIEL BISTER (35), *Institut für Betriebs- und Dialogsysteme, Universität Karlsruhe (TH), D-76128 Karlsruhe, GERMANY* [bister@ira.uka.de]

JACQUES BOUSQUET (43), *Institut de Recherche en Informatique de Nantes, B.P. 92208, 44322 Nantes cedex 3, FRANCE*

ALFRED M. BRUCKSTEIN (411), *CS Dept., Technion, Haifa 32000, ISRAEL* [freddy@cs.technion.ac.il]

PERE BRUNET (435), *Universitat Politècnica de Catalunya, Computer Graphics section, Software Department, E.T.S.E.I.B., Avda. Diagonal 647, Planta 8, 08028 Barcelona, SPAIN* [pere@turing.upc.es]

JIA-DING CAO (53), *Department of Mathematics, Fudan University, Shanghai, 200433, People's Republic of CHINA* [guch@bepc2.ihep.ac.cn]

DANIEL CÁRDENAS-MORALES (303), *Facultad de Ciencias Experimentales, Universidad de Jaén, 23071, Jaén, SPAIN* [cardenas@piturda.ujaen.es]

GIULIO CASCIOLA (61), *Dept. of Math., University of Bologna, Piazza di Porta S.Donato, 5, 40127 Bologna, ITALY* [casciola@dm.unibo.it]

ISABELLE CATTIAUX-HUILLARD (103), *Université de Valenciennes et du Hainaut-Cambrésis, ENSIMEV, Laboratoire IMAV B.P. 311, F-59304 Valenciennes, FRANCE* [cattiaux@univ-valenciennes.fr]

B. CHALMOND (135), *Cergy-Pontoise University and ENS Cachan (DIAM-CMLA), 2, av. A. Chauvin, 95302 Cergy Cedex, FRANCE* [Bernard.Chalmond@cmla.ens-cachan.fr]

JEAN-MARC CHASSERY (363), *TIMC-IMAG Laboratory, UMR CNRS 5525, Albert Bonniot Institut, Domaine de la Merci, 38706 La Tronche Cedex, FRANCE* [Jean-Marc.Chassery@imag.fr]

LAURENT D. COHEN (77), *CEREMADE, Universite Paris IX Dauphine, Place du Marechal de Lattre de Tassigny, 75775 Paris cedex 16, FRANCE* [Laurent.Cohen@ceremade.dauphine.fr]

PAOLO COSTANTINI (85), *Dipartimento di Matematica, Via del Capitano, 15, I 53100 Siena, ITALY* [costantini@unisi.it]

MARC DANIEL (43, 239), *Institut de Recherche en Informatique de Nantes, B.P. 92208, 44322 Nantes cedex 3, FRANCE* [Marc.Daniel@ec-nantes.fr]

PAUL DE FAGET DE CASTELJAU (69), *4 Avenue du Commerce, 78000 Versailles, FRANCE*

RON A. DEVORE (95), *Department of Mathematics, University of South Carolina, Columbia SC 29208* [devore@math.sc.edu]

J.M. DINTEN (135), *LETI (CEA - Technologies Avancées) DSYS - CEN/G, 17 avenue des Martyrs, 38054 Grenoble Cedex 9, FRANCE*

JEAN-CHARLES FIOROT (17, 103, 111), *Université de Valenciennes et du Hainaut Cambrésis, IUT- GEII, Laboratoire LIMAV, B.P. 311, 59304 Valenciennes Cedex, FRANCE* [fiorot@ univ-valenciennes.fr]

M. GARCÍA-ESNAOLA (119), *Departamento de Matemática Aplicada, Universidad de Zaragoza, 50009 Zaragoza, SPAIN* [mgesnaola@mcps.unizar.es]

THIERRY GENSANE (127), *Université du Littoral, L.M.A., bât. H. Poincaré, 50 rue F. Buisson B.P. 699, 62228 Calais Cedex, FRANCE* [gensane@lma.univ-littoral.fr]

S. GIRARD (135), *LETI (CEA - Technologies Avancées) DSYS - CEN/G, 17 avenue des Martyrs, 38054 Grenoble Cedex 9, FRANCE* [girard@dsys.ceng.cea.fr]

SONIA M. GOMES (143), *Department of Applied Mathematics, State University of Campinas, Caixa Postal 6065, 13081-970 Campinas SP, BRAZIL* [soniag@ime.unicamp.br]

HEINZ H. GONSKA (53), *Department of Mathematics, University of Duisburg, D - 47048 Duisburg, GERMANY* [gonska@informatik.uni-duisburg.de]

T. N. T. GOODMAN (151), *Dept. of Mathematics, Dundee University, Dundee, DD1 4HN, Scotland, U.K.* [tgoodman@mcs.dundee.ac.uk]

STEFANIE HAHMANN (159), *LMC-IMAG, Université Joseph Fourier, BP 53X, 38041 Grenoble, FRANCE* [Stefanie.Hahmann@imag.fr]

TIAN-XIAO HE (169), *Department of Mathematics, Illinois Wesleyan University, Bloomington, IL 61702-2900* [the@sun.iwu.edu]

WOLFGANG HEIDRICH (177), *Universität Erlangen, Graphische Datenverarbeitung, Am Weichselgarten 9, D-91058 Erlangen, GERMANY* [Heidrich@informatik.uni-erlangen.de]

JOSEF HOSCHEK (185), *Fachbereich Mathematik, Technische Hochschule Darmstadt, Schloßgartenstraße 7, D–64289 Darmstadt, GERMANY* [hoschek@mathematik.th--darmstadt.de]

PIERRE JEANNIN (111), *Université du Littoral, Centre Universitaire de la Mi-Voix, Bât. H. Poincaré, 50 rue F. Buisson, 62100 Calais, FRANCE*

BERT JÜTTLER (203), *Technische Hochschule Darmstadt, Fachbereich Mathematik, AG 3, Schloßgartenstraße 7, 64289 Darmstadt, GERMANY* [juettler@mathematik.th-darmstadt.de]

JOHANNES KAASA (211), *SINTEF, Institute for Applied Mathematics, P.O. BOX 124, Blindern, N-0314 Oslo, NORWAY* [Johannes.Kasa@si.sintef.no]

DANIELA P. KACSÓ (53, 217), *Department of Mathematics, University of Duisburg, D - 47048 Duisburg, GERMANY* [kacso@informatik.uni-duisburg.de]

PANAGIOTIS D. KAKLIS (345), *Ship-Design Laboratory, Dept. of Naval Architecture and Marine Engineering, National Technical University of Athens, Heroon Polytechneiou 9, Zografou 15773, Athens, GREECE* [kaklis@deslab.naval.ntua.gr]

Y. L. KERGOSIEN (223), *Université de Cergy-Pontoise, Informatique, 2 Av. Adolphe Chauvin, 95302 Cergy-Pontoise Cedex, FRANCE* [kergos@u-cergy.fr]

RON KIMMEL (411), *Mail-stop 50A-2152, LBNL, UC Berkeley, CA 94720* [ron@csr.lbl.gov]

STEFAN KONZ (159), *Universität Kaiserslautern, Fachbereich Informatik, D-67653 Kaiserslautern, GERMANY* [ko@nemetschek.de]

RIMVYDAS KRASAUSKAS (231), *Vilnius University, Department of Mathematics, Naugarduko 24, 2600 Vilnius, LITHUANIA* [rimvydas.krasauskas@maf.vu.lt]

ALEXANDER K. KUSHPEL (143), *University of Leicester, Department of Mathematics and Computer Science, Leicester LE1 7RH, ENGLAND* [ak99@mcs.le.ac.uk]

YEVGENIY KUZMIN (239), *Lab. for Computation Methods, Dept. of Mathematics, Moscow State University, Moscow, 119899 RUSSIA* [cls@online.ru]

BORIS I. KVASOV (247), *Institute of Computational Technologies, Russian Academy of Sciences, 6, Lavrentyev Avenue, 630090 Novosibirsk, RUSSIA* [boris@math.sut.ac.th]

GEORGE LABAHN (177), *Department of Computer Science, University of Waterloo, Waterloo, Ontario, CANADA, N2L 3G1* [glabahn@daisy.uwaterloo.ca]

PIERRE-JEAN LAURENT (255), *LMC-IMAG, Université Joseph Fourier, BP 53X, 38041 Grenoble, FRANCE* [pjl@imag.fr]

MICHEL LÉGER (263), *Institut Français du Pétrole, 1 & 4, avenue de Bois-Préau, 92506 Rueil-Malmaison, FRANCE* [Michel.LEGER@ifp.fr]

JEREMY LEVESLEY (143, 371), *University of Leicester, Department of Mathematics and Computer Science, Leicester LE1 7RH, ENGLAND* [jl1@mcs.le.ac.uk]

DANY LEVIATAN (95), *School of Mathematical Sciences, Sackler Faculty of Exact Sciences, Tel Aviv University, Tel Aviv 69978, ISRAEL* [leviatan@math.tau.ac.il]

M. C. LÓPEZ DE SILANES (329), *Departamento de Matemática Aplicada, C.P.S., Universidad de Zaragoza, María de Luna 3, 50015 Zaragoza, SPAIN* [mcsilanes@mcps.unizar.es]

J. LORENTE-PARDO (271), *Dpto. de Matemática Aplicada, Universidad de Granada, Granada, SPAIN* [lorente@goliat.ugr.es]

LUCIA MADDALENA (1), *Center for Research on Parallel Computing and Supercomputers (CPS), Via Cintia, 80126 Napoli, ITALY* [lucia@matna2.dma.unina.it]

EMMANUEL MALGRAS (279), *Institut de Recherche en Informatiques de Nantes, 2, rue de la Houssinière, 44072 Nantes cedex 03, FRANCE* [Emmanuel.Malgras@irin.univ-nantes.fr]

STEPHEN MANN (287), *Computer Science Department, University of Waterloo, 200 University Ave W, Waterloo, Ontario N2L 3G1, CANADA* [smann@cgl.uwaterloo.ca]

CHRISTOPH MÄURER (295), *Fachbereich Mathematik, Technische Hochschule Darmstadt, Schloßgartenstraße 7, D–64289 Darmstadt, GERMANY* [cmaeurer@mathematik.th-darmstadt.de]

TANYA MATSKEWICH (25), *Institute of Computer Science, Hebrew University, Jerusalem, 91904, ISRAEL* [fisa@cs.huji.ac.il]

MARIE-LAURENCE MAZURE (255), *LMC-IMAG, Université Joseph Fourier, BP 53X, 38041 Grenoble, FRANCE* [mazure@imag.fr]

ROSSANA MORANDI (403), *Dipartimento di Matematica, Universitá di Perugia, Via Vanvitelli 1, 06123 Perugia, ITALY* [morandi@gauss.dipmat.unipg.it]

SERENA MORIGI (61), *Dept. of Math., University of Bologna, Piazza di Porta S.Donato, 5, 40127 Bologna, ITALY* [serena@csr.unibo.it]

GÉRALDINE MORIN (255), *LMC-IMAG, Université Joseph Fourier, BP 53X, 38041 Grenoble, FRANCE*

FRANCISCO-JAVIER MUÑOZ-DELGADO (303), *Departamento de Mateáticas, Universidad de Jaén, Escuela Politechnica Superior, 23071, Jaén, SPAIN* [fdelgado@piturda.ujaen.es]

A. NIGRO (311), *LMC-IMAG, Université Joseph Fourier, BP53, F38041 Grenoble (cedex 9), FRANCE* [Abdelmalek.Nigro@imag.fr]

LYLE NOAKES (319), *Department of Mathematics, The University of Western Australia, Nedlands, WA 6907, AUSTRALIA* [lyle@maths.uwa.edu.au]

B. H. ONG (151), *School of Mathematical Sciences, Universiti Sains Malaysia, 11800 Penang, MALAYSIA* [bhong@cs.usm.my]

M. PASADAS (329), *Departamento de Matemática Aplicada, E.U.A.T., Universidad de Granada, Severo Ochoa s/n, 18001 Granada, SPAIN* [mpasadas@goliat.ugr.es]

J. M. PEÑA (119), *Departamento de Matemática Aplicada, Universidad de Zaragoza, 50009 Zaragoza, SPAIN* [jmpena@posta.unizar.es]

LES PIEGL (337), *Department of Computer Science & Engineering, University of South Florida, 4202 Fowler Avenue, ENG 118, Tampa, FL 33620* [piegl@babbage.csee.usf.edu]

KONSTANTINOS G. PIGOUNAKIS (345), *Ship-Design Laboratory, Dept. of Naval Architecture and Marine Engineering, National Technical University of Athens, Heroon Polytechneiou 9, Zografou 15773, Athens, GREECE* [kpig@deslab.naval.ntua.gr]

IOANNIS PITAS (363), *Department of Informatics, University of Thessaloniki, P.O. Box 451, 54006 Thessaloniki, GREECE* [pitas@zeus.csd.auth.gr]

N. PLA (435), *Universitat Politècnica de Catalunya, Computer Graphics section, Software Department, E.T.S.E.I.B., Avda. Diagonal 647, Planta 8, 08028 Barcelona, SPAIN* [nuria@turing.upc.es]

GISELA PÖPLAU (355), *Fachbereich Mathematik, Universität Rostock, D-18051 Rostock, GERMANY* [gisela.poeplau@mathematik.uni-rostock.de]

HELMUT POTTMANN (445), *Institut für Geometrie, Technische Universität Wien, Wiedner Hauptstraße 8–10/113, A-1040 Wien, AUSTRIA* [pottmann@geometrie.tuwien.ac.at]

HARTMUT PRAUTZSCH (35), *Institut für Betriebs- und Dialogsysteme, Universität Karlsruhe (TH), D-76128 Karlsruhe, GERMANY* [prau@ira.uka.de]

WILLIAM PUECH (363), *TIMC-IMAG Laboratory, UMR CNRS 5525, Albert Bonniot Institut, Domaine de la Merci, 38706 La Tronche Cedex, FRANCE* [William.Puech@imag.fr]

DAVID L. RAGOZIN (143, 371), *University of Washington, Department of Mathematics, Seattle, WA 98195-4350* [rag@math.washington.edu]

KLAUS-DIETER REINSCH (379), *Mathematisches Institut, Technische Universität, D-80290 München, GERMANY* [kladire@mathematik.tu-muenchen.de]

MLADEN ROGINA (387), *Dept. of Mathematics, University of Zagreb, Bijenička 30, 10000 Zagreb, CROATIA* [rogina@math.hr]

MALCOLM A. SABIN (395), *Numerical Geometry Ltd., 26 Abbey Lane, Lode, Cambridge CB5 9EP, ENGLAND* [malcolm@geometry.demon.co.uk]

PAUL SABLONNIÈRE (271), *Laboratoire L.A.N.S., I.N.S.A. de Rennes, 20, avenue des buttes de Coesmes, 35043 Rennes, FRANCE* [sablonni@perceval.univ-rennes1.fr]

PAIROTE SATTAYATHAM (247), *School of Mathematics, Suranaree University of Technology, Nakhon Ratchasima 30000, THAILAND* [pairote@sura1.sut.ac.th]

MIKE SCHNEIDER (185), *Fachbereich Mathematik, Technische Hochschule Darmstadt, Schloßgartenstraße 7, D–64289 Darmstadt, GERMANY* [mschneider@mathematik.th--darmstadt.de]

M. C. SERRANO-PÉREZ (271), *Dpto. de Matemática Aplicada, Universidad de Granada, Granada, SPAIN* [cserrano@goliat.ugr.es]

ALESSANDRA SESTINI (403), *Dipartimento di Energetica "Sergio Stecco", Universitá di Firenze, Via Lombroso 6/17, 50134 Firenze, ITALY* [sestini@ingfi1.ing.unifi.it]

I. A. SHEVCHUK (95), *Institute of Mathematics, National Academy of Sciences of Ukraine, Kyiv 252601, UKRAINE* [shevchuk@imat.gluk.apc.org]

FRAUKE SPRENGEL (355), *Fachbereich Mathematik, Universität Rostock, D-18051 Rostock, GERMANY* [frauke.sprengel@mathematik.uni-rostock.de]

AMI STEINER (411), *EE Dept., Technion, Haifa 32000, ISRAEL*
[steiner@tx.technion.ac.il]

HOLGER THEISEL (419), *University of Rostock, Computer Science Department, PostBox 999, 18051 Rostock, GERMANY*
[theisel@informatik.uni-rostock.de]

WAYNE TILLER (337), *GeomWare, Inc., 3036 Ridgetop Road, Tyler, TX 75703, USA* [76504.3045@compuserve.com]

J. J. TORRENS (329), *Departamento de Matemática e Informática, Universidad Pública de Navarra, Campus de Arrosadía s/n, 31006 Pamplona, SPAIN* [jjtorrens@upna.es]

KENJI UEDA (427), *Ricoh Company, Ltd., 1-1-17 Koishikawa, Bunkyo-ku, Tokyo 112, JAPAN* [ueda@src.ricoh.co.jp]

M. VIGO (435), *Universitat Politècnica de Catalunya, Computer Graphics section, Software Department, E.T.S.E.I.B., Avda. Diagonal 647, Planta 8, 08028 Barcelona, SPAIN* [marc@turing.upc.es]

OLEG VOLPIN (25), *Institute of Computer Science, Hebrew University, Jerusalem, 91904, ISRAEL* [oleg@cs.huji.ac.il]

JOHANNES WALLNER (445), *Institut für Geometrie, Technische Universität Wien, Wiedner Hauptstraße 8–10/113, A-1040 Wien, AUSTRIA* [hannes@geometrie.tuwien.ac.at]

HANS–JÖRG WENZ (217), *Dept. of Mathematics, University of Duisburg, D–47057 Duisburg, GERMANY* [wenz@informatik.uni-duisburg.de]

GEIR WESTGAARD (211), *SINTEF, Institute for Applied Mathematics, P.O. BOX 124, Blindern, N-0314 Oslo, NORWAY* [Geir.Westgaard@si.sintef.no]

JOAB R WINKLER (465), *Department of Computer Science, The University of Sheffield, Regent Court, 211 Portobello Street, Sheffield S1 4DP, ENGLAND* [j.winkler@dcs.shef.ac.uk]

RAINER ZEIFANG (473), *Hewlett Packard GmbH, Mechanical Design Division, Herrenberger Straße 130, 71034 Böblingen, GERMANY* [Rainer_Zeifang@hpbbn.bbn.hp.com]

Parallelization Strategies for the B-spline Curve Interpolation Problem

Marco D'Apuzzo and Lucia Maddalena

Abstract. In this paper we focus on the design of parallelization strategies for the problem of interpolating large data sets via parametric B-spline curves. It is well known [5] that, subject to appropriate hypotheses on the parametrization and on the knot vector of the B-spline curve [13], the classical formulation of the B-spline interpolation problem leads to the solution of a linear system whose coefficient matrix is banded and totally positive [5,8]. For such a system a suitable sequential solver is the Gaussian elimination algorithm for banded linear systems with no need for pivoting [6]. Unfortunately this algorithm is no longer very efficient in parallel environments [7]. We present two strategies that attempt to introduce a higher degree of parallelism into the considered problem. The first strategy is based on a suitable row block partitioning of the coefficient matrix; the second adopts a different approach to the interpolation problem, based on a domain decomposition strategy. Both approaches lead to linear systems whose coefficient matrices have particular almost block diagonal structures that are better suited for the development of efficient parallel algorithms for MIMD distributed memory machines, as is also shown by a few numerical experiments.

§1. Introduction

Scientific visualization is a fundamental tool in Scientific Computing, allowing the exploration of information and data deriving from long and complex numeric computations in order to obtain a deep comprehension and an intuitive interpretation of them. The visualization of time-varying images produced by long running numerical simulations, *data navigation*, and the interactive *steering* of computations are some examples of scientific visualization applications that require great computational power. A *natural* answer to such a need is given by the effective use of parallel architectures. For example there are many applications in which the *rendering* of an image (or of a sequence of images) is done by parallel computers (e.g. [10]). Several parallel algorithms have been developed for *ray tracing*, for *radiosity* computation, etc.

Curves and Surfaces with Applications in CAGD
A. Le Méhauté, C. Rabut, and L. L. Schumaker (eds.), pp. 1–8.
Copyright © 1997 by Vanderbilt University Press, Nashville, TN.
ISBN 0-8265-1293-3.
All rights of reproduction in any form reserved.

(e.g. [14]). Regarding curves and surfaces representation and manipulation, instead, only few parallel algorithms have been proposed (e.g. [2,11]).

In this paper we are concerned with this latter aspect of scientific visualization. Specifically we consider the *parametric interpolation problem* via *B-spline curves*. The classical approach to the problem leads to the solution of a linear system whose coefficient matrix is banded, with upper and lower bandwidth less than or equal to the degree of the B-spline curve, and totally positive [6,5,8]. Gaussian elimination is an efficient algorithm for its solution in a sequential environment, but this is no longer true in a distributed memory environment [7]. Therefore, we investigated different strategies that attempt to introduce a higher degree of parallelism into the considered problem.

In §2 we give a brief introduction to the interpolation problem via B-splines and to the classical approach to its solution. In §3 we present the parallel environment we are concerned with and the two parallelization strategies proposed. A few numerical experiments are presented in §4.

§2. The B-Spline Curve Interpolation Problem

In this paper we are concerned with the problem of interpolating a large set of prescribed data, a computationally intensive problem very frequent in Computer Aided Design, Computer Graphics, Image Processing, etc. [5,12]. In particular we will consider B-spline curves of degree h, whose parametric representation in $\mathbb{R}^2$ is

$$C(u) = \sum_{i=0}^{m} p_i N_{i,h}(u), \qquad u \in [a,b],$$

where the $N_{i,h}(u)$ are the *B-spline basis functions* of degree h defined on a nondecreasing *knot vector* $(t_0, \ldots, t_{m+h+1}) \in \mathbb{R}^{(m+h+2)}$, and the $p_i = (x_i, y_i) \in \mathbb{R}^2, i = 0, \ldots, m$, are the *control points* [12]. The results obtained can be easily extended to the case of B-spline curves in the space.

The *Lagrange interpolation problem* via parametric B-spline curves can be formulated as follows:

Problem 1. *Given $m + 1$ points $q_i = (a_i, b_i) \in \mathbb{R}^2$, $i = 0, \ldots, m$, assigned a parametrization $u_0, \ldots, u_m$, and a fixed knot vector $(t_0, \ldots, t_{m+h+1})$, determine a B-spline curve $C(u)$ of degree h that interpolates the given points:*

$$C(u_i) = q_i, \qquad i = 0, \ldots, m,$$

that is, determine the control points $p_j = (x_j, y_j) \in \mathbb{R}^2$, $j = 0, \ldots, m$, such that

$$C(u_i) = \sum_{j=0}^{m} p_j N_{j,h}(u_i) = q_i, \qquad i = 0, \ldots, m. \tag{1}$$

Necessary and sufficient conditions for the nonsingularity of the coefficient matrix N of the linear system (1), and therefore for the well-posedness of Problem 1, are given by the Schoenberg-Whitney conditions [13]:

$$t_i < u_i < t_{i+h+1}, \qquad i = 0, \ldots, m. \tag{2}$$

If (2) are satisfied, the matrix N is banded, with upper and lower bandwidth no greater than the degree h of the B-spline curve [5] and totally positive [8]. It follows that a suitable sequential solver for linear system (1) can be the Gaussian elimination algorithm for banded linear systems with no need for pivoting [6].

§3. Parallelization Strategies

The target environment considered for the design of parallelization strategies for the B-spline curves interpolation problem is a distributed memory message passing computing environment, consisting of P nodes, logically organized as a ring and numbered from 0 to $P - 1$. Each node consists of a CPU and a local memory directly accessible only by the node itself and accessible by other nodes only via a bidirectional communication channel.

In such an environment it is very hard to develop efficient parallel solvers for linear systems such as (1) that are based on the Gaussian elimination algorithm. In fact, in this case only a small portion of parallel work is available in updating the trailing submatrix, especially for small bandwidth [7]. We remark that in the problem of B-spline curves interpolation a low degree h (and therefore small upper and lower bandwidth of the coefficient matrix N of system (1)), ranging from two to five, is satisfactory for most applications. The above observations lead us to adopt different strategies in order to introduce a higher degree of parallelism into the solution of the B-spline curves interpolation problem.

First Strategy

A first approach to introduce parallelism into the computations consists in applying a suitable logical decomposition of the coefficient matrix N of (1) into blocks and in distributing them among the nodes, so that more parallel computation is available and less communication is needed.

Obviously, many ways to partition the banded coefficient matrix N into blocks can be considered. As an example, we can consider the block decomposition of Fig. 1(a), where odd (grey) blocks are orthogonal as well as even (black) blocks. Therefore, computations in odd blocks can be carried out in parallel by different nodes owning different blocks of the matrix, and the same applies for even blocks. Clearly communication is still needed in order to update components of the solution belonging to different nodes.

Another kind of block decomposition of the coefficient matrix is the one of Fig. 1(b), where we have tryed to minimize "intersections" between adjacent blocks. Here a bigger amount of computation can be carried out concurrently by different nodes owning different big (grey) blocks and, after that, communication is required in order to update components of the solution related to the small (black) blocks.

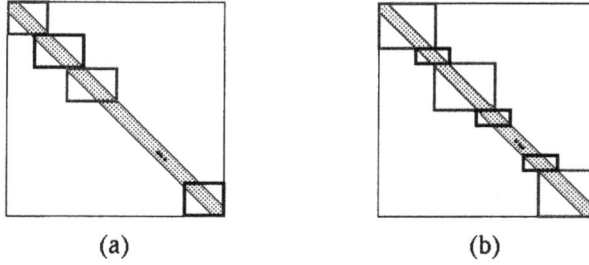

(a)   (b)

Fig. 1. Examples of block decomposition of the coefficient matrix N of (1).

Second Strategy

In order to introduce parallelism into the B-spline curves interpolation problem we investigated a second strategy, based on a domain decomposition approach to the problem.

The basic idea consists in subdividing the set of $m+1$ interpolation points into P subsets and in constructing concurrently in each subset the B-spline interpolant; in order to obtain the *global* interpolant as the *union* of *local* curves, suitable *boundary conditions* at the junction points of the subsets must be imposed [4]. Therefore a new formulation of the problem can be the following:

Problem 2. *Given $m+1$ points $q_i = (a_i, b_i) \in \mathbb{R}^2$, $i = 0, \ldots, m$, assigned a parametrization $u_0, \ldots, u_m$, and a fixed knot vector $(t_0, \ldots, t_{m+h+1})$, determine B-spline curves $C_1(u), C_2(u), \ldots, C_P(u)$ of degree h such that the curve*

$$C(u) = \begin{cases} C_1(u), & u \leq \overline{u}_1, \\ C_2(u), & \overline{u}_1 \leq u \leq \overline{u}_2, \\ \vdots & \vdots \\ C_P(u), & u \geq \overline{u}_{P-1}, \end{cases}$$

where $\overline{u}_1, \overline{u}_2, \ldots, \overline{u}_{P-1}$ are fixed parameter values, solves the B-spline interpolation problem

$$C(u_i) = q_i \qquad i = 0, \ldots, m.$$

As already mentioned, the crucial point is to impose suitable *boundary conditions* in suitable parameter values $\overline{u}_1, \overline{u}_2, \ldots, \overline{u}_{P-1}$. Focusing on the case of cubic B-spline curves, such issues are addressed by the following theorem.

Theorem 1. *[4] Suppose we are given the interpolation points $q_i = (a_i, b_i) \in \mathbb{R}^2$, $i = 0, \ldots, m$, assigned to a parametrization $u_0, \ldots, u_m$, along with a fixed knot vector $(t_0, \ldots, t_{m+4})$, such that*

(a) $t_i < u_i < t_{i+4}$, $i = 0, \ldots, m$,

(b) $t_{l_i-1} < t_{l_i} = u_{m_i} < t_{l_i+1}$, $i = 1, \ldots, P-1$,

where $m_i, l_i \in \mathbb{N}, i = 0, \ldots, P$, and

$$0 = m_0 < \ldots < m_P = m; \quad 3 = l_0 < \ldots < l_P = m + 1.$$

Then the linear system

$$
\begin{cases}
C_1(u_i) = \sum_{j=0}^{l_1-1} p_j N_{j,3}(u_i) = q_i, & i = m_0, \ldots, m_1, \\
\vdots & \vdots \\
C_P(u_i) = \sum_{j=l_{P-1}+3(P-2)}^{l_P+3(P-1)-1} p_j N_{j-3(P-1),3}(u_i) = q_i, & i = m_{P-1}, \ldots, m_P, \\
C_1^{(i)}(u_{m_1}) = C_2^{(i)}(u_{m_1}), & i = 1, 2, \\
\vdots & \\
C_{P-1}^{(i)}(u_{m_{P-1}}) = C_P^{(i)}(u_{m_{P-1}}), & i = 1, 2,
\end{cases}
\tag{3}
$$

has a unique solution of the form

$$(p_0, \ldots, p_{m+3P-3}) = (\overline{p}_0, \ldots, \underbrace{\overline{p}_{l_1-3}, \overline{p}_{l_1-2}, \overline{p}_{l_1-1}}, \underbrace{\overline{p}_{l_1-3}, \overline{p}_{l_1-2}, \overline{p}_{l_1-1}}, \cdots$$

$$\cdots, \underbrace{\overline{p}_{l_i-3}, \overline{p}_{l_i-2}, \overline{p}_{l_i-1}}, \underbrace{\overline{p}_{l_i-3}, \overline{p}_{l_i-2}, \overline{p}_{l_i-1}}, \cdots \tag{4}$$

$$\cdots, \underbrace{\overline{p}_{l_{P-1}-3}, \overline{p}_{l_{P-1}-2}, \overline{p}_{l_{P-1}-1}}, \underbrace{\overline{p}_{l_{P-1}-3}, \overline{p}_{l_{P-1}-2}, \overline{p}_{l_{P-1}-1}}, \ldots, \overline{p}_m),$$

where $(\overline{p}_0, \overline{p}_1, \ldots, \overline{p}_m)$ is the unique solution of the Lagrange interpolation problem via parametric cubic B-spline curves, *i.e.*,

$$C(u_i) = \sum_{j=0}^{m} p_j N_{j,3}(u_i) = q_i \quad i = 0, \ldots, m. \tag{5}$$

It follows from Th. 1 that the solution of the linear system (5) can be easily obtained from that of the linear system (3) by eliminating in (4) one of the triplets of each pair of coincident triplets of control points.

The coefficient matrix, say A, of linear system (3) is a particular almost block diagonal matrix that can be logically partitioned into two row blocks: A_1, consisting of P rectangular banded disjoint subblocks with upper and lower bandwidth no greater than the degree $h = 3$ of the curve, and A_2, consisting of $P - 1$ rectangular disjoint subblocks of dimension 2×6 (Fig. 2), see [4].

As in the case of the first strategy, parallelism can be enhanced, since a great part of the computation can be carried out concurrently by different nodes owning different subblocks of A_1 and A_2.

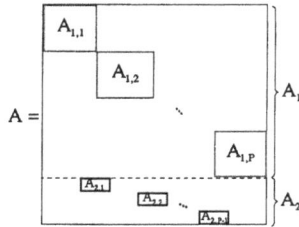

Fig. 2. Structure of the coefficient matrix of linear system (3).

§4. Numerical Experiments

We have developed two parallel algorithms for the solution of the cubic B-spline curves interpolation problem: Alg. 1, based on the first parallelization strategy using the block decomposition shown in Fig. 1(b), and Alg. 2, based on the second parallelization strategy. Both algorithms use a parallel iterative solver based on a block version of the *Kaczmarz algorithm*, SBRPK (*Symmetric Block Row-Projection Kaczmarz*) [3]. This solver has been proved to be well suited for the solution of linear systems with the described structure, see [4].

The testing environment used is the *Intel Touchstone Delta System* of the California Institute of Technology (CalTech), a MIMD distributed memory machine consisting of 512 compute nodes based on the Intel i860 microprocessor with 2D-mesh topology.

Numerical experiments have been carried out for a set of problems, where interpolation points are automatically generated as:

$$q_i = (a_i, b_i) \equiv \begin{cases} a_i = i^2, \\ \\ b_i = 2i * (-1)^i, \end{cases} \qquad i = 0, \dots, m,$$

with $m = 1024, 2016, 4000, 8000$, fixing the *centripetal* parametrization [9].

The model adopted for parallel efficiency evaluation is the classical one:

$$E_P = \frac{T_P^1(n)}{T_P(n)P},$$

where $T_P^1(n)$ is the execution time of our parallel algorithm on one node for a problem of dimension n subdivided into P subproblems, and $T_P(n)$ is the execution time of the parallel algorithm on P nodes for the same problem, fixing the number of iterations.

Efficiency values obtained with Alg. 1 and Alg. 2 varying the dimension m and the number P of nodes, and fixing the number of iterations equal to 10 are shown in Tab. 1. We observe that efficiency values obtained by both algorithms are generally satisfactory (greater than 0.77) and slightly better for Alg. 2; the results obtained show that the developed algorithms are scalable.

We would like to remark that these results have to be taken as a lower bound for results obtainable using parallel solvers other than SBRPK.

	Alg. 1			Alg. 2		
m	E_2	E_4	E_8	E_2	E_4	E_8
1024	0.97	0.90	0.79	0.97	0.89	0.77
2016	0.97	0.94	0.87	0.98	0.94	0.86
4000	0.98	0.95	0.93	0.99	0.96	0.96
8000	0.99	0.96	0.95	0.99	0.98	0.98

Tab. 1. Efficiency values (E_P) obtained with Alg. 1 and Alg. 2 varying the dimension m and the number P of nodes.

§5. Conclusions and Perspectives

We have presented two different strategies for introducing parallelism into the interpolation problem via parametric cubic B-spline curves, one based on a suitable row block partitioning of the coefficient matrix of the interpolation system, and the other based on a domain decomposition approach to the interpolation problem. These strategies lead to linear systems whose coefficient matrix has a particular almost block diagonal structure, well suited for the development of parallel algorithms. First numerical experiments carried out using a block Kaczmarz algorithm show that good efficiency values can be obtained with both strategies and, for the considered solver, they are slightly better when the domain decomposition approach to the problem is applied.

Starting from these results, future research will deal with the extension of the domain decomposition approach to tensor product B-spline surface interpolation, as well as to the B-spline curve and surface approximation problems.

Acknowledgments. This work was performed in part using the Intel Touchstone Delta System operated by CalTech on behalf of Concurrent Supercomputing Consortium. Access to this facility has been provided by CalTech.

References

1. Boehm, W., G. Farin, and J. Kahmann, A survey of curve and surfaces methods in CAGD, Comput. Aided Geom. Design **1** (1984), 1–60.

2. Chung, K. L., and L. J. Shen, Vectorized algorithm for B-spline curve fitting on CRAY X-MP EA/16se, IEEE Computer Society Press, 1992, 166–169.

3. D'Apuzzo, M., and M. Lapegna, A parallel row projection solver for large sparse linear systems, in *3rd EUROMICRO Workshop on Parallel and Distributed Processing*, IEEE Computer Society Press, 1995, 432–441.

4. D'Apuzzo, M., and L. Maddalena, A parallel row projection-based algorithm for parametric cubic B-spline curves interpolation, Tech. Rep. CPS-CNR, n. 3/96, 1996, submitted for publication.

5. de Boor, C., *A Practical Guide to Splines*, Springer-Verlag, 1978.

6. de Boor, C., and A. Pinkus, Backward error analysis for totally positive linear systems, Numer. Math. **27** (1977), 485–490.

7. Demmel, J. W., M. T. Heath, and H. A. van der Vorst, Parallel numerical linear algebra, Acta Numerica (1993), 138–197.

8. Karlin, S., *Total Positivity*, vol. I, Stanford University Press, 1968.

9. Lee, E. T. J., Choosing nodes in parametric curve interpolation, Computer Aided Design **21** (1989), n. 6, 363–370.

10. Messina, P., and T. Sterling (eds.), *Systems Software and Tools for High Performence Computing Environments*, SIAM, 1993.

11. Pham, B., and H. Schroder, Parallel algorithms and a systolic device for cubic B-spline curve and surface generation, Comput. & Graphics **15** (1991), n. 3, 349–354.

12. Rogers, D. F., and J. A. Adams, *Mathematical Elements for Computer Graphics*, II ed., McGraw-Hill, 1990.

13. Schoenberg, J. J., and A. Whitney, On Polya frequency functions. III. The positivity of translation determinants with an application to the interpolation problem by B-spline curves, Trans. Amer. Math. Soc. **74** (1953), 246–259.

14. Singh, J. P., A. Gupta, and M. Levoy, Parallel visualization algorithms: performance and architectural implications, IEEE Computer **27** (1994), n. 7, 45–55.

Marco D'Apuzzo
Center for Research on Parallel Computing and Supercomputers (CPS)
and University of Naples "Federico II"
Via Cintia, 80126 Napoli, ITALY
dapuzzo@matna2.dma.unina.it

Lucia Maddalena
Center for Research on Parallel Computing and Supercomputers (CPS)
Via Cintia, 80126 Napoli, ITALY
lucia@matna2.dma.unina.it

Connection-Matrix Splines
by Divided Differences

Richard H. Bartels and John C. Beatty

Abstract. We move the classical divided-difference construction for B-splines into a connection-matrix setting by establishing a one-sided basis for connection-matrix splines together with a differencing process that transforms the one-sided basis into a compact-support basis that partitions unity.

§1. Introduction and Motivation

One way of achieving geometric continuity in spline curves and tensor product spline surfaces is through the use of basis functions that automatically support specified continuity conditions across the knots. The initial work on such basis functions by Barsky, for G^1 and G^2 reparameterization continuity, produced the beta-splines. Generalizations of continuity conditions for curves by Barsky, DeRose, Goldman, Micchelli, Seidel, and others have led to the concept of connection-matrix splines. The more recent work, however, has concentrated on theoretical issues about the continuity conditions and representations of the spline as a whole, often from a control-point and polar-form approach rather than from the point of view of basis functions. Yet basis splines have an important role to play. Techniques of interpolation, approximation, constraint-based and direct manipulation, fairing, and finite-element design all may employ computations most conveniently expressed using values of basis splines.

We move the classical, divided-difference construction for B-splines into a connection-matrix setting by establishing a one-sided basis for connection-matrix splines and a differencing process that transforms collections of one-sided basis splines into basis splines that have compact support and partition unity. A recent paper by Seidel [3] has constructed connection-matrix splines as affine maps of a universal spline expressed in Bézier representation. The Seidel paper uses a symbolic algebra system and requires matrix inversion.

Curves and Surfaces with Applications in CAGD
A. Le Méhauté, C. Rabut, and L. L. Schumaker (eds.), pp. 9–16.
Copyright ⓒ 1997 by Vanderbilt University Press, Nashville, TN.
ISBN 0-8265-1293-3.
All rights of reproduction in any form reserved.

9

In practice that has limited the degree of splines that can be managed. The approach presented here, which extends a construction for beta-splines in [1], is computational rather than symbolic, is straightforward to program, and requires no inversion. The dark side of this approach is numerical cancellation when two or more knots are placed very close together relative to their distance from the remaining knots. By experience, however, we have found that the closeness must be unusually extreme before any effects are noticeable.

§2. Definitions

Connection-matrix splines are piecewise polynomials of common degree d satisfying certain continuity conditions between the pieces. They are a special case of the g-splines described in [2] for which each of the functionals γ_{ij} applied to pieces $i-1$ and i is the jth derivative.

The important terms for connection-matrix splines are as follows. A *breakpoint* is a domain point u_i at which two polynomial segments p_{i-1} and p_i of a spline join. The *derivative vector* is the vector of the first k derivatives of a polynomial

$$D^{(k)}p(u) = \left[p^{(1)}(u), \ldots, p^{(k)}(u) \right]^T$$

spline connectivity refers to whatever relationship exists between the derivative vectors of a spline $s(u)$ at the left and right of a breakpoint u_i

$$D_-^{(k_i)}s(u_i) = D^{(k_i)}p_{i-1}(u_i)$$
$$D_+^{(k_i)}s(u_i) = D^{(k_i)}p_i(u_i).$$

The connectivity of the splines under consideration are specified by a $(k_i \times k_i)$ *connection matrix*

$$D_+^{(k_i)}s(u_i) = C_i D_-^{(k_i)}s(u_i).$$

Every standard form of geometric continuity between curve segments is characterized by some connection matrix. In particular, reading from left to right, we have the matrices for *parametric continuity, geometric (reparameterization) continuity,* and *Frenet continuity:*

$$\begin{bmatrix} 1 & & \\ & 1 & \\ & & 1 \\ & & \cdots \end{bmatrix}, \begin{bmatrix} \beta_1 & & \\ \beta_2 & \beta_1^2 & \\ \beta_3 & 3\beta_1\beta_2 & \beta_1^3 \\ & \cdots & \end{bmatrix}, \begin{bmatrix} c_{11} & & \\ c_{21} & c_{11}^2 & \\ c_{31} & c_{32} & c_{11}^3 \\ & \cdots & \end{bmatrix}.$$

The matrices are required to have nonnegative minors. Each matrix is a special case of the one that follows. These matrix connections are also meaningful for tensor products, but they do not provide the fullest sense of geometric continuity for surfaces.

As with polynomial splines having parametric continuity, each breakpoint is the location of one or more *knots*, and the number of connection conditions specified (and hence, the size k_i of the connection matrix) is determined by the number of knots at the breakpoint. Generally the sequence of knots (the *knot vector*) will have clusters

$$[\cdots < \underbrace{t_j = \cdots = t_{j+\mu_i-1}}_{i^{th} \text{ cluster}} < \cdots] \ ,$$

and the ith breakpoint u_i corresponds to the ith knot cluster: $u_i \equiv t_j = \cdots = t_{j+\mu_i-1}$. The size of the connection matrix at this breakpoint will be $k_i = d - \mu_i$; that is, C_i must specify the first $d - \mu_i$ derivatives of a degree-d spline. Confluence of knots may occur, as in the case of familiar splines. At two nearby breakpoints there will be two connection matrices but if these breakpoints merge, the result will be a single breakpoint with a revised (smaller) connection matrix:

$$\begin{array}{cc} C_i & C_{i+1} \\ u_i & , & u_{i+1} \end{array} \longrightarrow \begin{array}{c} \bar{C}_i \\ \bar{u}_i \end{array} \ .$$

We shall not be concerned about connection-matrix splines with fluid knots, but we shall have to be concerned in our computational approach with knots that are positioned very nearly together.

§3. Truncated Powers

It is straightforward to honor the connectivity at a single breakpoint u_i by adding a Taylor correction to p_{i-1}. Since precisely the derivatives $1, \ldots, k_i$ are specified across breakpoint u_i, we add

$$a_{ik_i}(u - u_i)_+^{k_i} + \cdots + a_{i1}(u - u_i)_+^1$$

to p_{i-1}. The coefficients $a_{i\ell}$ are determined by the connection relation

$$\begin{bmatrix} p_{i-}^{(1)}(u_i) \\ \vdots \\ p_{i-}^{(k_i)}(u_i) \end{bmatrix} + \begin{bmatrix} a_{i1} \\ \vdots \\ k_i! a_{ik_i} \end{bmatrix} = \begin{bmatrix} c_{11} \\ \vdots & \ddots \\ c_{31} & \cdots & c_{11}^{k_i} \end{bmatrix} \begin{bmatrix} p_{i-}^{(1)}(u_i) \\ \vdots \\ p_{i-}^{(k_i)}(u_i) \end{bmatrix} \ .$$

In matrix terms, the vector A_i of coefficients $a_{i\ell}$ are produced by the matrix transformation

$$\left[F^{(k_i)}\right]^{-1} [C_i - I] \left[D^{(k_i)} p_{i-}(u_i)\right] = A_i,$$

where $F^{(k_i)}$ is the diagonal matrix of successive factorials and $C_i - I$ is triangular.

This observations leads to the definition of a number of one-sided basis functions $g_{ij}(u)$, the connection-matrix analog of a truncated power, by the following considerations. At a breakpoint u_i of multiplicity μ_i there is an unspecified jump in derivatives $k_i + 1 = d - \mu_i + 1, \ldots, d$ and a connection-matrix specified jump in derivatives $1, \ldots, d - \mu_i = k_i$. We begin the definition with $g_{ij}(u) \equiv 0$ for $u < u_i$, for which the connection-matrix conditions are satisfied trivially. The unspecified jumps are satisfied at u_i by $g_{ij}(u) = (u - u_i)_+^j$ for $d \geq j \geq d - \mu_i + 1$, where we associate $j = d$ with the leftmost knot, t_κ of the cluster defining u_i, $j = d - 1$ with the next sequential knot in the cluster, if it exists, and so on. At each successive breakpoint u_{i+r}, add in

$$\sum_{\ell=1}^{d - \mu_{i+r}} a_{i\ell j} (u - u_{i+r})_+^\ell \quad ,$$

and provide the coefficients $a_{i\ell j}$ as above, using the polynomial g_{ij} constructed thus far as p_{i+r-1} to obtain

$$g_{ij}(u) = (u - u_i)_+^j + \sum_{r \geq 1} \sum_{\ell=1}^{d - \mu_{i+r}} a_{i\ell j} (u - u_{i+r})_+^\ell$$

for each $j \in d, \ldots, \{d - \mu_i + 1\}$ in decreasing order. The sum in r ends with the breakpoint corresponding to $t_{\kappa+d+1}$. This is shown schematically in Figure 1.

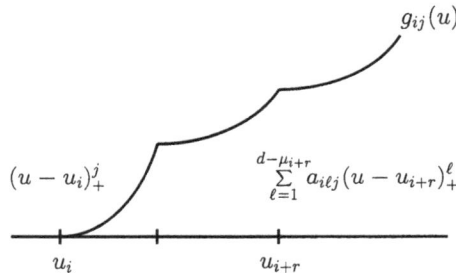

Fig. 1. Schematic view of $g_{ij}(u)$.

§4. Differencing

The functions $g_{ij}(u)$ yield to a differencing process that is patterned entirely after the one used for ordinary truncated powers to produce the B-splines. One g_{ij} is associated with each knot in sequence, as has been indicated:

$$\begin{aligned}
g_{i,d}(u) &\equiv (u - t_\kappa)^d & + \cdots & \leftrightarrow & t_\kappa & (t_\kappa > t_{\kappa-1}) \\
g_{i,d-1}(u) &\equiv (u - t_{\kappa+1})^{d-1} & + \cdots & \leftrightarrow & t_{\kappa+1} & (t_{\kappa+1} = t_\kappa)
\end{aligned}$$
etc.

Each $d + 2$-length subsequence of these functions is taken together to form a differencing cohort. They are brought to a common expression in powers of $(u - R)$ using the binomial expansion

$$\sum_{\gamma=0}^{j} \binom{j}{\gamma} (R - u_i)^\gamma (u - R)^{j-\gamma} + \sum_{r \geq 1} \sum_{\ell=1}^{d - \mu_{i+r}} a_{i\ell j} \sum_{\lambda=0}^{\ell} \binom{\ell}{\lambda} (R - u_{i+r})^\lambda (u - R)^{\ell-\lambda} .$$

Terms of like power are collected to provide the representation

$$(u-R)^j + \sum_{\ell=0}^{j-1} p_\ell (u-R)^\ell \ .$$

Finding the p_ℓ, which requires computing linear combinations of the coefficients $a_{i\ell j}$ based on binomial coefficients and powers of the expressions $(R-u_\kappa)$, is a preprocessing step that needs to be performed only once. Taking R so that $\min |R - u_\kappa|$ is maximized is equivalent to picking R as the midpoint of the largest breakpoint interval between t_κ and $t_{\kappa+d+1}$, and this has proven to be a numerically reasonable choice.

To evaluate a basis function, the resulting representations of the g_{ij} are differenced pairwise and normalized to cancel highest terms. All this takes place with respect to a differencing region past the rightmost knot of the cohort. Figure 2 shows this schematically.

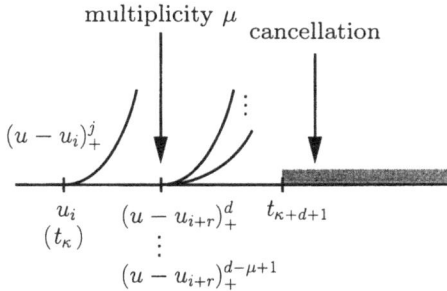

Fig. 2. Schematic view of differencing.

The differencing and renormalization pattern follows the outline

$$(u-R)^\delta + \sum_{\ell=0}^{\delta-1} p_\ell(u-R)^\ell - (u-R)^\delta + \sum_{\ell=0}^{\delta-1} q_\ell(u-R)^\ell$$

$$\rightarrow (u-R)^n + \left(\frac{1}{p_n - q_n}\right)\sum_{\ell=0}^{n-1}(p_\ell - q_\ell)(u-R)^\ell,$$

where n is the highest-power term not canceled by the subtraction.

A specific example may make this more clear. Figure 3 shows an example for $d = 2$ where breakpoint u_i has multiplicity 2, the next breakpoint has multiplicity 1, and the one after that has multiplicity 2.

The knot sequence is

$$t_\kappa = u_i, \ t_{\kappa+1} = u_i, \ t_{\kappa+2} = u_{i+1}, \ t_{\kappa+3} = u_{i+2}, \ t_{\kappa+4} = u_{i+2}$$

$$(u - u_i)_+^2 \qquad (u - u_{i+1})_+^2 \qquad (u - u_{i+2})_+^2$$
$$(u - u_i)_+^1 \qquad\qquad\qquad\qquad (u - u_{i+2})_+^1$$

$$u_i \qquad\qquad u_{i+1} \qquad\qquad u_{i+2}$$

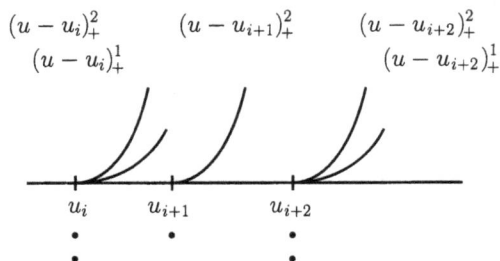

Fig. 3. Quadratic example.

The corresponding functions g_{ij} begin with the expressions

$$g_{i,2} = (u - u_i)^2 + \cdots, \; g_{i,1} = (u - u_i)^1 + \cdots$$
$$g_{i+1,2} = (u - u_{i+1})^2 + \cdots$$
$$g_{i+2,2} = (u - u_{i+2})^2 + \cdots, \; g_{i+2,1} = (u - u_{i+2})^1 + \cdots$$

Two successive differencing cohorts are present

$$g_{i,2}, g_{i,1}, g_{i+1,2}, g_{i+2,2}$$
$$g_{i,1}, g_{i+1,2}, g_{i+2,2}, g_{i+1,1}.$$

The first one provides the differencing pattern shown in Figure 4.

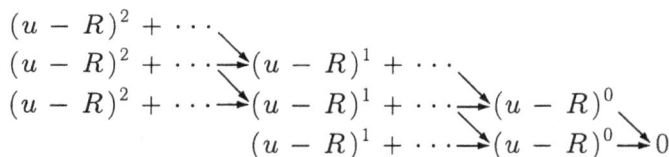

$$(u - R)^2 + \cdots$$
$$(u - R)^2 + \cdots \rightarrow (u - R)^1 + \cdots$$
$$(u - R)^2 + \cdots \rightarrow (u - R)^1 + \cdots \rightarrow (u - R)^0$$
$$(u - R)^1 + \cdots \rightarrow (u - R)^0 \rightarrow 0$$

Fig. 4. Quadratic example difference table.

This difference table gives the evaluation algorithm for the connection-matrix basis spline having the support $t_\kappa \leq u \leq t_{\kappa+d+1}$. The table need only be used for values of u in this interval, and the basis spline is simply set to zero outside.

§5. Examples and Closing Remarks

In Figures 5 and 6 we show two basis spline examples. Both are cubic examples on 5 breakpoints of multiplicity 1. For Figure 5 $u_i = 0$, $u_{i+1} = 1$, $u_{i+2} = 2$, $u_{i+3} = 3$, $u_{i+4} = 4$, and for Figure 6 $u_i = 0$, $u_{i+1} = 2.0001$, $u_{i+2} = 2.00011$, $u_{i+3} = 3$, $u_{i+4} = 4$. Both examples use the same choice of connection matrices

$$C_i = \begin{bmatrix} 1 & 0 \\ 0 & 1 \end{bmatrix}, \; C_{i+1} = \begin{bmatrix} 4 & 0 \\ 6 & 8 \end{bmatrix}, \; C_{i+2} = \begin{bmatrix} 3 & 0 \\ 12 & 9 \end{bmatrix}$$

$$C_{i+3} = \begin{bmatrix} 1 & 0 \\ 4 & 1 \end{bmatrix}, \; C_{i+4} = \begin{bmatrix} 1 & 0 \\ 0 & 1 \end{bmatrix}$$

The second example has a pair of knots extremely close, and the example is chosen to show that even in this unusual situation the results are quite good. Compared to a symbolic evaluation of the same basis function, the computations have a relative error of only 10^{-6} in double-precision (10-decimal) arithmetic.

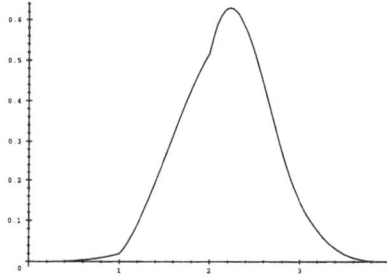

Fig. 5. Connection-matrix basis spline.

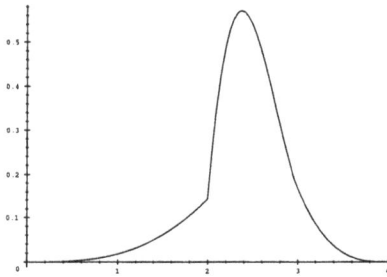

Fig. 6. Connection-matrix basis spline with nearly confluent knots.

Under the usual requirements for the connection matrices, the nonnegativity of the basis functions produced can be established inductively by carefully following the differencing and normalization process, observing at every step that one subtracts a more quickly growing polynomial from a more slowly growing one and normalizes the result to be positive. The partition of unity that the functions provide can be established by concentrating on an interval and observing that the sum of the basis functions on that interval is the same as the sum of the last pairwise differences. The normalization at this last pairwise difference stage turns out to be one, the sum of the pairwise differences collapses, and the only remaining nonzero function on the interval has the constant value one over the interval.

In closing, it is interesting to note what happens in cases when a connection matrix fails to have its required nonnegativity property. The process we have described may still produce linearly independent splines of compact support, but they cannot be expected to be nonnegative or partition unity. An example for beta-splines is shown in Figures 18.8 and 18.9 of [1].

References

1. Bartels, R. H., J. C. Beatty, and B. A. Barsky, *Splines for Use in Computer Graphics and Geometric Modeling*, Morgan Kaufmann, Los Altos, California, 1987.
2. Schumaker, L. L., *Spline Functions: Basic Theory*, Wiley, New York, 1981.
3. Seidel, H.-P., Polar forms for geometrically continuous spline curves of arbitrary Degree, ACM Trans. on Graphics **12** (1993), 1–34.

Richard Bartels
Computer Science Department
University of Waterloo
Waterloo, Ontario N2L 3G1
CANADA
rhbartel@uwaterloo.ca

John Beatty
Computer Science Department
University of Waterloo
Waterloo, Ontario N2L 3G1
CANADA
jcbeatty@uwaterloo.ca

Massic-Vector-Based Conics Modelling

Jean-Paul Bécar and Jean-Charles Fiorot

Abstract. From three linearly independent massic vectors describing a proper conic (BR-form), we determine its geometric features: F the focus, d the directrix and e the eccentricity. Conversely, from a (F, d, e) triple of data, we obtain a symmetric BR-form of the corresponding conic.

§1. Introduction and General Background

Let us consider $\mathcal{E}$ an affine space ($\mathbb{R}^2$, $\mathbb{R}^3$ in general), $\vec{\mathcal{E}}$ its associated linear space, and the massic vectors linear space $\hat{\mathcal{E}} = (\mathcal{E} \times \mathbb{R}^*) \cup \vec{\mathcal{E}}$ ($\alpha \in \hat{\mathcal{E}}$ a massic vector is either a weighted point: $\alpha = (A; a)$ or a pure vector: $\alpha = \vec{U}$).

Any rational curve $C(t)$ of $\mathcal{E}$ or $\tilde{\mathcal{E}}$ ($\tilde{\mathcal{E}}$ is the completion of $\mathcal{E}$ i.e. $\mathcal{E}$ with its point at infinity) is defined by $\alpha = (\alpha_0, \alpha_1, \ldots, \alpha_n)$ a set of massic vectors and is written $C(t) = BR[\alpha_0, \alpha_1, \ldots, \alpha_n](t)$ (or $BR[\alpha](t)$ in short). $BR[\alpha](t)$ is called the BR-form of $C(t)$ [3, 4]. This BR-form includes the rational Bézier curves [1, 2].

Let $\Pi = \hat{\mathcal{E}} - \{\vec{0}\} \to \tilde{\mathcal{E}}$ be the natural projection defined by $\Pi(A; a) = A$, $\Pi(\vec{U}) = (\vec{U})_\infty$ where $(\vec{U})_\infty$ designates the infinity point of $\mathcal{E}$ in the direction $\vec{U}$. We have the property: $\forall \lambda \neq 0, \forall \beta \in \hat{\mathcal{E}}, \Pi(\lambda * \beta) = \Pi(\beta)$.

We shall use the property that $BR[\alpha_0, \alpha_1, \ldots, \alpha_n](t) = \Pi(\sum_{i=0}^{n} B_i^n(t) * \alpha_i)$ i.e. a BR-curve of $\mathcal{E}$ is the Π-projection of a Bézier-de Casteljau curve of the space $\hat{\mathcal{E}}$. $\hat{\mathcal{E}}$ is a linear space with an addition operator denoted by $\oplus$ and an external multiplication denoted by $*$. For instance $(A; a) \oplus (B; b) = \left(\dfrac{aA + bB}{a + b}; a + b \right)$ if $a + b \neq 0$; $(A; a) \oplus (B; b) = a\overrightarrow{BA} = b\overrightarrow{AB}$ if $a + b = 0$; $(A; a) \oplus \vec{U} = (A + \dfrac{\vec{U}}{a}; a)$; $\lambda * (A; a) = (A; \lambda a)$ for $\lambda \in \mathbb{R}$.

Let us also recall that $\alpha_0, \alpha_1, \alpha_2$ being three linearly independent massic vectors of $\hat{\mathcal{E}}$ ($\mathcal{E}$ may be $\mathbb{R}^2$) the support of $BR[\alpha_0, \alpha_1, \alpha_2]$ is a proper conic bitangential at $\Pi(\alpha_0)$ and $\Pi(\alpha_2)$ respectively to the straight lines $\Pi(\alpha_0)\Pi(\alpha_1)$

Curves and Surfaces with Applications in CAGD
A. Le Méhauté, C. Rabut, and L. L. Schumaker (eds.), pp. 17–24.

Copyright ℗ 1997 by Vanderbilt University Press, Nashville, TN.
ISBN 0-8265-1293-3.
All rights of reproduction in any form reserved.

and $\Pi(\alpha_2)\Pi(\alpha_1)$. Conversely, any proper conic in a plane is a BR-curve support: $BR[\alpha_0, \alpha_1, \alpha_2]$ where $\alpha_0, \alpha_1, \alpha_2$ being three linearly independent massic vectors (Proposition 5.1.3 in [4]).

In a plane $\mathcal{E}$ let be given a straight line d (the directrix) a point F (the focus) and a positive scalar e (the eccentricity). In elementary books of geometry (one of the oldest book is by Pappus (4th Century) [7]), a conic is defined as the set of points M satisfying $\dfrac{MF}{MH} = e$, where H is the orthogonal projection of M on line d *.

In this paper, for CAGD purposes, we establish the link between the two previous descriptions of a conic. From the (F, d, e) triple data we determine a symmetric BR-form of the conic. Conversely, suppose we are given three linearly independent massic vectors defining a conic. By using adequate homographic parameter changes studied subsequently three new massic vectors are determined in such a way that they give the (F, d, e) geometric features of the conic directly. In [6] the author has had this preoccupation, from a second degree rational Bézier curve: the foci and the vertices of an ellipse (resp. hyperbola) were determined in an analytical way. In this work, from the BR-form of a conic including the Bézier rational form, we propose a more geometrical and more exhaustive presentation of the problem.

We will use the following result about the homographic parameter change (h.p.c.) for BR-curves. Consider

$$t = h(u) = \frac{a(1 - u) + bu}{c(1 - u) + du}, \qquad ad \neq bc. \qquad (1)$$

From Proposition 5.2 and Remark 3.1 in [5] we deduce that

$$BR[\alpha_0, \alpha_1, \alpha_2] \circ h(u) = BR[\beta_0, \beta_1, \beta_2](u),$$

with

$$\begin{aligned}
\beta_0 &= (c - a)^2 \alpha_0 \oplus 2a(c - a)\alpha_1 \oplus a^2 \alpha_2 \\
\beta_1 &= (c - a)(d - b)\alpha_0 \oplus (bc - 2ab + ad)\alpha_1 \oplus ab\alpha_2 \qquad (2) \\
\beta_2 &= (d - b)^2 \alpha_0 \oplus 2b(d - b)\alpha_1 \oplus b^2 \alpha_2
\end{aligned}$$

For the particular case $t = h(u) = \dfrac{bu}{c(1 - u) + bu}$, $(h(0) = 0, h(1) = 1)$ we obtain $\beta_0 = c^2 \alpha_0$, $\beta_1 = bc\alpha_1$, $\beta_2 = b^2 \alpha_2$. When $\alpha_0, \alpha_1, \alpha_2$ are weighted points, this h.p.c is used in [8].

By Proposition 5.1.6 in [4], we know that $\alpha_0, \alpha_1, \alpha_2$ being three massic vectors and $BR[\alpha_0, \alpha_1, \alpha_2]$ the associated proper conic, then if $\chi(\alpha_1)^2 - \chi(\alpha_0)\chi(\alpha_2) > 0$(resp. 0, resp. < 0), $BR[\alpha_0, \alpha_1, \alpha_2]$ is a hyperbola (resp. a parabola, resp. an ellipse) where χ is the linear form $\chi : \hat{\mathcal{E}} \to \mathbb{R}$ with $\chi(A; a) = a, \chi(\vec{U}) = 0$.

---

* This definition was already given by Euclid (3rd Century B.C.) and Archimedes (c. 287-212 B.C.)

§2. Determination of a BR-form of a Conic Given by (F, d, e)

Proposition 1. *Let (C) be a conic defined by a focus F, a directrix d and an eccentricity e. Consider A_1 the orthogonal projection of F on d and points A_0, A_2 on (C) symmetrically from F.*
Then, defining the three massic vectors $\alpha_0 = (A_0; 1), \alpha_1 = (A_1; e), \alpha_2 = (A_2; 1)$, the support (γ) of $BR[\alpha_0, \alpha_1, \alpha_2](t)$, $t \in \tilde{\mathbb{R}}$ is the conic (C).

Proof: By elementary results of geometry we know that:
(i) the straight lines $A_1 A_0$ and $A_1 A_2$ are respectively the tangent lines on (C) at A_0 and A_2,
(ii) S the vertex relative to the focus F satifies $SF = eSA_1$,
(iii) $e = \tan \widehat{FA_1 A_0}$.
We have $BR[\alpha_0, \alpha_1, \alpha_2](0) = \Pi(\sum_{i=0}^{2} B_i^2(0) * \alpha_i) = \Pi(\alpha_0) = A_0$ and similarly $BR[\alpha_0, \alpha_1, \alpha_2](1) = A_2$. As recalled in Section 1, straight lines $A_0 A_1 = \Pi(\alpha_0)\Pi(\alpha_1)$ and $A_2 A_1 = \Pi(\alpha_2)\Pi(\alpha_1)$ are tangent to (γ) respectively at A_0 and A_2. By some calculations we obtain successively $BR[\alpha_0, \alpha_1, \alpha_2](\frac{1}{2}) = \Pi(\frac{1}{4}\alpha_0 \oplus \frac{1}{2}\alpha_1 \oplus \frac{1}{4}\alpha_2) = \frac{F + eA_1}{1 + e} = S$. As the BR-conic (γ) and the conic (C) have five common elements (points A_0, A_2, tangent lines $A_0 A_1, A_2 A_1$ and point S), they are identical. ∎

Remark 2.1. The general BR-form (with three massic vectors) of the conic (F, d, e) is given by $BR[\beta_0, \beta_1, \beta_2](u) = BR[\alpha_0, \alpha_1, \alpha_2] \circ h(u)$ with h defined by (1) and where the massic polygon $(\alpha_0, \alpha_1, \alpha_2)$ (resp. $(\beta_0, \beta_1, \beta_2)$) is given by Proposition 1 (resp. formulae (2)).

In the next sections we deal with the reciprocal problem: from a BR-form of a conic, using successive h.p.c. we determine the corresponding geometric features (F, d, e).

§3. The Parabola

As previously recalled, a BR-parabola is defined by three massic vectors $\alpha_0, \alpha_1, \alpha_2$ satisfying $\chi(\alpha_1)^2 - \chi(\alpha_0)\chi(\alpha_2) = 0$. Then its representation can be summed up in three cases (1): $BR[(P_0; 1), (P_1; w), (P_2; w^2)]$, $w \in \mathbb{R}^*$, (2): $BR[(P_0; 1), \overrightarrow{A_1}, \overrightarrow{A_2}]$, (3): $BR[\overrightarrow{A_0}, \overrightarrow{A_1}, (P_2; 1)]$. By an affine parameter change $h(u) = 1 - u$, (3) gives $BR[\overrightarrow{A_0}, \overrightarrow{A_1}, (P_2; 1)] \circ h = BR[(P_2; 1), \overrightarrow{A_1}, \overrightarrow{A_0}]$. Thus, we only consider cases (1) and (2).

Proposition 2. *Let $BR[\alpha_0, \alpha_1, \alpha_2]$ be a parabola where $\alpha_0 = (P_0; 1)$, $\alpha_1 = (P_1; w)$, $\alpha_2 = (P_2; w^2)$, $w \in \mathbb{R}^*$. At the most, there are h_1, h_2, h_3 three h.p.c. successively giving*
(i) If $w = 1$, $h_1(u) = \dfrac{u}{1 - u}$, $BR[\alpha_0, \alpha_1, \alpha_2] \circ h_1 = BR[(P_0; 1), \overrightarrow{B_1}, \overrightarrow{B_2}]$ with $\overrightarrow{B_1} = \overrightarrow{P_0 P_1}, \overrightarrow{B_2} = \overrightarrow{P_1 P_0} + \overrightarrow{P_1 P_2}$,

if $w \neq 1, h_1(u) = \dfrac{1}{1-w}u, BR[\alpha_0, \alpha_1, \alpha_2] \circ h_1 = BR[(P_0; 1), \overrightarrow{B_1}, \overrightarrow{B_2}]$ with

$\overrightarrow{B_1} = \dfrac{w}{1-w}\overrightarrow{P_0 P_1}, \overrightarrow{B_2} = (\dfrac{w}{1-w})^2(\overrightarrow{P_1 P_0} + \overrightarrow{P_1 P_2}).$

In both cases $\overrightarrow{B_2}$ is the axis direction of the parabola. If $\overrightarrow{B_1}.\overrightarrow{B_2} = 0$, P_0 is the vertex of the parabola, $\overrightarrow{B_1}$ is collinear to the tangent at P_0.

(ii) If $\overrightarrow{B_1}.\overrightarrow{B_2} \neq 0, h_2(u) = \dfrac{ru}{1-u+ru}, r = \dfrac{\overrightarrow{B_1}.\overrightarrow{B_2}}{\| \overrightarrow{B_2} \|^2}, BR[(P_0; 1), \overrightarrow{B_1}, \overrightarrow{B_2}] \circ h_2 = $

$BR[(P_0; 1), \overrightarrow{C_1}, \overrightarrow{C_2}]$ with $\overrightarrow{C_1} = r\overrightarrow{B_1}, \overrightarrow{C_2} = r^2\overrightarrow{B_2}$ and $\overrightarrow{C_1}.\overrightarrow{C_2} = \| \overrightarrow{C_2} \|^2.$

(iii) If $\overrightarrow{C_1}.\overrightarrow{C_2} \neq 0$ and $\overrightarrow{C_1}.\overrightarrow{C_2} = \| \overrightarrow{C_2} \|^2, h_3(u) = \dfrac{-(1-u)+u}{u},$

$BR[(P_0; 1), \overrightarrow{C_1}, \overrightarrow{C_2}] \circ h_3 = BR[(D_0; 1), \overrightarrow{D_1}, \overrightarrow{D_2}]$ with $D_0 = P_0 - 2\overrightarrow{C_1} + \overrightarrow{C_2}$ is the vertex of the parabola, $\overrightarrow{D_1} = \overrightarrow{C_1} - \overrightarrow{C_2}, \overrightarrow{D_2} = \overrightarrow{C_2}, \overrightarrow{D_1}.\overrightarrow{D_2} = 0$; $\overrightarrow{D_1}$ is collinear to the tangent line at D_0 ; $\overrightarrow{D_2}$ is the axis direction.

Proof: The proof is straightforward by definition of h_1, h_2, h_3, formulae (2), properties of Π and operators $\oplus, *$. We only give some details for cases (i) and (iii). $BR[(P_0; 1), \overrightarrow{B_1}, \overrightarrow{B_2}](1) = \Pi(\overrightarrow{B_2}) = (\overrightarrow{B_2})_\infty$, then $\overrightarrow{B_2}$ is the axis direction of the parabola. $BR[(P_0; 1), \overrightarrow{B_1}, \overrightarrow{B_2}](0) = \Pi(P_0; 1) = P_0$. By Proposition 4.1.1 in [4] we have $\dfrac{d}{dt}(BR[(P_0; 1), \overrightarrow{B_1}, \overrightarrow{B_2}])(0) = 2\overrightarrow{B_1}$. If $\overrightarrow{B_1}$ is orthogonal to $\overrightarrow{B_2}$ (the axis direction) then P_0 is the vertex of the parabola. ∎

Remark 3.1. In Proposition 2 (ii) (hypothesis $\overrightarrow{B_1}.\overrightarrow{B_2} \neq 0$) taking P_0 (for instance) as the common origin of vectors $\overrightarrow{B_1}, \overrightarrow{B_2}, \overrightarrow{C_1}, \overrightarrow{C_2}$ and denoting by b_1 (resp. b_2) the end point of $\overrightarrow{B_1}$ (resp. $\overrightarrow{B_2}$), we determine (after some calculations) the end point c_1 (resp. c_2) of $\overrightarrow{C_1}$(resp. $\overrightarrow{C_2}$) by the following geometric drawing. The parallel line to (b_1, b_2) passing through point b_1' (the orthogonal projection of b_1 on (P_0, b_2)) cuts (P_0, b_1) at c_1 and c_2 is the orthogonal projection of c_1 on (P_0, b_2).

Proposition 3. Let $BR[(D_0; 1), \overrightarrow{D_1}, \overrightarrow{D_2}]$ be a parabola with $\overrightarrow{D_1}.\overrightarrow{D_2} = 0$. Let us define points $M = D_0 + \overrightarrow{D_1}, N = D_0 + \overrightarrow{D_2}, D$ on line $(D_0, \overrightarrow{D_2})$ such that the orthogonal projection of D on line (MN) is M and finally $F = D_0 - \overrightarrow{D_0 D}$. Then F is the focus of the parabola, $d = (D, \overrightarrow{D_1})$ is the directrix and $(D, \overrightarrow{D_2})$ is the axis.

Proof: (i) Let us define $\rho = \dfrac{\| \overrightarrow{D_1} \|}{\| \overrightarrow{D_2} \|}, \overrightarrow{E_1} = \rho\overrightarrow{D_1}, \overrightarrow{E_2} = \rho^2\overrightarrow{D_2}$ (hence $\| \overrightarrow{E_1} \| = \| \overrightarrow{E_2} \| = \| \overrightarrow{D_0 D} \|$) and $h_4(u) = \dfrac{\rho u}{(1-u)+\rho u}$, then $BR[(D_0; 1), \overrightarrow{D_1}, \overrightarrow{D_2}] \circ h_4 = $

$BR[(D_0; 1), \overrightarrow{E_1}, \overrightarrow{E_2}]$.

(ii) Consider $h_5(u) = \dfrac{-(1 - u) + u}{2u}$ we have $BR[(D_0; 1), \overrightarrow{E_1}, \overrightarrow{E_2}] \circ h_5$

$= BR[(A_0; 1), (D; 1)(A_2; 1)]$ with $A_0 = D_0 - 2\overrightarrow{E_1} + \overrightarrow{E_2}$,$(D = D_0 - \overrightarrow{E_2})$, $A_2 = D_0 + 2\overrightarrow{E_1} + \overrightarrow{E_2}$. Let (C) be the parabola $(e = 1)$ defined by the focus $F = D_0 + \overrightarrow{E_2}(= D_0 - \overrightarrow{D_0 D} = \dfrac{A_0 + A_2}{2})$, the directrix $d = (D, \overrightarrow{A_0 A_2}) = (D, \overrightarrow{D_1})$. Proposition 1 implies that the support of $BR[(A_0; 1), (D; 1)(A_2; 1)]$ is identical to (C). ■

Finally following the five steps (at the most) of Propositions 1 and 2 in this order we determine the focus and the directrix of a BR-parabola.

§4. The Ellipse

A BR-ellipse is defined by three massic vectors $\alpha_0, \alpha_1, \alpha_2$ satisfying $\chi(\alpha_1)^2 - \chi(\alpha_0)\chi(\alpha_2) < 0$. We have six possibilities for the sign of the triplet $(\chi(\alpha_0), \chi(\alpha_1), \chi(\alpha_2))$: (+,+,+) (+,-,+) (-,+,-) (-,-,-) (+,0,+) (-,0,-). By taking the opposite massic polygon we come back to three cases (+,+,+) (+,-,+) and (+,0,+). And considering a h.p.c. $h(u) = \dfrac{u}{-(1 - u) + u}$ we have $BR[\alpha_0, \alpha_1, \alpha_2] \circ h = BR[\alpha_0, -\alpha_1, \alpha_2]$, hence the second case also comes back to the first. In conclusion, they are two cases left to be examined: (+,+,+) and (+,0,+).

Proposition 4. *Let* $BR[(A_0; a_0), (A_1; a_1), (A_2; a_2)]$ *be a BR ellipse* $(a_i > 0, a_1^2 - a_0 a_2 < 0)$. *At the most, there are* h_1, h_2, h_3 *three h.p.c. successively giving :*

(i) $BR[(A_0; a_0), (A_1; a_1), (A_2; a_2)] \circ h_1 = BR[(A_0; a_0), \overrightarrow{B_1}, (B_2; b_2)]$ *with* $h_1(u) =$
$\dfrac{ru}{(1 - u) + (1 + r)u}$, $r = -\dfrac{a_0}{a_1}$, $\overrightarrow{B_1} = a_0 \overrightarrow{A_1 A_0}$, $B_2 = \dfrac{a_0 A_0 + 2a_1 r A_1 + a_2 r^2 A_2}{a_0 + 2a_1 r + a_2 r^2}$,
$b_2 = a_0 + 2a_1 r + a_2 r^2 > 0$.

(ii) $BR[(A_0; a_0), \overrightarrow{B_1}, (B_2; b_2)] \circ h_2 = BR[(A_0; 1), \overrightarrow{C_1}, (B_2; 1)]$
with $h_2(u) = \dfrac{(b_2)^{-1/2} u}{a_0^{-1/2}(1 - u) + (b_2)^{-1/2} u}$, $\overrightarrow{C_1} = (a_0 b_2)^{-1/2} \overrightarrow{B_1}$.

(iii) Define $k_1 = \overrightarrow{A_0 B_2} . \overrightarrow{C_1}$, $k_2 = \| \overrightarrow{C_1} \|^2 - \dfrac{1}{4} \| \overrightarrow{A_0 B_2} \|^2$,

if $k_1 = k_2 = 0$, $BR[(A_0; 1), \overrightarrow{C_1}, (B_2; 1)]$ *is the circle of center* $\dfrac{1}{2}(A_0 + B_2)$ *and radius* $\dfrac{1}{2} \| \overrightarrow{A_0 B_2} \|$, *else*

define $h_3(u) = \dfrac{\sin \theta .(1 - u) + \cos \theta .u}{(\cos \theta + \sin \theta).(1 - u) + (-\sin \theta + \cos \theta).u}$,

with $\theta = \dfrac{\pi}{8}$ *(resp.* $\theta = \dfrac{1}{4} \text{Arctan} \dfrac{k_1}{k_2})$ *if* $k_1 \neq 0, k_2 = 0$ *(resp.* $\forall k_1 \in \mathbb{R}, k_2 \neq 0)$,

$BR[(A_0; 1), \overrightarrow{C_1}, (B_2; 1)] \circ h_3 = BR[(D_0; 1), \overrightarrow{D_1}, (D_2; 1)]$ *with* $\overrightarrow{D_0 D_2} . \overrightarrow{D_1} = 0$,

$D_0 = \cos^2\theta . A_0 + \sin^2\theta . B_2 + \sin 2\theta . \overrightarrow{C_1}, \; \overrightarrow{D_1} = \cos 2\theta . \overrightarrow{C_1} + \dfrac{1}{2}\sin 2\theta . \overrightarrow{A_0 B_2}, \; D_2 = \sin^2\theta . A_0 + \cos^2\theta . B_2 - \sin 2\theta . \overrightarrow{C_1}.$

Proof: By some calculations. Remember the circle is characterized by Proposition 5.4.2 in [4]. ∎

After the three steps (at the most) of Proposition 4 we determine the (F, d, e) geometric features of the ellipse.

Proposition 5. *Let us consider the BR-ellipse* $BR[(D_0; 1), \overrightarrow{D_1}, (D_2; 1)]$ *with* $\overrightarrow{D_0 D_2} . \overrightarrow{D_1} = 0$. *Then point* $O = \dfrac{1}{2}(D_0 + D_2)$ *is the center.*

(i) If $\| \overrightarrow{D_0 D_2} \| > 2 \| \overrightarrow{D_1} \|$, *define* $a.\vec{i} = \overrightarrow{OD_2}$, $b.\vec{j} = \overrightarrow{D_1}$ *($\vec{i}$ and $\vec{j}$ are orthonormal vectors). If* $\| \overrightarrow{D_0 D_2} \| < 2 \| \overrightarrow{D_1} \|$ *define* $a.\vec{i} = \overrightarrow{D_1}$, $b.\vec{j} = \overrightarrow{OD_0}$. *Then* $c = (a^2 - b^2)^{1/2}$, *the eccentricity* $e = \dfrac{c}{a}$, *the foci are given by* $\overrightarrow{OF} = -\overrightarrow{OF'} = c\vec{i}$ *and the marks on* $(O, \vec{i})$ *of the associated directrices are* $\overrightarrow{OH} = -\overrightarrow{OH'} = \dfrac{a^2}{c}\vec{i}$.

(ii) If $\| \overrightarrow{D_0 D_2} \| = 2 \| \overrightarrow{D_1} \|$ *the ellipse is a circle.*

Proof: Considering the Cartesian frame $(O, \vec{i}, \vec{j})$ the equation of the BR-ellipse is as follows: $\dfrac{x^2}{a^2} + \dfrac{y^2}{b^2} = 1$. ∎

§5. The Hyperbola

Proposition 6. *Let* $BR[\alpha_0, \alpha_1, \alpha_2]$ *be a BR-hyperbola. Let* $a_0 = \chi(\alpha_0), a_1 = \chi(\alpha_1), a_2 = \chi(\alpha_2)$ $(a_1^2 - a_0 a_2 > 0)$.
(i) (a) $a_0 - 2a_1 + a_2 \neq 0$, *let* t_1, t_2 *be the two real roots of the second degree equation (denominator of the BR)*

$$t^2(a_0 - 2a_1 + a_2) + 2t(a_1 - a_0) + a_0 = 0 \qquad (3)$$

and $h_1(u) = (1 - u)t_2 + ut_1$, *then*

$$BR[\alpha_0, \alpha_1, \alpha_2] \circ h_1 = BR[\dfrac{1}{\chi(\beta_1)}\overrightarrow{B_0}, (B_1; 1), \dfrac{1}{\chi(\beta_1)}\overrightarrow{B_2}] \text{ where}$$

$\overrightarrow{B_0} = (1 - t_2)^2 \alpha_0 \oplus 2t_2(1 - t_2)\alpha_1 \oplus t_2^2 \alpha_2,$
$B_1 = \Pi(\beta_1), \; \beta_1 = (1 - t_1)(1 - t_2)\alpha_0 \oplus (t_1 - 2t_1 t_2 + t_2)\alpha_1 \oplus t_1 t_2 \alpha_2$
$\overrightarrow{B_2} = (1 - t_1)^2 \alpha_0 \oplus 2t_1(1 - t_1)\alpha_1 \oplus t_1^2 \alpha_2 ,$

(b) $a_0 - 2a_1 + a_2 = 0$, $t_1 = \dfrac{a_0}{2(a_0 - a_1)}$, $h_1(u) = \dfrac{t_1(1 - u) + u}{1 - u},$

$$BR[\alpha_0, \alpha_1, \alpha_2] \circ h_1 = BR[\dfrac{1}{\chi(\beta_1)}\overrightarrow{B_0}, (B_1; 1), \dfrac{1}{\chi(\beta_1)}\overrightarrow{B_2}] \text{ where}$$

$\overrightarrow{B_0} = (1 - t_1)^2 \alpha_0 \oplus 2t_1(1 - t_1)\alpha_1 \oplus t_1^2 \alpha_2,$

$B_1 = \Pi(\beta_1), \quad \beta_1 = -(1 - t_1)\alpha_0 \oplus (1 - 2t_1)\alpha_1 \oplus t_1\alpha_2,$

$\overrightarrow{B_2} = \alpha_0 \ominus 2\alpha_1 \oplus \alpha_2 ,$

(ii) $h_2(u) = \dfrac{bu}{\dfrac{1}{b}(1 - u) + bu}$, define $b = \left(\dfrac{\parallel \overrightarrow{B_0} \parallel}{\parallel \overrightarrow{B_2} \parallel}\right)^{1/4}$ then

$BR[\dfrac{1}{\chi(\beta_1)}\overrightarrow{B_0}, (B_1; 1), \dfrac{1}{\chi(\beta_1)}\overrightarrow{B_2}] \circ h_2 = BR[\overrightarrow{C_0}, (B_1; 1), \overrightarrow{C_2}]$

with $\overrightarrow{C_0} = \dfrac{1}{b^2}\dfrac{\overrightarrow{B_0}}{\chi(\beta_1)}, \overrightarrow{C_2} = b^2\dfrac{\overrightarrow{B_2}}{\chi(\beta_1)}$ $(\parallel \overrightarrow{C_0} \parallel = \parallel \overrightarrow{C_2} \parallel)$.

Proof: (i) (a), $\overrightarrow{B_0}$ (resp. $\overrightarrow{B_2}$) is a pure vector because t_2 (resp. t_1) is a root of (3). The support of the entire BR-hyperbola is not altered by an affine parameter change or a h.p.c then $(\chi(\beta_1))^2 - \chi(\overrightarrow{B_0})\chi(\overrightarrow{B_2}) > 0$ implies $\chi(\beta_1) \neq 0$: β_1 is a weighted point. (i)(b), we have $a_0 \neq a_1$ else $a_1^2 - a_0 a_2 = 0$. $\overrightarrow{B_0}$ is a pure vector because t_1 is a root of (3), $\overrightarrow{B_2}$ is a pure vector according to the hypothesis $a_0 - 2a_1 + a_2 = 0$. (ii) With the choice of b we verify that $\parallel \overrightarrow{C_0} \parallel = \parallel \overrightarrow{C_2} \parallel$. ∎

Now after two steps at the most, we can obtained the desired geometric features of the hyperbola.

Proposition 7. *Consider* $BR[\overrightarrow{C_0}, (B_1; 1), \overrightarrow{C_2}]$ *with* $\parallel \overrightarrow{C_0} \parallel = \parallel \overrightarrow{C_2} \parallel$ *then* B_1 *is the center of the hyperbola,* $(B_1, \overrightarrow{C_0})$ *and* $(B_1, \overrightarrow{C_2})$ *are the asymptotic lines. Define* $a\vec{i} = \dfrac{1}{2}(\overrightarrow{C_0} + \overrightarrow{C_2}), b\vec{j} = \dfrac{1}{2}(\overrightarrow{C_0} - \overrightarrow{C_2})$ ($\vec{i}$ *and* $\vec{j}$ *are orthonormal vectors), then* $c = (a^2 + b^2)^{1/2}$, *the eccentricity* $e = \dfrac{c}{a}$, *the foci are given by* $\overrightarrow{B_1 F} = -\overrightarrow{B_1 F'} = c\vec{i}$, *and the marks on* $(B_1, \vec{i})$ *of the associated directrices are* $\overrightarrow{B_1 H} = -\overrightarrow{B_1 H'} = \dfrac{a^2}{c}\vec{i}$.

Proof: Relative to the Cartesian frame $(B_1, \vec{i}, \vec{j})$, the equation of the BR-hyperbola is $\dfrac{x^2}{a^2} - \dfrac{y^2}{b^2} = 1$. ∎

References

1. Farin, G. E., Algorithms for rational Bézier curves, Computer-Aided Design **15** (1983), 73-77.

2. Faux, I. E., and M. J. Pratt, *Computational Geometry for Design and Manufacture*, Ellis Horwood, Chichester, 1979.

3. Fiorot, J.-C., and P. Jeannin, Nouvelle description des courbes rationnelles à l'aide de points et vecteurs de contrôle, Comptes-Rendus de l'Académie des Sciences de Paris **305**, Ser. I, (1987), 435-440.

4. Fiorot, J.-C., and P. Jeannin, *Courbes et Surfaces Rationnelles, Applications à la CAO*, Masson, RMA 12, Paris, 1989. English translation: *Rational Curves and Surfaces. Applications to CAD*, J. Wiley, Chichester, 1992.

5. Fiorot, J.-C., and P. Jeannin, and S. Taleb , New control massic polygon of a B-Rational curve resulting from a homographic change of parameter, Numerical Algorithms **6** (1994), 379-418.

6. Lee, E. T. Y., The rational Bezier representation for conics in *Geometric Modelling: Algorithms and New Trends* ,G. Farin (ed.) SIAM, Philadelphia, 1987, 3-19.

7. Pappus, *Collection Mathématique (livre 7)*, A. Blanchard, Paris, 1982.

8. Patterson, R. R. , Projective transformations of the parameter of a Bernstein-Bézier curve, ACM Trans. on Graphics **6** (1985), 276-290.

Jean-Paul Bécar
Université de Valenciennes et du Hainaut Cambrésis
IUT- GEII, Laboratoire LIMAV,
B.P. 311, 59304 Valenciennes Cedex, FRANCE
becar@ univ-valenciennes.fr

Jean-Charles Fiorot
Université de Valenciennes et du Hainaut Cambrésis
ENSIMEV, Laboratoire LIMAV,
B.P. 311, 59304 Valenciennes Cedex, FRANCE
fiorot@ univ-valenciennes.fr

Globally G^1 Free Form Surfaces Using Kirchhoff-Love Plate Energy Methods

Michel Bercovier, Oleg Volpin and Tanya Matskewich

Abstract. A new paradigm for free form surfaces based on quadrilateral patches and a "physical" (i. e. parametrically independent) plate energy is introduced. The elementary patch energy construction is detailed. Inter-patch G^1 continuity is constructed, and for the case of Bézier patches the corresponding minimization problem under constraints is introduced. The present method is very effective for large collections of patches and/or difficult situations including sharp junctions and n-patch constructions. This is illustrated by several free form surfaces with vertex interpolation constraints.

§1. Introduction

Multi-patch surface building/editing by means of a plate energy type method is a classical approach in *Computer Aided Geometric Design* (CAGD). Thus Celniker and Gossard [5] apply the finite element method, using Zienkiewicz [10] rational C^1 triangular patches. On quadrilateral patches the so-called "plate energy" is usually based on the minimization of a functional defined in term of local patch parametrization (Bercovier and Volpin [2] and the references given therein). The definition of a "physical" plate quadrilateral is difficult due to the necessity of constructing Hermite type basis functions over arbitrary meshes. Hence the FEM literature is limited to cubic type or classical spline elements as given in [6] and [4] over regular grids, or to Clough-Tocher macro-elements [10]. In the present work an entirely new paradigm for "plate type" quadrilateral patches well suited for free form surfaces is presented. This approach has numerous applications such as reverse engineering, data interpolation over arbitrary quadrilateral meshes, invariant n-patches, hierarchical data reduction and so on. The present work centers on the construction of this new type of patches. Due to lack of space, details of the implementation on arbitrary meshes and actual algorithmic implementations are omitted. They will appear elsewhere [3].

Curves and Surfaces with Applications in CAGD
A. Le Méhauté, C. Rabut, and L. L. Schumaker (eds.), pp. 25–34.
Copyright ©1997 by Vanderbilt University Press, Nashville, TN.
ISBN 0-8265-1293-3.
All rights of reproduction in any form reserved.

§2. Problem Definition

Consider an arbitrary mesh consisting of non-overlapping quadrilateral elements. The number of elements sharing a common vertex is not restricted. The purpose of the present work is to build a global G^1 surface interpolating the given mesh vertices. The elementary patches consist of $(n+1) \times (m+1)$ control points based surfaces (from Bézier to NURBS). Hence the free control points are computed by the global minimization of a "plate " type functional. Here the implementation is given for Bézier patches [9].

2.1 Plate Energy Functional

Let Ω_k be a quadrilateral in a plane, say Oxy, the corresponding Kirchhoff-Love plate bending energy of this element is given by

$$E_k = \int\int_{\Omega_k} \left(\frac{\partial^2 z}{\partial x^2} + \frac{\partial^2 z}{\partial y^2}\right)^2 - 2(1-\nu)\left(\frac{\partial^2 z}{\partial x^2}\frac{\partial^2 z}{\partial y^2} - \left(\frac{\partial^2 z}{\partial x \partial y}\right)^2\right) dxdy. \quad (1)$$

Here $z(x,y)$ is the normal displacement relative to the plate initial plane $(0 \le \nu \le 0.5)$. If this plane were actually the tangent plane of a surface and Ω_k were small enough, equation (1) would be a good approximation to its Gaussian curvature energy [8]. The quadratic functional is independent of the actual position of the plane Oxy and of any underlying parameterization. To compute it, one introduces the notion of reference element.

2.2 Reference Element

Consider the planar reference element $\Omega' = [0,1]^2$ and the bilinear mapping defined by

$$\varphi_1(u,v) = (1-u)(1-v) \qquad \varphi_2(u,v) = u(1-v)$$
$$\varphi_3(u,v) = uv \qquad \varphi_4(u,v) = (1-u)v \qquad (2)$$

and

$$x(u,v) = \sum_{i=1}^{4} x_i\varphi_i(u,v) \qquad y(u,v) = \sum_{i=1}^{4} y_i\varphi_i(u,v), \quad (3)$$

where (x_i, y_i) are the coordinates of the vertices of element Ω_k. Following the FEM, the bending function $z(x,y)$ over Ω_k will be by definition such that

$$z(x(u,v), y(u,v)) = \tilde{z}(u,v), \quad (4)$$

where

$$\tilde{z}(u,v) = \sum_{ij} Z_{ij}\psi_{ij}(u,v). \quad (5)$$

(In the considered case $\psi_{ij}(u,v) = B_i^n(u)B_j^m(v)$). Of course, the Z_{ij} depend also on k - pointer of the initial patch in the mesh. The Z_{ij} are the element degrees of freedom, and the $\psi_{ij}(u,v)$ the shape functions defined on the reference element. Let J be the Jacobian matrix of the bilinear transformation. Then

$$J = \begin{pmatrix} \frac{\partial x}{\partial u} & \frac{\partial x}{\partial v} \\ \frac{\partial y}{\partial u} & \frac{\partial y}{\partial v} \end{pmatrix}; \qquad J^{-1} = \frac{1}{\det J} \begin{pmatrix} \frac{\partial y}{\partial v} & -\frac{\partial y}{\partial u} \\ -\frac{\partial x}{\partial v} & \frac{\partial x}{\partial u} \end{pmatrix} = \begin{pmatrix} \frac{\partial u}{\partial x} & \frac{\partial u}{\partial y} \\ \frac{\partial v}{\partial x} & \frac{\partial v}{\partial y} \end{pmatrix}. \qquad (6)$$

Define for any functional over Ω_k,

$$\tilde{F}(\tilde{z}(u,v)) = F(z(x,y)). \qquad (7)$$

Thus its integral can be evaluated on the reference element, since

$$\int\int_{\Omega_k} F(z(x,y))dxdy = \int\int_{\Omega'} \tilde{F}(\tilde{z}(u,v))|\det J| \, dudv. \qquad (8)$$

§3. Surface Construction

In Mechanical Engineering, shells have been approximated by assemblies of flat plates [1], and the present paradigm relies on and extends this approach to surface construction. The global energy functional for the whole mesh is taken as the sum of the local functionals (E_k) over all mesh elements:

$$\mathbf{E} = \sum_k E_k. \qquad (9)$$

The minimization energy is constructed element (i.e. patch) by element in two steps. First a classical plate formulation is introduced on a local plane related to the underlying mesh. Assembling all the local energies in a global one is then done by introducing the constrained "in plane" degrees of freedom and adding a least square correction to the classic plate energy. In the present work the patch surface functions $\tilde{z}$ (and even the parameterization $\tilde{x}, \tilde{y}$) are represented as Bézier patch of degree 4 (note that for a regular grid cubic patches would do also). For each patch the control points Z_{ij} of the bending function are defined so as to minimize in some way the energy functional.

3.1 Definition of the Local Coordinate System

Let us consider one element of the mesh - generally it is a non-planar quadrilateral with vertices A, B, C, D. (In this section subscribing indices, corresponding to the element number in the mesh are omitted). The first purpose is to approximate this non-planar element by a planar one as close as possible to the original one, and such that the normal to the constructed quadrilateral is roughly a "normal" to the initial element. It is relative to this new plane quadrilateral that the plate energy will be defined.

As shown in Figure 1, the origin O' of the local coordinate system is defined as:

$$\vec{OO'} = (\vec{OA} + \vec{OB} + \vec{OC} + \vec{OD})/4. \qquad (10)$$

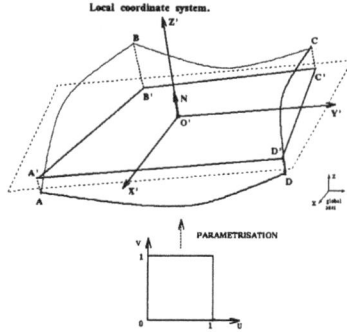

Fig. 1. Local Coordinate System and Initial Parameterization.

Here O is the origin of the global coordinate system. The normal (i.e. Z' axis) of the local coordinate system is chosen to be

$$\vec{N}' = (\vec{N}_{\Delta ABC} S_{\Delta ABC} + \vec{N}_{\Delta BCD} S_{\Delta BCD} + \vec{N}_{\Delta CDA} S_{\Delta CDA} + \vec{N}_{\Delta DAB} S_{\Delta DAB}) \tag{11}$$

where $\vec{N}$ and S describe the normal and the relative surface of the corresponding triangle. (Note that such $\vec{N}'$ is the mean value of the normal in the bilinear interpolation of quadrilateral $ABCD$). $\vec{N}'$ defines the Z' axis of the local coordinate system, and the direction of the two other axes can be arbitrarily chosen in the perpendicular plane. Their choice is discussed below. The resulting plane P passing through the origin of the local coordinate system O' with normal $\vec{N}'$ (equation (11)) will be the initial plane for the patch plate energy. Obviously, this plane depends on the actual mesh element.

3.2 Bilinear Interpolation

Next the vertices of the initial quadrilateral element are projected on the plane P. This defines four nodes in that plane: A', B', C', D'. The directions of the X' and Y' axis are then defined by the Gram-Schmidt orthogonalization of the vectors

$$\vec{N}', \quad \frac{1}{2}(\vec{O'A'} + \vec{O'B'}), \quad \frac{1}{2}(\vec{O'A'} + \vec{O'D'}). \tag{12}$$

Since the vertices A', B', C', D' are in the plane $X'Y'$,

$$A' = \begin{pmatrix} x_1 \\ y_1 \end{pmatrix} \quad B' = \begin{pmatrix} x_2 \\ y_2 \end{pmatrix} \quad C' = \begin{pmatrix} x_3 \\ y_3 \end{pmatrix} \quad D' = \begin{pmatrix} x_4 \\ y_4 \end{pmatrix}. \tag{13}$$

The resulting planar quadrilateral $A'B'C'D'$ is defined by the bilinear interpolation equation (3).

3.3 Initial Energy Functional in the u, v Coordinates

As is done in the FEM, to compute the initial energy functional (equation (1)) for each mesh element in the corresponding u, v coordinates the following steps are performed:

- Express all partial derivatives from the equation (1) in terms of u, v.
- Compute the Jacobian of the transformation from local coordinates x, y to the reference domain coordinates u, v.
- Compute the integrals over the reference domain $\Omega' = \{(u, v) \mid 0 \leq u, v \leq 1\}$ instead of over the initial domain Ω.

Consider the computation of one of the partial derivatives. It is easily seen that

$$\frac{\partial^2 z}{\partial x^2} = \frac{\partial^2 z}{\partial u^2}\left(\frac{\partial u}{\partial x}\right)^2 + 2\frac{\partial^2 z}{\partial u \partial v}\frac{\partial u}{\partial x}\frac{\partial v}{\partial x} + \frac{\partial^2 z}{\partial v^2}\left(\frac{\partial v}{\partial x}\right)^2 + \frac{\partial z}{\partial u}\frac{\partial^2 u}{\partial x^2} + \frac{\partial z}{\partial v}\frac{\partial^2 v}{\partial x^2} \quad (14)$$

where $\frac{\partial u}{\partial x}$ and $\frac{\partial v}{\partial x}$ are obtained from equation (6).
For the second order derivatives $\frac{\partial^2 u}{\partial x^2}$ and $\frac{\partial^2 v}{\partial x^2}$,

$$\begin{pmatrix} \frac{\partial^2 u}{\partial x^2} \\ \frac{\partial^2 v}{\partial x^2} \end{pmatrix} = 2\frac{\partial u}{\partial x}\frac{\partial v}{\partial x}\begin{pmatrix} \alpha\frac{\partial u}{\partial x} + \beta\frac{\partial u}{\partial y} \\ \alpha\frac{\partial v}{\partial x} + \beta\frac{\partial v}{\partial y} \end{pmatrix}, \quad (15)$$

where $\alpha = x_1 - x_2 + x_3 - x_4$ and $\beta = y_1 - y_2 + y_3 - y_4$.

The other second order partial derivatives are obtained in the same way. In the following steps the Jacobian can be easily computed from the bilinear interpolation (equation (3)) and the resulting integral over the reference square domain can be evaluated by numerical techniques.

3.4 Generalization of Bilinear Interpolation and Energy Functional

At this point, the patch is defined in the local reference frame $O'x'y'z'$ with $Z'(u, v)$ a Bézier function of order $n \times m$ and $X'(u, v)$ and $Y'(u, v)$ bilinear only. Now one must relate the *local* degrees of freedom $\mathbf{b}'_{i,j}$, $0 \leq i \leq n$, $0 \leq j \leq m$, to the global ones $\mathbf{b}_{i,j}$. Since the Bézier surface is invariant, by rotation one computes the coordinates of the local Bézier points in the global system and sums up the element quadratic functionals. Such a procedure is done in the FEM to define the resulting deformed surface [5,4]. Even so the resulting surface is "poor", since in every local plane the corresponding Bézier polynomials $X'(u, v)$ and $Y'(u, v)$ are bilinear only. To make full use of the Bézier patch one would have to allow functions of higher degree than bilinear over the domain of definition Ω_k. Suppose that $x'(u, v)$, $y'(u, v)$ are Bézier Bernstein polynomials with unknown coefficients to be obtained as a result of the minimization problem. Denote these new polynomials by $\tilde{x}(u, v)$, $\tilde{y}(u, v)$ and their control points by $\tilde{bx}_{ij}$, $\tilde{by}_{ij}$. Since one would like to stay as close as possible to the plate energy formulation, these polynomials must stay close to the initial bilinear description, so as to preserve the expressions for the partial derivatives in the initial energy functional. (Another option is to recompute the energy functional, according to the $\tilde{x}(u,v)$, $\tilde{y}(u,v)$, but then this energy functional will contain unknown coefficients of these polynomials and lead to non-linear equations.) Here the initial energy functional (equation (1)) is modified so as to minimize the deviation from the bilinear description. Let n, m - be the degrees of polynoms $\tilde{x}(u,v)$, $\tilde{y}(u,v)$, denote by u_i, v_j, $0 \leq i \leq n+1$, $0 \leq j \leq m+1$), the points in the reference square domain such that $u_i = \frac{i}{n+1}$, $v_j = \frac{j}{m+1}$. The new energy is then

$$\tilde{E}_k = E_k + w \sum_{i,j=0}^{n,m} \left[(x(u_i, v_j) - \tilde{x}(u_i, v_j))^2 + (y(u_i, v_j) - \tilde{y}(u_i, v_j))^2 \right], \quad (16)$$

where $x(u,v)$ and $y(u,v)$ denote the initial bilinear interpolation.

§4. G^1 Continuity Constraints

By extending the bending plate energy and allowing some "in plane" displacement and hence reparametrization, the problem of getting C^0 connections was solved. Consider now the smooth connection problem between adjacent patches where G^1 continuity properties are required. Two neighboring patches are G^1 continuous if they have the same tangent plane along their common boundary edge [9]. Hence G^1 continuity for a pair of adjacent patches sharing a common edge is easy to implement (see Figure 2, where two adjacent patches control points are shown). In the present model, the control points for each patch are described in the patches own local system. G^1 continuity between neighboring patches is enforced by imposing the following relation in term of global coordinates:

$$R \begin{pmatrix} bx_{ij} - bx_{kl} \\ by_{ij} - by_{kl} \\ bz_{ij} - bz_{kl} \end{pmatrix} + \begin{pmatrix} t_x \\ t_y \\ t_z \end{pmatrix} = \alpha \left[-R' \begin{pmatrix} bx'_{i'j'} - bx'_{k'l'} \\ by'_{i'j'} - by'_{k'l'} \\ bz'_{i'j'} - bz'_{k'l'} \end{pmatrix} + \begin{pmatrix} t'_x \\ t'_y \\ t'_z \end{pmatrix} \right]. \quad (17)$$

Here R, R' are orthogonal matrices of rotation and $\vec{t}, \vec{t'}$ shift vectors defining the respective transformations from the local coordinate systems of the two patches to the global one. The choice of $\alpha = 1$ linearizes the G^1 continuity condition. The control of G^1 continuity between several patches with a common corner is more complex. In Figure 2 the tangent vectors for the boundaries of the respective patches are denoted by $\tau_1, \ldots, \tau_n$. Each of these vectors can be expressed as a difference of control Bézier points. To set G^1 constraints the following non-linear constraint equation must be satisfied: There exists n non-null coefficients $\alpha_1, \ldots, \alpha_n$ such that

$$\alpha_1 (\vec{\tau}_1 \times \vec{\tau}_2) = \alpha_2 (\vec{\tau}_2 \times \vec{\tau}_3) = \ldots = \alpha_n (\vec{\tau}_n \times \vec{\tau}_1). \quad (18)$$

One of the methods to linearize this constraint is to fix a common plane for all vectors $\tau_1, \ldots, \tau_n$. This plane must include the common vertex V (Figure 2) and its direction is defined by a normal at the vertex ($\vec{N}_{vert}$) chosen in an

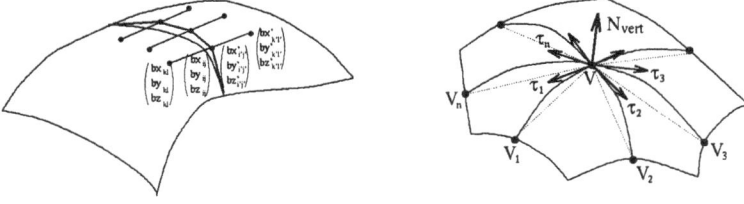

Fig. 2. G^1 connection.

appropriate way. Several methods for the choice of that normal were studied. The best results were obtained by fixing the normal $\vec{N}_{vert}$ (in the global coordinate system) according to the following equation (19):

$$\vec{N}_{vert} = \frac{V\vec{V}_1 \times V\vec{V}_2}{\| V\vec{V}_1 \times V\vec{V}_2 \|} + \frac{V\vec{V}_2 \times V\vec{V}_3}{\| V\vec{V}_2 \times V\vec{V}_3 \|} + \ldots + \frac{V\vec{V}_n \times V\vec{V}_1}{\| V\vec{V}_n \times V\vec{V}_1 \|} \tag{19}$$

Finally, once the normal is chosen, the linear equations for control points of the patches with a common vertex V are obtained by using the transformation from local to global coordinate system as in equation (17).

§5. Minimization Problem under Constraints and Implementation

Let p be the number of patches in the mesh. For simplicity, let $d = (m+1) \times (n+1)$ be the degrees of each Bézier polynomial. Then there are $3pd$ unknowns - (Bézier control points) - $bx_{ij}^{(k)}, by_{ij}^{(k)}, bz_{ij}^{(k)}$, ($1 \leq k \leq p$; $0 \leq i \leq m$;, $0 \leq j \leq n$.) These unknowns can be grouped into three vectors (each vector has $q \overset{def}{=} pd$ components):

$$\vec{\mathbf{bx}} = \left\{ bx_{ij}^{(k)} \right\}_{k,i,j} \qquad \vec{\mathbf{by}} = \left\{ by_{ij}^{(k)} \right\}_{k,i,j} \qquad \vec{\mathbf{bz}} = \left\{ bz_{ij}^{(k)} \right\}_{k,i,j} \tag{20}$$
$$1 \leq k \leq p, \ 0 \leq i \leq m, \ 0 \leq j \leq n$$

and the vector $\mathbf{b}$ containing all unknowns is defined by $\vec{\mathbf{b}} = \left(\vec{\mathbf{bx}} \ \ \vec{\mathbf{by}} \ \ \vec{\mathbf{bz}} \right)^t$.

The vector $\vec{\mathbf{b}}$ is the solution of the minimization problem under constraints, $\tilde{E}(\vec{\mathbf{b}})$, defined in equation (16) and the constraints are given by the linear equations ($C\vec{\mathbf{b}} = \vec{g}$) of G^1 global geometric continuity (C being an $m \times q$ matrix). Thus, the problem to be solved is the following:

$$\tilde{E}(\vec{\mathbf{b}}) \longrightarrow min$$
$$\text{s.t.} \qquad C\vec{\mathbf{b}} = \vec{g} \tag{21}$$

By the method of Lagrange multipliers it is equivalent to a saddle point problem [2], leading to the following system:

$$\mathbf{A}\vec{\mathbf{b}} + C^T \vec{\lambda} - \vec{\mathbf{f}} = \vec{0}$$
$$C\vec{\mathbf{b}} = \vec{g}. \tag{22}$$

Here the square matrix $\mathbf{A}_{m \times m}$ and the vector $\vec{f}_m$ are obtained by differentiation of the energy functional with respect to all unknown variables. Solution of this system is unique if the constraints are not redundant (i.e., $rank \ C = q$). Effective solution procedures can be implemented by taking into account the special structure of matrices in the equation (22) in the present method [3].

§6. Examples

6.1 G^1 surface over four patches meeting at a sharp angle

The example show how the G^1 connection over a mesh with a sharp corner is built (Figure 3). The resulting G^1 surface illustrates that the normal at the common corner vertex is chosen in a "natural" way by ((17), (18), (19)).

6.2 G^1 Surface over "step" like mesh.

Figure 4 shows that, in the present algorithm, the directions of the tangent vectors along a common edges of adjacent patches agree with these of the tangent vectors at the vertices of the patches.

6.3 G^1 surface over an arbitrary quadrilateral mesh.

Consider a set of points over an arbitrary quadrilateral mesh (Figure 5). It represents results from a FE computation using the CFD code FIDAP ([7]). The mesh vertices are defined as interpolation points. The result is a G^1 surface where each element is a Bézier patch of degree 4.

6.4 G^1 Surface over part of the Car's Chair.

The example shown in the Figure 6 is a difficult one, the initial surface was only C^0 and included overlaps and gaps.

§7. Conclusion

Starting from the classic FEM approximation of shells by flat plates, a new surface paradigm has been constructed. Its main features are:

- Usage of an "objective" minimization functional;
- Construction using control points for the elementary patches (i.e. well adapted for assembly of Bézier or NURBS patches);
- G^1 continuity over arbitrary quadrilateral meshes;
- Proper choice of normal directions to linearize the G^1 constraints;
- Assembly of classic basic patches whether Bézier or NURBS;
- Interpolation at the vertices and automatic deduction of the remaining control points. These features can be used in a very effective manner in constructing surfaces from data sets, in re-editing/simplifying very large surfaces. The FEM gives as a side result a global parameterization of the complex of patches, thus the composite surface can be viewed as an abstract spline over a FEM type quadrilateral mesh (not limited to the

Fig. 3. Four patches connection with sharp angles between patches.

Fig. 4. Mesh and G^1 Surface over "step" figure.

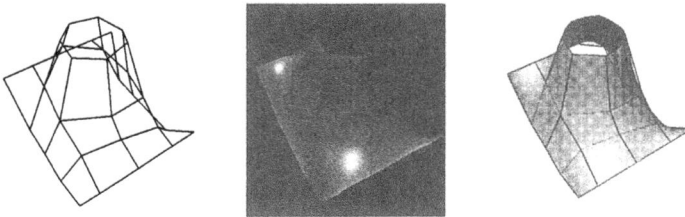

Fig. 5. G^1 Surface over given arbitrary mesh.

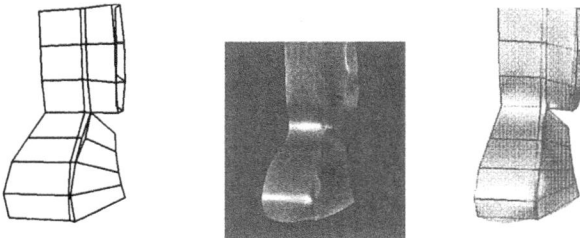

Fig. 6. G^1 surface over part of the car's chair.

map of a regular grid). Complex and time consuming mapping definitions such as a mixture of Voronoy triangulation and conformal mapping are avoided.

Acknowledgments. We would like to thank Dr. M. Engelman and FIDAP company for Example 6.5, and Prof. J. Hoschek who suggested Example 6.4.

References

1. I. Babuŝka and M. Suri, The P and H-P versions of the finite element method, basic principles and properties, Society for Industrial And Applied Mathematics 4(1994), 578–632.

2. M. Bercovier and O. Volpin, G^1 Hierarchical Bézier surface over irregular mesh, Technical Report, Leibnitz, Jerusalem, 1996.

3. M. Bercovier O. Volpin and T. Matskewich, The approximation of shells by "nearly" flat quadrilateral plates applied to the construction of free form surfaces over irregular meshes, to appear in Technical Report, Leibnitz, Jerusalem 1996.

4. M. Bernadou, *Finite Element Method for Thin Shell Problems*, Masson Publisher, Paris, 1996.

5. G. Celniker and D. Gossard, Deformable curve and surface finite elements for free-form shape design, ACM Computer Graphics 25(1991), 157–266.

6. P. G. Ciarlet, *Introduction à L'Analyse Numérique Matricielle et à L'Optimisation*,Masson, Paris, Cambridge University Press, 1982.

7. M. Engelman, *FIDAP 7.5* , FDI, 1995.

8. G. Greiner, Surface construction based on variational principles, in *Curve and Surfaces II*, P.-J. Laurent, A. Le Méhauté, and L. L. Schumaker (eds.), A. K. Peters, Wellesley MA, 1994.

9. J. Hoshek and D. Lasser, *Computer Aided Geometric Design*, A. K. Peters, 1993.

10. O. C. Zienkiewicz, *The Finite Element Method in Engineering Science*, McGrawHill Publishing, London, 1971.

Michel Bercovier, Tanya Matskewich and Oleg Volpin
Institute of Computer Science,
Hebrew University, Jerusalem, 91904, ISRAEL
berco@cs.huji.ac.il
oleg@cs.huji.ac.il
fisa@cs.huji.ac.il

A New Approach to Tchebycheffian B-Splines

Daniel Bister and Hartmut Prautzsch

Abstract. Originally, Tchebycheffian B-splines have been defined by
generalized divided differences. In this paper, we define Tchebycheffian
B-splines by integration. Based upon this definition, all basic algorithms
for Tchebycheffian splines can be derived in a straightforward manner.
As an example, a knot insertion algorithm for Tchebycheffian splines is
constructed.

§1. Introduction

The class of Tchebycheffian splines contains many different kinds of splines:
for example B-splines, exponential splines, and hyberbolic splines, see [11].
Algorithms for Tchebycheffian splines have been constructed by generalized
divided differences, see e.g. [5], by generalized polar forms [8,6], and by generalized de-Boor-Fix dual functionals [1]. A fourth possibility based upon
a new construction method for Tchebycheffian B-splines is presented in this
paper. This construction method, which can be considered as a generalized
convolution having its origin in the derivative formula for B-splines, makes it
possible to derive all basic algorithms for Tchebycheffian splines in a straightforward elementary manner [2]. In this paper, we will present a knot insertion
algorithm for Tchebycheffian splines to illustrate the method.

§2. Definition of Basis Splines

One can introduce ordinary B-splines by their derivative formula and derive
all further properties from this definition [9]. It is also possible to construct
exponential B-splines of arbitray order by this method [4]. We will use this
approach with a simple modification and get a much more general class of
splines. In the following section, we will show that this class contains Tchebycheffian splines.

First, let us recall some basic concepts from analysis. A function $f: \mathbb{R} \to
\overline{\mathbb{R}} := \mathbb{R} \cup \{\pm\}$ is called *locally integrable*, abbreviated $f \in L_{loc}$, if f is Lebesgue
integrable over every compact interval J with $J \subset \mathbb{R}$, see [12]. The space L_{loc}

Curves and Surfaces with Applications in CAGD
A. Le Méhauté, C. Rabut, and L. L. Schumaker (eds.), pp. 35–41.
Copyright © 1997 by Vanderbilt University Press, Nashville, TN.
ISBN 0-8265-1293-3.
All rights of reproduction in any form reserved.

is a function algebra: with $f, g \in L_{loc}$ and $\lambda \in \mathbb{R}$, the functions $f + g$, λf, and $f \cdot g$ belong to L_{loc}. A locally integrable function is called *integral-positive* if the integral

$$\int_J f(x)\, dx$$

is positive for every compact interval J with positive length. Further, a *knot sequence* is a non-decreasing sequence of numbers. With these concepts we are able to construct certain basis splines:

Definition 1. *Let* $\mathbf{t} = (t_i)_{i \in \mathbb{Z}}$ *be a knot sequence and* $\mathbf{w} = (w_0, \ldots, w_n)$ *be a sequence of integral-positive functions. Then the basis splines* $A_i^n(x) = A_i^n(x; \mathbf{t}, \mathbf{w})$ *of order* $n + 1$ *over* $\mathbf{t}$ *with respect to* $\mathbf{w}$ *are recursively defined by*

(0)

$$A_i^0(x) := \begin{cases} w_0(x), & \text{if } t_i \leq x < t_{i+1}, \\ 0, & \text{otherwise,} \end{cases}$$

(1)

$$A_i^n(x) := w_n(x) \int_{-\infty}^{x} \left(A_i^{n-1}(y)/\alpha_i^{n-1} - A_{i+1}^{n-1}(y)/\alpha_{i+1}^{n-1} \right) dy$$

where $\alpha_j^{n-1} := \int_{-\infty}^{\infty} A_j^{n-1}(y)\, dy$ *is the area of* A_j^{n-1} *and the following rule is used if* $t_j = t_{j+n}$:

$$\int_{-\infty}^{x} A_j^{n-1}(y)/\alpha_j^{n-1}\, dy := \begin{cases} 0, & \text{if } x < t_j, \\ 1, & \text{if } x \geq t_j. \end{cases}$$

Example 2. *If we choose* $w_0(x) = w_1(x) = \cdots = w_n(x) = 1$, *we obtain B-splines, see* [9].

Next, we state some properties of the basis splines A_i^n which can be verified by straightforward induction. The details and further properties are given in [2].

Positivity. For $t_i < t_{i+n+1}$ the area α_i^n of the basis spline A_i^n is positive. Hence, the basis splines in Definition 1 are well-defined. Moreover, the integral

$$\int_J A_i^n(x)\, dx$$

is positive for every interval J with positive length and with $J \subseteq [t_i, t_{i+n+1}]$. For $t_i = t_{i+n+1}$ the basis spline A_i^n is zero.

Local Support. If $t_i < t_{i+n+1}$, the support of A_i^n is the interval $[t_i, t_{i+n+1}]$.

Basis Property. The basis splines $A_i^n, A_{i+1}^n, \ldots, A_{i+n}^n$ are linearly independent over any non-empty interval (t_{i+n}, t_{i+n+1}).

Remark 3. *It is possible to replace the Lebesgue integral in Definition 1 by a Lebesgue-Stieltjes integral*

$$A_i^n(x) := w_n(x) \int_{-\infty}^{x} (\cdots)\, d\sigma_n(y),$$

where σ_n *is a locally bounded, strictly increasing, and right-continuous function.*

§3. Tchebycheffian Splines

Let us now examine what kinds of basis splines can be constructed by Definition 1. To do this, we repeat the definition of Tchebycheffian splines given in the book of Schumaker [11].

Let $I = [a, b]$ be a compact subinterval of $\mathbb{R}$ and let $(u_0, \ldots, u_n)$ be a sequence of functions in $C^n(I)$. Then $(u_0, \ldots, u_n)$ is called an Extended Complete Tchebycheff system on I, short ECT-system, if for all $k = 0, \ldots, n$ and and each non-decreasing sequence $(t_0, \ldots, t_k)$ of numbers in I the determinant

$$\det \left(\left[D^{d_i} u_j(t_i) \right]_{i,j=0}^{k} \right)$$

is positive, where

$$d_i := \max \left\{ r \mid t_i = \cdots = t_{i-r} \right\}.$$

A linear space is called an *ECT-space* on I if it has a basis forming an ECT-system on I.

Definition 4. *Let $I = [a, b]$ be a compact interval, let U be an $(n + 1)$-dimensional ECT-space on I, and let $\mathbf{t} = (t_0, \ldots, t_{m+n+1})$ be a knot sequence. Suppose $t_0 = t_n = a$, $t_{m+1} = t_{m+n+1} = b$, and $\ell_i \leq n+1$ for $n < i \leq m$, where ℓ_i denotes the multiplicity of the knot t_i in $\mathbf{t}$. Then a function $s\colon [a, b) \to \mathbb{R}$ is called a Tchebycheffian spline, abbreviated $s \in \mathcal{S}(U, \mathbf{t})$ if s agrees on every non-empty knot interval $[t_i, t_{i+1})$ with a function in U and if $s \in C^{n-\ell_i}(t_i)$ for any knot t_i, where $n < i \leq m$.*

Theorem 5. *Every space $\mathcal{S}(U, \mathbf{t})$ of Tchebycheffian splines has a basis of basis splines $A_0^n, \ldots, A_m^n$ where the A_i^n are constructed by Definition 1.*

Proof: Let $(u_0, \ldots, u_n)$ be an ECT-system for U. A theorem in [3, p. 379] says that every ECT-system $(u_0, \ldots, u_n)$ can be written as iterated integrals of positive weight functions $w_i \in C^i$:

$$u_0(x) = w_n(x)$$

$$u_1(x) = w_n(x) \int_a^x w_{n-1}(s_{n-1}) \, ds_{n-1}$$

$$\vdots$$

$$u_n(x) = w_n(x) \int_a^x w_{n-1}(s_{n-1}) \cdots \int_a^{s_1} w_0(s_0) \, ds_0 \cdots ds_{n-1}.$$

Hence, the basis splines $A_0^n, \ldots, A_m^n$ over $\mathbf{t}$ with respect to $\mathbf{w} = (w_0, \ldots, w_n)$ belong to $\mathcal{S}(U, \mathbf{t})$. Since the dimension of $\mathcal{S}(U, \mathbf{t})$ is $m + 1$, see [11, p. 378], the assertion follows from the basis property of the basis splines A_i^n. ∎

It is also possible to produce non-Tchebycheffian B-splines with Definition 1. Consider the functions $u_0(x) = 1$, $u_1(x) = 2x^{1/2}$, $u_2(x) = \frac{2}{3}x^{3/2}$, and $u_3(x) = \frac{1}{5}x^{5/2}$ with $x \in [0, 1]$ as examined in [10]. They do not span an

ECT-system since u_1 is not differentiable at $x = 0$. However, the corresponding weight functions $w_3(x) = w_1(x) = w_0(x) = 1$ and $w_2(x) = x^{-1/2}$ are
Lebesgue integrable and positive on $[0, 1]$, so the construction of basis splines
with Definition 1 is feasible.

§4. Knot Insertion

Let **t** be a knot sequence, and let **w** be a sequence of integral-positive functions. A spline **s** over **t** with respect to **w** is defined as a linear combination
of the basis splines $A_i^n(x) = A_i^n(x; \mathbf{t}, \mathbf{w})$, i.e.,

$$\mathbf{s}(x) = \sum_i \mathbf{c}_i A_i^n(x), \quad \text{where } \mathbf{c}_i \in \mathbb{R}^d.$$

The points $\mathbf{c}_i$ are called *control points*. They form the *control polygon* of **s**.
    We want to construct a knot insertion algorithm for these splines. Let
$\hat{t} \in \mathbb{R}$ be a number occuring with multiplicity ℓ in $\mathbf{t} = (t_i)_{i \in \mathbb{Z}}$. If $\hat{t}$ is not
contained in **t**, we set $\ell := 0$. Let r be the number with $t_r < \hat{t} \le t_{r+1}$. If $\hat{t}$ is
inserted in **t**, we obtain the refined knot sequence $\hat{\mathbf{t}} = (\hat{t}_i)_{i \in \mathbb{Z}}$ where

$$\hat{t}_i := \begin{cases} t_i & \text{if } i < r+1, \\ \hat{t} & \text{if } i = r+1, \\ t_{i-1} & \text{if } i > r+1. \end{cases}$$

We write $\hat{\mathbf{t}} = \mathbf{t}[\hat{t}]$ to indicate that $\hat{\mathbf{t}}$ is obtained by inserting $\hat{t}$ into **t**.

Theorem 6. *Let A_i^n be the basis splines over **t** with respect to **w**, and let
$B_i^n(\cdot) = A_i^n(\cdot; \hat{\mathbf{t}}, \mathbf{w})$ be the basis splines over the refined knot sequence $\hat{\mathbf{t}} = \mathbf{t}[\hat{t}]$
with respect to **w**. Then there exist numbers $\lambda_i^n \in \mathbb{R}$ and $\mu_i^n \in \mathbb{R}$ with*

$$A_i^n = \lambda_i^n B_i^n + \mu_i^n B_{i+1}^n. \tag{1}$$

Proof: We show the theorem by induction. Let r be such that $t_r < \hat{t} \le t_{r+1}$,
and let ℓ be the multiplicity of $\hat{t}$ in **t**. For $n \le \ell$, we obtain from Definition 1

$$A_i^n = \begin{cases} B_i^n & \text{if } i < r, \\ B_i^n + B_{i+1}^n & \text{if } i = r, \\ B_{i+1}^n & \text{if } i > r. \end{cases}$$

Thus equation (1) holds for

$$\lambda_i^n := \begin{cases} 1 & \text{if } i \le r, \\ 0 & \text{if } i > r \end{cases} \quad \text{and} \quad \mu_i^n := \begin{cases} 0 & \text{if } i < r, \\ 1 & \text{if } i \ge r. \end{cases}$$

Suppose $n > \ell$. Let α_j^{n-1} and β_j^{n-1} be the areas of A_j^{n-1} and B_j^{n-1} respectively, and assume $t_i < t_{i+n}$ and $t_{i+1} < t_{i+n+1}$ so that α_i^{n-1} and α_{i+1}^{n-1} do

not vanish. Suppose for the induction that there are numbers λ_j^{n-1} and μ_j^{n-1} such that

$$A_j^{n-1} = \lambda_j^{n-1} B_j^{n-1} + \mu_j^{n-1} B_{j+1}^{n-1}. \tag{2}$$

Using this expression (2) for a substitution in the definition of A_i^n gives

$$A_i^n(x) = w_n(x) \int_{-\infty}^{x} \left[\left(\lambda_i^{n-1} B_i^{n-1}(y) + \mu_i^{n-1} B_{i+1}^{n-1}(y) \right) / \alpha_i^{n-1} \right.$$
$$\left. - \left(\lambda_{i+1}^{n-1} B_{i+1}^{n-1}(y) + \mu_{i+1}^{n-1} B_{i+2}^{n-1}(y) \right) / \alpha_{i+1}^{n-1} \right] dy.$$

Applying Definition 1 to B_i^n and B_{i+1}^n, we obtain

$$A_i^n(x) = \lambda_i^{n-1} \frac{\beta_i^{n-1}}{\alpha_i^{n-1}} B_i^n(x) + \mu_{i+1}^{n-1} \frac{\beta_{i+2}^{n-1}}{\alpha_{i+1}^{n-1}} B_{i+1}^n(x)$$
$$+ \left(\frac{\mu_i^{n-1}}{\alpha_i^{n-1}} + \frac{\lambda_i^{n-1} \beta_i^{n-1}}{\alpha_i^{n-1} \beta_{i+1}^{n-1}} - \frac{\lambda_{i+1}^{n-1}}{\alpha_{i+1}^{n-1}} - \frac{\mu_{i+1}^{n-1} \beta_{i+2}^{n-1}}{\alpha_{i+1}^{n-1} \beta_{i+1}^{n-1}} \right) w_n(x) \int_{-\infty}^{x} B_{i+1}^{n-1}(y) \, dy.$$

The last term in this equation vanishes since integrating equation (2) gives

$$\alpha_j^{n-1} = \lambda_j^{n-1} \beta_j^{n-1} + \mu_j^{n-1} \beta_{j+1}^{n-1}.$$

Hence equation (1) is valid for

$$\lambda_i^n := \lambda_i^{n-1} \frac{\beta_i^{n-1}}{\alpha_i^{n-1}} \quad \text{and} \quad \mu_i^n := \mu_{i+1}^{n-1} \frac{\beta_{i+2}^{n-1}}{\alpha_{i+1}^{n-1}}.$$

A similar computation gives

$$\lambda_i^n := 1 \quad \text{and} \quad \mu_i^n := \mu_{i+1}^{n-1} \frac{\beta_{i+2}^{n-1}}{\alpha_{i+1}^{n-1}} \qquad \text{for } t_i = t_{i+n} < t_{i+n+1}$$

and

$$\lambda_i^n := \lambda_i^{n-1} \frac{\beta_i^{n-1}}{\alpha_i^{n-1}} \quad \text{and} \quad \mu_i^n := 1 \qquad \text{for } t_i < t_{i+1} = t_{i+n+1}.$$

The case $t_i = t_{i+n+1}$ is trivial. ∎

The proof of Theorem 6 shows that the numbers λ_i^n and μ_i^n are as follows:

Corollary 7. *Let α_i^m and β_i^m be the areas of A_i^m and B_i^m, respectively. Then the numbers λ_i^n and μ_i^n in Theorem 6 can be computed by*

$$\lambda_i^n = \begin{cases} 1 & \text{if } i \leq r - n + \ell, \\ \prod_{m=\ell+r-i}^{n-1} (\beta_i^m / \alpha_i^m) & \text{if } r - n + \ell < i \leq r, \\ 0 & \text{if } i > r, \end{cases}$$

$$\mu_i^n = 1 - \lambda_{i+1}^n$$

where r is such that $t_r < \hat{t} \leq t_{r+1}$ and ℓ is the multiplicity of $\hat{t}$ in **t**.

If we apply Theorem 6 to linear combinations of the basis splines A_i^n, we obtain the following knot insertion algorithm:

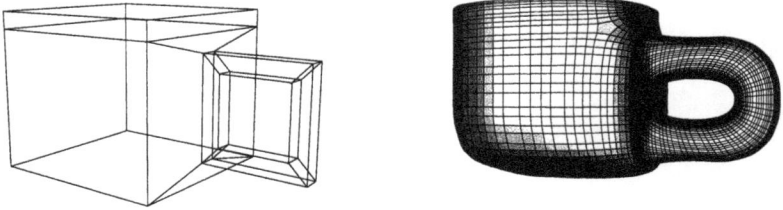

Fig. 1. Generating a cup by subdivision.

Algorithm 8 (Knot Insertion). *Every spline* $s = \sum_i c_i A_i^n$ *over* t *can be written as a spline* $s = \sum_i d_i B_i^n$ *over* $t[\hat{t}]$ *where the control points* d_i *are given by*

$$d_i = (1 - \lambda_i^n)\, c_{i-1} + \lambda_i^n c_i, \quad \lambda_i^n \text{ as in Corollary 7.}$$

Algorithm 8 implies that knot insertion is a corner cutting algorithm, see [7] for a detailed description of corner cutting algorithms. With a knot insertion algorithm it is easy to derive subdivision algorithms. For example, if the functions $w_0(x) = x(1 - x)$ and $w_1(x) = w_2(x) = 1$ defined on the interval $[0, 1]$ are periodically continued to $\mathbb{R}$, we can construct a local corner cutting algorithm by repeated knot insertion which produces C^2-curves with flat points, see [2] for full details. Also, by forming tensor products and introducing special rules for non-quadrilateral meshes, we can extend this local corner cutting algorithm to control nets of arbitrary topology, see Fig. 1 for an illustration.

References

1. Barry, P. J., de-Boor-Fix dual functionals and algorithms for Tchebycheffian B-spline curves, preprint, University of Minnesota, 1995.

2. Bister, D., Ein neuer Zugang für eine verallgemeinerte Klasse von Tschebyscheff-Splines, dissertation, Universität Karlsruhe (TH), 1996.

3. Karlin, S. and W. J. Studden, *Tchebycheff Systems: With Applications in Analysis and Statistics*, Interscience, New York, 1966.

4. Koch, P. E. and T. Lyche, Construction of exponential B-Splines of arbitrary order, in *Curves and Surfaces*, P.-J. Laurent, A. Le Méhauté, and L. L. Schumaker (eds.), Academic Press, New York, 1991, 255–258.

5. Lyche, T., A recurrence relation for Chebyshevian B-splines, Constr. Approx. **1** (1985), 155–173.

6. Mazure, M.-L. and H. Pottmann, Tchebycheff curves, in *Total Positivity and its Applications*, M. Gasca and C. A. Micchelli (eds.), Kluwer Academic Publishers, Dordrecht, 1996, 187–218.

7. Paluszny, M., H. Prautzsch, and M. Schäfer, A geometric look at corner cutting, to appear in Comput. Aided Geom. Design.

8. Pottmann, H., The geometry of Tchebycheffian splines, Comput. Aided Geom. Design **10** (1993), 181–210.

9. Prautzsch, H., Unterteilungsalgorithmen für Bézier- und B-Spline-Flächen, master thesis, Technische Universität Braunschweig, 1983.

10. Reddien, G. W. and L. L. Schumaker, On a collocation method for singular two-point boundary value problems, Numer. Math. **25** (1976), 427–432.

11. Schumaker, L. L, *Spline Functions: Basic Theory*, Wiley, New York, 1981.

12. Walter, W., *Analysis II*, Springer, Berlin, 1990.

Daniel Bister
Institut für Betriebs- und Dialogsysteme
Universität Karlsruhe (TH)
D-76128 Karlsruhe, GERMANY
bister@ira.uka.de

Hartmut Prautzsch
Institut für Betriebs- und Dialogsysteme
Universität Karlsruhe (TH)
D-76128 Karlsruhe, GERMANY
prau@ira.uka.de

Flaw Removal on Surfaces

Jacques Bousquet and Marc Daniel

Abstract. Two approaches for automatic removal of flaws on surfaces are proposed. They both work by correction of control points for different classes of surfaces. The first approach deals with "locally nearly convex" surfaces and functions by rebuilding the convex hull of the control points. It can be extended to locally nearly convex regions on surfaces. The second concerns a larger variety of surfaces and minimizes a discrete surface strain energy.

§1. Introduction

The problem of how to detect and correct flaws in designed surfaces in CAD systems remains essential in everyday use in industry. The first step is obviously the detection of potential imperfections. This can be difficult when they appear as oscillations of very low amplitude; surface inspection techniques are required. Once flaws are clearly identified, the designer has to use all his know-how and his experience to try to correct them. From an experimental approach, he will have to adjust interpolation points or control points in an interactive process until he obtains a satisfactory result. The difficulty of estimating the direction and the size of the shift for each point makes the correction process laborious and demonstrates the usefulness techniques which shift points globally. Concerning the problem of locally fairing surfaces, references are scarce; five main ones can be referred to:

- Kaufmann and Klass [11] developed a fairing method based on reflection lines used for inspecting the surface.

- Andersson et al. [2] improved the convexity of a surface using a constant sign condition on normal curvature. Using adequate linearization, the problem is solved by linear programming.

- Sapidis [15] developed an algorithm to locally fair B-Spline curves by subsequently removing and reinserting knots.

- The idea was expanded by Hadenfeld who introduced the criteria of minimal strain energy and generalized it to surfaces [8]. One control point is

Curves and Surfaces with Applications in CAGD
A. Le Méhauté, C. Rabut, and L. L. Schumaker (eds.), pp. 43–52.
Copyright © 1997 by Vanderbilt University Press, Nashville, TN.
ISBN 0-8265-1293-3.
All rights of reproduction in any form reserved.

moved at each step. Its new location is chosen in order to minimize the criterion.

- Hamann and Jean [9] proposed a method which works by projecting a Coons's patch on the region of the surface to be corrected.

The search for systematic and, if possible, automatic methods to eliminate irregularities on B-Spline surfaces is the aim of this paper. We first study in Section 2 the restrictive but no less important case of a *nearly convex surface*. By *nearly convex surface*, we mean a surface which should have been convex but which is not perfectly convex. The proposed heuristic method can be extended to surfaces having nearly convex regions. In both cases, it reconstructs (globally or locally) the convex hull of control points. In Section 3 we propose an optimization process in order to compare our heuristic method with more classical approaches. Moreover, the process allows us to enlarge the spectrum of corrected surfaces. The optimization is based on a discrete strain energy.

§2. A Heuristic Method for Flaw Removal

2.1. Setting-up a Heuristic Method

Let us consider convex surface regions, a category which includes substantial proportion of the cases to be dealt with: a local flaw is shown by a local loss of the convexity property. In the case of surfaces, the variation-diminishing property does not exist and the convexity of the control polyhedron does not imply the convexity of the designed surface. Another condition must be added. In practice and by experience, one knows that the more regular the spacing of the control points on the convex hull, the better the result. This also corresponds to current industrial requirements for control nets which concern the quality of surface definitions [13]. Having a convex control polyhedron with proper point spacing is the practical basis of our heuristic method for achieving a better surface. Our surface convexity correction corresponds first to the correction of control polyhedron convexity, sometimes completed by improvement of the regular spacing of the control points *on* the convex hull.

This heuristic method can be justified by the sufficient condition stated by H. Schelske in 1984 [10] : A tensor-product Bézier or B-Spline surface is convex provided all control points are on the boundary of the convex hull of the control points, and any set of four control points of the form $B_{i,j}$, $B_{i+1,j}$, $B_{i,j+1}$, $B_{i+1,j+1}$ must be the corners of a parallelogram. This is a very stringent condition, but it corresponds to the limit case of what is expected in CAD context. We should note that weaker conditions are currently being studied, and results already exist for non-parametric surfaces [6].

2.2. Description of the Algorithm

Our algorithm [3] fairs a nearly convex region. It is divided into four main steps:

1) The decomposition of the surface control net into facets.

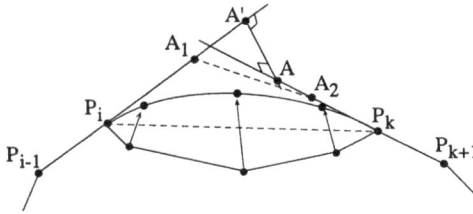

Fig. 1. Correction of a row of control points.

2) Row-by-row scanning in two steps:
   - determination of the group of points which are not located on the convex hull.
   - calculation of the new position of these points in order to bring them back onto the convex hull.
3) Column-by-column scanning with the same two steps.
4) Improving the display regularity of the control points by shifting them along the convex hull.

The decomposition of the surface control net into facets will be done first; each quadrilateral of the control polyhedron is divided into two triangular facets, then the normal facet vectors will be calculated all turned in the same direction.

Determining the group of points which are not located on the convex hull works by scanning them row by row then column by column. The algorithm for a row is an extension of the planar Jarvis's March algorithm [14]. It provides two extreme points P_i and P_k on the same row (line or column) in between intermediate "wrong" control points which have to be shifted.

The fairing algorithm constructs a space cubic correction curve joining the two extreme points, defined by points P_i, A_1, A_2, P_k (Figure 1). Points A and A', representing the shortest distance from the two tangent lines at the extreme points of the curve, are computed. Control points A_1 and A_2 are placed on segments $[P_i A']$ and $[A P_k]$. The exact location of these points is defined by the ratio

$$r = \frac{P_i A_1}{P_i A'} = \frac{P_k A_2}{P_k A}.$$

This ratio is a very important parameter of the algorithm, allowing it to deal with more or less flat or bulging correction curves. The wrong control points are mapped on the cubic curve in accordance with their control-length parametrization. A simple variation rule of r for the different rows (columns), taking into account the shape of the surface around the hole, is used. It does not ensure that the convexity will be achieved in one step "study by row then by column". In practice, a few iterations may be required. In difficult cases, first applying a regularization of the control net, as explained in the following paragraph, improves the result.

The regularization of the control net tends to organize the control point rows and columns "as parallel as possible". To do this, we first calculate the average space between each pair of following lines, then each pair of following columns. The regulation process in itself will then operate in an iterative manner. For each iteration, the rows and then the columns are taken in groups of three neighboring rows (columns), where the middle row (column) is moved *on* the convex hull according to the two others in a distance ratio defined by the previously computed average values. To measure the improvement of the control polyhedron regularity composed of $(nu+1)$ rows and $(nv+1)$ columns, we define *the regularity index*. Let us consider the Euclidean distance d and note the standard deviation sd. For each pair of neighboring rows of control points we define

$$D_i = d(B_{i,j}, B_{i+1,j}) \ / \ \sum_{i=0}^{nu-1} d(B_{i,j}, B_{i+1,j})$$

$$ECT_i = sd(D_j) \quad , \quad ECT_j = sd(D_i) \quad 0 \leq i \leq nu - 1 \ , \ 0 \leq j \leq nv - 1.$$

Finally, the regularity index I_r of the control polyhedron is

$$\frac{\sum_{i=0}^{nu-1} ECT_i \ + \ \sum_{j=0}^{nv-1} ECT_j}{nu + nv}.$$

This index is equal to zero when the control net is perfectly regular. The regulation process continues until the index reaches a sufficient low value set by the designer. It provides very promising results. As it corresponds to a heuristic method, we did not study the influence of inverting the "row then column" order of study.

Our algorithm can be applied to any surface having nearly convex regions after preprocessing in order to identify these regions. The discrete Gaussian curvature, as described in Section 3.2, is computed for each vertex of the control net. The vertices are considered as vertices of a graph. Classical techniques of graph exploration allow the detection of regions whose control points form a nearly convex polyhedron. Local correction of the control net is finally proposed for each region.

2.3. Examples

The first example corresponds to a surface including different types of flaws: undulations, holes and bumps perturbed in a disordered manner. Each figure shows the surface (left) and its control polyhedron (right). The strain energy is computed. Figure 2 shows the initial surface and Figure 3 the surface obtained after a few iterations of steps 1 to 3 of our algorithm, completed by one iteration of step 4. The strain energy decreases from 36 to 6.7 (before step 4) and 6.17 (after step 4). The study of the Gaussian curvature map shows that the final surface is convex. The second example, Figure 4, illustrates the regularization of the control polyhedron (step 4) which is initially convex but not regular (the surface is not convex) and become regular by "parallelization" of control points lines. This visibly improves the quality of the surface.

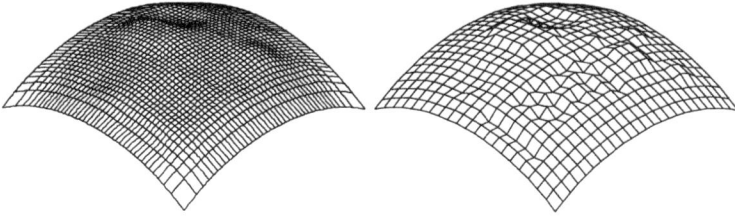

Fig. 2. Inital surface: Strain energy $= 36.0$.

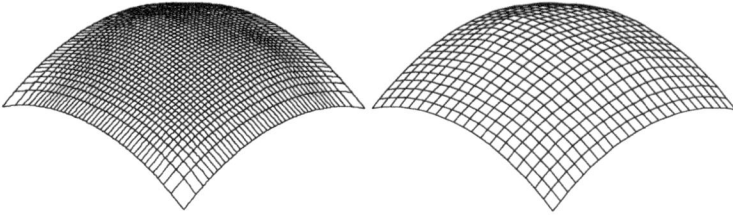

Fig. 3. Convex surface, Strain energy $= 6.17$.

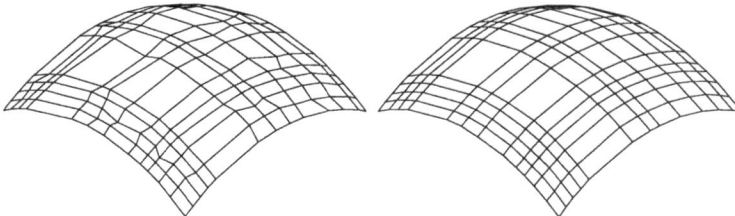

Fig. 4. Regularity index of the control polyhedron: 0.16 before, 0.004 after.

§3. A Correction Algorithm Based on Global Optimization

3.1. Global Optimization Methods

The method presented above corrects particular surfaces. We decided to compare our heuristic method with more classical approaches, which are generally global optimization methods. Moreover, it allows us to enlarge the spectrum of corrected surfaces. In point of fact, global optimization methods, if they could be used, would treat any surface. In this section we propose a correction based on optimization of a discrete strain energy.

The principle of these methods consists in optimizing the value of an *objective function* generally dependent in a non-linear manner on N parameters. In our case, the objective function to be minimized is the strain energy of the surface and the variables, the position of free control points which are able to move in an interval dl around their initial position. The set of control points are split into two categories: fixed control points and free control points. These methods can to be applied for fairing any surface but require substantial computer time. In fact the problem is rather posed as follows: how many free control points can be accepted in order to keep processing times compatible with the interactive environment of CAD/CAM systems? Satisfactory performance must be demontrated for problems which may contain hundreds of variables. To this end, we first limit the number of variables by moving each free control point in only one direction "perpendicular" to the surface, lateral moves being achieved by the regularization method studied above.

But the main problem lies in evaluating the objective function, the surface strain energy, which requires too much computing time to be calculated with the classical "true" formula (where H and K denote the mean and Gaussian curvature and ds denotes the surface element):

$$E = \int_S (4H^2 - 2K)\, ds$$

3.2. Strain Energy of the Control Polyhedron

We propose to compute the "strain energy of the control polyhedron" instead of the surface strain energy. A fundamental part of the theory of intrinsic geometry of non-regular surfaces developed by Alexandrov [1] is the application of the notion of polyhedral metrics. Discrete analogues of well-known concepts of classical differential geometry as Gaussian and mean curvatures can be deduced [4]. They are easily and quickly computed and can be directly applied on the control net of surfaces. Let us consider F_k and F_{k+1}, consecutive triangular facets around control point $P_{i,j}$. $Angle(F_k, F_{k+1})$ is the oriented angle between the plane of facet F_k and F_{k+1}. Using all the notations illustrated in Figure 5 for the mean curvature and Figure 6 for the Gaussian curvature, these two discrete curvatures and the strain energy are computed at each control point by the following formulas:

$$H_{i,j} = \frac{1}{4} \sum_k (\pi - angle(F_k, F_{k+1})).l_{k,k+1}$$

$$K_{i,j} = 2\pi - \sum_k \alpha_k$$

$$E_{i,j} = 4H_{i,j}^2 - 2K_{i,j}.$$

Finally, the strain energy of the complete control polyhedron is

$$E = \sum_{i=0}^{nu} \sum_{j=0}^{nv} E_{i,j}$$

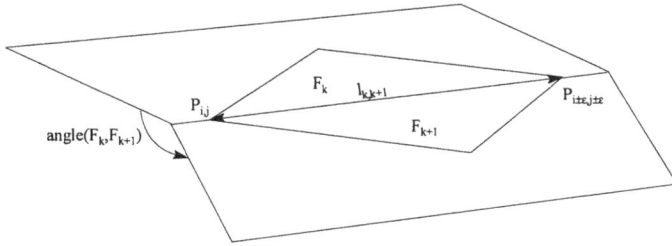

Fig. 5. Computation of mean curvature.

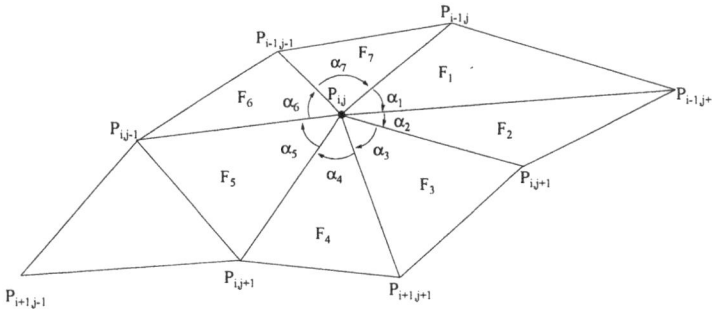

Fig. 6. Computation of Gaussian curvature.

The Gauss-Bonnet Theorem posits that the integral of the Gaussian curvature over the surface can be expressed using an integral along the boundaries (see [5] for example). If we consider that the boundaries of the surface can be modified, we cannot suppress this term in our definition of the energy. Otherwise, its suppression greatly decreases computing time.

This function may possess local optima among which it is necessary to choose; a stochastic approach revents us from being trapped in a local optimum, assuming that proper adjustment of its parameters is provided. One example of such a minimum is proposed in Figure 7. Let us consider this well-spaced control polygon. The free points are the points in the hole (black dots). The current position of the free control points corresponds to a local minimum of energy.

Three main stochastic methods are known: the *Simulated annealing method* [12] which reproduces the physical phenomenon of the slow cooling down of a solid in fusion state, leading to a low energy state, the *Genetic algorithms* which are based on a natural selection mechanism in species evolution and the *Tabu search method* [7] which is a heuristic method developed to treat discrete problems from operational research. We obtain the best results with the Simulated annealing method.

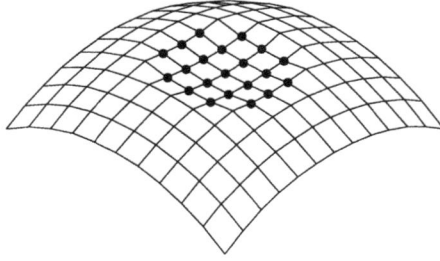

Fig. 7. A local minimum of the strain energy.

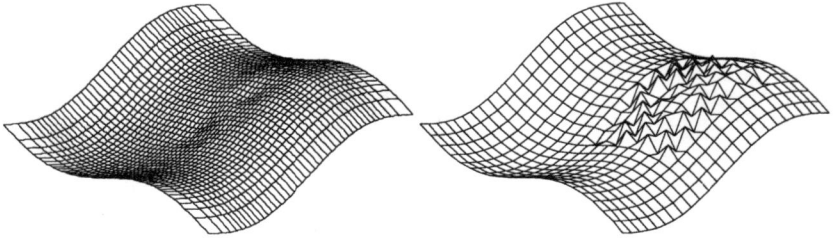

Fig. 8. Initial surface.

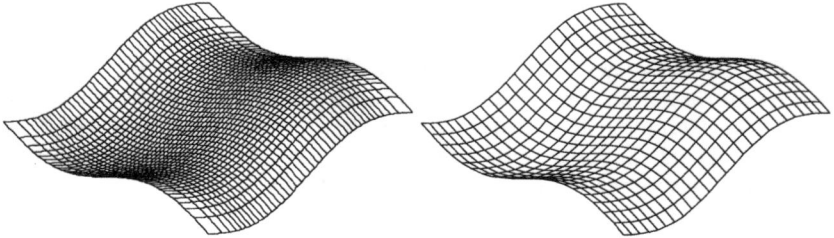

Fig. 9. Faired surface using simulated annealing algorithm.

3.3. Examples

The initial surface is shown in Figure 8. The strain energy of the surface is 43.1
and 14,500 when computed on its control polyhedron. The arbitrary number
of free control points is 97. The final strain energy of the surface (Figure
9) is 17.6 and 100.2 computed on its control polyhedron. Using classical
computation of the strain energy, the time required for our computer is 3
hours, and with our discrete formula, it falls to 2 minutes with fixed boundaries
and 6 minutes with free boundaries. For other examples, using this discrete
technique always divides the computing time by a factor of 50 to 60.

In addition, the surface proposed in Section 2.3 has been corrected with this approach (204 free points and 321 fixed points). The final energy is equivalent to the previous one, but no better. The result is obtained in 7 minutes, compared to the 10 seconds required with our heuristic method.

§4. Conclusion

The study described in this paper was implemented on a SUN Sparc 2 workstation with CADDS 5 CAD software by Computervision. Its CV-DORS programming interface allows us, by direct action on the graphic data base, to analyze and modify the curve and surface characteristics.

The automatic correction algorithm for nearly convex surfaces gives excellent results with execution times of a few seconds. A complementary module makes it possible to determine nearly convex regions on any surface to fair them. This approach is not completely general, but the number of cases for which it can be applied is sufficient to justify its interest.

We propose a discrete formula for computing the strain energy of surfaces. It allows the use of global optimization methods to improve the fairing of surfaces in CAD software packages with acceptable computation times. This technique is suitable for any surface, but it is better to apply our faster, first algorithm when possible.

Acknowledgments. The authors would like to thank the referees for their invaluable comments and suggestions.

References

1. Alexandrov, A. D., and V. A. Zalgaller, *Intrinsic Geometry of Surfaces*, edited by AMS, Rhode Island, 1967.

2. Andersson, E., M. Boman, T. Elmroth, B. Dahlberg and B. Johansson, Automatic construction of surfaces with prescribed shape, Computer-Aided Design **20**(6) (1988), 317–324.

3. Bousquet, J., Automatic fairing methods for surfaces, IDMME'96 (Integrated Design and Manufacturing in Mechanical Engineering), Nantes, France, April 1996, 965–974.

4. Bousquet, J., and P. Aimé, Une application de la géométrie différentielle discrète à la modélisation de surfaces, Research Report IRIN, forthcoming, Nantes, 1997.

5. Do Carmo, M., *Differential Geometry of Curves and Surfaces*, Prentice-Hall, 1976.

6. Floater, M., A weak condition for the convexity of tensor-product Bézier and B-splines surfaces, Advances in Computational Mathematics **2** (1994), 67–80.

7. Glover, F., Tabu search, Center of Applied Artificial Intelligence Report 88-3, University of Colorado, 1988.

8. Hadenfeld, J. , Local Energy Fairing of B-Spline Surfaces, in *Mathematical Methods for Curves and Surfaces*, M. Dæhlen, T. Lyche and L. L. Schumaker (eds.), Vanderbilt University Press, 1995, 203–212.

9. Hamann, B., and B. A. Jean, Interactive surface correction based on a local approximation scheme, Comput. Aided Geom. Design **13**(4) (1996), 351–358.

10. Hoschek, J., and D. Lasser, *Fundamentals of Computer-Aided Geometric Design*, A.K. Peters, 1993.

11. Kaufmann, E., and R. Klass, Smoothing surfaces using reflection lines for families of splines, Computer-Aided Design **20**(6) (1988), 312–316.

12. Kirkpatrick, S., C. D. Gelatt, and M. P. Vecchi, Optimization by simulated annealing. *Science* **220**(4598) (1982), 671–680.

13. Lichah, T., Développement de commandes d'analyse et de contrôle de la définition mathématiques des modèles surfaciques sous CADDS4X, DEA ENSIMEV, Valenciennes, France, in collaboration with Peugeot S.A., 1994.

14. Preparata, F. P., and M. I. Shamos, *Computational Geometry: an Introduction*, Springer Verlag, 1985.

15. Sapidis, N., and G. Farin, Automatic fairing algorithm for B-Spline curves, Computer-Aided Design **22**(2) (1990), 121–129.

Jacques Bousquet, Marc Daniel
Institut de Recherche en Informatique de Nantes
B.P. 92208
44322 Nantes cedex 3, FRANCE
Marc.Daniel@ec-nantes.fr

On Some Polynomial Curves
Derived from Trigonometric Kernels

Jia-ding Cao, Heinz H. Gonska, and Daniela P. Kacsó

Abstract. We present a class of polynomial curves which are based on operators giving the best degree of approximation. These curves possess a number of properties shared by Bézier curves, but give a much better closeness of fit.

§1. Introduction

The classical Bernstein operators play a central role in CAGD. However, from the point of view of Approximation Theory, their degree of approximation is rather slow. As was shown by the first two authors several years ago, there are more powerful approximation methods which are quite similar to the Bernstein operators. This statement is also true as far as ease of computation is concerned. In this note we will discuss some of these methods and evaluate them according to the needs of CAGD. In doing so we will be guided by Goldman's catalogue (dating from 1985) in which he listed several desirable properties of curves for CAGD.

The approximation operators underlying the curves to be discussed will be transformed trigonometric convolution–type operators based on positive even kernels (details to be given in Section 3). Using appropriate quadrature formulae these can be discretized and corrected by a Boolean sum approach to arrive at the final design tools

$$\Lambda_{m,N}^+(f,x) \ = \ \sum_{j=0}^{N+1} f(x_{j,N}) \cdot A_{j,m,N}^+(x).$$

Replacing the numbers $f(x_{j,N})$ by control points $\vec{b}_j \in \mathbb{R}^d$, $0 \le j \le N+1$, yields polynomial curves which have many properties relevant to CAGD.

Curves and Surfaces with Applications in CAGD   53
A. Le Méhauté, C. Rabut, and L. L. Schumaker (eds.), pp. 53–60.
Copyright © 1997 by Vanderbilt University Press, Nashville, TN.
ISBN 0-8265-1293-3.
All rights of reproduction in any form reserved.

§2. Some Pros and Cons Regarding Bernstein Polynomials

In the present section we motivate our investigation of the curves briefly described in the introduction by recalling some properties of Bernstein polynomials and by giving some additional historical information.

One feature to be mentioned is the existence of a two term recursive evaluation algorithm, colloquially known as de Casteljau's algorithm. This is probably the most fundamental one in the field of curve and surface design. While the historical root mentioned in regard to it is usually a 1959 technical report by de Casteljau (see, e.g., [9, p.25]), it should be noted that the algorithm was available in quite an explicit form as early as 1933 in the work of Popoviciu (see [11, p.38]).

This is, however, only one reason for which Bernstein–Bézier curves play such an important role in CAGD. A more complete list of reasons is that they possess most of the important properties of CAGD curves as listed in two papers by Goldman [10] and Barry and Goldman [1].

From the point of view of Approximation Theory, the Bernstein operators are even best possible in a certain sense. This is explained in the following result due to Berens and DeVore [2]:

Theorem 1. *Let $\mathcal{L}_n$ denote the class of all operators L_n mapping $C[0,1]$ into itself with*

(i) $L_n(f) \in \Pi_n$ for all $f \in C[0,1]$,
(ii) $L_n(l) = l$ for all $l \in \Pi_1$,
(iii) $[L_n(f)]^{(j)} \geq 0$, if $f^{(j)} \geq 0$, $j = 0, 1, \ldots, n$.

Then

$$\frac{x(1-x)}{n} = B_n[(\cdot - x)^2, x] = \inf_{L_n \in \mathcal{L}_n} L_n[(\cdot - x)^2, x], \qquad 0 \leq x \leq 1.$$

In other words, among all polynomial operators reproducing linear functions and possessing shape preservation properties as expressed by (iii), from a quantitative point of view the Bernstein operators are best possible. Unfortunately, here "best possible" only means that they are best possible in a class of operators having "bad" approximation potential. To make this more clear, we recall a result in a paper by Cao [4], where it was shown that the Bernstein operators satisfy

$$|B_n(f, x) - f(x)| \leq c \cdot \omega_2\left(f, \sqrt{\frac{x(1-x)}{n}}\right) \qquad (2.1)$$

for $n \in \mathbb{N}, f \in C[0,1], x \in [0,1]$. Here ω_2 denotes the standard second order modulus of smoothness (see [12] for the definition and some properties). So uniformly, one only has $\|B_n f - f\|_\infty = \mathcal{O}(\frac{1}{n})$ for functions in C^2. Furthermore, inequality (2.1) is best possible for Bernstein operators and Lipschitz classes defined with respect to ω_2; this was shown in an article of Berens and Lorentz [3].

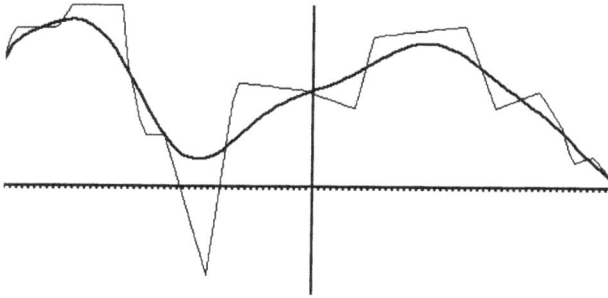

Fig. 1. Approximation of a polygon by a Bernstein polynomial of degree 48 .

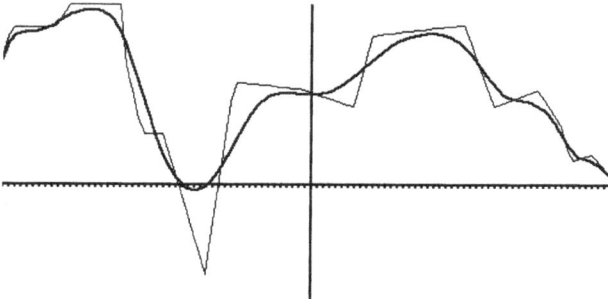

Fig. 2. Approximation of the same polygon by the polynomial $\Lambda^+_{48,26}(f, \cdot)$.

A visual consequence of this fact is a frequently observed drawback of Bézier curves: especially for higher degree curves, the similarity between the control polygon and its Bézier curve is rather poor. Approximating polynomials have more to offer. In a series of papers Cao and Gonska investigated certain operators L_n satisfying estimates of the form

$$|L_n(f, x) - f(x)| \le c \cdot \omega_2 \left(f; \frac{\sqrt{x(1-x)}}{n} \right),$$ (2.2)

where n, f and x are as above. This means $\|L_n f - f\| = \mathcal{O}(\frac{1}{n^2})$ for $f \in C^2[0,1]$. Besides this, an estimate as (2.2) also guarantees that the operators L_n provide endpoint interpolation, reproduce linear functions and, most of all, have a speed twice that of Bernstein operators. For technical details and further references, see [6] (especially in regard to computational aspects) and [7] (for shape preservation).

The difference in the visual behaviour of operators satisfying (2.1) in comparison to such verifying (2.2) is maybe demonstrated best by showing how they approximate piecewise linear functions.

It is obvious that the polynomial $\Lambda^+_{48,26}(f, \cdot)$ gives a much closer approximation to the polygon, although it is based on only 26+2 point evaluations as opposed to 49 for the Bernstein case. The next section will show how to construct it and other related polynomials.

§3. Polynomial Curves derived from Trigonometric Kernels

3.1 The construction

Step 0: Write the (formal) Chebyshev-Fourier expansion of $f \in \mathbb{R}^{[-1,1]}$, i.e.,

$$f(x) \approx \frac{1}{\pi} \int_{-1}^{1} \frac{f(t)}{\sqrt{1-t^2}} \, dt + \sum_{k=1}^{\infty} \frac{2}{\pi} \int_{-1}^{1} \frac{f(t) \cdot T_k(t)}{\sqrt{1-t^2}} \, dt \cdot T_k(x),$$

where T_k is the k–th Chebyshev polynomial.

Step 1: (Summation) Choose an even and positive trigonometric kernel of the form

$$K_m(t) = \frac{1}{2} + \sum_{k=1}^{m} \rho_{k,m} \cdot \cos kt \geq 0, \quad t \in \mathbb{R};$$

then define

$$G[K_m](f;x) := \frac{1}{\pi} \int_{-1}^{1} \frac{f(t)}{\sqrt{1-t^2}} \, dt + \sum_{k=1}^{m} \rho_{k,m} \cdot \left\{ \frac{2}{\pi} \int_{-1}^{1} \frac{f(t) \cdot T_k(t)}{\sqrt{1-t^2}} \, dt \right\} \cdot T_k(x).$$

The operator $G_m := G[K_m]$ can already be used to approximate a function. However, it does not provide endpoint interpolation and is not discretely defined.

Step 2: (Discretization) Choose a quadrature formula of the form

$$Q_N(g) = \sum_{j=0}^{N+1} B_{j,N} \cdot g(x_{j,N}) \approx \int_{-1}^{1} \frac{g(t)}{\sqrt{1-t^2}} \, dt,$$

where $1 = x_{N+1,N} > x_{N,N} > \ldots > x_{1,N} > x_{0,N} = -1$, and $B_{j,N} \geq 0$, $0 \leq j \leq N+1$. The discretized operator $G[K_m]$ then attains the form (cf. [6])

$$\Lambda_{m,N}(f,x) := \Lambda[K_m; Q_N](f,x) = \sum_{j=0}^{N+1} f(x_{j,N}) \cdot A_j[K_m, Q_N](x)$$

with non–negative fundamental functions $A_{j,m,N} := A_j[K_m, Q_N]$.

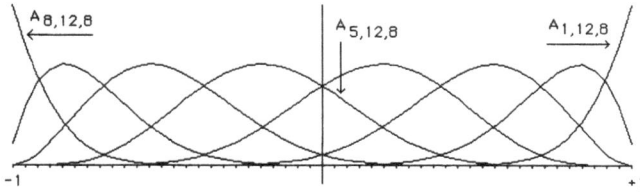

Fig. 3. Fundamental functions of $\Lambda_{12,8}$.

Step 3: (Correction = enforcement of endpoint interpolation) For $L(f;x) = \frac{1}{2}[f(1)(x+1) + f(-1)(1-x)]$ and $\Lambda_{m,N}$ as above consider the Boolean sum $\Lambda_{m,N}^+(f;x) := (L + \Lambda_{m,N} - L \circ \Lambda_{m,N})(f;x)$. Then $\Lambda_{m,N}^+(f;\pm1) = f(\pm1)$, and we have

$$\Lambda_{m,N}^+(f,x) := \Lambda_{m,N}^+[K_m, Q_N](f,x) = \sum_{j=0}^{N+1} f(x_{j,N}) \cdot A_j^+[K_m, Q_N](x).$$

Note that $\Lambda_{m,N}^+(f,x)$ can also be written as a linear combination of Chebyshev polynomials (cf. [6]).

The next question is how to choose a good candidate for K_m. This can be judged with the aid of $G[K_m]$. A brief list of "first choice" kernels includes Fejér–Korovkin kernels of order $p \geq 2$, and Jackson–Matsuoka kernels of order $s \geq 3$ (see [8] and [5], respectively). This list is justified by the fact that the operators G_m based on these kernels exhibit very good approximation behaviour.

As "second choice" kernels we mention Bohman-Zheng kernels, Fejér-Korovkin kernels, and Jackson-de La Vallée Poussin kernels. For "second-choice" kernels the approximation behaviour of the related operators G_m is slightly weaker, but still much better than that of Bernstein operators.

Natural choices of the quadrature formula Q_N are generalized Lobatto formula of type (r,s), in particular those for $r = s = 0$ (Gauss-Chebyshev quadrature), and for $r = s = 1$ (Lobatto-Markov quadrature). The latter uses endpoint information at ±1; so this appears to be most natural in this context.

Next we give two examples showing the fundamental functions in two particular cases.

(i) The 8 fundamental functions $A_{r,m,N}$ of $\Lambda_{m,N}, 1 \leq r \leq N$ for the case $m = 12$ (Jackson-Matsuoka kernel of order 3) and Gauss integration with $N = 8$ are shown in Fig. 3.

(ii) The 6 fundamental functions $A_{j,m,N}^+$ of $\Lambda_{m,N}^+$, $0 \leq j \leq N$, for $m = 6$ and $N = 5$ (Jackson-Matsuoka kernel of order 3 and Lobatto-Markov integration):

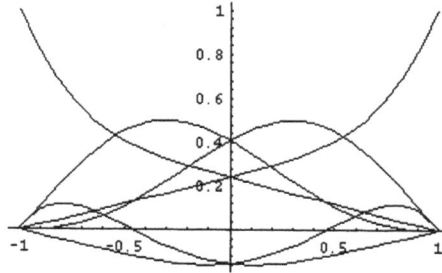

Fig. 4. Fundamental functions of $\Lambda_{6,5}^+$.

3.2. The properties of $\Lambda_{m,N}$ and $\Lambda_{m,N}^+$.

As further general rules for the construction of $\Lambda_{m,N}$ and $\Lambda_{m,N}^+$ we mention the following: It is often easier to check the analytical properties of G_m than those of operators derived from it. This suggests to use a quadrature formula that makes sure that $\Lambda_{m,N} = \Lambda[K_m, Q_N]$ will inherit relevant properties from $G_m = G[K_m]$. A "good choice" for Q_N means to make sure that Q_N is exact for polynomials of degree $\leq m+2$. See the related discussion in [6] to confirm this fact. "Safe choices" are

- for the Gauss case ($r = s = 0$), $N = m + 1$, and
- for the Lobatto case ($r = s = 1$, i.e., endpoint information included), $N = m - 1$.

These choices also warrant that the number of fundamental functions generated are equal to the dimension of $\Pi_m \supseteq G_m(C[-1,1])$.

We are now ready to discuss the criteria listed in the catalogues of Barry and Goldman mentioned in the above. The following statements are all with respect to Gauss quadrature ($r = s = 0$).

1. *"Well defined" (affine invariance)*: Fullfilled for both $\Lambda_{m,N}$ and $\Lambda_{m,N}^+$.

2. *Convex hull property*: Fullfilled for $\Lambda_{m,N}$ (by construction), slightly violated for $\Lambda_{m,N}^+$.

2'. *Behavior for straight lines*: $\Lambda_{m,N}$ does not reproduce all linear functions; $\Lambda_{m,N}^+$ does. This implies: In the parametric case, if all control points lie on a straight line, then $\Lambda_{m,N}$ will create part of the convex hull of the points, $\Lambda_{m,N}^+$ will create all of it.

3. *Smoothness*: Given trivially, since the operators are polynomial operators.

4. *Interpolation at the endpoints*: Not true for $\Lambda_{m,N}$ (and impossible for positive kernels); true for $\Lambda_{m,N}^+$ by construction.

5. *Extension to surfaces*: For the rectangular case, use tensor products; for the triangular case a general method appears to be unknown.

6. *Symmetry*: Fullfilled for both $\Lambda_{m,N}$ and $\Lambda^+_{m,N}$.

7. *Geometric construction algorithm*: At present it is unclear what it would look like; research on this topic is going on.

8. *Exact reproduction of points (consistency) and lines*: For points this is clear; for lines see 2'.

9. *Nondegeneracy (linear independence of the fundamental functions)*: This is true for $\Lambda_{m,N}$ if $N \leq m+1$, but it is not true for $\Lambda^+_{m,N}$ based on Gauss quadrature.

10. *Subdivision algorithm*: It exists if the fundamental functions are a basis for Π_m, the latter is true if one picks $N = m+1$ (for the Gauss case) for the construction of $A_{j,m,N}$. The existence of an algorithm is unknown for $\Lambda^+_{m,N}$, even for $N = m+1$ (the $A^+_{j,m,N}$ do not form a polynomial basis, see 9.).

11. *Augmentation algorithm (degree elevation)*: It exists for both $\Lambda_{m,m+1}$ and $\Lambda^+_{m,m+1}$.

12. *Variation diminishing property*: There is hope to have this for "nice" kernels K_m and $\Lambda_{m,N}$; but there are problems for $\Lambda^+_{m,N}$, at least in the Gauss case. Research in this direction is likewise going on.

13. *Local control*: This is not available due to the fact that all operators considered here are polynomial.

Acknowledgments. The authors gratefully acknowledge the technical assistance and valuable contributions of Rita Hülsbusch, Jörg Simon and Hans–Jörg Wenz, as well as the support by the National Natural Science Foundation of China under project code 19171020.

References

1. P. J. Barry and R. N. Goldman, What is the natural generalization of a Bézier curve?, in *Mathematical Methods in Computer Aided Geometric Design*, T. Lyche and L. Schumaker (eds.), Academic Press, New York, 1989, 71–85.

2. H. Berens and R. A. DeVore, A characterization of Bernstein polynomials, in *Approximation Theory III*, E. W. Cheney (ed.), Acad. Press, New York, 1980, 213–219.

3. H. Berens and G. G. Lorentz, Inverse theorems for Bernstein polynomials. Indiana Univ. Math. J. **21** (1972), 693–708.

4. J.-d. Cao, On linear approximation methods (Chinese). Acta Sci. Natur. Univ. Fudan **9** (1964), 43–52.

5. J.-d. Cao and H. H. Gonska, Approximation by Boolean sums of positive linear operators II: Gopengauz-type estimates. J. Approx. Theory **57** (1989), 77–89.

6. J.-d. Cao and H. H. Gonska, Approximation by Boolean sums of positive linear operators III: estimates for some numerical approximation schemes. Numer. Funct. Anal. Optim. **10** (1989), 643–672.

7. J.-d. Cao and H. H. Gonska, Pointwise estimates for higher order convexity preserving polynomial approximation. J. Austral. Math. Soc. Ser. B **36** (1994), 213–233.

8. J.-d. Cao and H. H. Gonska, H.-J. Wenz, Approximation by Boolean sums of positive linear operators VII: Fejér–Korovkin kernels of higher order. Acta Math. Hungar. **73** (1996), 71–85.

9. G. Farin, *Curves and Surfaces for Computer Aided Geometric Design*, Academic Press, New York, 1988.

10. R. N. Goldman, Polya's urn model and Computer Aided Geometric Design. SIAM J. Algebraic Discrete Meth. **6** (1985), no. 1, 1-28.

11. T. Popoviciu, *Despre cea mai bună aproximaţie a funcţiilor continue prin polinoame.* Cluj: Institutul de arte grafice "Ardealul" 1937.

12. L. L. Schumaker, *Spline Functions: Basic Theory*, Wiley, New York, 1981.

Jia-ding Cao
Department of Mathematics
Fudan University
Shanghai, 200433, CHINA
guch@bepc2.ihep.ac.cn

Heinz H. Gonska and Daniela P. Kacsó
Department of Mathematics
University of Duisburg
D - 47048 Duisburg, GERMANY
gonska@informatik.uni-duisburg.de
kacso@informatik.uni-duisburg.de

Spline Curves in Polar and Cartesian Coordinates

Giulio Casciola and Serena Morigi

Abstract. A new class of spline curves in polar coordinates has been presented in [11] and independently considered in [5] . These are rational trigonometric curves in Cartesian coordinates and can be represented as NURBS. An alternative way to derive some useful tools for modelling splines in polar coordinates is provided. Moreover, an ad hoc algorithm of degree elevation for splines in polar coordinates is presented, and its efficiency and stability is proved.

§1. Introduction

Recently, in [11] a class of spline curves in polar coordinates was proposed. We refer to these curves as *p-splines*. They have proved to be a generalization of those considered in [10], which we call *p-Bézier curves*.

The p-splines were independently considered in [5], and called Focal splines. These classes of curves are interesting because they allow for modelling and interpolation of free forms in polar coordinates with the same facilities as Cartesian splines.

In [11], Sánchez-Reyes emphasizes the fact that the p-spline curves are piecewise rational Bézier in Cartesian coordinates but they are not rational splines. Actually, this last assertion is neither proved nor supported by any justification. In this paper we will provide the algorithm that leads to a representation of these curves as NURBS.

In addition to knot insertion, knot removal and subdivision, another known result from Sánchez-Reyes' papers is the possibility of carrying out degree elevation from degree n to degree kn. In [2] we proposed an algorithm for degree elevation for p-Bézier curves. In this paper we will suggest how to use it for p-splines, together with tests on efficiency and numerical stability.

These two results, together with their generalization to surfaces, have convinced us of the usefulness of extending our NURBS-based modelling system by supplying it with a modelling environment for p-spline curves and surfaces in polar, spherical, and mixed polar-Cartesian coordinates. This allows us to manage polar and spherical models as NURBS [3].

Curves and Surfaces with Applications in CAGD
A. Le Méhauté, C. Rabut, and L. L. Schumaker (eds.), pp. 61–68.
Copyright © 1997 by Vanderbilt University Press, Nashville, TN.
ISBN 0-8265-1293-3.
All rights of reproduction in any form reserved.

Fig. 1. p-spline curve of degree $n = 3$ together with its control polygon - $\{t_i\} = \{0, 0, 0, 0, 0.75, 1.5, 2.25, 3, 3, 3, 3\}$, $\underline{d_i} = \{(0, 0.3), (0.75, 0.3), (2.25, 0.22), (4.5, 0.3), (6.75, 0.22), (8.25, 0.3), (9, 0.3)\}$.

§2. P-spline Curves

A p-spline curve $\underline{c}(t)$ of degree n is defined as

$$\underline{c}(t) = \begin{pmatrix} \rho(t) \\ \theta(t) \end{pmatrix} = \begin{pmatrix} \frac{1}{\sum_{i=0}^{K+n} \delta_i M_{i,n}(t)} \\ nt \end{pmatrix},$$

where $\theta(t)$ denotes the polar angle and $\rho(t)$ is the radius defined as the reciprocal of a trigonometric spline. Without loss of generality, we consider $t \in [-\Delta, \Delta]$. The functions $M_{i,n}(t)$ are normalized trigonometric B-splines [8] and are defined by the following recurrence relation

$$M_{i,n}(t) = \frac{\sin(t - t_i)}{\sin(t_{i+n} - t_i)} M_{i,n-1}(t) + \frac{\sin(t_{i+n+1} - t)}{\sin(t_{i+n+1} - t_{i+1})} M_{i+1,n-1}(t) \quad (1)$$

$$M_{i,0}(t) = \begin{cases} 1, & \text{if} \quad t_i \le t < t_{i+1}, \\ 0, & \text{otherwise}, \end{cases}$$

on a non-decreasing knot sequence $\{t_i\}_{i=0}^{K+2n+1}$ satisfying the constraint $t_{i+n} - t_i < \pi$, for all i. Note that each trigonometric spline piece is a trigonometric polynomial belonging to the space

$$T_n = \begin{cases} span\{1, \cos 2t, \sin 2t, \cos 4t, \sin 4t, ..., \cos nt, \sin nt\}, & n \text{ even} \\ span\{\cos t, \sin t, \cos 3t, \sin 3t, ..., \cos nt, \sin nt\}, & n \text{ odd}. \end{cases}$$

The coefficients δ_i^{-1} and the Greville radial directions $\xi_i = \sum_{j=i+1}^{i+n} t_j$ define, in polar coordinates, the control points $\underline{d_i} = (\xi_i, \delta_i^{-1})$ of the p-spline $\underline{c}(t)$. The knot constraint implies that, in polar coordinates, $\xi_i - \xi_{i-1} < \pi$ holds. Figure 1 illustrates an example of a p-spline curve and relative control polygon.

    P-spline curves enjoy properties of local control, linear precision, convex hull, and variation diminishing inherited from splines in Cartesian coordinates. Moreover, p-splines of degree 2 are conic sections with foci at the origin of the coordinates.

§3. NURBS Representation of p-splines

It is known [11] that the class of p-spline restricted to a single segment (p-Bézier curves) represents a subclass of rational Bézier curves in Cartesian coordinates. Therefore, we can state that p-splines represent a subclass of NURBS.

A first approach to obtain a NURBS representation of a p-spline curve has been suggested in [11]. Given a p-spline curve over an arbitrary knot sequence, this can be converted by subdivision into a piecewise curve whose individual pieces are p-Bézier curves, so that every p-Bézier curve can be represented in terms of rational Bézier curves in Cartesian coordinates.

In the alternative approach proposed here, a non-piecewise Bézier representation of a p-spline $\underline{c}(t)$ as a NURBS curve $\underline{q}(v)$ will be provided.

Let $\underline{c}(t)$ be the p-spline represented as a scalar function

$$\rho(\frac{\theta}{n}) = \frac{1}{\sum_{i=0}^{K+n} \delta_i M_{i,n}(\frac{\theta}{n})}, \tag{2}$$

where $\theta \in [-n\Delta, n\Delta]$. Then the correspondent curve of (2) in Cartesian coordinates will be obtained by a simple change of coordinates:

$$\rho(\frac{\theta}{n}) \begin{pmatrix} \cos\theta \\ \sin\theta \end{pmatrix}. \tag{3}$$

Applying the identities [6]

$$\cos\theta = \sum_{i=0}^{K+n} \cos\xi_i M_{i,n}(\tfrac{\theta}{n}), \qquad \sin\theta = \sum_{i=0}^{K+n} \sin\xi_i M_{i,n}(\tfrac{\theta}{n}),$$

relation (3) assumes the following trigonometric rational form

$$\frac{\sum_{i=0}^{K+n} \binom{\cos\xi_i}{\sin\xi_i} M_{i,n}(\frac{\theta}{n})}{\sum_{i=0}^{K+n} \delta_i M_{i,n}(\frac{\theta}{n})}. \tag{4}$$

In [7] the important transformation $\gamma_n : P_n -> T_n$; $(\gamma_n f)(x) = \cos^n x \cdot f(\tan x)$, was provided; more precisely, if $p \in P_n$ on $[\tan\alpha, \tan\beta]$, then $\gamma_n p \in T_n$ on $[\alpha, \beta]$ when $-\frac{\pi}{2} < \alpha < \beta < \frac{\pi}{2}$. From this assertion it follows that a polynomial B-spline is proportional to a trigonometric B-spline. In particular, the following important relation can easily be proved:

$$M_{i,n}(t) = \frac{\cos^n t}{\prod\limits_{j=i+1}^{i+n} \cos t_j} N_{i,n}(\varphi(t)) \tag{5}$$

$$\varphi(t) = \tan t, \qquad -\frac{\pi}{2} < t < \frac{\pi}{2},$$

where the $N_{i,n}$ are the polynomial B-spline functions defined on the knot sequence $\{\varphi(t_i)\}$. In virtue of (5), relation (4) becomes

$$q(v) = \frac{\sum_{i=0}^{K+n} \begin{pmatrix} \cos \xi_i \\ \sin \xi_i \end{pmatrix} \left(\prod_{j=i+1}^{i+n} \cos t_j \right)^{-1} N_{i,n}(v)}{\sum_{i=0}^{K+n} \left(\prod_{j=i+1}^{i+n} \cos t_j \right)^{-1} \delta_i N_{i,n}(v)},$$

where

$$v = \varphi(t) = \frac{1}{2} \left[1 + \frac{\tan t}{\tan \Delta} \right]. \tag{6}$$

Thus, we can conclude that a p-spline in Cartesian coordinates has the following NURBS representation:

$$q(v) = \frac{\sum_{i=0}^{K+n} P_i w_i N_{i,n}(v)}{\sum_{i=0}^{K+n} w_i N_{i,n}(v)}, \qquad v \in [0,1], \tag{7}$$

with weights $w_i = \delta_i / (\prod_{j=i+1}^{i+n} \cos t_j)$ and control points $P_i = \delta_i^{-1} \begin{pmatrix} \cos \xi_i \\ \sin \xi_i \end{pmatrix}$; the $N_{i,n}(v)$ functions are defined over a knot sequence $\{v_i\}$ obtained applying relation (6) to the knots t_i. Note that the P_i are given by the transformation in Cartesian coordinates of the p-spline control points $\underline{d_i}$.

For example, the NURBS representation of the p-spline curve illustrated in Figure 1 has knot vector $\{v_i\} = \{0,0,0,0,0.4\bar{6},0.5,0.5\bar{3},1,1,1,1\}$, and weights $\{w_i\} = \{1, 0.09668, 0.00933, 0.00066, 0.00933, 0.09668, 1\}$.

If $2\Delta \geq \pi$, in order to satisfy the applicability conditions of relation (5), it will be necessary to subdivide the p-spline curve into piecewise p-splines defined on intervals whose size is less than π.

§4. Tools for p-splines

Knot insertion, subdivision, knot removal, and degree elevation are some of the many tools that play an important role in a spline-based modelling system. For p-splines, algorithms for knot insertion, subdivision, and knot removal can be obtained from analogous algorithms for trigonometric splines, as can be easily deduced from the definition of $\underline{c}(t)$.

Alternative algorithms can be obtained using relation (5) and the analogous algorithms for polynomial splines. For example, the knot insertion algorithm for p-splines may be schematized through the following steps:

Let $t_\ell < \hat{t} \leq t_{\ell+1}$ be the knot to be inserted.

1) Compute $c_i = \delta_i / (\prod_{j=i+1}^{i+n} \cos t_j) \qquad i = \ell - n, \cdots, \ell$;

2) Insert knot $\varphi(\hat{t})$ by means of the polynomial spline algorithm on c_i coefficients to achieve $\hat{c}_i$ over the $\{\hat{v}_i\}$ knot partition;

3) Compute $\hat{\delta}_i = (\prod_{j=i+1}^{i+n} \cos \hat{t}_j) / \hat{c}_i \qquad i = \ell - n, \cdots, \ell + 1.$

In fact, applying (5) to $\underline{c}(t)$, we have

$$\frac{1}{\sum_{i=0}^{K+n} \delta_i M_{i,n}(t)} = \frac{1}{cos^n t \; \sum_{i=0}^{K+n} c_i N_{i,n}(v)} \qquad (8)$$

executing knot insertion for polynomial splines, and applying relation (5) once again, we obtain

$$= \frac{1}{cos^n t \; \sum_{i=0}^{K+n+1} \hat{c}_i \hat{N}_{i,n}(v)} = \frac{1}{\sum_{i=0}^{K+n+1} \hat{\delta}_i \hat{M}_{i,n}(t)}$$

with c_i, $\hat{c}_i$ and $\hat{\delta}_i$ as indicated in 1), 2) and 3) above.

Analogously, relation (5) can be used in order to evaluate the p-spline $\underline{c}(t)$, referring the evaluation of a trigonometric spline to a polynomial spline. It should be noted that these tips can improve the efficiency of a p-spline-based modelling system.

Unlike the above-considered tools, the algorithm for the degree elevation of a p-spline is not achievable either from the degree elevation algorithm for a trigonometric polynomial considered in [1], or from the degree elevation algorithm for polynomial splines. In fact, the application of such algorithms does not modify the parametric interval size, as results from the definition of $\underline{c}(t)$. From this the need emerges for an ad hoc algorithm to determine the degree elevated p-spline curve.

4.1. Degree elevation for p-splines

From the expression of a p-spline in terms of the Fourier basis, one can deduce that this subset of curves is closed with respect to degree elevation from degree n to degree kn, for any natural value k [10].

Following the idea in [9] for polynomial splines, we propose a degree elevation technique for p-splines that consists in the following steps:

1) decompose the p-spline into piecewise p-Bézier curves (subdivision);

2) apply degree elevation to each p-Bézier curve;

3) remove unnecessary knots until the continuity of the original curve is guaranteed (knot removal).

In order to realize step 2., the following result is exploited [2].

Degree elevation formula for p-Bézier curves

Let $p(t) = \sum_{j=0}^{n} c_j A_{j,n}(t)$, $t \in [-\Delta, \Delta]$ be a generic trigonometric polynomial of degree n in the Bernstein trigonometric basis. Then

$$p(t) = \sum_{r=0}^{kn} \overline{c_r} A_{r,kn}(s), \qquad s = t/k, \qquad s \in \left[-\frac{\Delta}{k}, \frac{\Delta}{k} \right],$$

where $\qquad \overline{c_r} = \frac{n!}{\sin^n(2\Delta)} \binom{kn}{r}^{-1} \sum_{j=0}^{n} c_j \sum_{\Gamma_j} \gamma_j \eta_{j,r}$

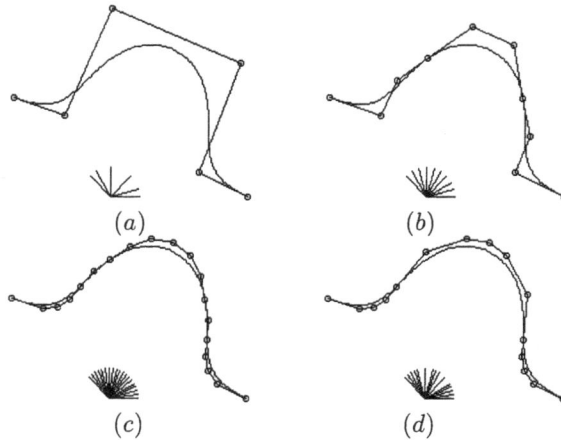

Fig. 2. Degree elevation steps; (a) original cubic p-spline curve , (b) subdivision in 3 p-Bézier curves, (c) degree elevation of each p-Bézier curve , (d) degree-elevated curve after the knot removal step.

and $\qquad \gamma_j = \tan^{I+J}\left(\frac{2\Delta}{k}\right) \cdot \prod_{d=1, d\ odd}^{D} \frac{\left[(-1)^{d\,div\,2}\binom{k}{d}\right]^{i_d+j_d}}{i_d!j_d!}$

$$\eta_{j,r} = \sum_{h=max(0,r-kj)}^{min(k(n-j)-I,r-J)} \binom{k(n-j)-I}{h}\binom{kj-J}{r-J-h} \cos^{k(n-j)+r-2h}\left(\frac{2\Delta}{k}\right).$$

D = max odd less than or equal to k,

$$\begin{aligned}\Gamma_j \;=\; & \{(i_1,i_3,..,i_D,j_1,j_3,..,j_D)\,;\, i_1,i_3,..,i_D,j_1,j_3,..,j_D \geq 0\,, \\ & i_1+i_3+..+i_D = n-j,\; j_1+j_3+..+j_D = j\}\end{aligned}$$

$$I = i_1 + 3i_3 + ... + Di_D, \qquad \text{and} \qquad J = j_1 + 3j_3 + ... + Dj_D.$$

Our implementation of the above formula provides a preprocessing phase that performs the entities that recur many times in the given formula; in particular, the coefficients γ_j are precomputed.

Note that step 2) uses an optimized version of this algorithm, in fact it requires only one preprocessing phase for the degree elevation of all the p-Bézier curves, and this also contributes to the efficiency of the degree elevation algorithm for p-splines.

In Figure 2, the three main algorithm steps are tested on an initial p-spline curve of degree $n = 3$ with 2 single interior knots, to obtain a p-spline of degree $kn = 6$ with 2 interior knots, both having a multiplicity of 4.

Computational results

The algorithm has been implemented in Pascal (BORLAND 7.0), carried out in double precision (15-16 significant figures), and tested on a Pentium 90 PC.

The test curves considered, without loss of generality, have been chosen with $\theta \in [0, \pi]$, the coefficients $\delta_i^{-1} = 1$, $i = 0, \cdots, n$, and randomly distributed knots.

The computational results reported here summarize the outcomes given in [3]. The tests were performed on numerical stability and execution time of the algorithm. Numerical stability was assessed by $MAXERR := \|\underline{c}(t) - \underline{c}(s)\|_{\infty D}$ on a uniformly-spaced set of points, where $\underline{c}(s)$ denotes the degree-elevated curve, resulting in $10^{-16} \leq MAXERR \leq 10^{-15}$ with $kn \leq 64$. This reveals the accuracy of our algorithm.

$k \backslash n$	1	2	3	4
2	0.036	0.073	0.147	0.253
3	0.073	0.187	0.340	0.807
4	0.090	0.220	0.587	1.250
5	0.107	0.440	1.500	4.510

$k \backslash n$	1	2	3	4
2	0.180	0.547	1.313	2.640
3	0.326	1.420	3.880	8.127
4	0.620	2.900	8.453	18.57
5	0.993	5.380	16.07	36.50

(a) (b)

Table 1: Execution time $(10^{-2} sec)$ results of degree elevation.

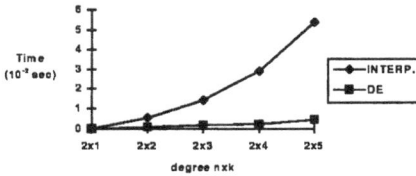 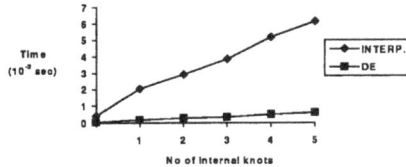

(a) (b)

Fig. 3. Degree elevation results.

In order to evaluate performance, the algorithm for the degree elevation of p-splines was compared with the interpolation technique, the only means at our disposal for degree-elevating a p-spline. Table 1 reports a comparison of execution times required by our algorithm (a) and by the interpolation technique (b) for 3 interior knots.

The graphs in Figure 3 provide a clearer understanding of these results. The graph (a) shows timings in the second column in Table 1 (a) and (b); note that while the execution times of our algorithm present a quasi linear growth rate, the interpolation has an exponential growth rate. The graph (b) illustrates the execution times as functions of the number of internal knots, while the degree kn remains unchanged at value 9.

All of the tests considered used p-spline curves with single interior knots. It is clear that our algorithm performs better, compared with the interpolation technique, when the knot multiplicity is increased.

Remark. After this paper was submitted, Sánchez-Reyes and the authors found a more efficient algorithm for p-Bézier degree elevation (see [4]). This also involves an improvement of the results here presented.

Acknowledgments. Supported by CNR-Italy, Contract n.95.00730.CT01.

References

1. Alfeld, P., M. Neamtu and L. L. Schumaker, Circular Bernstein-Bézier Polynomials, in *Mathematical Methods for Curves and Surfaces*, Morten Dæhlen, Tom Lyche, Larry L. Schumaker (eds.), Vanderbilt University Press, Nashville & London, 1995, 11–20.

2. Casciola, G., M. Lacchini and S. Morigi, Degree elevation for single-valued curves in polar coordinates, Dept. of Math., Bologna, Italy, 13, 1996.

3. Casciola, G. and S. Morigi, Modelling of curves and surfaces in polar and Cartesian coordinates, Dept. of Math., Bologna, Italy, 12, 1996.

4. Casciola, G., S. Morigi and J. Sánchez-Reyes J., Degree elevation for p-Bézier curves, submitted to Comput. Aided Geom. Design, October 1996.

5. de Casteljau, P., Splines Focales, *Curves and Surfaces in Geometric Design*, P.-J. Laurent, A. Le Méhauté, and L. L. Schumaker (eds.), A. K. Peters, Wellesley MA, 1994, 91–103.

6. Goodman, T. N. T. and S. L. Lee, B-Splines on the circle and trigonometric B-Splines, in *Approximation Theory and Spline Functions*, S. P. Singh, J. H. W. Burry, and B. Watson (eds.), Reidel (Dordrecht), 1984, 297–325.

7. Koch, P. E., Multivariate trigonometric B-splines, J. Approx. Theory **54** 1988, 162–168.

8. Lyche, T. and R. Winther, A stable recurrence relation for trigonometric B-splines, J. Approx. Theory **25** 1979, 266–279.

9. Piegl, L. and W. Tiller, Software-engineering approach to degree elevation of B-Spline curves, Computer-Aided Design **26** 1994, 17–28.

10. Sánchez-Reyes, J., Single-valued curves in polar coordinates, Computer-Aided Design **22** 1990, 19–26.

11. Sánchez-Reyes J., Single-valued spline curves in polar coordinates, Computer Aided Design **24** 1992, 307–315.

Giulio Casciola and Serena Morigi
Dept. of Math., University of Bologna
P.zza di Porta S.Donato, 5
40127 Bologna, ITALY
`casciola@dm.unibo.it`
Dept. of Math. Bologna webb page `http://dm.unibo.it`

La Tolérance d'usinage
chez Citroën dans les années (19)60

Paul de Faget de Casteljau

Abstract. The purpose of this paper is discuss what was done at Citroën during the sixties in order to solve certain manufacturing problems, and to fill the gap between mathematics and milling machines. Particular emphasis is placed on milling tolerance. Concerning machine finishing of a given surface, the milling tolerance should take into account first of all the normal (to the surface) deviation, then the geodesic deviation, and finally the deviation between furrows. The paper also gives an outline of some constraints that have to be satisfied because of the properties of milling machines (shape of the milling tool, shape of its stamp, continuity of the sweeping path, etc.)

§0. Introduction

Dans les années (19)60, au Laboratoire de Fraisage Numérique de la Société Citroën, créé et dirigé par Monsieur de la Boixière, il fallût étudier une foule de problèmes, indispensables au passage de la théorie mathématique au stade industriel. En particulier il convient de limiter l'écart entre l'usinage obtenu et la forme théorique; c'est le rôle de la tolérance, calcul de grande importance. Si l'analyse d'un problème n'est pas assez poussée, l'expérience se charge immédiatement de vous infliger de bien cruelles désillusions. Or, pour balayer correctement une portion de surface par une succession de déplacements linéaires d'un outil, il est nécessaire de bien sérier les difficultés. On peut répartir les principes à respecter, tels qu'ils sont analysés dans cet exposé, en trois catégories distinctes:

En premier lieu, comment passer des tendances aux extrémités d'un segment, à une formulation d'écart cubique.

En second, les trois types de tolérance rencontrés lors d'un balayage d'une portion de surface: Ecart normal, Ecart géodésique, Ecart entre sillons.

En troisième partie, les précautions nécessitées par les contraintes imposées par le comportement des machines outils, fraiseuses, rectifieuses, etc...

Curves and Surfaces with Applications in CAGD
A. Le Méhauté, C. Rabut, and L. L. Schumaker (eds.), pp. 69–76.
Copyright © 1997 by Vanderbilt University Press, Nashville, TN.
ISBN 0-8265-1293-3.

All rights of reproduction in any form reserved.

69

§1. Ecarts et Tendances.

On considère l'équation du troisième degré:

$$y = \varphi(x) \equiv a + bx + cx^2 + dx^3.$$

L'écart avec la corde C_{ij}, qui relie les points d'abscisse x_i et x_j, s'écrit

$$E_{ij}(x) = \varphi - C_{ij} = (x - x_i)(x - x_j)[c + d(x_i + x + x_j)].$$

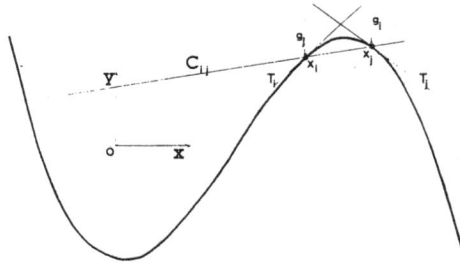

Fig. 1. Tendances au troisième degré.

Cette formule s'applique aussi aux tangentes $T_i = C_{ii}$ et $T_j = C_{jj}$, et on nommera tendances les valeurs des écarts $g_1 = E_{11}(x_2)$ et $E_{22}(x_1) = g_2$

$$g_1 = E_{11}(x_2) = (x_2 - x_1)^2[c + d(2x_1 + x_2)],$$

$$g_2 = E_{22}(x_1) = (x_2 - x_1)^2[c + d(x_1 + 2x_2)],$$

d'où l'on tire les termes parabolique $g_1 + g_2 = (x_2 - x_1)^2 \, \varphi''(\frac{x_1 + x_2}{2})$ et cubique $g_2 - g_1 = (\frac{x_2 - x_1}{6})^3 \, \varphi'''$. Pour raccourcir ou allonger un arc, en multipliant l'intervalle $\Delta x = x_2 - x_1$ par p, ainsi que pour passer à l'arc suivant ($p = -1$), de façon à ajuster l'écart à la tolérance admise, il convient de poser $x_i = 0$, et en banalisant l'autre variable, il vient:

$$g_0 = x^2(c + dx) \qquad\qquad g_2 = x^2(c + 2dx)$$

Le tableau ci-dessous, où chaque ligne, à partir de la troisième, est la somme des deux précédentes, donne les résultats recherchés:

$2g_0 - g_1 = -cx^2$	$2G_0 - G_1 = -cp^2x^2 = p^2[2g_0 - g_1]$	$2G_0 - G_1 = 2g_0 - g_1$
$g_1 - g_0 = \quad dx^3$	$G_1 - G_0 = d.p^3x^3 = p^3[g_1 - g_0]$	$G_1 - G_0 = g_0 - g_1$
$g_0 = cx^2 + dx^3$	$G_0 = p^2[g_0 + (1-p)(g_0 - g_1)]$	$G_0 = 3g_0 - 2g_1$
$g_1 = cx^2 + 2dx^3$	$G_1 = p^2[g_1 + 2(1-p)(g_0 - g_1)]$	$G_1 = 4g_0 - 3g_1$

La troisième colonne, obtenue pour $p = -1$, mise sous la forme matricielle $G = Mg$ met en œuvre une matrice

$$M = \begin{pmatrix} 3 & -2 \\ 4 & -3 \end{pmatrix}$$

racine carrée de l'unité $M^2 = 1$ $M = M^{-1}$

En reprenant la fonction d'écart, écrite sous une forme réduite ($0 \leq x \leq 1$), on peut en calculer la dérivée, et de là en tirer les racines.

$$f \equiv x(1-x)[(1-x)g_1 + xg_2]$$

$$f' \equiv (1-x)(1-3x)g_1 + x(2-3x)g_2$$

Suivant que g_1 et g_2 sont de mêmes signes, ou de signes contraires, il y a un ou deux maxima dans l'intervalle, et dans ce dernier cas on reportera notre attention sur le plus grand des deux.

Il est remarquable que l'écriture qui utilise une valeur absolue et la détermination positive d'un radical, donne dans tous les cas et sans ambiguïté le plus grand écart possible à l'intérieur de l'arc.

Posons $S = |g_1 + g_2|$, terme parabolique, et $R = \sqrt{g_1^2 - g_1g_2 + g_2^2}$ ou encore $R = \frac{1}{2}\sqrt{(g_1+g_2)^2 + 3(g_2-g_1)^2}$ ce radical tient compte de la valeur du terme cubique.

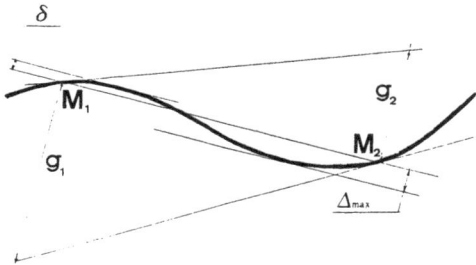

Fig. 2. Ecart Maximum.

L'écart maximum dans l'intervalle $]x_1, x_2[$ est alors:

$$\Delta\ max = \frac{(S+R)^2}{9(S+2R)} = \frac{1}{9}\left[S + \frac{R^2}{S+2R}\right].$$

Cette formule n'est pas mise en défaut par une inflexion, ou le balayage d'une surface le long d'une ligne asymptotique.

Bien entendu, il reste à ajuster la longueur de l'arc à la valeur imposée à la tolérance. On vient d'apprendre à le faire par des calculs approchés, ainsi que de prédéterminer l'intervalle suivant.

§2. Tolérance et Empreinte d'usinage.

Un usinage laisse presque toujours un excédent de matière. Le travail de finition consistera à l'enlever avec des outils appropriés. De cet excès de matière, on tire une notion intituitive de la tolérance, avec le risque d'accumuler des recettes vite incompatibles, ce qui se traduira par un résultat bien peu satisfaisant.

Afin de préciser cette notion de tolérance, rien ne vaut l'analogie avec un fruit, orange par exemple, dont on enlève la peau (croûte, écorce, coquille, bogue, etc.), que l'on supposera d'épaisseur constante, bien que ce ne soit pas une condition indispensable.

Définitions.

1) La tolérance est donnée par l'épaisseur T d'une couche, limitée à deux surfaces parallèles, l'une interne S, l'autre externe Σ.

2) A un instant donné de l'usinage, l'outil qui vient d'enlever une portion de cette couche intermédiaire, est en contact avec cette couche. Cette surface de contact est appelée l'empreinte.

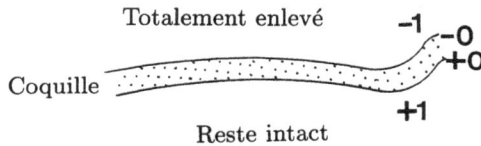

Fig. 3. Traces d'usinage.

Il existe donc trois zones dans le bloc brut de matière à usiner que l'on peut colorer par la pensée:

1) L'intérieur de la surface S, coloré en bleu (+1) ne doit jamais être endommagé par l'outil. C'est le but final de l'usinage.

2) L'extérieur de la surface Σ, coloré en rouge (−1) qui doit être enlevé dans sa totalité par l'usinage.

3) La couche intermédiaire, colorée en vert, subira un enlèvement partiel (−0) pendant l'usinage, le reste (+0) ne sera enlevé qu'en finition.

Dans la pratique, le centre de l'outil décrit une ligne polygonale, tangente en quelques points à la surface S, sans jamais s'en écarter au delà de la surface Σ. Les points de contact avec S, dépendent du signe de la concavité.

Forme de l'outil.

En théorie toutes les fantaisies sont possibles, comme le fil "à couper le beurre", possible par électro-érosion.

En pratique on a un outil en rotation rapide (fraise lors d'un fraisage) défini par la méridienne de son enveloppe.

Chez Citroën, il y eût aussi une machine à rectifier les cames ou un rapide va et vient d'une meule, définissait un plan de rectification, donnant un résultat de qualité exceptionelle.

Le cas le plus classique est celui d'une fraiseuse 3-axes: Si on calcule l'angle avec la verticale de la normale d'un point à usiner, il est facile de positionner n'importe quelle forme de fraise, dont la méridienne serait définie par enveloppe de droites ; son équation est donnée sous la forme normale d'Euler:

$$xcos\varphi + ysin\varphi = p(\varphi)$$

(Le cas des fraises cylindriques ou coniques $\varphi = c^t$, est exclu, et nécessite un calcul à part).

Le plus souvent on en reste au cas $p = C^{te} = R$ de la fraise sphérique, encore appelée "cul d'oeuf" d'usage très courant, mais qui ne fait pas un travail excellent au voisinage de son axe.

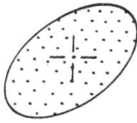

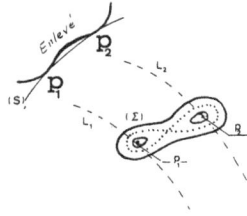

Fig. 4. Ellipse de contact   **Fig. 5.** le Bicontact.

En première approximation, l'empreinte est une ellipse, qui dépend des courbures respectives des formes en contact (indicatrice d'Euler, pression de Hertz). Le cas d'une forme hyperbolique est exclu, mais on peut avoir le cas de droites parallèles, qui remplace le cas parabolique. Ainsi pour l'usinage de l'arête rectiligne d'une surface développable par une fraise cylindrique (Ô combien spectaculaire !).

On comprend mieux cette notion d'empreinte d'un contact, dans le cas toujours possible du bi-contact. Un outil de forme torique peut ainsi être en contact avec une forme, le long de deux lignes voisines, et il serait théoriquement possible d'éliminer ainsi les crêtes d'usinage, permettant ainsi d'atteindre une quasi perfection. La forme de l'empreinte ressemble alors à un haricot. On peut encore prendre conscience de la possibilité de ce bi-contact, en considérant le tore comme enveloppe d'une sphère en rotation excentrée autour de son axe: on prend alors deux positions voisines de la sphère, et on détermine les conditions pour qu'elles définissent un tore.

Le rayon de la fraise "cul d'oeuf" ne peut descendre en dessous de la valeur T de la tolérance, et alors il suffit de considérer le contour de l'empreinte, qui est aussi l'intersection de la fraise avec la surface Σ. Pour réaliser l'usinage, il suffit de déplacer cette empreinte de façon à couvrir complètement Σ.

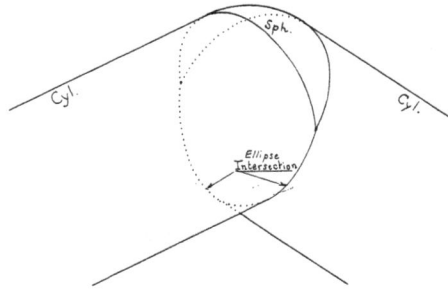

Fig. 6. Correction de concavité.

Correction de concavité.

Si l'on usine une zone concave (points voisins de la surface du même coté que le centre de fraise par rapport au plan tangent), ce sont les sommets de la ligne polygonale du parcours, qui correspondront à un contact tangent avec la surface S. On porte sur la normale le rayon R de la fraise pour en calculer le centre. Sinon le contact se fait le long d'un segment rectiligne du parcours. Pour rester en excès de matière, le contact aux sommets du contour polygonal est reporté au niveau de la surface Σ. La trace d'un sillon a donc l'allure d'un genou: deux cylindres tangents à une même sphère, dont l'empreinte ne se remarque que dans les parties concaves.

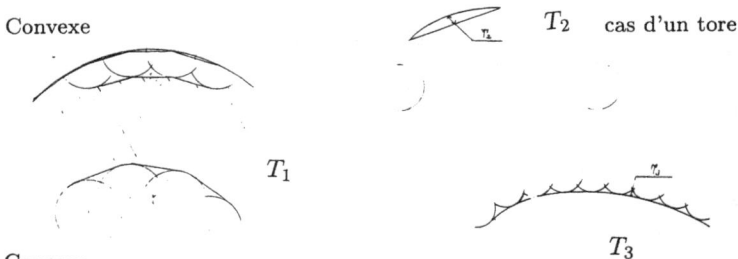

Fig. 7. les trois tolérances.

Principes régissant la tolérance

Il en existe trois et dans le cas d'une courbe plane, la première suffit:

1) *Tolérance normale* T_1. A tout instant de son parcours, une fraise parallèle à $T_1(R + T_1$ ou $p + T_1)$ serait sécante avec la surface S à usiner, sans que la fraise le soit. Ceci suffit à prouver l'existence d'une empreinte.

2) *Tolérance géodésique* T_2. A tout instant de son parcours, une fraise parallèle à $T_2(T_2 \gg T_1$ en pratique) couperait la courbe, décrite sur la surface. Indispensable pour suivre une asymptotique, ou la ligne de contact entre un plan tangent constant et une surface, cette tolérance évite des catastrophes.

3) *Tolérance entre sillons* T_3. En fin de balayage, la surface parallèle à $+T_3$ a totalement disparu.

Calculs pratiques

On utilise pour les deux premières, la notion de tendance, permettant de calculer l'écart maximum intermédiaire Δmax.

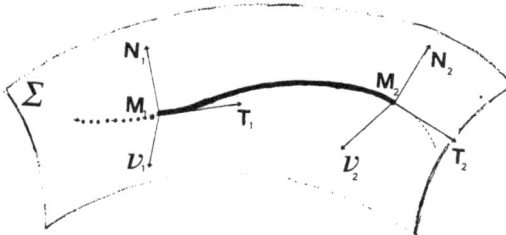

Fig. 8. calcul des "tendances".

1. Ecart normal. On détermine la concavité ; si $M_1 M_2 . \overrightarrow{N}_1 > 0$

$$P_1 = M_1 + R\overrightarrow{N}_1 \quad \text{et} \quad g_1 = \overrightarrow{M_1 M_2} . \overrightarrow{N}_1$$

Sinon on aura:

$$P_1 = M_1 + (R+T)\overrightarrow{N}_1 \quad \text{et} \quad g_1 = \overrightarrow{P_1 P_2} . \overrightarrow{N}_1 .$$

S'il y a changement de concavité entre M_1 et M_2, la plus grande valeur absolue impose sa loi.

2. Ecart géodésique. Les tendances sont alors calculées , à partir de la normale géodésique, perpendiculaire à la tangente à la courbe, située dans le plan tangent à la surface. $\overrightarrow{\nu}_1$ est le vecteur unitaire de $\overrightarrow{T}_1 \wedge \overrightarrow{N}_1$.

$$\gamma_1 = \overrightarrow{M_1 M_2} . \overrightarrow{\nu}_1 \quad \gamma_2 = \overrightarrow{M_1 M_2} . \overrightarrow{\nu}_2$$

La valeur de T_2 est moins critique que T_1, de l'ordre de la moitié de T_3.

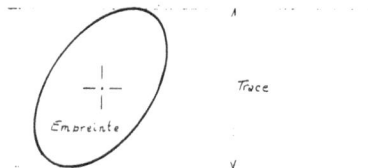

Fig. 9. Largeur d'une trace.

3. Ecart en sillons. C'est la valeur minima de la distance entre les deux tangentes à l'empreinte, parallèles à la tangente à la courbe, le long du parcours.

Ces calculs utilisent les deux formes quadratiques $\overrightarrow{dM^2}$ et $\overrightarrow{N}\overrightarrow{d^2 M}$ de la théorie différentielle des surfaces. Déjà résumé, le document Citroën relatif à la fraise sphérique dépasse la vingtaine de pages que l'on ne peut reproduire ici.

Pour être complet, il faut signaler les interférences entre ces trois tolérances qu'il convient de limiter. Ainsi le déport latéral, du à la tolérance géodésique, se superpose à l'écart entre sillons, au point qu'un balayage peut quelquefois recouper le précédent.

Fig. 10. Balayage aller et retour   **Fig. 11** Balayage "labyrinthe".

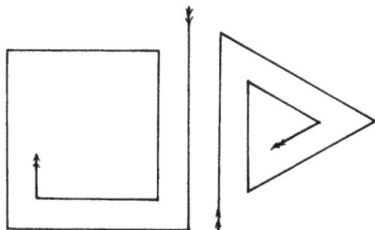

De manière à ce que la fraise travaille toujours de la même façon, le balayage aller et retour était proscrit au bénéfice du balayage "labyrinthe" en spirale convergeant vers le centre du carreau.

Il faut aussi, à tout prix, éviter les ralentissements aux coins du labyrinthe et au dégagement de fraise ; sinon la fraise "cire" et laisse une marque profonde. Il avait été prévu quatre paliers de vitesse, des moteurs pas à pas, dont il était préférable d'éviter la résonnance (à une vitesse assez basse) ; la règle était de $\Delta v^2 \leq 0,28$. Il était donc nécessaire de prévoir un coin, au moins trois ou quatre segments à l'avance. Il restait donc un entier n, suivi d'une fraction f de segments à faire. Ainsi au lieu de faire un segment de $0,1$ qui aurait imposé un fort ralentissement, si l'on avait prévu deux segments avant, on pouvait faire $\frac{1}{3}(2,1) = 0,7$ ce qui est déjà mieux.

L'expérience n'a pas su départager entre les valeurs:

$$n+1; \qquad n,f + \frac{1-f}{n}; \qquad n,f + (1-f)^2; \qquad \text{ou} \qquad n,f + \frac{(1-f)^2}{n}.$$

La dernière étant la plus élaborée. Même précaution entre les sillons.

Ces calculs sont adaptés à une interpolation linéaire. Certaines estimations semblent donner un réel avantage, dans le sens de la réduction du codage, à une interpolation parabolique, où un Δ_2 serait ajouté à un Δ_1, dont on reprendrait les premiers chiffres pour les additionner à la fonction; Toute retenue au niveau des unités est envoyée aux moteurs. On imagine le spectacle que pourrait procurer une telle interpolation appliquée sur une machine cinq-axes usinant avec une fraise torique.

Paul de Faget de Casteljau
4 Avenue du Commerce
78000 Versailles, FRANCE

Avoiding Local Minima for Deformable Curves in Image Analysis

Laurent D. Cohen

Abstract. We present an overview of part of our work over the past few years on snakes, balloons, and deformable models, with applications to image analysis. The main drawbacks of the active contour model being its initialization and minimization, we present three approaches that help to avoid being trapped in a local minimum of the energy. We introduced the balloon model to extract a contour being less demanding on the initial curve. In a more recent approach, based on minimal paths and geodesics, we find the global minimum of the energy between two points. A third approach is defined by a hybrid region-based energy taking into account homogeneity of the region inside the contour.

§1. Introduction

Active contour models, introduced by Kass, Witkin and Terzopoulos [15], and many variations on these deformable models have been studied for almost a decade and used for many applications. Using deformable models and templates, the extraction of a shape is obtained by giving an initial guess and through minimization of an energy composed of an internal regularization term and an external attraction potential (data fitting term), illustrated for example in [4,20,15,14,2,1]. Since the relevant shapes in medical images are usually smooth, the use of such models is particularly interesting for locating structures found in biomedical images and tracking their nonrigid deformation. We began using snakes for detecting the contour of Left Ventricular cavity of the heart in MR Images [10].

The main drawbacks of the active contour model being its initialization and minimization, we present three approaches that help to avoid being trapped in a local minimum of the energy, making initialization easier. We defined the balloon model [10,11] to extract the closed contour of an object being less demanding on the initial curve. This model was used as a first approach to make 3D reconstruction from cross sections [11]. A simplified 3D model was introduced as a stack of 2D balloons deforming simultaneously

Curves and Surfaces with Applications in CAGD
A. Le Méhauté, C. Rabut, and L. L. Schumaker (eds.), pp. 77–84.
Copyright © 1997 by Vanderbilt University Press, Nashville, TN.
ISBN 0-8265-1293-3.
All rights of reproduction in any form reserved.

[14]. To recover more general 3D surfaces in 3D medical images, the snake and balloon model was generalized to 3D and implemented with a finite element method in [14]. The normal inflation force appeared later as the basic evolution equation for many models with a moving front [5,18,25]. We also used recently a front evolution approach to find efficiently the global minimum of the snake energy [9]. A third approach making active contours less sensitive to initialization takes into account homogeneity of the region defined inside the curve. It combines edge and region terms, similar to Mumford-Shah energy [13]. Another way to make the algorithms more robust is to use parametric models like superquadrics or hyperquadrics. Once a global match with a rough shape is obtained, we have introduced ways to refine the shape given by the parametric model [8,3]. We also introduced a new mathematical formulation of some two-step iterative algorithms for deformable models, like deformable templates or B-spline snakes [17,2], using auxiliary variables [12].

§2. Active Contour Models or Snakes

We are looking for a plane curve $v(s) = (x(s), y(s))$ minimizing energy:

$$v \mapsto E(v) = \int_0^1 [w_1\|v'(s)\|^2 + w_2\|v''(s)\|^2 + P(v(s))]ds. \tag{1}$$

This energy models mechanical properties that are between an elastic string (first order) and a more rigid rod or spline (second order). The minimization of the potential P attracts the curve to the interesting features in the image. In the original model [15], it is based on the gradient of image I : $P(v) = -\|\nabla I(v)\|^2$. In [11], we defined an attraction potential from a binary image of previously extracted edge points, using Gaussian convolution or a distance map. In the latter case, $P(v) = f(d(v))$ is a function of the Chamfer distance to edges. The attraction acts as a zero length string that links a point of the curve and a data point [14]. Related to robust statistics, f can be chosen bounded, in order to allow the string linked to the data to break [12]. We see in Figure 2 an example of distance map.

Starting from an initial estimate, we solve the evolution equation with fixed, free or periodic boundary conditions:

$$\frac{\partial v}{\partial t} - \frac{\partial}{\partial s}(w_1 \frac{\partial v}{\partial s}) + \frac{\partial^2}{\partial s^2}(w_2 \frac{\partial^2 v}{\partial s^2}) = F(v)$$

where $F = -\nabla P$ is the attraction force towards contours. This equation is equivalent to making a gradient descent of the energy converge to a minimum of (1).

In the 3D case [14], we model a surface that has physical properties between an elastic membrane (first order, like food wrapping plastic paper) and a thin plate (second order, for example projection transparencies).

The equations are solved using finite differences or finite elements. In any case, we can always write the iterative scheme in matrix form as

$$(Id + \tau A)V^t = (V^{t-1} + \tau F(V^{t-1})),$$

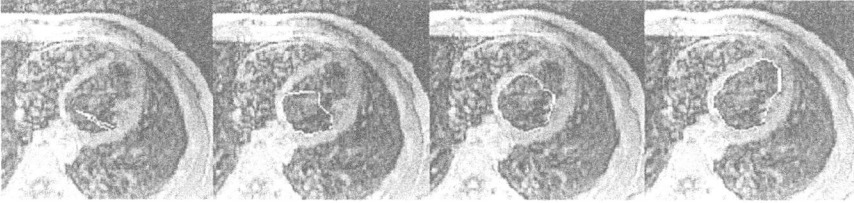

Fig. 1. MRI Image of the heart: Evolution of the balloon.

where V^t represents the vector of unknowns (nodes or degrees of freedom) of the discrete curve at iteration t.

§3. Balloon Model: Inflation Pressure Force

Usually, to make the curve converge to the right solution, the user has to provide an initial guess that is rather close to it. To make the curve converge to the solution, even when it is not close to it, we introduced the "Balloon model" [10]. We add a *pressure force* pushing outwards like inflating a *"balloon"*. This gives the curve a more dynamic behavior

$$F_{balloon} = k\vec{n}(s), \tag{2}$$

where $\vec{n}(s)$ is the unit normal vector to the curve at point $v(s)$. The curve behaves like a balloon which is inflated. It is stopped by a strong edge but avoids the curve being "trapped" by spurious isolated edge points. This makes the result much less sensitive to the initial conditions. Since by inflating the model, the size of the curve increases, it may be necessary to increase the number of discretization nodes. This may be obtained by making reparametrization every few iterations, defining regularly spaced nodes.

We remark that this force derives from the inside area energy

$$E_{area} = -k \int_{inside region} dA$$

that measures the area inside the region bounded by the curve. Minimization of such energy corresponds to get a region as large as possible. This is obtained by a pressure force in the outward normal direction.

We show an example of the balloon evolution in Figure 1. Starting from almost any curve inside the object permits the recovery of the whole boundary by inflating the curve like a balloon. This avoids the need for an initialization close to the solution.

Recently, much work has been done based on the evolution of a plane curve subject to a normal force. This was either in a purely mathematical framework or for various applications in image processing [21,22,5,18,6]. We also note the similarity between the evolution of a plane curve subject to a pressure force (2) and a dilatation in mathematical morphology.

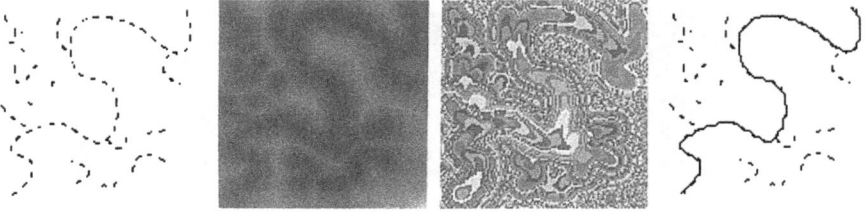

Fig. 2.   Line image. From left to right: original, potential, minimal action (random look up table to show the level set propagation starting from the bottom left), minimal path between bottom left and top right .

The current trend to define a deformable curve or surface is to use an intrinsic geometric model [5,18]. The surface deforms as a front evolution in the normal direction of the zero level set of a 3D function. This function is "deformed" in order to make its zero level set follow the minimization of the potential. These are called either geometric or geodesic deformable models [6,7], implicit deformable surface [26], or bubbles [25]. This is an efficient way to change the topology. Other models also permit the curve or surface to change topology [24,19].

§4. Global Minimum using Geodesics

We now present an approach introduced in [9] which is based on normal front evolution of level sets to find the global minimum of the energy.

The minimization problem we are trying to solve is slightly different from (1). The reason we modified the energy is that we now have an expression where the internal regularization energy is included in the potential term. We can then solve the energy minimization in a similar way to that of finding the shortest path between two points on a surface using the method developed in [16]. The energy of the new model has the following form:

$$E(v) = \int_0^L [w\|v'(s)\|^2 + P(v(s))]ds = wL + \int_0^L P(v(s))ds = \int_0^L \tilde{P}(v(s))ds \quad (3)$$

where $\tilde{P} = P + w$. Here v is in the space of all curves connecting two given points (restricted by boundary conditions): $v(0) = p_0$ and $v(L) = p_1$, where L is the length of the curve. Contrary to the classical snake energy, here s represents the arc-length parameter, which means that $\|v'(s)\|^2 = 1$. This makes the energy depend only on the geometric curve $\mathcal{C}$ and not on the parameterization. The regularization term with w now exactly measures the length of the curve. To solve this minimization problem, we first search for the *surface of minimal action* U starting at $p_0 = v(0)$. At each point p of the image plane, the value of this surface U corresponds to the minimal energy integrated along a path starting at p_0 and ending at p:

$$U(p) = \inf_{v(L)=p} \left\{ \int_{\mathcal{C}} \tilde{P} ds \right\}. \quad (4)$$

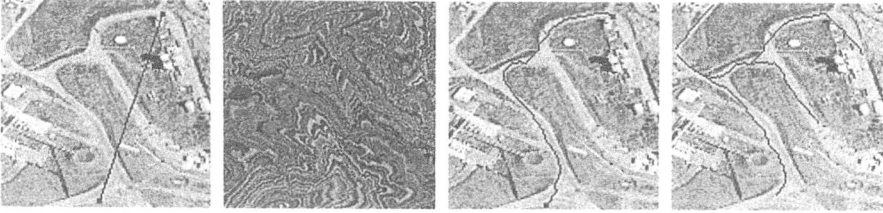

Fig. 3. Road Image. From left to right: initial data; minimal action level sets; path of minimal action connecting the two black points; many paths are obtained simultaneously connecting the start point on the upper left to 4 points.

Applying ideas of [16] to minimize our energy (3), it is possible to formulate a partial differential evolution equation describing the level set curves $\mathcal{L}$ of U:

$$\frac{\partial \mathcal{L}(s,t)}{\partial t} = \frac{1}{\tilde{P}} \vec{n}(s,t), \tag{5}$$

where $\tilde{P} = P + w$ and $\vec{n}(s,t)$ is the normal to the closed curve $\mathcal{L}(.,t)$: $S^1 \to \mathbb{R}^2$. This evolution equation is initialized by a curve $\mathcal{L}(s,0)$ which is a small circle surrounding the point p_0. It corresponds to a null energy. This evolution equation (5) is similar to a balloon evolution [11] with an inflation force depending on the potential.

This equation is solved using the Eulerian formulation for curve evolution introduced in [21] to overcome numerical difficulties and handle topological changes. Minimal action U can also be found efficiently using the fast marching method recently introduced by Sethian [23].

The algorithm is thus composed of two steps. First, minimal action U from p_0 is computed using front propagation starting from an infinitesimal circle centered at p_0. Then a backpropagation is made, tracking the minimal path by gradient descent on U starting from p_1 ending at p_0. A synthetic example is presented in Figure 2. We demonstrate the performance of the proposed algorithm by applying it to an aerial road image in Figure 3. Since road areas are lighter and correspond to higher gray levels, we chose $P = -I$. Observe the way the level curves propagate faster along the path. We need only two points while with a classical snake, a very close initialization is necessary to avoid local minima (see [9]). Our approach can be used for the minimization of many paths emerging from the same point in one single calculation of the minimal action (Figure 3).

§5. Region-based Energy

Active contour models only take into account the information along the curve. This often stops the evolution of the curve, trapped in a local minimum. We take into account here the fact that the contour is the boundary of an homogeneous region by the introduction of a hybrid energy composed of region-based

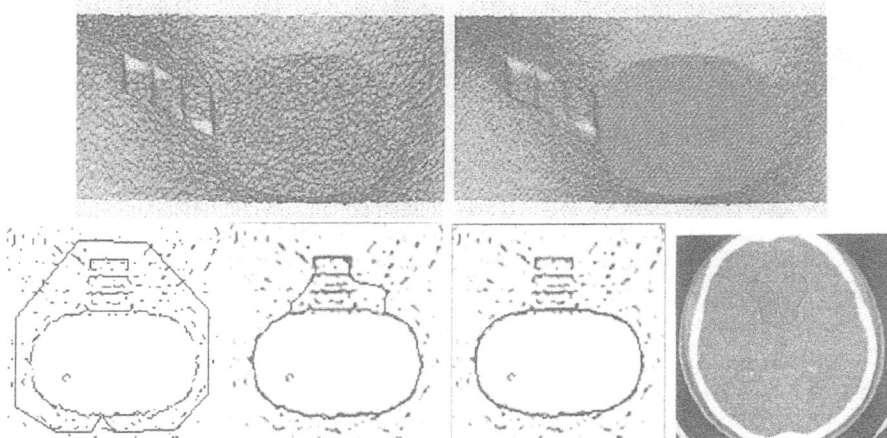

Fig. 4. Synthetic digital terrain model: above: original $z = I(x,y)$ and reconstruction $z = u(x,y)$ with our energy; below: On the left, initial curve is superimposed on an edge image obtained from I above. On the two middle images, we see the snake superimposed on the potential obtained from the edges. With the same initial curve on the left, a classic snake is stopped (middle-left) while the region term pushes the curve to the right boundary (middle-right). On the right, this is applied to two constant regions simultaneously to detect brain ventricles.

energy and boundary energy [13]. In [13] we used an active contour to detect a curve of discontinuities in a digital terrain model of a mountain and lake. The surface has to be smoothed and at the same time, we wish to recover the lake as a constant elevation region bounded by discontinuities.

We thus defined the reconstruction of a surface ($z = u(x,y)$) composed of two different kinds of regions. One has a constant elevation u_0 inside region L, the other is the background ($R - L$). The boundary curve B between them is obtained with an active contour model v. This combines the two problems of surface reconstruction with discontinuities and contour detection. We minimize an energy that is function of the couple of unknowns (u, v):

$$E_g(u,v) = E_{snake}(v) + \int_L (u_0 - I(x,y))^2 dx dy + \int_{R-L} ((u-I)^2 + \lambda^2 \|\nabla u\|^2) \quad (6)$$

The algorithm successively minimizes energy E_g with respect to each of the two variables u and v. When u is given, minimization in v corresponds to active contour evolution where new external forces are added that derive from the surface terms since L depends on v. This allows the model to take into account the fact that the level inside the region has to be homogeneous. This avoids the curve being trapped by spurious edges (see Figure 4). Initialization of the contour is made easier (see [13]). This property is not satisfied in classical active contours, where the curve "sees" only what happens locally along the curve and this may stop its evolution due to local minima.

Acknowledgments. Some of the presented work was done in collaboration with Nicholas Ayache, Eric Bardinet and Ron Kimmel. I would like to thank them, and express my wish that these fructuous collaborations go on.

References

1. Ayache, N., Medical computer vision, virtual reality and robotics, Image and Vision Computing, **13** (4) (1995), 295–313.

2. Ayache, N., P. Cinquin, I. Cohen, L. D. Cohen, F. Leitner, and O. Monga, Segmentation of complex 3D medical objects: a challenge and a requirement for computer assisted surgery planning and performing, in *Computer Integrated Surgery*. MIT Press, 1995, 59–74.

3. Bardinet, E., L. D. Cohen, and N. Ayache. Tracking medical 3D data with a deformable parametric model. in *Proc. ECCV'96*, U. K., I:317-328.

4. Blake, A. and A. Zisserman, *Visual Reconstruction*, MIT Press, 1987.

5. Caselles, V., F. Catté, T. Coll, and F. Dibos, A geometric model for active contours, Numer. Math. **66** (1993), 1–31.

6. Caselles, V., R. Kimmel, and G. Sapiro, Geodesic active contours, in *Proc. ICCV'95*, Cambridge, USA, June 1995, 694–699.

7. Caselles, V., R. Kimmel, G. Sapiro, and C. Sbert, Minimal surfaces: a 3D segmentation approach, in *Proc. ECCV'96*, Cambridge, U. K., 97-106.

8. Cohen, I., and L. D. Cohen. A hybrid hyperquadric model for 2-D and 3-D data fitting. in *Proc. 12th ICPR*, Jerusalem, 1994, B:403-405.

9. Cohen, L. D., and R. Kimmel. Global minimum for active contour models: A minimal path approach. in *Proc. CVPR'96*, San Francisco, June 1996.

10. Cohen, L. D., On active contour models. in *Active Perception and Robot Vision*. Springer, July 1989, 599–613.

11. Cohen, L. D., On active contour models and balloons, Computer Vision and Image Understanding **53**(2) (1991), 211–218.

12. Cohen, L. D., Auxiliary variables and two-step iterative algorithms in computer vision problems, in Journal of Mathematical Imaging and Vision **6**(1) (1996), 61–86. See also *Proc. IEEE ICCV'95*, Boston.

13. Cohen, L. D., E. Bardinet, and N. Ayache, Surface reconstruction using active contour models, in *Proc. Conference on Geometric Methods in Computer Vision*, San Diego, CA, July 1993.

14. Cohen, L. D., and I. Cohen, Finite element methods for active contour models and balloons for 2-D and 3-D images, in IEEE Trans. Pattern Anal. and Machine Intelligence, **15** (11) (1993), 1131-1147..

15. Kass, M., A. Witkin, and D. Terzopoulos, Snakes: Active contour models, International Journal of Computer Vision, **1**(4) (1988), 321–331.

16. Kimmel, R., A. Amir, and A. Bruckstein, Finding shortest paths on surfaces, *Curves and Surfaces in Geometric Design*, P.-J. Laurent, A. Le Méhauté, and L. L. Schumaker (eds.), A. K. Peters, Wellesley MA, 1994, 259–268.

17. Leitner, F., I. Marque, S. Lavallée and P. Cinquin, Dynamic segmentation: finding the edge with snake-splines, in *Curves and Surfaces*, P.-J. Laurent, A. Le Méhauté, and L. L. Schumaker (eds.), Academic Press, New York, 1991, 1–4.

18. Malladi, R., J. A. Sethian, and B. C. Vemuri, Shape modeling with front propagation: A level set approach. IEEE Trans. Pattern Anal. and Machine Intelligence **17**(2) (1995), 158–175.

19. McInerney, T. and D. Terzopoulos, Medical image segmentation using topologically adaptable snakes, In Springer, editor, *Proc. CVRMed'95*, Nice, France, April 1995, 92–101. See also ICCV'95.

20. Mumford, D. and J. Shah, Boundary detection by minimizing functionals, in *Proc. CVPR'85*, San Francisco, June 1985, 22-26.

21. Osher, S. J. and J. A. Sethian. Fronts propagation with curvature dependent speed: Algorithms based on Hamilton-Jacobi formulations, J. Comput. Phys. **79** (1988), 12–49.

22. Sapiro G. and A. Tannenbaum, On invariant curve evolution and image analysis, Indiana Univ. Math. J. **42**(3) (1993), 985–1009.

23. Sethian, J. A., A fast marching level set method for monotonically advancing fronts, in Proc. Nat. Acad. Sci. **93**(4) (1996), 1591–1595.

24. Szeliski, R., D. Tonnessen, and D. Terzopoulos, Curvature and continuity control in particle-based surface models, in *Proc. SPIE 93 Conf. on Geometric Methods in Computer Vision*, San Diego, CA, July 1993, 172-181.

25. Tek, H. and B. Kimia, Image segmentation by reaction-diffusion bubbles, in *Proc. ICCV'95*, Cambridge, USA, June 1995, 156–162.

26. Whitaker, R., Algorithms for implicit deformable models, in *Proceedings ICCV'95*, Cambridge, USA, June 1995, 822–827.

Laurent D. Cohen
CEREMADE, Université Paris IX Dauphine
Place du Maréchal de Lattre de Tassigny
75775 Paris cedex 16, FRANCE
Laurent.Cohen@ceremade.dauphine.fr
http://www.ceremade.dauphine.fr/~cohen

Variable Degree Polynomial Splines

Paolo Costantini

Abstract. The purpose of this paper is to give a first description of a class of spline functions made up by polynomials whose shape can be very efficiently and locally controlled using their variable degrees. Applications to constrained interpolation and CAGD will be considered.

§1. Introduction

The importance of controlling the shape of curves, both for classical functional interpolation and for CAGD applications, is today well-known and completely accepted by the scientific community.

The first, and probably the most used approach in interpolation is given by *tension methods*, based on piecewise exponential or rational functions. The main drawback of these kinds of tension functions – with the exception of particular choices of rationals – is that they have no geometrical counterpart (like the control net for Bézier polynomials) and this, in turn, has made the computation of shape–constraints a difficult task – almost impossible for 2-D interpolation – and has inhibited their direct use in CAGD applications.

However, in the last few years, polynomial alternatives to rationals and exponentials have been introduced and have been successfully applied in several shape-preserving problems. The basic fact is that the n_i-degree ith polynomial piece, which is a cubic for $n_i = 3$ and tends to be linear as n_i tends to infinity, has a Bézier net as simple as the cubic one. This has allowed very easy formulations of the shape constraints, and has therefore opened the way to several 2-D extensions. The additional features of numerical stability and low computational cost (practically equivalent to the cubic case for any n_i), have made this approach useful for functional shape-preserving interpolation. We refer to [4] for a survey on these applications and for technical details. In addition, the simplicity of the net has made possible to use these polynomials as basic elements for curve design. We refer again to [4, Sect. 5] for details and graphical examples, and note only that it is possible to obtain C^2 curves

Curves and Surfaces with Applications in CAGD
A. Le Méhauté, C. Rabut, and L. L. Schumaker (eds.), pp. 85–94.
Copyright © 1997 by Vanderbilt University Press, Nashville, TN.
ISBN 0-8265-1293-3.
All rights of reproduction in any form reserved.

and to locally control the shape, using the degrees to stretch the polynomial pieces in their central part (like the $\beta2$ parameter in the Beta–splines of [1]).

In this paper we introduce a more general class of splines in which, for any polynomial piece, we assign *two* tension parameters which control the behavior of the curve independently in the left and in right part of the interval, thus providing a more stringent control of the shape (the curves easily get cusp or linear-like appearances). The interesting fact is that there is no price to pay, neither in computational cost nor in geometric simplicity, because we have again a very simple Bézier net (or *pseudo-net* as we might better say). These polynomial tension splines were first introduced in [10] for C^2 interpolation, and then briefly recalled in [9]; here we want to analyze them from the *geometric* point of view, relating their features in shape-preserving interpolation and in curve design with their underlying pseudo-net.

As explained in the last section, this new class of *two-degree* polynomial splines has all the potential uses in constrained approximation of the previous *one-degree* one and, perhaps more importantly, it seems to have promising CAGD applications, especially if used to define C^2 NURBS-like curves or surfaces. The aim of the present paper is therefore to begin the analysis of its basic properties and its applications in these two fields.

This paper is divided into five sections: in the next we introduce the polynomial element and its pseudo-net; in Section 3 we discuss the possibility of an automatic choice of the tension parameters when used in shape-preserving interpolation and, in Section 4 we show their application in curve construction. Section 5 is then devoted to final remarks and anticipations of other results which, due to space limitations will be reported elsewhere.

§2. The Polynomial Element

Let integers $n, m \geq 3$ be given, and in the interval $[0,1]$ define the functions

$$\Phi(t) = \Phi(t;n) := (1-t)^n; \ \Psi(t) = \Psi(t;m) := t^m, \tag{1}$$

which can also be expressed as $\Phi = \mathcal{B}^{(n)}\ell_\Phi$ and $\Psi = \mathcal{B}^{(m)}\ell_\Psi$, where $\mathcal{B}^{(n)}, \mathcal{B}^{(m)}$ are the Bernstein–Bézier operators and ℓ_Φ, ℓ_Ψ are piecewise linear functions such that $\ell_\Phi(0) = 1$, $\ell_\Phi(k/n) = 0$, $k = 1, 2, \ldots, n$, $\ell_\Psi(k/m) = 0$, $k = 0, 1, \ldots, m-1$, and $\ell_\Psi(1) = 1$. We define

$$BP_{n,m} := span\{t, (1-t), \Phi, \Psi\} \tag{2}$$

which is clearly a four-dimensional linear space such that $BP_{3,3} = \mathbb{P}_3$. Let us suppose $m > n$ (an analogous theory can be developed for $n > m$). Since Bernstein operators are linear and reproduce linear functions, we can express any $b(t) = b(t;n,m) = \alpha(1-t) + \beta t + \gamma\Phi(t) + \delta\Psi(t) \in BP_{n,m}$ as

$$b(t) = \alpha(1-t) + \beta t + \gamma\mathcal{B}^{(n)}(\ell_\Phi(t)) + \delta\mathcal{B}^{(m)}(\ell_\Psi(t)). \tag{3}$$

Obviously we cannot extract the control net of b because it is made up of polynomials of different degrees, but we will immediately see that the piecewise linear function

$$\ell(t) = \alpha(1 - t) + \beta t + \gamma \ell_\Phi(t) + \delta \ell_\Psi(t); \tag{4}$$

(hereafter called the *control pseudo-net*) plays the same role of the control net of cubic Bézier polynomials. Let $\mathcal{B}^{(n,m)}$ be the linear operator given by (3) and (4), i.e. $b(t) = \mathcal{B}^{(n,m)}\ell(t)$. We have the following variation-diminishing property.

Theorem 1. *The polynomial $b = \mathcal{B}^{(n,m)}\ell$ has no more intersections with any straight line than the piecewise linear function ℓ given in (4).*

Proof: Let us define the one dimensional degree elevation operator $\mathcal{E}_p^q$ (see, e.g. [5]) such that, for any control net $\tilde{\ell}$ of p degree (that is defined by $p + 1$ control points), $\mathcal{B}^{(p)}\tilde{\ell} = \mathcal{B}^{(q)}\mathcal{E}_p^q\tilde{\ell}$, and let

$$\lambda_m(t) = \alpha(1 - t) + \beta t + \gamma \mathcal{E}_n^m \ell_\Phi(t) + \delta \ell_\Psi(t) . \tag{5}$$

Obviously $b = \mathcal{B}^{(m)}\lambda_m$. Now, let μ be any integer such that $n \le \mu \le m$. Since $\Phi = \mathcal{B}^{(n)}\ell_\Phi = \mathcal{B}^{(\mu)}\mathcal{E}_n^\mu \ell_\Phi$ and since (see (1)) $\Phi^{(j)}(1) = 0$, $j = 0, 1, \ldots, n - 1$, it follows that $\mathcal{E}_n^\mu \ell_\Phi(k/\mu) = 0$, $k = \mu - n + 1, \cdots, \mu$, or, in other words,

$$\mathcal{E}_n^\mu \ell_\Phi(t) = \begin{cases} \mathcal{E}_n^\mu \ell_\Phi(t), & t \in [0, \frac{\mu-n+1}{\mu}] \\ \ell_\Phi(t), & t \in (\frac{\mu-n+1}{\mu}, 1]. \end{cases}$$

This implies that the modifications of the net (5) with respect to the pseudo-net (4) have no effect on the right hand side of the interval, where this is formed taking the only contribution of t, $(1 - t)$, ℓ_Ψ. Since the operator $\mathcal{E}_n^\mu$ is nothing more that a sequence of repeated convex combinations, we may write $\alpha(1 - t) + \beta t + \mathcal{E}_n^\mu(\gamma \ell_\Phi)(t) = \mathcal{E}_n^\mu(\alpha(1 - t) + \beta t + \gamma \ell_\Phi(t))$, and therefore λ can be obtained from ℓ using the following *forward* degree-elevation scheme (see also Fig. 1). We define $\lambda_n := \ell$ and then, for $\nu = n + 1, \cdots, m$, we set

$$\lambda_\nu(t) := \begin{cases} \mathcal{E}_{\nu-1}^\nu \lambda_{\nu-1}(t), & t \in [0, \frac{\nu-n+1}{\nu}] \\ \lambda_{\nu-1}, & t \in (\frac{\nu-n+1}{\nu}, 1] \end{cases} .$$

Since any step is made up by convex interpolations, we have that λ_m has no more intersections with any straight line than ℓ, and the proof follows immediately from the variation diminishing property of Bernstein polynomials. ∎

In the theorem above we have seen that the shape of a polynomial of the form (2) can be controlled by the pseudo-net (4) which is as simple as the cubic one, being defined by only four control points, located at 0, $1/n$, $1 - 1/m$, 1. The effect of increasing the degrees n and m will be explained later; now we want to recall that $b(t) = [\alpha(1 - t) + \beta t] + \gamma(1 - t)^n + \delta t^m$, and therefore the computational cost for evaluating b is practically independent of the degrees.

We conclude this section with the obvious observation that the choice $n = m$ puts this class of functions in the simpler one discussed in [4].

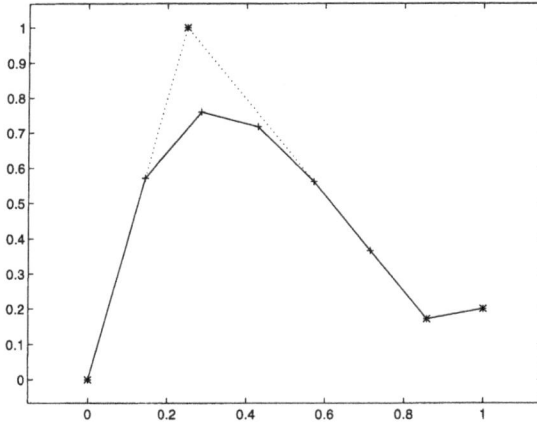

Fig. 1. The pseudo-net ℓ (dotted) and the net λ_m (solid) for $n = 4$, $m = 7$.

§3. Shape–preserving Interpolation

The aim of this section is to discuss the shape–preserving properties of interpolating splines made up of polynomials of the form (3). Because of space
limitations, we will confine ourselves to local C^1 interpolation at knots (see
[10] for an idea of global, cubic-like, C^2 interpolation), we start with an analysis of the monotonicity and convexity domains.

The polynomial $b = \mathcal{B}^{(n,m)}\ell$ can be conveniently expressed as the unique
solution of the following interpolation problem:

$$b(0; n, m) = y_0;\ \frac{db}{dt}(0; n, m) = d_0;\ b(1; n, m) = y_1;\ \frac{db}{dt}(1; n, m) = d_1.$$

Let $\Delta = (y_1 - y_0)/(1 - 0)$ be the data slope. An easy computation leads to

$$b(t; n, m) = \alpha(1 - t) + \beta t + \gamma(1 - t)^n + \delta t^m , \qquad (6)$$

and to

$$b'(t; n, m) = -\alpha + \beta - n\gamma(1 - t)^{n-1} + m\delta t^{m-1} , \qquad (7)$$

$$b''(t; n, m) = n(n - 1)\gamma(1 - t)^{n-2} + m(m - 1)\delta t^{m-2} , \qquad (8)$$

where

$$\alpha = \frac{(nm - n)y_0 - my_1 + (m - 1)d_0 + d_1}{nm - n - m};$$

$$\beta = \frac{-ny_0 + (nm - m)y_1 - d_0 - (n - 1)d_1}{nm - n - m};$$

$$\gamma = \frac{-my_0 + my_1 - (m - 1)d_0 - d_1}{nm - n - m};\ \delta = \frac{ny_0 - ny_1 + d_0 + (n - 1)d_1}{nm - n - m} .$$

Let us recall that the data are said to be *increasing* if $d_0, \Delta, d_1 > 0$ and *convex* if $d_0 < \Delta < d_1$. A simple inspection of (8) reveals that b is convex if and only if $b''(0), b''(1) > 0$, that is if and only if

$$(d_0, d_1) \in D_{CVX}(\Delta, n, m) :=$$
$$\{(u, v) \in \mathbb{R}^2 : m\Delta - (m-1)u - v > 0; \; n\Delta - (n-1)v - u < 0\} . \tag{9}$$

Since (9) are equivalent to requiring that the pseudo-net be convex, we have that b is convex if and only if ℓ is convex, thus restating a condition given in [2] for the case $n = m$ (obviously concavity conditions are obtained reversing the sign of inequalities (9)). Note also that for convexity we have

$$D_{CVX}(\Delta, n, m) \subset D_{CVX}(\Delta, n+1, m) , \; D_{CVX}(\Delta, n, m+1). \tag{10}$$

Let us now move to monotonicity conditions. It is an obvious consequence of Theorem 1 that b is increasing if ℓ is increasing, that is if

$$d_0 > 0; \; d_1 > 0; \; mn\Delta - nd_1 - md_0 > 0 ,$$

but, even if the computation of "necessary and sufficient" domains is an impossible algebraic task in this arbitrary case, we would at least improve the sufficient domain, and not to be obliged to choose useless large degrees. In [3], for $n = m$, it was proven that b is increasing if (d_0, d_1) is contained in $D_{INC}(\Delta, n, n) := ConvHull\{(0,0), (n\Delta, 0), (\rho(n)\Delta, \rho(n)\Delta), (0, n\Delta)\}$, where $\rho(n) := \frac{n(1-2^{(n-2)})}{n-2^{(n-1)}}$. Despite the algebraic computations, $\rho(n)$ was simply computed assuming $d_0 = d_1$, proving that in this case b'' given in (8) has the unique zero at $t = 1/2$ and then finding the threshold value of d_0 (that is $\rho(n)\Delta$) such that, for any $d_0 > \rho(n)\Delta$, $b'(1/2; n, n) < 0$. In the present case, where $n \neq m$, it is impossible to analytically locate the root of b'' and the previous steps can be performed only numerically. We have the following result.

Theorem 2. *The polynomial b given in (6) is increasing if*

$$(d_0, d_1) \in D_{INC}(\Delta, n, m) :=$$
$$ConvHull\{(0,0), (n\Delta, 0), (\rho(n,m)\Delta, \rho(n,m)\Delta), (0, m\Delta)\} ,$$

where the points above are on the boundary of the monotonicity region and the values of $\rho(n, m)$ can be numerically computed.

Proof: Obviously $(0,0)$ is on the boundary of the monotonicity domain. Let us take $d_0 = 0$, $d_1 > m\Delta$. Then we have immediately from (8) and (9) that b'' is negative in a right neighborhood of 0 and therefore b' must be negative. Similar considerations can be done for $d_0 > n\Delta$, $d_1 = 0$. Now assume $d_0 = d_1$ in (6)–(8) (that is we are moving on the first bisector of uv-plane) and $d_0, d_1 > \Delta$ (that is we are not in the convexity domain). From (8) we see that b'' has a single root $\xi \in (0, 1)$, and from (7) that $b'(\xi; n, m)$ is linear in d_0, having substituted $d_1 = d_0$. Then, solving the equation $b'(\xi; n, m) = 0$ and calling $\rho(n, m)\Delta$ its solution, we have

$$\rho(n, m) = \frac{mn(1 - \xi^{m-1} - (1 - \xi)^{n-1})}{n + m - mn(\xi^{m-1} + (1 - \xi)^{n-1})}.$$

Obviously $(\rho(n, m)\Delta, \rho(n, m)\Delta)$ is on the boundary of monotonicity domain. ∎

Another useful result proven in [3] for $n = m$ was that $D_{INC}(\Delta, n, n) \subset D_{INC}(\Delta, n + 1, n + 1)$. In the present case we would expect some property like

$$D_{INC}(\Delta, n, m) \subset D_{INC}(\Delta, n + 1, m) \,, \; D_{INC}(\Delta, n, m + 1) \,, \qquad (11)$$

but we have been able to prove only the following result.

Theorem 3. *For any $n, m \geq 3$, $D_{INC}(\Delta, n, m) \subset D_{INC}(\Delta, n + 1, m + 1)$.*

Proof: It suffices to show that $(\rho(n, m)\Delta, \rho(n, m)\Delta) \in D_{INC}(\Delta, n+1, m+1)$, that is $b(\cdot; n + 1, m + 1)$ is increasing. This follows immediately by observing that (see (4)) since the end derivatives are the same, $\ell(\cdot; n + 1, m + 1)$ has the slopes of the first and third segment equal to, and the (negative) slope of the second segment greater than the corresponding ones of $\ell(\cdot; n, m)$. Now, consider $\lambda_{m+1}(\cdot)$ and $\lambda_m(\cdot)$ obtained applying the algorithmic scheme of the proof of Theorem 1 to $\ell(\cdot; n + 1, m + 1)$ and $\ell(\cdot, n, m)$ respectively; then the slopes of $\lambda_{m+1}(\cdot)$ are greater or equal to those of $\mathcal{E}_m^{m+1}\lambda_m(\cdot)$. The proof is then completed by observing that $b(\cdot; n, m) = \mathcal{B}^{(m+1)}\mathcal{E}_m^{m+1}\lambda_m(\cdot)$ and $b(\cdot; n + 1, m + 1) = \mathcal{B}^{(m+1)}\lambda_{m+1}(\cdot)$. ∎

On the contrary, numerical computations of ρ show that $D_{INC}(\Delta, 3, 3) \not\subset D_{INC}(\Delta, 3, 4)$ but that (11) is valid for $n, m \geq 4$. We can thus state the following conjecture.

Proposition 4. *Relations (11) are satisfied for any $n, m \geq 4$.*

As an experimental proof we have computed $\rho(n, m)$ for $3 \leq n, m \leq 40$; some of these values are

$$\{\rho(3, m); \; m = 3, 4, 5, 6\} = \{3 \, , \; 2.943 \, , \; 2.912 \, , \; 2.894\} \,;$$
$$\{\rho(4, m); \; m = 4, 5, 6\} = \{3 \, , \; 3.055 \, , \; 3.106\} \,;$$
$$\{\rho(5, m); \; m = 5, 6\} = \{3.182 \, , \; 3.294\} \,; \; \{\rho(6, 6)\} = \{3.462\} \,.$$

We observe that (10) and Proposition (4) above are nothing more than a restatement – useful for applications – of a simple property inherent in the structure of the pseudo-net: for large values of n and m, ℓ tends to the straight line interpolating y_0 and y_1. Therefore, taking also into account Theorem 1, if the derivatives d_0, d_1 and the slope Δ agree with the definitions of increase and convexity, it is always possible to find degrees such that the corresponding ℓ and $\mathcal{B}^{(n,m)}\ell$ are increasing and/or convex.

Suppose now we are given a set of data (x_i, y_i); $i = 0, 1, \ldots, N$, and the corresponding derivatives d_i; $i = 0, 1, \ldots, N$ are either given or computed somehow (e.g. by Bessel-like formulas). Set, for any i, $t := (x - x_i)/(x_{i+1} - x_i)$ and $\Delta_i := (y_{i+1} - y_i)/(x_{i+1} - x_i)$. Then, the ith piece of a C^1 piecewise polynomial interpolating function can be easily computed taking, for any interval, a polynomial of the form (6). This can then be forced to be shape–preserving selecting the corresponding degrees n_i, m_i in such a way that (d_i, d_{i+1}) belongs to $D_{INC}(\Delta_i, n_i, m_i)$ or $D_{DEC}(\Delta_i, n_i, m_i)$ and/or $D_{CVX}(\Delta_i, n_i, m_i)$ or $D_{CNC}(\Delta_i, n_i, m_i)$.

§4. Curve Construction

In this section we discuss the use of variable degree polynomials in free form design. For reasons of space, we will limit ourselves to C^1 parametric curves, deferring applications to B-spline or NURBS like, C^2, parametric curves and surfaces construction to forthcoming papers. We start by expressing the space (2) using the Bernstein-Bézier pseudo-basis:

$$BP_{n,m} := span\left\{B_0(\cdot; n, m), B_1(\cdot; n, m), B_2(\cdot; n, m), B_3(\cdot; n, m)\right\},$$

where B_0, B_1, B_2, B_3 are obtained substituting for (y_0, d_0, y_1, d_1) in (6) respectively the values $(1, -1/n, 0, 0)$; $(0, 1/n, 0, 0)$; $(0, 0, 0, -1/m)$; $(0, 0, 1, 1/m)$. From geometric considerations we easily see they form a partition of unity. Now, taking in $\mathbb{R}^3$ the control points $\mathbf{b}_0, \mathbf{b}_1, \mathbf{b}_2, \mathbf{b}_3$, we can form the parametric Bézier polynomial

$$\mathbf{b}(\cdot; n, m) = \sum_{k=0}^{3} \mathbf{b}_k B_k(\cdot; n, m) . \tag{12}$$

We note that the properties of the convex hull, affine invariance, end-point interpolation, linear precision, pseudo-local control, hold as in the "classical" cubic case, and for symmetry we have

$$\sum_{k=0}^{3} \mathbf{b}_k B_k(t; n, m) = \sum_{k=0}^{3} \mathbf{b}_{3-k} B_k(1 - t; m, n) .$$

The degrees can of course be used for modifying the polynomial shape; in fact, since Bézier polynomials tend to their net when the degree increases, we see immediately, taking in (3) $\alpha, \beta, \gamma, \delta \in \mathbb{R}^3$, that the curve is attracted by $\mathbf{b}_1$ when n increases, by $\mathbf{b}_2$ when m increases and tends to become linear between $\mathbf{b}_1$ and $\mathbf{b}_2$ when both increase. See Fig. 2 for some graphical examples.

Now let $\mathbf{d}_i$; $i = 0, 1, \ldots, N$ be a given set of points, let u_i; $i = 0, 1, \ldots, N$ be any sequence of knots and let h_i; $i = 0, 1, \ldots, N-1$ be the lengths of knot intervals. A C^1 curve can be easily constructed connecting Bézier polynomials of the form (12), with $t = (u - u_i)/h_i$, $n = n_i$, $m = m_i$, which interpolate the data and whose tangent directions, $\mathbf{s}_i$; $i = 0, 1, \ldots, N$, are given by any

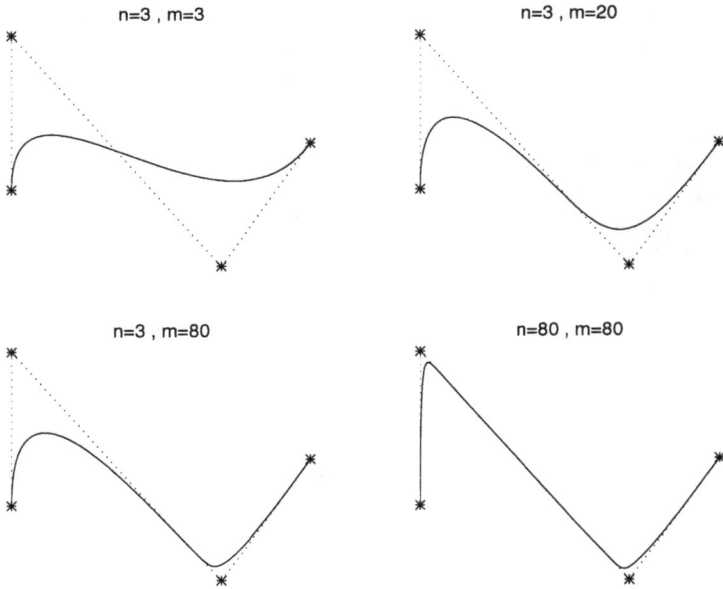

Fig. 2. The effect of increasing the degree on the polynomial shape.

estimating formula, like FMILL, Catmull-Rom or Bessel (see, e.g. [5]). The control points of the ith Bézier polynomial are then given by

$$\mathbf{b}_{3i} = \mathbf{d}_i; \ \mathbf{b}_{3i+1} = \mathbf{d}_i + \frac{1}{n_i}\mathbf{s}_i; \ \mathbf{b}_{3i+2} = \mathbf{d}_{i+1} - \frac{1}{m_i}\mathbf{s}_{i+1}; \ \mathbf{b}_{3i+3} = \mathbf{d}_{i+1} .$$

In this C^1 composite case, increasing n_i (m_i) pushes the curve towards $\mathbf{d}_i$ $(\mathbf{d}_{i+1})$; therefore, we obtain a cusp-like behavior around $\mathbf{d}_i$ if m_{i-1} and n_i are increased, and a linear shape between $\mathbf{d}_i$ and $\mathbf{d}_{i+1}$ when n_i and m_i are increased. In Fig. 3 and Fig. 4 we show some graphical examples obtained using Bessel estimates and chord length parametrization.

§ 5. Conclusions and Remarks

We have seen in the previous sections some basic properties and applications of variable degree polynomial splines, and shown how these could be used as a powerful tool both in functional constrained approximation and in curve design. As we have already anticipated in the introduction, there are several, and sometimes more interesting, applications we cannot discuss in this paper.

In [4] are reported several extensions of variable degree splines made up by polynomials *with a single tension parameter* (that is generated by (1) with $n_i = m_i$) to bivariate constrained interpolation. Obviously, all these extensions can be obtained starting with our new polynomials, which provide

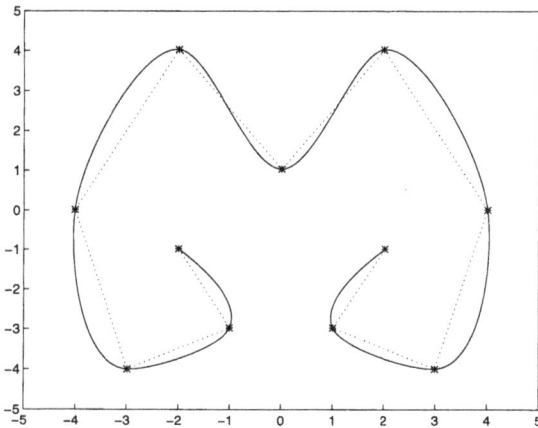

Fig. 3. A C^1 interpolating curve with $\boldsymbol{n} = \{3, \cdots, 3\}$, $\boldsymbol{m} = \{3, \cdots, 3\}$.

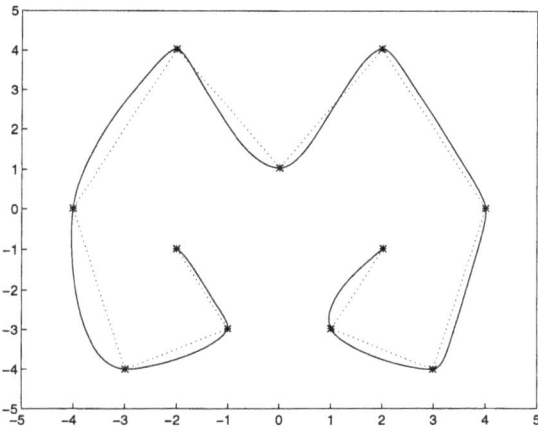

Fig. 4. $\boldsymbol{n} = \{15, 3, 3, 3, 15, 3, 15, 15, 3, 3\}$, $\boldsymbol{m} = \{15, 3, 3, 15, 3, 3, 15, 15, 3, 3\}$.

a more flexible handling of the constraints. This greater facility in controlling the shape seems to be particularly promising in the case of scattered data interpolation.

Although the results of the previous section open the way to useful applications, for an effective utilization in CAGD we need something equivalent to B–spline curves. Recent results of [6] allow us to say we can define a variable degree C^2 polynomial spline space, and use a B–spline like basis (with all the properties of the cubic ones). It is also possible, in addition, to directly compute the pseudo-Bézier nets from the B-spline control polygon, thus repeating one of the most used tool in CAGD. In addition, all the classical extensions of B-spline curves are possible: tensor-product, Boolean sums and so on. The rational counterparts, that is variable degree curve and surface NURBS, seem to be extremely powerful, because the control provided by the weights and the degrees (which have a different kind of influence) makes possible almost any change on the shape.

Finally, it is worthwhile to say that generalizations to other function spaces are possible; we refer to the papers [7,8] and the references therein.

References

1. Barsky, B., *Computer Graphics and Geometric Modeling Using Beta–splines*. Springer Verlag, Berlin, 1988.

2. Costantini, P., On monotone and convex spline interpolation, Math. Comp. **46** (1986), 203–214.

3. Costantini, P., Co–monotone interpolating splines of arbitrary degree. A local approach, SIAM J. Sci. Statist. Comput. **8** (1987), 1026–1034.

4. Costantini, P., Shape-preserving interpolation with variable degree polynomial splines, in *Advanced Course on FAIRSHAPE*, J. Hoschek and P. D. Kaklis (eds.), B. G. Teubner, Stuttgart, 87–114.

5. Farin, G., *Curves and Surfaces for Computer Aided Geometric Design*, 3rd ed. Academic Press, Boston, 1993.

6. Kvasov, B. I., GB–splines and algorithms of shape–preserving approximation, Preprint, 1994, to appear in Int. J. on CAGD.

7. Kvasov, B. I. and P. Sattayathan, Generalized tension B–splines, in *Curves and Surfaces with Applications in CAGD*, A. Le Méhauté, C. Rabut, and L. L. Schumaker (eds.), Vanderbilt Univ. Press, Nashville, 1997, 247–254.

8. Laurent, P. J., M. L. Mazure and G. Morin, Shape effects with polynomial Chebyshev splines, in *Curves and Surfaces with Applications in CAGD*, A. Le Méhauté, C. Rabut, and L. L. Schumaker (eds.), Vanderbilt Univ. Press, Nashville, 1997, 255–262.

9. Schmidt, J. W., Staircase algorithm and construction of convex spline interpolants up to the continuity C^3, Comput. Math. Appl. **31**, 67–79.

10. Soanes, R. W., VP–splines, an extension of twice differentiable interpolation, in *Proceedings of 1976 Army Numerical Analysis and Computer Conf.*, ARO Report 76-3, US Army Research Office, Research Triangle Park, NC, 1976, 141–152.

Paolo Costantini
Dipartimento di Matematica
Via del Capitano, 15
I 53100 Siena, ITALY
`costantini@unisi.it`

Approximation of Monotone Functions:
A Counter Example

R. A. DeVore, D. Leviatan and I. A. Shevchuk

Abstract. When we approximate a continuous nondecreasing function f in $[-1, 1]$, we wish sometimes that also the approximating polynomials be nondecreasing. However, this constraint restricts very much the degree of approximation that the polynomials can achieve, namely, only the rate of $\omega_2(f, 1/n)$. It turns out as we will prove somewhere else that relaxing the monotonicity requirement in intervals of length $1/n^2$ near the endpoints allows the polynomials to achieve the rate of ω_3. On the other hand, we show in this paper, that even when we relax the requirement of monotonicity of the polynomials on sets of measures approaching 0, (no matter how slowly or how fast), ω_4 is not reachable.

§1. Introduction

Let $f \in C[-1, 1]$ be nondecreasing on $I := [-1, 1]$. Then DeVore [1] proved that there exist nondecreasing polynomials such that

$$\|f - P_n\|_{C(I)} \leq c\,\omega_2(f, 1/n), \tag{1.1}$$

where c is an absolute constant and $\omega_k(f; \cdot)$ denotes the modulus of smoothness of order k, of f. (See also pointwise estimates by DeVore and Yu [2] and similar estimates involving the second Ditzian- Totik modulus of smoothness by Leviatan [3].)

On the other hand it is known (see Shvedov [4]) that in (1.1) one cannot replace ω_2 by ω_k with any $k \geq 3$.

It is quite natural to ask whether one can strengthen (1.1) in the sense of being able to replace ω_2 by moduli of smoothness of higher order, if one is willing to allow P_n not to be monotone on a rather "small" subset of I. This indeed is the case. If we allow the polynomials not to be monotone in intervals of length $1/n^2$ near the end points, then it is possible to achieve the estimates

$$\|f - p_n\| \leq c\,\omega_3(f, 1/n). \tag{1.2}$$

Curves and Surfaces with Applications in CAGD
A. Le Méhauté, C. Rabut, and L. L. Schumaker (eds.), pp. 95–102.
Copyright ℗ 1997 by Vanderbilt University Press, Nashville, TN.
ISBN 0-8265-1293-3.
All rights of reproduction in any form reserved.

We will prove that as a special case of a more general result (pointwise estimates for comonotone approximation) in another paper. However, even this
improvement comes to a halt; it cannot be extended to ω_4, and thus not to
ω_k for any $k > 3$.

In order to state our theorem we need some notation. Given $\epsilon > 0$ and a
nondecreasing function $f \in C[-1, 1]$, we denote

$$E_n^{(1)}(f; \epsilon) := \inf_{P_n} \|f - P_n\|_{C(I)},$$

where the infimum is taken over all polynomials P_n of degree not exceeding
n satisfying

$$\text{meas}(\{x \,:\, P_n'(x) \geq 0\} \cap I) \geq 2 - \epsilon.$$

Theorem. *For each sequence* $\bar{\epsilon} = \{\epsilon_n\}_{n=1}^{\infty}$, *of nonnegative numbers tending
to 0, there exists a nondecreasing function* $f := f_{\bar{\epsilon}} \in C[-1, 1]$ *such that*

$$\limsup_{n \to \infty} \frac{E_n^{(1)}(f; \epsilon_n)}{\omega_4(f, 1/n)} = \infty. \qquad (1.3)$$

Remarks 1. *A weaker version of our theorem, where* P_n *is required to be
monotone in* I, *that is, under the stronger condition that* $\epsilon_n = 0$, $n = 1, 2, \ldots$,
is due to Wu and Zhou [5]. *One should note that Wu and Zhou have a single
function while Shvedov* [4] *obtained for an arbitrarily large prescribed* c, *and
for each* n, *a different function* $f_{n,c}$, *which violates* (1.1).

2. *Note that we allow the relaxing of monotonicity (on small sets) anywhere
in* I, *not necessarily near the endpoints, and* ω_4 *cannot be had.*

§2. A Counter Example (Proof of the Theorem)

While above we have used c as an absolute constant which may differ on
different occurrences, in this section we will have to keep track of the constants,
therefore we denote them by $C_1, C_2, \ldots$. We begin by recalling some simple
properties of the Chebyshev polynomials for the interval $[-2, 2]$. For $\nu > 1$,
let

$$t_\nu(x) := \cos \nu \arccos \frac{x}{2}, \quad x \in [-2, 2],$$

denote the Chebyshev polynomial and let $z_j := 2\cos \frac{j\pi}{\nu}$, $j = 0, \ldots, \nu$, be
its extrema. Given $0 < b < \frac{1}{2}$, we take two points on both sides of z_j,
$j = 1, \ldots, \nu - 1$, namely, we set $z_{j,l} := 2\cos(\frac{(j+b)\pi}{\nu})$ and $z_{j,r} := 2\cos(\frac{(j-b)\pi}{\nu})$.
Note that

$$|t_\nu(z_{j,l})| = |t_\nu(z_{j,r})| = \cos \pi b,$$

and

$$z_{j,r} - z_{j,l} = 4\sin \frac{j\pi}{\nu} \cdot \sin \frac{\pi b}{\nu} < 4\pi \frac{b}{\nu}. \qquad (2.1)$$

We truncate the Chebyshev polynomial by setting

$$t_\nu^*(x) := t_{\nu,b}^*(x) := \begin{cases} \cos \pi b, & t_\nu(x) > \cos \pi b \\ -\cos \pi b, & t_\nu(x) < -\cos \pi b \\ t_\nu(x), & \text{otherwise.} \end{cases}$$

Since

$$1 - \cos \pi b = 2 \sin^2 \frac{\pi b}{2} < 5b^2, \tag{2.2}$$

for any $x \in I$, it follows by the monotonicity of the areas as we go away from the origin, and the alternation in sign of these areas, that

$$\left| \int_0^x (t_\nu(u) - t_\nu^*(u)) \, du \right| \le \left| \int_{z_{[\frac{\nu}{2}],l}}^{z_{[\frac{\nu}{2}],r}} (t_\nu(u) - t_\nu^*(u)) \, du \right| < 4\pi \frac{b}{\nu} 5b^2 = C_1 \frac{b^3}{\nu}, \tag{2.3}$$

where we have applied also (2.1).

Now, given $n \ge 1$ and $0 < b < \frac{1}{2}$, let $\nu := [b^{\frac{3}{4}}n] + 2$, where $[a]$ denotes the largest integer not exceeding a. Put

$$t_{\nu,b} := t_\nu + \cos \pi b, \quad \text{and} \quad \tilde{t}_{\nu,b} := t_{\nu,b}^* + \cos \pi b.$$

Finally,

$$T_{\nu,b}(x) := \int_0^x t_{\nu,b}(u) \, du \quad \text{and} \quad f_{n,b}(x) := \int_0^x \tilde{t}_{\nu,b}(u) \, du, \quad x \in I.$$

Obviously $f_{n,b}$ is a nondecreasing function on I and it readily follows by (2.3) that

$$\|f_{n,b} - T_{\nu,b}\| \le C_1 \frac{b^3}{\nu} \le C_1 \frac{b^{\frac{9}{4}}}{n}, \tag{2.4}$$

where we denote by $\|\cdot\|_J$ the max-norm taken on the interval J, and when the norm is on I, we suppress the subscript.

If we set $\tilde{z}_{j,l} := 2 \cos \left(\frac{(j+b/2)\pi}{\nu} \right)$, and $\tilde{z}_{j,r} := 2 \cos \left(\frac{(j-b/2)\pi}{\nu} \right)$, then we have for all j for which $z_j \in I$,

$$\tilde{z}_{j,r} - z_j = 4 \sin \frac{(j - \frac{b}{4})\pi}{\nu} \sin \frac{b\pi}{4\nu} > \frac{b}{\nu}, \tag{2.5}$$

and similarly,

$$z_j - \tilde{z}_{j,l} > \frac{b}{\nu}. \tag{2.5'}$$

Let j be odd. Since $\sin b\pi/4 > 3b/4$ for b satisfying $b\pi/4 < \pi/6$, we have

$$T_{\nu,b}'(x) = t_{\nu,b}(x) \le -\cos b\pi/2 + \cos b\pi = -2 \sin b\pi/4 \sin 3b\pi/4$$

$$< -2 \frac{3b}{4} \frac{3b}{2} = -\frac{9b^2}{4}, \quad x \in [\tilde{z}_{j,l}, \tilde{z}_{j,r}]. \tag{2.6}$$

Then since $I \subset [-2,2]$, it follows by the Bernstein inequality that

$$\|T_{\nu,b}^{(4)}\| = \|t_{\nu,b}^{(3)}\| = \|t_\nu^{(3)}\| \leq \frac{C_2 \nu^3}{(1-(\frac{1}{2})^2)^{3/2}} \|t_\nu\|_{[-2,2]} = C_3 \nu^3.$$

Hence, by (2.4),

$$\begin{aligned}
\omega_4(f_{n,b}, \frac{1}{n}) &\leq \omega_4(f_{n,b} - T_{\nu,b}, \frac{1}{n}) + \omega_4(T_{\nu,b}, \frac{1}{n}) \\
&\leq 2^4 \|f_{n,b} - T_{\nu,b}\| + \frac{1}{n^4}\|T_{\nu,b}^{(4)}\| \qquad (2.7)\\
&\leq 2^4 C_1 \frac{b^{9/4}}{n} + \frac{C_3 \nu^3}{n^4} \leq C_4 \frac{b^{9/4}}{n},
\end{aligned}$$

by the relation between ν and n.

Next we need a simple lemma.

Lemma 1. *There exists a constant C_5 such that for any interval $J \subseteq I$, we have the following. For any measurable sets $E \subseteq I$, if*

$$P_n'(x) \geq 0, \qquad x \in J \setminus E, \qquad (2.8)$$

then

$$\|f_{n,b} - P_n\|_J \geq \frac{b^2|J|}{n} - \frac{C_5}{n}(b^{9/4} + b|E| + \frac{b^{5/4}}{n}). \qquad (2.9)$$

Proof: Let J_0 denote the middle third of J. We consider two cases. First we assume that J_0 contains at most one of the z_j's. Then by the definition of ν we get

$$|J| < C_6 \frac{1}{\nu} < C_6 \frac{b^{-3/4}}{n}.$$

Hence

$$\|f_{n,b} - P_n\|_J \geq 0 > \frac{b^2|J|}{n} - C_6 \frac{b^{5/4}}{n^2}. \qquad (2.10)$$

On the other hand, if J_0 contains at least two extrema z_j, then it contains at least $2C_7\nu|J|$ extrema, for some constant C_7. These extrema satisfy (2.5) and (2.5'), and about half of them (and at least one) have odd indices, then together with (2.6) we conclude that

$$\text{meas}\,(J_0 \cap \{x : T_{\nu,b}'(x) < -\frac{9b^2}{4}\}) \geq \frac{1}{2}\frac{2b}{\nu}C_7\nu|J| = C_7 b|J|. \qquad (2.11)$$

Now, if $C_7 b|J| \leq |E|$, then

$$\|f_{n,b} - P_n\|_J \geq 0 \geq \frac{b^2|J|}{n} - \frac{b|E|}{nC_7}. \qquad (2.12)$$

Otherwise, $C_7 b |J| > |E|$. Then by (2.11) there is a point $x_0 \in J_0 \setminus E$, for which

$$T'_{\nu,b}(x_0) < -\frac{9b^2}{4}.$$

Hence, (2.8) yields,

$$\frac{9b^2}{4} \le P'_n(x_0) - T'_{\nu,b}(x_0) \le \frac{2}{|J|} \frac{n}{\sqrt{1 - (1/3)^2}} \|P_n - T_{\nu,b}\|_J,$$

where we have used the Bernstein inequality. Therefore from (2.4),

$$\begin{aligned}
\frac{b^2 |J|}{n} &\le \frac{3\sqrt{2}}{4} \frac{b^2 |J|}{n} \le \|P_n - T_{\nu,b}\|_J \\
&\le \|P_n - f_{n,b}\|_J + \|f_{n,b} - T_{\nu,b}\|_J \qquad (2.13) \\
&\le \|P_n - f_{n,b}\|_J + C_1 \frac{b^{9/4}}{n}.
\end{aligned}$$

Taking $C_5 := \max\{C_6, \frac{1}{C_7}, C_1\}$, (2.9) now follows by combining (2.10), (2.12) and (2.13). ∎

We are now in a position to define $f_{\bar{\epsilon}} := f$, for a given sequence $\bar{\epsilon} = \{\epsilon_n\}$. Let $b_n := (\max\{\epsilon_n^2, \frac{1}{n}\})^{2/5}$, and set $d_0 := 1$, and

$$d_j := \frac{b_{n_j}^{9/4}}{n_j} d_{j-1} = \prod_{\nu=1}^{j} \frac{b_{n_\nu}^{9/4}}{n_\nu}, \quad j \ge 1,$$

where the sequence $\{n_\nu\}$ is defined by induction as follows. First, we choose n_1 so large that $b_{n_1}^{1/8} < \frac{1}{12}$ (as needed in (2.15) below) and $J_0 := I$. Suppose that $\{n_1, \dots, n_{\sigma-1}\}$ and $J_{\sigma-2} \subseteq J_{\sigma-3} \subseteq \cdots \subseteq J_0$, $\sigma \ge 2$, have been defined. Then put

$$F_{\sigma-1} := \sum_{j=1}^{\sigma-1} d_{j-1} f_{n_j, b_{n_j}},$$

and let $J_{\sigma-1}$ be an interval such that $J_{\sigma-1} \subseteq J_{\sigma-2}$ and

$$F'_{\sigma-1}(x) = 0, \quad x \in J_{\sigma-1}. \qquad (2.14)$$

(The induction process will guarantee the existence of such intervals.) Let $N_{1,\sigma}$ be such that

$$|J_{\sigma-1}| \ge b_n^{1/8}, \quad n \ge N_{1,\sigma}, \qquad (2.15)$$

and let

$$N_{2,\sigma} := \left(\frac{\|F_{\sigma-1}^{(2)}\|}{d_{\sigma-1}} \right)^{10}. \qquad (2.16)$$

Finally, we take

$$n_\sigma > \max\{n_{\sigma-1}, N_{1,\sigma}, N_{2,\sigma}\}$$

so big that the function $f'_{n_\sigma, b_{n_\sigma}}$ oscillates a few times inside the interval $J_{\sigma-1}$ and since it vanishes on some interval in each oscillation, that is, inside $J_{\sigma-1}$, there exists an interval $J_\sigma \subset J_{\sigma-1}$ as required in (2.14).

Now denote

$$\Phi_\sigma := \sum_{j=\sigma}^{\infty} d_{j-1} f_{n_j, b_{n_j}},$$

where the convergence of the series is justified by the definition of the d_j's and the fact that $\|f_{n,b_n}\| \leq 2$, for all n. In fact

$$\|\Phi_\sigma\| \leq 2d_{\sigma-1}(1 + \frac{b_{n_\sigma}^{9/4}}{n_\sigma} + \frac{b_{n_\sigma}^{9/4} b_{n_{\sigma+1}}^{9/4}}{n_\sigma n_{\sigma+1}} + \ldots)$$

$$\leq 2d_{\sigma-1} \sum_{j=0}^{\infty} 2^{-j} = 4d_{\sigma-1}. \tag{2.17}$$

So we define

$$f := f_{\bar{\epsilon}} := \sum_{j=1}^{\infty} d_{j-1} f_{n_j, b_{n_j}},$$

and we prove

Lemma 2. For each $\sigma \geq 1$ we have

$$\omega_4(f, 1/n_\sigma) \leq C_8 d_\sigma. \tag{2.18}$$

Proof: First, by (2.17)

$$\omega_4(\Phi_{\sigma+1}, 1/n_\sigma) \leq 2^4 \|\Phi_{\sigma+1}\| \leq 2^6 d_\sigma. \tag{2.19}$$

At the same time, (2.7) yields

$$\omega_4(d_{\sigma-1} f_{n_\sigma, b_{n_\sigma}}, 1/n_\sigma) \leq d_{\sigma-1} C_4 \frac{b_{n_\sigma}^{9/4}}{n_\sigma} = C_4 d_\sigma. \tag{2.20}$$

Finally,

$$\omega_4(F_{\sigma-1}, 1/n_\sigma) \leq 4\omega_2(F_{\sigma-1}, 1/n_\sigma)$$

$$\leq \frac{4}{n_\sigma^2} \|F_{\sigma-1}^{(2)}\|$$

$$= 4\frac{\|F_{\sigma-1}^{(2)}\|}{d_{\sigma-1}} n_\sigma^{-1/10} \left(\frac{1}{n_\sigma^{2/5} b_{n_\sigma}}\right)^{9/4} d_\sigma \tag{2.21}$$

$$\leq 4d_\sigma,$$

by virtue of (2.16) and the definitions of b_{n_σ}, d_σ and n_σ. Lemma 2 follows by combining (2.19), (2.20) and (2.21). ∎

The last lemma that we need is

Lemma 3. *There is an absolute constant C_9 such that whenever $E \subset I$ is a measurable set satisfying*

$$|E| \leq \epsilon_{n_\sigma}, \tag{2.22}$$

and P_{n_σ} is a polynomial satisfying

$$P'_{n_\sigma}(x) \geq 0, \quad x \in I \setminus E, \tag{2.23}$$

then

$$\|f - P_{n_\sigma}\| \geq (b_{n_\sigma}^{-1/8} - C_9)d_\sigma. \tag{2.24}$$

Proof: Since $F_{\sigma-1}$ is constant on $J_{\sigma-1}$, we may write

$$f(x) = d_{\sigma-1}f_{n_\sigma, b_{n_\sigma}}(x) + \Phi_{\sigma+1}(x) + M, \quad x \in J_{\sigma-1}. \tag{2.25}$$

Let

$$Q_{n_\sigma} := \frac{1}{d_{\sigma-1}}(P_{n_\sigma} - M).$$

Then it follows from (2.23),

$$Q'_{n_\sigma}(x) \geq 0, \quad x \in J_{\sigma-1} \setminus E.$$

Thus by virtue of Lemma 1,

$$\|Q_{n_\sigma} - f_{n_\sigma, b_{n_\sigma}}\|_{J_{\sigma-1}} \geq \frac{b_{n_\sigma}^2 |J_{\sigma-1}|}{n_\sigma} - \frac{C_5}{n_\sigma}(b_{n_\sigma}^{9/4} + b_{n_\sigma}|E| + \frac{b_{n_\sigma}^{5/4}}{n_\sigma}). \tag{2.26}$$

The definition of n_σ and (2.15) yield,

$$b_{n_\sigma}^2 |J_{\sigma-1}| = b_{n_\sigma}^{17/8}\left(\frac{|J_{\sigma-1}|}{b_{n_\sigma}^{1/8}}\right) \geq b_{n_\sigma}^{17/8}.$$

On the other hand, (2.22) and the definition of b_{n_σ} imply

$$b_{n_\sigma}|E| \leq b_{n_\sigma}\epsilon_{n_\sigma} \leq b_{n_\sigma}^{9/4},$$

and

$$\frac{b_{n_\sigma}^{5/4}}{n_\sigma} \leq b_{n_\sigma}^{15/4} < b_{n_\sigma}^{9/4}.$$

Hence (2.26) implies

$$\|Q_{n_\sigma} - f_{n_\sigma, b_{n_\sigma}}\|_{J_{\sigma-1}} \geq \frac{1}{n_\sigma}(b_{n_\sigma}^{17/8} - 3C_5 b_{n_\sigma}^{9/4}) = \frac{b_{n_\sigma}^{9/4}}{n_\sigma}(b_{n_\sigma}^{-1/8} - 3C_5).$$

In other words,

$$\|P_{n_\sigma} - M - d_{\sigma-1}f_{n_\sigma, b_{n_\sigma}}\|_{J_{\sigma-1}} \geq d_{\sigma-1}\frac{b_{n_\sigma}^{9/4}}{n_\sigma}(b_{n_\sigma}^{-1/8} - 3C_5) = d_\sigma(b_{n_\sigma}^{-1/8} - 3C_5).$$

In view of (2.25), it follows from (2.17) that,

$$\|f - P_{n_\sigma}\| \geq \|f - P_{n_\sigma}\|_{J_{\sigma-1}} \geq \|P_{n_\sigma} - M - d_{\sigma-1} f_{n_\sigma, b_{n_\sigma}}\|_{J_{\sigma-1}} - \|\Phi_{\sigma+1}\|$$
$$\geq (b_{n_\sigma}^{-1/8} - (3C_5 + 4))d_\sigma,$$

and Lemma 3 is proved with $C_9 := 3C_5 + 4$. ∎

The proof of (1.3) now follows from Lemmas 2 and 3 since

$$\limsup_{n\to\infty} \frac{E_n^{(1)}(f;\epsilon_n)}{\omega_4(f,1/n)} \geq \limsup_{\sigma\to\infty} \frac{E_{n_\sigma}^{(1)}(f;\epsilon_{n_\sigma})}{\omega_4(f,1/n_\sigma)}$$
$$\geq \limsup_{\sigma\to\infty} \frac{1}{C_8}(b_{n_\sigma}^{-1/8} - C_9) = \infty. \quad \blacksquare$$

Acknowledgments. Part of this work was done while the second and third authors were visiting the University of South Carolina. All three authors acknowledge partial support by ONR grant N00014-91-1076 and by DoD grant N00014-94-1-1163.

References

1. DeVore, R. A., Degree of approximation, in *Approximation Theory II*, G. G. Lorentz, C. K. Chui, and L. L. Schumaker (eds.), Academic Press, New York, 1976, 117–162.
2. DeVore, R. A. and X. M. Yu, Pointwise estimates for monotone polynomial approximation, Constr. Approx. 1 (1985), 323–331.
3. Leviatan D., Monotone and comonotone approximation revisited, J. Approx. Theory 53 (1988), 1-16.
4. Shvedov, A. S., Orders of coapproximation of functions by algebraic polynomials, Mat. Zametki 29 (1981), 117–130.
5. Wu, X. and S. P. Zhou, A problem on coapproximation of functions by algebraic polynomials in *Progress in Approximation Theory*, A. Pinkus and P. Nevai (eds.), Academic Press, New York, 1991, 857–866.

R. A. DeVore
Department of Mathematics
University of South Carolina
Columbia SC 29208
USA
devore@math.sc.edu

D. Leviatan
School of Mathematical Sciences
Sackler Faculty of Exact Sciences
Tel Aviv University
Tel Aviv 69978, ISRAEL
leviatan@math.tau.ac.il

I. A. Shevchuk
Institute of Mathematics
National Academy of Sciences of Ukraine
Kyiv 252601, UKRAINE
shevchuk@imat.gluk.apc.org

Uniform Point Distribution on a Circle

J-C. Fiorot and I. Cattiaux-Huillard

Abstract. Usually rational parametric curves, with the exception of straight lines, are not uniformly parameterized, i.e. the points which are images of evenly spaced values of the parameter are unevenly spaced on the curve. A remedy is proposed for a circle segment defined by three massic vectors. From the standard rational parameterization of the circle on $\widetilde{\mathbb{R}} = \mathbb{R} \cup \{\infty\}$, the full circle as a C^1 (respectively C^3) smoothly joined BR-curve image of $[0,1]$ can be achieved by a rational quadratic (respectively cubic) change of parameter. In both cases, two degrees of freedom remain. We take these two degrees of freedom into account by giving a satisfactory uniform parameterization.

§1. Introduction

A rational curve $BR[\theta](t)$, $t \in [0,1]$, different from a straight line, can never be parameterized by its arc length [1]. This implies that the speed $\left\| \frac{dBR[\theta](t)}{dt} \right\|$ is never a constant function.

But a "constant-speed" representation of a curve is very useful for visual representation and robotic applications, for example, to minimize the number of points needed for a satisfactory plot of the curve or to define tool-paths. Here we propose to find a rational parameterization $BR[\theta]$ of a unit circle segment or, more importantly, of an entire circle, which gives images $BR[\theta](t_i)$ of evenly spaced values $t_i = \frac{i}{n}$ for $i \in \{0, \ldots, n\}$ of $[0,1]$, as evenly spaced as possible on the circle.

In Section 2, we give a degree 2/degree 2 representation for a unit circle segment of length 2θ, and we notice that this parameterization is uniform enough for $\theta \leq \frac{\pi}{3}$. The main part of this paper is devoted to the full circle. It is represented in [3] as a C^1 (resp. C^3) smoothly joined BR-curve image of [0,1]. In Section 3, we first pose the problem. We propose degree 4/degree 4 and degree 6/degree 6 solutions respectively in Section 4 and in Section 5.

Curves and Surfaces with Applications in CAGD
A. Le Méhauté, C. Rabut, and L. L. Schumaker (eds.), pp. 103–110.
Copyright © 1997 by Vanderbilt University Press, Nashville, TN.
ISBN 0-8265-1293-3.

All rights of reproduction in any form reserved.

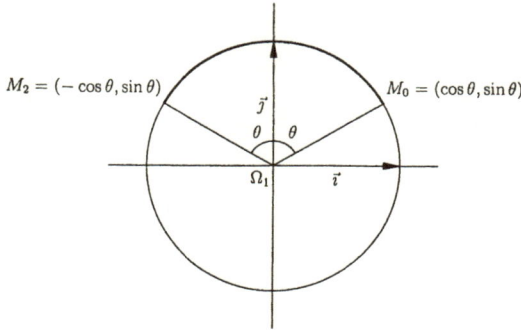

Fig. 1. The thick segment represents $BR[\omega](t)$, $t \in [0,1]$.

§2. Smooth Parameterization of a Circle Segment

Let M_0 and M_2 be two points of the trigonometric circle defining a segment of length 2θ. A cartesian frame $(\Omega_1, \vec{\imath}, \vec{\jmath})$ can be determined so as to obtain $M_0 = (\sin\theta, \cos\theta)$ and $M_2 = (-\sin\theta, \cos\theta)$ as described in Figure 1.

Lemma 1. *The circle segment $M_0 M_2$ is represented when $\theta \neq \frac{\pi}{2}$, by the massic polygon $\omega = \{(M_0; 1), (M_1; \rho\cos\theta), (M_2; \rho^2)\}$ where $M_1 = \left(0, \frac{1}{\cos\theta}\right)$:*
$$BR[\omega](t) = \frac{(1-t)^2 M_0 + 2t(1-t)\rho\cos\theta M_1 + t^2\rho^2 M_2}{(1-t)^2 + 2t(1-t)\rho\cos\theta + t^2\rho^2}, \; t \in [0,1]. \text{ When } \theta = \frac{\pi}{2} \text{ the semi-}$$
circle has the following representation:

$$BR[(M_0; 1), \rho\vec{\jmath}, (M_2; \rho^2)](t) = \frac{(1-t)^2 M_0 + t^2\rho^2 M_2}{(1-t)^2 + t^2\rho^2} + \frac{\rho\vec{\jmath}}{(1-t)^2 + t^2\rho^2}, t \in [0,1] \text{ (see [2])}.$$

In both cases, we take $\rho > 0$.

This parameterization admits one degree of freedom to be adjusted in order to obtain a distribution of points on the circle segment as uniform as possible.

Proposition 1. *For a given n, the parameter ρ (depending on θ) which determines the most uniform distribution of $n+1$ points on the circle segment of length 2θ is a solution of the following problem:*

$$(P_{1,n}) : Minimize \left\{\eta + 0.\rho \; \middle| \; (\forall i \in \{0, 1 \ldots, n-1\}, \; \tfrac{n}{2}.\Phi_{n,\rho,\theta}(\tfrac{i}{n}) - \theta \leq \eta,\right.$$

$$\left. -\tfrac{n}{2}.\Phi_{n,\rho,\theta}(\tfrac{i}{n}) + \theta \leq \eta \right), \; \eta \geq 0, \; \rho \geq 0\},$$

where $\Phi_{n,\rho,\theta}(t) = 2\arctan\left(\left(\tfrac{1}{n}\rho\sin\theta\right) / \left(1 - \tfrac{1}{n} + \tfrac{1}{n}\rho\cos\theta\right.\right.$

$$+ t(-2 + \tfrac{1}{n} + \tfrac{1}{n}\rho^2 + 2\rho\cos\theta - 2\tfrac{1}{n}\rho\cos\theta) + t^2(1 - \rho^2 - 2\rho\cos\theta)\Big)\Big).$$

Proof: Let be $n \in \mathbb{N}^*$, and let us denote $t_i = \frac{i}{n}$, $i \in \{0, \ldots, n\}$ and by $M(t) = BR[\omega](t)$ the current point on the circle segment. We want to determine ρ such that the segments $M(t_i)M(t_{i+1})$ are as near of the ideal segment $\frac{2\theta}{n}$ as possible.

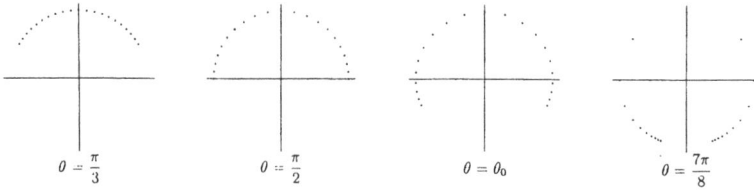

Fig. 2. Example of distribution of $n = 20$ points for different values of θ.

Then we want to determine the real $\rho > 0$ minimizing

$$\max_{i \in \{0,\ldots,n-1\}} \left| M(t_i) M(t_i + \frac{1}{n}) - \frac{2\theta}{n} \right| \text{, or, } n \text{ being a constant:}$$

$$\max_{i \in \{0,\ldots,n-1\}} \left| \frac{n}{2} . M(t_i) M(t_i + \frac{1}{n}) - \theta \right|. \text{ We denote by } \eta \text{ this previous expression.}$$

After some calculations we find that the arc length $M(t_i) M(t_i + \frac{1}{n})$ is equal to $\Phi_{n,\rho,\theta}(t_i)$. Therefore we obtain $(P_{1,n})$. ∎

Definition of an Asymptotic Version of $(P_{1,n})$.

$(P_{1,\infty})$: *Minimize* $\{\eta + 0.\rho \mid (\forall\, t \in [0,1]\,,\ g_{\rho,\theta}(t) \leq \eta\,,\ -g_{\rho,\theta}(t) \leq \eta)$,

$$\eta \geq 0\,,\ \rho \geq 0\}$$,

where: $g_{\rho,\theta}(t) = \frac{\rho \sin \theta}{(1-t)^2 + 2\rho t \cos \theta + t^2 (\rho^2 - 2\rho \cos \theta)} - \theta$.

Justification. $(P_{1,\infty})$ has an infinite number of constraints. Let us consider $\Phi_{n,\rho,\theta}(t)$ as a function of $\frac{1}{n}$. These constraints are obtained from the power series expansion of $\Phi_{n,\rho,\theta}(t)$ about 0, with respect to $\frac{1}{n}$, up to order 1:
$\frac{n}{2} . \Phi_{n,\rho,\theta}(t) - \theta = g_{\rho,\theta}(t) + n.o\left(\frac{1}{n}\right)$.

Function $g_{\rho,\theta}(t)$ has a unique maximum on $\mathbb{R}$ at $t_{\rho,\theta} = \frac{1 - \rho \cos \theta}{\rho^2 - 2\rho \cos \theta + 1}$. We define: $M(\rho, \theta) = g_{\rho,\theta}(t_{\rho,\theta})$ if $t_{\rho,\theta} \in [0,1]$, 0 otherwise. Then we prove that $(P_{1,\infty})$ is equivalent to:

$(P'_{1,\infty})$: *Minimize* $\{\eta + 0.\rho \mid (\text{for } t \in \{0,1\}\,,\ g_{\rho,\theta}(t) \leq \eta\,,\ -g_{\rho,\theta}(t) \leq \eta)$,
$$M(\rho, \theta) \leq \eta\,,\ \eta \geq 0\,,\ \rho \geq 0\}.$$

This problem has only 7 constraints. We prove that $(P_{1,\infty})$, i.e. $(P'_{1,\infty})$ has the following solution:

Proposition 2. *For any value of $\theta \in\,]0, \pi[$, the solution of $(P_{1,\infty})$ is $\bar{\rho} = 1$, $\bar{\eta} = \theta - \sin \theta$ if $\theta \in\,]0, \theta_0]$, and $\bar{\eta} = 2 \tan \frac{\theta}{2} - \theta$ if $\theta \in\,]\theta_0, \pi[$, where $\theta_0 \simeq 1,97475$ radian is the only root in $]0, \pi[$ of the equation $\theta - \sin \theta = 2 \tan \frac{\theta}{2} - \theta$.*

So, for any value of n, we *propose* this solution: $\bar{\rho} = 1$, for $(P_{1,n})$. According to this result, the value of $\bar{\eta}$ depends on the value of θ. The point distribution is sufficiently uniform for $\theta \leq \frac{\pi}{3}$ (i.e. $\bar{\eta} \leq 0.55$). But for larger values of θ, as shown in Figure 2., the distribution obtained is too erratic to be useful. Then we cannot have a good point distribution for an almost complete circle with an angle θ very close to π. For an entire circle, we will use higher degree parameterizations.

§3. A Complete Circle: the Problem

A circle (C) in the Cartesian frame $(\Omega_1, \vec{i}, \vec{j})$ has a rational parameterization
$$C(u): \left(\frac{1-u^2}{1+u^2}, \frac{2u}{1+u^2} \right), \text{ for } u \in \widetilde{\mathbb{R}} = \mathbb{R} \cup \{\infty\}.$$
This parameterization is obtained from the trigonometric parameterization of the circle $(\cos \lambda; \sin \lambda)$ by setting: $u = \tan \frac{\lambda}{2}$, where λ, the azimuth of $C(u)$, is the angle between $\vec{i}$ and $\overrightarrow{\Omega_1 C(u)}$.

We obtain the full circle when u describes $\widetilde{\mathbb{R}}$. To represent it as an image of $[0,1]$, we use a change of parameter $\varphi : [0,1] \longrightarrow \widetilde{\mathbb{R}}$. Let $M(t) = C(\varphi(t))$. We have
$$t \in [0,1] \longrightarrow u = \varphi(t) \in \widetilde{\mathbb{R}} \longrightarrow M(t) \in (C).$$

Let $t_i = \frac{1}{n}$ for $i \in \{0 \ldots, n-1\}$. The images $M(t_i)$ of evenly spaced points t_i of $[0,1]$ have to be as evenly spaced as possible on the circle.

Lemma 2. *With φ, M and t_i above defined, a change of parameter φ leads to a distribution of points $M(t_i)$ which is as uniform as possible if it minimizes the value η such that:*

$$\left| n. \arctan \left(\frac{\varphi(t_{i+1}) - \varphi(t_i)}{1 + \varphi(t_{i+1})\varphi(t_i)} \right) - \pi \right| \leq \eta, \quad \forall \; i \in \{0, \ldots, n-1\}.$$

Proof: We want to determine a change of parameter $\varphi(t)$ such that $M(t) = C(\varphi(t))$ gives a uniform point distribution. The relation between λ and t, $\varphi(t) = \tan \frac{\lambda}{2}$ allows us to write λ as a function of t: $\lambda(t) = 2 \arctan(\varphi(t))$.

We want to obtain n circle segments $M(t_i)M(t_{i+1})$ with the same length. As the circle radius is 1, on the one hand, this length must be $\frac{2\pi}{n}$ and, on the other hand: $M(t_i)M(t_{i+1}) = \lambda(t_{i+1}) - \lambda(t_i)$. We seek φ minimizing the following values: $\max_{i \in \{0,\ldots,n-1\}} \left| \lambda(t_{i+1}) - \lambda(t_i) - \frac{2\pi}{n} \right|$. We have $\lambda(t_{i+1}) - \lambda(t_i) = 2 \left(\arctan(\varphi(t_{i+1})) - \arctan(\varphi(t_i)) \right) = 2 \arctan \left(\frac{\varphi(t_{i+1})-\varphi(t_i)}{1+\varphi(t_{i+1})\varphi(t_i)} \right)$. So we look for φ to minimize $\max_{i \in \{0,\ldots,n-1\}} \left| 2. \arctan \left(\frac{\varphi(t_{i+1})-\varphi(t_i)}{1+\varphi(t_{i+1})\varphi(t_i)} \right) - \frac{2\pi}{n} \right|$ or, n being a constant: $\max_{i \in \{0,\ldots,n-1\}} \left| n. \arctan \left(\frac{\varphi(t_{i+1})-\varphi(t_i)}{1+\varphi(t_{i+1})\varphi(t_i)} \right) - \pi \right|$. Denoting this value by η we obtain the result. ∎

§4. Solution for a C^1 Circle

In [3] we have proved that a C^1 smoothly joined circle can be obtained with the following quadratic changes of parameter: $q = \dfrac{\alpha B_0^2 + \beta B_1^2 - \alpha B_2^2}{B_1^2}$,

with $\alpha \in \mathbb{R}^* = \mathbb{R} - \{0\}$ and $\beta \in \mathbb{R}$.

We will determine α and β in order to obtain the best point distribution on the circle for a degree 4/degree 4 rational parameterization.

Proposition 3. *The parameters α and β which determine the most uniform distribution of n points on the full circle are solutions of problem $(P_{2,n})$:*

$$Minimize\,\{\eta + 0.\alpha + 0.\beta \mid (\forall\, i \in \{0, 1 \ldots, n-1\}\,,\ n.Q_{\alpha,\beta,n}(t_i) - \pi \leq \eta,$$
$$-n.Q_{\alpha,\beta,n}(t_i) + \pi \leq \eta)\,,\ \eta \geq 0\,\},$$

where $Q_{\alpha,\beta,n}(t) = \arctan\left(\frac{q(t+\frac{1}{n})-q(t)}{1+q(t+\frac{1}{n})q(t)}\right).$

Proof: This is a direct reformulation of Lemma 2. ∎

Definition of the Asymptotic Problem of $(P_{2,n})$.

$(P_{2,\infty})$: $Minimize\,\{\eta + 0.\alpha + 0.\beta \mid (\forall\, t \in [0, 1]\,,\ g_{\alpha,\beta}(t) - \pi \leq \eta\quad (D_t),$
$$-g_{\alpha,\beta}(t) + \pi \leq \eta\quad (D'_t))\,,\ \eta \geq 0\,\},$$

with: $g_{\alpha,\beta}(t) = \frac{-2\alpha(1-2t+2t^2)}{(2t^2-2t)^2+(\alpha(2t-1)+\beta(2t^2-2t))^2}.$

Justification: Considering $Q_{\alpha,\beta,n}(t)$ as a function of $\frac{1}{n}$, the power series expansion of $Q_{\alpha,\beta,n}(t)$, about 0, with respect to $\frac{1}{n}$, up to order 1 is: $Q_{\alpha,\beta,n}(t) = \frac{1}{n}g_{\alpha,\beta}(t) + o(\frac{1}{n})$, with $g_{\alpha,\beta}(t)$ given above.

Proposition 4. *The solution of $(P_{2,\infty})$ is $(\bar{\eta}, \bar{\alpha}, \bar{\beta})$ with $\bar{\eta} = \pi - 2\sqrt{2} \simeq 0.3132$, $\bar{\alpha} = -\frac{\sqrt{2}}{2}$ and $\bar{\beta} = 0$.*

Proof: We can verify that $(\bar{\eta}, \bar{\alpha}, \bar{\beta})$ is a feasible solution of $(P_{2,\infty})$. Let us determine the other feasible solutions (η, α, β) verifying $\eta \leq \bar{\eta}$. We deduce from $\eta \leq \bar{\eta}$ and from (D_0) and (D'_0) that: $-\pi + 2\sqrt{2} \leq -\frac{2}{\alpha} - \pi \leq \pi - 2\sqrt{2}$,

then:
$$-\frac{\sqrt{2}}{2} \leq \alpha \leq \frac{1}{\sqrt{2}-\pi}. \tag{1}$$

From $(D'_{\frac{1}{2}})$ and $\eta \leq \bar{\eta}$, we deduce: $\beta^2 \leq -1 - \alpha\sqrt{2}$. But (1) implies that $-1 - \alpha\sqrt{2} \leq 0$ then $\beta = 0$. Therefore $-g_{\alpha,0}(\frac{1}{2}) - \pi \leq \bar{\eta}$, which implies $\alpha \leq -\frac{\sqrt{2}}{2}$. And from (1) we deduce $\alpha = -\frac{\sqrt{2}}{2}$. From (D'_0) we obtain: $-g_{-\frac{\sqrt{2}}{2},0}(0) + \pi \leq \eta$ yields $\eta \geq \pi - 2\sqrt{2}$. But by assumption $\eta \leq \bar{\eta}$ then $\eta = \bar{\eta}$. Consequently, all the feasible solutions satisfying $\eta \leq \bar{\eta}$ satisfy: $(\eta, \alpha, \beta) = (\bar{\eta}, \bar{\alpha}, \bar{\beta})$, and thus $(\bar{\eta}, \bar{\alpha}, \bar{\beta})$ is the solution of $(P_{2,\infty})$. ∎

So, for any value of n, we *propose* this solution: $\bar{\alpha} = -\frac{\sqrt{2}}{2}$, $\bar{\beta} = 0$ for $(P_{2,n})$.

Corollary 1. *With $\alpha = -\frac{\sqrt{2}}{2}$ and $\beta = 0$, the massic polygon corresponding to $M(t)$ is the following:*

$$\theta_0 = ((-1, 0); \tfrac{1}{2}) \qquad \theta_1 = \left(\left(0, -\tfrac{\sqrt{2}}{2}\right); 0\right) \qquad \theta_2 = ((\tfrac{5}{3}, 0)\,;\tfrac{1}{2})$$

$$\theta_3 = \left(\left(0, \tfrac{\sqrt{2}}{2}\right); 0\right) \qquad \theta_4 = ((-1, 0); \tfrac{1}{2})$$

Proof: Using the formulae given in [3] with the propose values, we obtain the above massic polygon. ∎

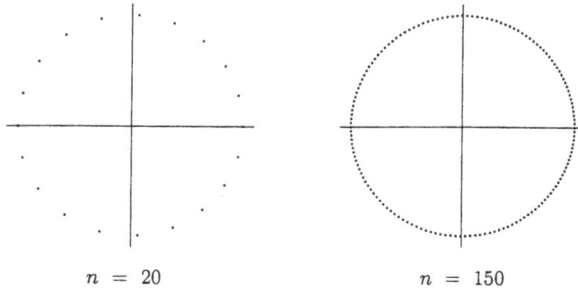

$n = 20$ $n = 150$

Fig. 3. Examples of point distributions on a C^1 circle.

Remark. It can be raised as an objection that the change of parameter q is not the most general quadratic change of parameter (the denominator of q can have three terms instead of one), and there may be another degree 4/degree 4 parameterization for the full circle which gives a better point distribution. This is not true: the best point distribution of other degree 4/degree 4 circles have the same point distribution as the proposed solution. We do not give the proof of this result here.

§5. Solution for a C^3 Circle

We obtain a C^3 smoothly joined circle [3] with a change of parameter of degree $\frac{3}{2}$ which can be rewritten as follows: $\varphi(t) = p.\dfrac{1 - 2t - 2t^2 + 2t^3}{t(1 - t)} + q$, where $p \in \mathbb{R}^*$ and $q \in \mathbb{R}$.

As in Section 4, Lemma 2 allows us to formalize the problem as

Proposition 5. To find the values of p and q which define a uniform parameterization φ, we solve the problem $(P_{3,n})$:

$Minimize \; \{\eta + 0.p + 0.q \mid (\; \forall \, i \in \{0, \ldots, n-1\}, \; n.\psi_{p,q,n}(t_i) - \pi \leq \eta,$

$$-n.\psi_{p,q,n}(t_i) + \pi \leq \eta) \; , \; \eta \geq 0 \; \},$$

where $\psi_{p,q,n}(t) = \arctan\left(\dfrac{\varphi(t+\frac{1}{n})-\varphi(t)}{1+\varphi(t+\frac{1}{n})\varphi(t)}\right)$.

To find a solution, we use a reduced problem:

Definition of an Approximating Problem $(P'_{3,n})$.

$Minimize \; \{\eta + 0.p + 0.q \mid (\; \forall \, i \in \{0, \ldots, n-1\}, \; f_{p,q}(t_i) - \pi \leq \eta \quad (C_i),$

$$-f_{p,q}(t_i) + \pi \leq \eta \quad (C'_i)) \; , \; \eta \geq 0 \; \},$$

with $f_{p,q}(t) = p.(1-2t+4t^2-4t^3+2t^4)/(-p^2+t(4p^2-2pq)+t^2(-1+6pq-q^2)$

$$+t^3(2-12p^2+2q^2)+t^4(-1+4p^2-8pq-q^2)+t^5(8p^2+4pq)-4p^2t^6).$$

In the same way as for $(P_{2,\infty})$, we prove that $\psi_{p,q,n}(t) = \frac{1}{n}f_{p,q}(t) + o(\frac{1}{n})$.

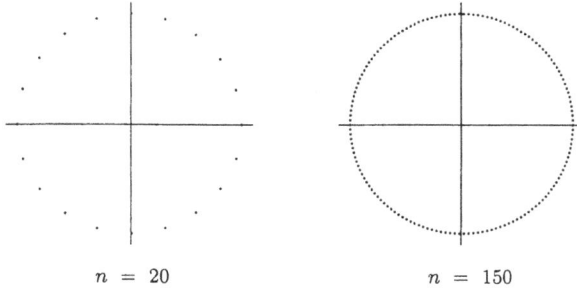

$n = 20$ $n = 150$

Fig. 4. Example of point distributions of a C^3 circle.

Proposition 6. *A solution of* $(P'_{3,10})$, *for* $n = 10$, *is* $(\bar\eta, \bar p, \bar q)$ *with:* $\bar\eta = 0.022$, $\bar p = -0.3142$ *and* $\bar q = -0.3138$.

This proposition is more tedious than difficult to prove, so we present only some elements of the proof.

Proof: $(\eta_0, p_0, q_0) = (0.05, -\frac{1}{\pi}, -\frac{1}{\pi})$ is a feasible solution of $(P'_{3,10})$. Let us prove that all the feasible solutions (η, p, q) of $(P'_{3,10})$ verifying $\eta \le \eta_0$ are in domain $D = [0, 0.05] \times [-0.3235, -0.3133] \times [-0.3814, -0.2769]$:

1) $\eta \in [0, 0.05]$ is a straightforward consequence of $\eta \le \eta_0$.
2) $p \in [-0.3235, -0.3133]$ is a consequence of constraints (C_0) and (C'_0), and of the bounds of η.
3) Studying the constraint (C_4), (respectively (C'_4) and (C'_5)) for (η, p) in $[0, 0.05] \times [-0.3235, -0.3133]$, we obtain: $q \le -0.2769$ or $q \ge 0.2978$ (resp. $-0.3814 \le q \le 0.4030$ and $-0.5386 \le q \le -0.1083$). From these previous bounds we deduce: $q \in [-0.3814, -0.2769]$.

The constraints of $(P'_{3,10})$ are nonlinear and nonconvex. But the domain D is restricted enough to allow us to suppose that the constraints are almost convex. To solve $(P'_{3,10})$, we use a linearisation algorithm: the method of the secant planes of Kelley [4]. Via two iterations, we obtain the solution $\bar\eta = 0.022$, $\bar p = -0.3142$ and $\bar q = -0.3138$. ∎

So, we *propose* this solution: $(p, q) = (-0.3142, -0.3138)$ for $(P_{3,n})$, for any value of $n \ge 10$.

Corollary 2. *With* $p = -0.3142$ *and* $q = -0.3138$ *the massic polygon corresponding to* $M(t)$ *is the following:*

$\theta_0 = ((-1, 0); 0.3949)$ $\theta_1 = ((-1, -1.5924); 0.2631)$

$\theta_2 = ((1.0015, -1.8858); 0.2665)$ $\theta_3 = ((2.9495, 0.0016); 0.2026)$

$\theta_4 = ((0.9985, 1.8846); 0.2669)$ $\theta_5 = ((-1, 1.5903); 0.2634)$

$\theta_6 = ((-1, 0); 0.3949)$

Proof: Using the formulae given in [3] with the proposed values ($a = 2p = -0.6284$, $b = \frac{2}{3}(p + q) = -0.4187$), we obtain the above massic polygon. ∎

Remark. *In the case $n \geq 10$, for $p = \bar{p}$ and $q = \bar{q}$, graphically we obtain* $|n.\psi_{\bar{p},\bar{q},n}(t) - \pi| \leq 0.042$, *then* $\left|2.\psi_{\bar{p},\bar{q},n}(t) - \frac{2\pi}{n}\right| \leq \frac{0.084}{n}$.

This result means that the difference between the lengths of circle segment $M(t_i)M(t_{i+1})$ and an ideal segment is less than $\frac{0.084}{n}$. For instance, for $n = 20$ (Figure 4.), it is less than 4.2×10^{-3}. Even for $n < 10$ this solution gives a satisfactory result.

References

1. Farrouki, R. T. and T. Sakkalis, Real rational curves are not 'unit-speed', Comput. Aided Geom. Design 8 (1991), 151–157.

2. Fiorot, J. C. and P. Jeannin, *Courbes et Surfaces Rationnelles, Applications à la CAO*, RMA 12, Masson, Paris, 1989. English translation: *Rational Curves and Surfaces, Applications to CAD*, Wiley, Chichester, 1992.

3. Fiorot, J. C. , P. Jeannin and I. Cattiaux–Huillard, The circle as a smoothly joined BR-curve on [0,1], Comput. Aided Geom. Design(1996), to appear. Details in Research Report 96-1 Limav, Université de Valenciennes F-59304, (1996).

4. Kelley, J. E., The cutting plane method for solving convex programs, SIAM J. Industrial and Applied Mathematics 8 (1960), 703-771.

Jean-Charles Fiorot and Isabelle Cattiaux-Huillard
Université de Valenciennes et du Hainaut-Cambrésis
ENSIMEV, Laboratoire IMAV B.P. 311
F-59304 Valenciennes - cedex, FRANCE
fiorot@univ-valenciennes.fr
cattiaux@univ-valenciennes.fr

BR-form of an Entire Rational Curve with a Preassigned Point

J-C. Fiorot and P. Jeannin

Abstract. Any rational curve of degree n/degree n is determined by $(n + 1)$ massic vectors (the BR-form) relative to a finite interval. The entire rational curve is the image of $\bar{\mathbb{R}} = \mathbb{R} \cup \{\infty\}$. For geometric modelling purposes, we determine by means of a quadratic change of parameter a new control massic polygon of $(2n + 1)$ massic vectors in terms of the original ones such that the entire curve is the image of $[0,1]$ and such that an arbitrary point of this curve is now the image of the origin.

§1. Introduction

In Computer Aided Geometric Design, many surfaces are defined from curves by means of elementary mappings. To obtain surfaces of revolution, ruled surfaces, surfaces with cross sections of constant shape but varying size [5], or more generally surfaces determined as variable combination of curves [8], we need to reparametrize curves. Here we deal with the following problem.

Any rational curve C can be written in the BR-form [6,7,8], i.e., controlled by a set of massic vectors (see Section 2) relative to an interval $[a, b]$. Likewise, a polynomial curve can be written in the Bézier-de Casteljau form [1,2], i.e. controlled by a set of points. We want to use the entire curve which is unfortunately the image of $\bar{\mathbb{R}} = \mathbb{R} \cup \{\infty\}$. To remedy this drawback, we determine via a quadratic change of parameter a new control massic polygon in order that the entire curve is the image of a finite interval, for instance $[0,1]$, with the supplementary requirement that an arbitrary point of the curve is now the image of the origin.

Since a Bézier-de Casteljau curve [2,1] and a rational Bézier curve [5,3,4] are particular cases of a BR-curve, we propose a solution for the latter.

Curves and Surfaces with Applications in CAGD
A. Le Méhauté, C. Rabut, and L. L. Schumaker (eds.), pp. 111–118.
Copyright © 1997 by Vanderbilt University Press, Nashville, TN.
ISBN 0-8265-1293-3.
All rights of reproduction in any form reserved.

§2. Notation and Preliminaries

Let $F(t)$ be a polynomial curve, $P = (P_0, P_1, \ldots, P_n)$ its Bézier-de Casteljau polygon (called control polygon) relative to the interval $[a, b]$,

$$F(t) = \sum_{i=0}^{n} B_i^n \left(\frac{t-a}{b-a} \right) P_i \text{ denoted by } BP[P_0, P_1, \ldots, P_n; [a, b]](t),$$

where $B_i^n \left(\frac{t-a}{b-a} \right) = \binom{n}{i} \left(\frac{b-t}{b-a} \right)^{n-i} \left(\frac{t-a}{b-a} \right)^i$, $i = 0, 1, \ldots, n$, are the Bernstein polynomials of degree n relative to $[a, b]$. When $a = 0$ and $b = 1$, $B_i^n \left(\frac{t-a}{b-a} \right)$ is usually denoted by $B_i^n(t)$ and $BP[P_0, P_1, \ldots, P_n; [a, b]](t)$ by $BP[P_0, P_1, \ldots, P_n](t)$.

We denote by $\text{DECAST}(P_0, P_1, \ldots, P_n; a, b; t_0)$ the procedure calculating the points P_i^j of the de Casteljau's scheme at t_0 from the control polygon $(P_0, P_1, \ldots, P_n)$ relatively to $[a, b]$: $P_i^0 = P_i$, $i = 0, \ldots, n$; for $j = 1, \ldots, n$, $i = 0, \ldots, n-j$: $P_i^j = \frac{b-t_0}{b-a} P_i^{j-1} + \frac{t_0-a}{b-a} P_{i+1}^{j-1}$.

By Section 4.6 in [4] or Proposition 1.3.3 in [9], we have

$$\forall t \in \mathbb{R}, \quad BP[P_0, P_1, \ldots, P_n; [a, b]](t) = BP[P_0^n, P_1^{n-1}, \ldots, P_n^0; [t_0, b]](t)$$
$$= BP[P_0^n, P_0^{n-1}, \ldots, P_0^0; [t_0, a]](t),$$

$(P_0^n, P_1^{n-1}, \ldots, P_n^0)$ (resp. $(P_0^n, P_0^{n-1}, \ldots, P_0^0)$) is the downward lower (resp. upward upper) diagonal of the de Casteljau's scheme.

We denote by $|\Gamma|_J$ the support of the curve Γ relative to the interval J i.e. $|\Gamma| = \{\Gamma(t); \ t \in J\}$.

Any rational curve of an affine space $\mathcal{E}$ ($\mathbb{R}^2, \mathbb{R}^3, \ldots$ for instance) or $\tilde{\mathcal{E}}$ (the completion of $\mathcal{E}$ i.e. $\mathcal{E}$ with its points at infinity) is written as a BR-curve relative to an interval $[a, b]$ (Theorem 1 in [7] or Proposition 2.2.2. in [8]). A BR-curve is determined by a control massic polygon $\theta = (\theta_0, \theta_1, \ldots, \theta_n)$ and is denoted $BR[\theta; [a, b]](t) = BR[\theta_0, \theta_1, \ldots, \theta_n; [a, b]](t)$. The θ_i are called massic vectors, they belong to the linear space $\widehat{\mathcal{E}} = (\mathcal{E} \times \mathbb{R}^*) \cup \vec{\mathcal{E}}$: θ_i is either a weighted point $(P_i; \beta_i) \in \mathcal{E} \times \mathbb{R}^*$ for $i \in I$ or a pure vector $\vec{U}_i \in \vec{\mathcal{E}}$ for $i \in \overline{I}$ with $I \cup \overline{I} = \{0, 1, \ldots, n\}$, $I \cap \overline{I} = \emptyset$ ($\vec{\mathcal{E}}$ is the associated linear vector space to $\mathcal{E}$).

We denote by $(\vec{U})_\infty$ the point at infinity of $\tilde{\mathcal{E}}$ in the direction of $\vec{U} \in \vec{\mathcal{E}}$. The explicit form of $BR[\theta_0, \theta_1, \ldots, \theta_n; [a, b]](t)$ is

$$BR[\theta; [a, b]](t) = \frac{1}{\beta(t)} \Big(\sum_{i \in I} \beta_i B_i^n \Big(\frac{t-a}{b-a}\Big) P_i \Big) + \frac{1}{\beta(t)} \Big(\sum_{i \in \bar{I}} B_i^n \Big(\frac{t-a}{b-a}\Big) \vec{U}_i \Big) \quad (1)$$

if $\beta(t) \neq 0$ and where $\beta(t) = \sum_{i \in I} \beta_i B_i^n \Big(\frac{t-a}{b-a}\Big)$

$$BR[\theta; [a, b]](t) = \Big(\sum_{i \in I} \beta_i B_i^n \Big(\frac{t-a}{b-a}\Big) P_i + \sum_{i \in \bar{I}} B_i^n \Big(\frac{t-a}{b-a}\Big) \vec{U}_i \Big)_\infty \qquad (2)$$

if $\beta(t) = 0$ and $\vec{V}(t) = \sum_{i \in I} \beta_i B_i^n \Big(\frac{t-a}{b-a}\Big) P_i + \sum_{i \in \bar{I}} B_i^n \Big(\frac{t-a}{b-a}\Big) \vec{U}_i \neq \vec{0}.$

$$BR[\theta; [a, b]](t_0) = \lim_{t \to t_0} BR[\theta; [a, b]](t) \text{ if } \beta(t_0) = 0 \text{ and } \vec{V}(t_0) = \vec{0}. \qquad (3)$$

We illustrate the way to obtain the representation of rational curves by BR-curves. The basic material comes from [8].

For simplicity, we consider the affine space $\mathcal{E} = \mathbb{R}^2$ (it could be $\mathbb{R}^3$) with the cartesian frame $(\Omega_1; \vec{\imath}, \vec{\jmath})$. Let Ω be a point not belonging to $\mathcal{E}$. We denote by $\mathcal{F}$ the 3-dimensional affine space with the frame $(\Omega; \vec{\imath}, \vec{\jmath}, \overrightarrow{\Omega\Omega_1})$ and $\vec{\mathcal{F}}$ its associated vector space. A point $P \in \mathcal{E}$ (resp. $Q \in \mathcal{F}$) is identified with the vector $\overrightarrow{\Omega_1 P} \in \vec{\mathcal{E}}$ (resp. $\overrightarrow{\Omega Q} \in \vec{\mathcal{F}}$).

Let $\Pi\Omega : \vec{\mathcal{F}} - \{\vec{0}\} \to \tilde{\mathcal{E}}$ be the conic projection of apex Ω. It associates with $Q \in \mathcal{F}$ the point of intersection P of the straight line (ΩQ) with $\tilde{\mathcal{E}}$. Similarly we define $\Pi : \vec{\mathcal{E}} - \{\vec{0}\} \to \tilde{\mathcal{E}}$ the natural projection : $\Pi(P; \beta)=P$, $\Pi(\vec{U})=(\vec{U})_\infty$. We have $\dim \widehat{\mathcal{E}} = \dim \vec{\mathcal{F}} = \dim \mathcal{E} + 1$.

The one-to-one mapping $\widehat{\Omega} : \widehat{\mathcal{E}} \to \vec{\mathcal{F}}$ defined by $\widehat{\Omega}(P; \alpha) = \alpha\overrightarrow{\Omega P}, \widehat{\Omega}(\vec{U}) = \vec{U}$ induces an addition operator and an external multiplication in $\widehat{\mathcal{E}}$, denoted by $\oplus$ and $*$ respectively such that $\widehat{\mathcal{E}}$ is a linear space and $\widehat{\Omega}$ is an isomorphism.

For instance : $(A; \alpha) \oplus (B; \beta)=\big((\alpha A + \beta B)/(\alpha + \beta); \alpha + \beta\big)$ if $\alpha + \beta \neq 0$ else $(A; \alpha) \oplus (B; \beta)=\alpha\overrightarrow{BA}$; $(A; \alpha) \oplus \vec{U} = \big(A+(\vec{U})/(\alpha); \alpha\big)$; $\lambda*(A; \alpha) = (A; \lambda\alpha)$ if $\lambda \in \mathbb{R}$.

The projections Π and $\Pi\Omega$ are linked by the relation $\Pi = \Pi\Omega \circ \widehat{\Omega}$ (Proposition 1.2.2.3 in [8]). We have $\forall \lambda \neq 0, \Pi(\lambda * \theta) = \Pi(\theta)$.

Consider a rational curve $C(t)$ of $\mathcal{E}$ given by its coordinates $x_i(t) = \frac{a_i(t)}{d(t)}$, $i = 1, 2$, where a_i and d are polynomials of degree $\leq n$. $C(t)$ is the $\Pi\Omega$-projection of the polynomial curve $Q(t) = a_1(t)\vec{\imath}+a_2(t)\vec{\jmath}+d(t)\overrightarrow{\Omega\Omega_1}$ of $\mathcal{F}$. $Q(t)$ can be represented by its BP-form : $Q(t) = BP[Q_0, Q_1, \ldots, Q_n; [a, b]](t) = \sum_{i=0}^n B_i^n \Big(\frac{t-a}{b-a}\Big) Q_i, Q_i \in \mathcal{F}$. Then

$$C(t) = \Pi\Omega\Big(\sum_{i=0}^n B_i^n \Big(\frac{t-a}{b-a}\Big) Q_i \Big). \qquad (4)$$

The relation $\Pi\Omega = \Pi \circ \widehat{\Omega}^{-1}$ implies $C(t) = \Pi\left(\sum_{i=0}^n B_i^n\left(\frac{t-a}{b-a}\right) * \widehat{\Omega}^{-1}(Q_i)\right)$.

Let us consider $I = \{i|\overrightarrow{\Omega Q_i} \notin \vec{\mathcal{E}}\}$ (the third coordinate of Q_i is different from zero)} and $\bar{I} = \{i|\overrightarrow{\Omega Q_i} \in \vec{\mathcal{E}}\}$. For $i \in I$ we define the massic vector $\theta_i = (P_i; \beta_i) \in \mathcal{E} \times \mathbb{R}^*$ such that $\theta_i = \widehat{\Omega}^{-1}(\overrightarrow{\Omega Q_i})$ i.e. $\beta_i \overrightarrow{\Omega P_i} = \overrightarrow{\Omega Q_i}$ and for $i \in \bar{I}$, $\theta_i = \widehat{\Omega}^{-1}(\overrightarrow{\Omega Q_i}) = \overrightarrow{\Omega Q_i} = \vec{U}_i \in \vec{\mathcal{E}}$.

Then we obtain the BR-form of $C(t)$

$$C(t) = \Pi\left(\sum_{i=0}^n B_i^n\left(\frac{t-a}{b-a}\right) * \theta_i\right) \quad (= BR[\theta_0, \theta_1, \ldots, \theta_n; [a,b]](t)). \quad (5)$$

This result means that a BR-curve of $\widetilde{\mathcal{E}}$, i.e. a rational curve of $\widetilde{\mathcal{E}}$, is the Π-projection of a Bézier-de Casteljau curve of the space $\widehat{\mathcal{E}}$. Equivalently we have seen in (4) that a BR-curve of $\widetilde{\mathcal{E}}$ is the $\Pi\Omega$-projection of a Bézier-de Casteljau curve of $\mathcal{F}$.

From (5) and Proposition 1.3 in [8] we obtain the explicit form (1), (2), (3) of a BR- curve. Equation (1) gives a point at finite distance, equation (2) gives an asymptotic or a parabolic point at infinity and equation (3) gives a point obtained by continuity (may be a point at infinity).

The rational Bézier curves [5, 3, 4] are a strict subset of the set of BR-curves. They are obtained from (1) when $\bar{I} = \emptyset$.

§3. The Problem and the Solution

Let $C(t)$ be a rational curve image of $\widetilde{\mathbb{R}} = \mathbb{R} \cup \{\infty\}$ written in a BR-form relative to the interval $[a,b] : C(t) = BR[\theta_0, \theta_1, \ldots, \theta_n; [a,b]](t)$. $M_0 = C(t_0)$ is a particular point $(t_0 \in \mathbb{R})$. We want to define the entire curve C as a BR-curve image of a finite interval for instance $[0,1]$ and such that M_0 is the value at the origin 0.

Algorithm

Data : $C(t) = BR[\theta_0, \theta_1, \ldots, \theta_n; [a,b]](t)$, $t_0 \in \mathbb{R}$, $M_0 = C(t_0)$.

1a) $\overrightarrow{\Omega P_i} = \widehat{\Omega}(\theta_i)$ $i = 0, 1, \ldots, n$,

1b) DECAST$(P_0, P_1, \ldots, P_n; a, b; t_0)$ gives

$$\begin{aligned}
(Q_0, Q_1, \ldots, Q_n) &= (P_0^n, P_1^{n-1}, \ldots, P_n^0) &\quad if \ |t_0 - b| \geq \frac{1}{2}|b-a| \\
(Q_0, Q_1, \ldots, Q_n) &= (P_0^n, P_0^{n-1}, \ldots, P_0^0) &\quad else
\end{aligned}$$

1c) $\omega_i = \widehat{\Omega}^{-1}(\overrightarrow{\Omega Q_i})$, $i = 0, 1, \ldots, n$.

2) Define the desired control polygon $(\tau_0, \tau_1, \ldots, \tau_{2n})$: for $j = 0, 1, \ldots, n$,

$$\tau_j = \frac{(2n-j)!j!}{2n!} \sum_{s:0 \leq 2s \leq j} (-1)^s \binom{n}{j-2s}\binom{n-j+2s}{s} \Delta^{j-2s}\omega_0,$$

for $j = 0, 1, \ldots, n-1$, $\tau_{2n-j} = (-1)^{n-j}\tau_j$.

3) Result:
$$M_0 = BR[\tau_0, \tau_1, \ldots, \tau_{2n}; [0,1]](0) = \Pi(\tau_0),$$
$$|BR[\theta_0, \theta_1, \ldots, \theta_n; [a,b]]|_{\widetilde{\mathbb{R}}} = |BR[\tau_0, \tau_1, \ldots, \tau_{2n}; [0,1]]|_{[0,1]}.$$

§4. Justification of the Algorithm

From Section 2 and point 1b above we have

$$\forall t, \quad BP[P_0, P_1, \ldots, P_n; [a,b]](t) = BP[Q_0, Q_1, \ldots, Q_n; [t_0, t_1]](t), \qquad (6)$$

where the values involved are defined by
if $|t_0 - b| \geq \frac{1}{2}|b-a|$: $t_1 = b$ and $(Q_0, Q_1, \ldots, Q_n) = (P_0^n, P_1^{n-1}, \ldots, P_n^0)$
else : $t_1 = a$ and $(Q_0, Q_1, \ldots, Q_n) = (P_0^n, P_0^{n-1}, \ldots, P_0^0)$.

Point t_1 is chosen in order to have an interval $[t_0, t_1]$ sufficiently large. From (6) we deduce

$$\Pi \circ \widehat{\Omega}^{-1}(BP[P_0, P_1, \ldots, P_n; [a,b]](t))$$
$$= \Pi \circ \widehat{\Omega}^{-1}(BP[Q_0, Q_1, \ldots, Q_n; [t_0, t_1]](t)),$$
$$\Pi(BP[\widehat{\Omega}^{-1}(P_0), \widehat{\Omega}^{-1}(P_1), \ldots, \widehat{\Omega}^{-1}(P_n); [a,b]](t))$$
$$= \Pi(BP[\widehat{\Omega}^{-1}(Q_0), \widehat{\Omega}^{-1}(Q_1), \ldots, \widehat{\Omega}^{-1}(Q_n); [t_0, t_1]](t)),$$

and by 1a and 1c,

$$BR[\theta_0, \theta_1, \ldots, \theta_n; [a,b]](t) = BR[\omega_0, \omega_1, \ldots, \omega_n; [t_0, t_1]](t). \qquad (7)$$

Moreover

$$BR[\omega_0, \omega_1, \ldots, \omega_n; [t_0, t_1]](t) = BR[\omega_0, \omega_1, \ldots, \omega_n; [0,1]](t') \text{ with } t' = \frac{t - t_0}{t_1 - t_0}, \qquad (8)$$

$$M_0 = BR[\theta_0, \theta_1, \ldots, \theta_n; [a,b]](t_0) = BR[\omega_0, \omega_1, \ldots, \omega_n; [0,1]](0) = \Pi(\omega_0). \qquad (9)$$

By (7) and (8) we deduce

$$|BR[\theta_0, \theta_1, \ldots, \theta_n; [a,b]]|_{\widetilde{\mathbb{R}}} = |BR[\omega_0, \omega_1, \ldots, \omega_n; [0,1]]|_{\widetilde{\mathbb{R}}}. \qquad (10)$$

Now we introduce the quadratic change of parameter $t' = q(v)$, $q(v) = \frac{v(1-v)}{1-2v}$, q is a one-to-one (increasing) correspondence between $[0,1]$ and $\widetilde{\mathbb{R}}$. When v sweeps interval $[0,1]$, then $BR[\omega_0, \omega_1, \ldots, \omega_n; [0,1]]\left(\frac{v(1-v)}{1-2v}\right)$ follows $|BR[\omega_0, \omega_1, \ldots, \omega_n; [0,1]]|_{\widetilde{\mathbb{R}}}$ and we obtain successively

$$BR[\omega_0, \omega_1, \ldots, \omega_n; [0,1]](t) = \Pi\left(\sum_{i=0}^{n} B_i^n(t)\omega_i\right) = \Pi\left(\sum_{i=0}^{n} \binom{n}{i} t^i \Delta^i \omega_0\right)$$

$$BR[\omega_0, \omega_1, \ldots, \omega_n; [0,1]](q(v)) = \Pi\left(\sum_{i=0}^{n} \binom{n}{i} \frac{v^i(1-v)^i}{(1-2v)^i} \Delta^i \omega_0\right)$$

$$= \Pi\left(\sum_{i=0}^{n} \binom{n}{i} v^i(1-v)^i (1-2v)^{n-i} \Delta^i \omega_0\right).$$

Let us define $V_1 = 1 - v, V_2 = v, 1 - 2v = V_1^2 - V_2^2$ then

$$BR[\omega_0, \omega_1, \ldots, \omega_n; [0, 1]](q(v)) = \Pi\left(\sum_{i=0}^{n} \binom{n}{i} V_1^i V_2^i (V_1^2 - V_2^2)^{n-i} \Delta^i \omega_0\right). \quad (11)$$

The binomial formula gives

$$(V_1^2 - V_2^2)^{n-i} = \sum_{s=0}^{n-i} \binom{n-i}{s} (-1)^s V_1^{2(n-i-s)} V_2^{2s},$$

then for $i = 0, 1, \ldots, n$,

$$V_1^i V_2^i (V_1^2 - V_2^2)^{n-i} = \sum_{s=0}^{n-i} (-1)^s \binom{n-i}{s} V_1^{2n-i-2s} V_2^{2s+i}. \quad (12)$$

The homogeneous polynomial $V_1^i V_2^i (V_1^2 - V_2^2)^{n-i}$ of degree $2n$ in V_1, V_2 can be written $V_1^i V_2^i (V_1^2 - V_2^2)^{n-i} = \sum_{j=0}^{2n} a_i^j V_1^{2n-j} V_2^j$. An identification with (12) yields

for $i = 1, 2, \ldots, n : a_i^0 = a_i^1 = \cdots = a_i^{i-1} = 0,$ (13)
(for $i = 1, 2, \ldots, n : a_i^{2n-i+1} = a_i^{2n-i+2} = \cdots = a_i^{2n} = 0$),
for $i \leq j \leq 2n - i$ and $j - i$ odd $: a_i^j = 0,$ (14)

for $i \leq j \leq 2n - i$ and $j - i$ even $: a_i^j = (-1)^{\frac{j-i}{2}} \binom{n-i}{\frac{j-i}{2}}.$ (15)

We have

$$\sum_{i=0}^{n} \binom{n}{i} V_1^i V_2^i (V_1^2 - V_2^2)^{n-i} \Delta^i \omega_0 = \sum_{i=0}^{n} \binom{n}{i} \sum_{j=0}^{2n} a_i^j V_1^{2n-j} V_2^j \Delta^i \omega_0$$

$$= \sum_{j=0}^{2n} \left(\sum_{i=0}^{n} \binom{n}{i} a_i^j \Delta^i \omega_0\right) V_1^{2n-j} V_2^j$$

$$= \sum_{j=0}^{2n} \left(\frac{1}{\binom{2n}{j}} \sum_{i=0}^{n} \binom{n}{i} a_i^j \Delta^i \omega_0\right) \binom{2n}{j} V_1^{2n-j} V_2^j. \quad (16)$$

As $\binom{2n}{j} V_1^{2n-j} V_2^j = B_j^{2n}(v)$, (11) and (16) imply

$$BR[\omega_0, \omega_1, \ldots, \omega_n; [0, 1]](q(v)) = \Pi\left(\sum_{j=0}^{2n} B_j^{2n}(v) \frac{1}{\binom{2n}{j}} \sum_{i=0}^{n} \binom{n}{i} a_i^j \Delta^i \omega_0\right).$$

Defining $\tau_j = \binom{2n}{j}^{-1} \sum_{i=0}^{n} \binom{n}{i} a_i^j \Delta^i \omega_0, j = 0, 1, \ldots, 2n$, we obtain

$$BR[\omega_0, \omega_1, \ldots, \omega_n; [0, 1]](q(v)) = \Pi\left(\sum_{j=0}^{2n} B_j^{2n}(v) \tau_j,\right),$$

$$BR[\omega_0, \omega_1, \ldots, \omega_n; [0, 1]](q(v)) = BR[\tau_0, \tau_1, \ldots, \tau_{2n}; [0, 1]](v). \quad (17)$$

Let us prove that for $j = 0, 1, \ldots, n - 1 : \tau_{2n-j} = (-1)^{n-j}\tau_j$. This property is called property of ρ reciprocity in [10], and $\rho = -1$ here. Defining $\varphi(V_1, V_2) = \sum_{i=0}^{n} \binom{n}{i} V_1^i V_2^i (V_1^2 - V_2^2)^{n-i} \Delta^i \omega_0$, φ is a homogeneous polynomial of degree 2n in V_1, V_2. By (16) and the definition of τ_j we may write

$$\varphi(V_1, V_2) = \sum_{j=0}^{2n} \binom{2n}{j} V_1^{2n-j} V_2^j \tau_j. \tag{18}$$

The definition of φ implies

$$\varphi(-V_1, V_2) = (-1)^n \varphi(V_2, V_1). \tag{19}$$

Substituting (18) into (19) yields

$$\sum_{j=0}^{2n} \binom{2n}{j} (-1)^{2n-j} V_1^{2n-j} V_2^j \tau_j = (-1)^n \sum_{j=0}^{2n} \binom{2n}{j} V_2^{2n-j} V_1^j \tau_j$$

$$= (-1)^n \sum_{j=0}^{2n} \binom{2n}{j} V_1^{2n-j} V_2^j \tau_{2n-j}.$$

Then by identification we obtain the property of ρ-reciprocity:

$$(-1)^{n-j}\tau_j = \tau_{2n-j}, \qquad j = 0, 1, \ldots, n - 1.$$

Hence it is sufficient to calculate τ_j, $j = 0, 1, \ldots, n$.
For $j = 0, 1, \ldots, n$ and $i > j$ $(i \leq n)$, by (13) $a_i^j = 0$, then

$$\tau_j = \binom{2n}{j}^{-1} \sum_{i=0}^{j} \binom{n}{i} a_i^j \Delta^i \omega_0.$$

By (14) and (15)

$$\tau_j = \binom{2n}{j}^{-1} \sum_{s:0 \leq 2s \leq j} \binom{n}{j - 2s} a_{j-2s}^j \Delta^{j-2s} \omega_0,$$

$$\tau_j = \binom{2n}{j}^{-1} \sum_{s:0 \leq 2s \leq j} (-1)^s \binom{n}{j - 2s} \binom{n - j + 2s}{s} \Delta^{j-2s} \omega_0, \quad j = 0, 1, \ldots, n.$$

It remains to justify (3) in the algorithm. By (17) and the property of q,

$$|BR[\omega_0, \omega_1, \ldots, \omega_n; [0, 1]]|_{\widetilde{\mathbf{R}}} = |BR[\tau_0, \tau_1, \ldots, \tau_{2n}; [0, 1]]|_{[0,1]}. \tag{20}$$

By (10) and (20),

$$|BR[\theta_0, \theta_1, \ldots, \theta_n; [a, b]]|_{\widetilde{\mathbf{R}}} = |BR[\tau_0, \tau_1, \ldots, \tau_{2n}; [0, 1]]|_{[0,1]}$$

and by (9), (17) and the property $q(0) = 0$,

$$M_0 = BR[\theta_0, \theta_1, \ldots, \theta_n; [a, b]](t_0) = BR[\tau_0, \tau_1, \ldots, \tau_{2n}; [0, 1]](0).$$

References

1. Bézier, P., Définition numérique des courbes et surfaces, Automatique **11** (1966), 625-632.

2. de Casteljau, P., Outillage, méthode de calcul, André Citroën Automobile S.A., Paris, 1959.

3. Farin, G.E., Algorithm for rational Bézier curves, Comp. Aided Design 15, (1983), 73-77.

4. Farin, G.E., *Curves and Surfaces for Computer Aided Design*, 3rd ed., Academic Press, San Diego. CA, 1993.

5. Faux, I.E. and M.J. Pratt, *Computational Geometry for Design and Manufacture*, Ellis Horwood, Chichester, UK, 1979.

6. Fiorot, J.C. and P. Jeannin, Courbes Bézier rationnelles, XIXème Congrès National d'Analyse Numérique, Port-Barcarès, France, 1986.

7. Fiorot, J.C. and P. Jeannin, Nouvelle description et calcul des courbes rationnelles à l'aide de points et vecteurs de contrôle, Comptes-Rendus à l'Académie des Sciences de Paris 305, Sér. 1 (1987), 435-440.

8. Fiorot, J.C. and P. Jeannin, *Courbes et Surfaces rationnelles, Applications à la CAO*, Masson, Paris, 1989 ; english translation : *Rational Curves and Surfaces, Applications to CAD*, Wiley, Chichester, UK, 1992.

9. Fiorot, J.C. and P. Jeannin, *Courbes splines rationnelles, Application à la CAO*, RMA 24, Masson, Paris, 1992.

10. Fiorot, J.C. and P. Jeannin and S. Taleb, B-rational curves and reparametrization : the quadratic case, RAIRO, Mathematical Modelling and Numerical Analysis (M^2AN) **27** (1993), 289-311.

J-C. Fiorot
Université de Valenciennes et du Hainaut Cambrésis
ENSIMEV, Laboratoire LIMAV, B.P. 311
F - 59304 Valenciennes cedex, FRANCE
Jean-Charles.Fiorot@univ-valenciennes.fr

P. Jeannin
Université du Littoral
Centre Universitaire de la Mi-Voix
Bâtiment H. Poincaré
50 rue F. Buisson, F - 62100 Calais, FRANCE

Optimal Convexity Preserving Bases

M. García-Esnaola and J. M. Peña

Abstract. Among all totally positive bases of a space of functions, bases with optimal shape preserving properties have recently been characterized. They are called *B-bases*. Here we give new characterizations of B-bases, and generalize this concept to spaces with a nonnegative basis. As an application, we show that B-bases have optimal convexity preserving properties among the nonnegative bases.

§1. Introduction and Motivation

Let us start by introducing some basic definitions.

Definitions 1.1. *Given a system* $(u_0, \ldots, u_n)$ *of functions defined on* $[a, b]$ *and* $t_0 < t_1 < \cdots < t_m$ *in* $[a, b]$, *the collocation matrix of* $u_0, \ldots, u_n$ *at* $t_0 < \cdots < t_m$ *is the matrix*

$$M \begin{pmatrix} u_0, \ldots, u_n \\ t_0, \ldots, t_m \end{pmatrix} := (u_j(t_i))_{i=0,\ldots,m;j=0,\ldots,n}, \quad a \leq t_0 < t_1 < \cdots < t_m \leq b.$$

A matrix is called TP_r, $1 \leq r \leq n+1$, *if all its* $k \times k$ *minors,* $k = 1, \ldots, r$, *are nonnegative. If all the minors are nonnegative then* A *is called totally positive. A system of functions* $(u_0, \ldots, u_n)$ *is called* TP_r *(resp.,* TP*) if all its collocation matrices are* TP_r *(resp.,* TP*).*

Now we shall introduce three preorder relations among nonnegative bases (i.e. bases of nonnegative functions) of a space.

Definitions 1.2. *Let* $\boldsymbol{u} = (u_0, \ldots, u_n)$, $\boldsymbol{v} = (v_0, \ldots, v_n)$ *be two nonnegative bases of a vector space* $\mathcal{U}$ *of functions defined on* $\Omega \subseteq \mathbb{R}^s$. *Then* $\boldsymbol{u} \preceq \boldsymbol{v}$ *if there exists a nonnegative matrix* A *such that* $\boldsymbol{v} = \boldsymbol{u}A$. *If* A *is a* TP_k *matrix (resp.,* TP *matrix) we shall write* $\boldsymbol{u} \preceq_{TP_k} \boldsymbol{v}$ *(resp.,* $\boldsymbol{u} \preceq_{TP} \boldsymbol{v}$*).*

It is well-known (see for instance Lemma 3.2 (ii) of [7]) that a nonnegative matrix with nonnegative inverse must be a generalized permutation matrix, i.e. a matrix with the same zero pattern as a permutation matrix. With a

Curves and Surfaces with Applications in CAGD   119
A. Le Méhauté, C. Rabut, and L. L. Schumaker (eds.), pp. 119–126.
Copyright © 1997 by Vanderbilt University Press, Nashville, TN.
ISBN 0-8265-1293-3.
All rights of reproduction in any form reserved.

slight modification of the proof of Lemma 3.2 of [7] one can also prove that, for $k > 1$, a TP_k matrix with TP_k inverse must be a diagonal matrix. Thus the fact that $\boldsymbol{u} \preceq \boldsymbol{v}$ and $\boldsymbol{v} \preceq \boldsymbol{u}$ is equivalent to saying that $\boldsymbol{u}$ and $\mathbf{v}$ are identical up to permutation and scaling. For $k > 1$, the relations of the previous definitions are partial orders among the nonnegative bases up to scaling.

Nonnegative bases which are minimal for $\preceq$ have been considered in [7] and [8] due to their optimal stability properties. Totally positive bases which are minimal for $\preceq_{TP}$ have been studied in some recent papers ([4,5]) due to their optimal shape preserving properties among all totally positive bases and have been called B-bases. Here we shall see that B-bases are also minimal for $\preceq_{TP_k}$, and we shall show that this fact implies for $k = 3$ that B-bases have also optimal convexity preserving properties among nonnegative bases.

Definition 1.3. *A totally positive basis $\boldsymbol{b}$ of a vector space of functions $\mathcal{U}$ is said to be a B-basis if for any totally positive basis $\boldsymbol{u}$ of $\mathcal{U}$ there exists a totally positive matrix A such that $\boldsymbol{u} = \boldsymbol{b}A$.*

In [5], B-bases were constructed and characterized. In particular, in Proposition 3.12 of that paper it was proved that a totally positive basis $(b_0, \ldots, b_n)$ is a B-basis if and only if it satisfies

$$\inf \left\{ \frac{b_i(t)}{b_j(t)} \, \Big| \, t \in I, b_j(t) \neq 0 \right\} = 0, \forall i \neq j. \tag{1.1}$$

We shall weaken this condition in Theorem 2.4. Examples of B-bases are the Bernstein basis in the case of the space of polynomials of degree less than or equal to n on a compact interval, and the B-spline basis in the case of the corresponding polynomial spline space.

Let us comment now the interpretation of these concepts in Computer Aided Geometric Design. Given $u_0, \ldots, u_n$ functions defined on $[a, b]$ and $P_0, \ldots, P_n \in \mathbb{R}^k$, a curve $\gamma(t)$ may be defined by $\gamma(t) = \sum_{i=0}^{n} u_i(t) P_i$. The points $P_0, \ldots, P_n$ are called *control points* and the polygon $P_0 \cdots P_n$ with vertices $P_0, \ldots, P_n$ is called *control polygon* of γ. In Computer Aided Geometric Design the functions $u_0, \ldots, u_n$ are usually nonnegative and $\sum_{i=0}^{n} u_i(t) = 1$, $\forall t \in [a, b]$, and in this case we say that $(u_0, \ldots, u_n)$ is a *blending system*. A system is blending if and only if it satisfies the *convex hull property*: the generated curve always lies in the convex hull of the control polygon. Another relevant property is the *linear precision*: if the control polygon coincides with a line, then the graph of the function generated also coincides with that line. A system of functions satisfies the linear precision property if

$$\sum_{i=0}^{n} \frac{i}{n} u_i(t) = \frac{t-a}{b-a}. \tag{1.2}$$

It is well-known that Bernstein basis and B-spline basis satisfy the linear precision property. An important consequence of (1.2) (see Section 6 of [2]) is the *endpoint interpolation property*, that is, the function interpolates its own control polygon at the end points: $c_0 = \sum_{i=0}^{n} c_i u_i(a)$, $c_n = \sum_{i=0}^{n} c_i u_i(b)$.

It is well known (cf. [9,4]) that blending totally positive systems lead to shape preserving representations, in the sense that many shape properties of the control polygon are inherited by the corresponding curve. Furthermore, a blending B-basis has optimal shape preserving properties among all totally positive bases of a space, i.e. the curve γ imitates better the shape of the control polygon with respect to a B-basis than the shape of the control polygon with respect to any other totally positive basis (see [4,5,7]).

On the other hand, given a system of functions $(u_0, \ldots, u_n)$ and given positive weights $w_0, \ldots, w_n$, we may define for any control polygon $P_0 \cdots P_n$ the corresponding rational curve $\gamma(t) = \frac{1}{\sum_{i=0}^{n} w_i u_i(t)} \sum_{i=0}^{n} w_i P_i u_i(t)$, which can be seen as a curve generated by the blending system

$$\left(\frac{w_0 u_0}{\sum_{i=0}^{n} w_i u_i}, \ldots, \frac{w_n u_n}{\sum_{i=0}^{n} w_i u_i} \right) \tag{1.3}$$

and the control points $P_0, \ldots, P_n$.

A system of functions $(u_0, \ldots, u_n)$ is *monotonicity preserving* if for any increasing sequence $c_0, \ldots, c_n$ the function $\sum_{i=0}^{n} c_i u_i$ is also increasing. In Theorem 2.6 of [3] it was proved that $(u_0, \ldots, u_n)$ is TP_2 if and only if all the systems (1.3) are monotonicity preserving for all $w_0, \ldots, w_n$.

In [3] it was introduced the concept of *geometrically convexity preserving* system of functions. These systems can be characterized by the following property: if the control polygon of a planar curve is convex, then the curve is also convex. From Corollary 4.6 of [3] we have that a blending system of functions satisfying (1.2) is TP_3 if and only if all the systems (1.3) are geometrically convexity preserving for all $w_0, \ldots, w_n$.

Let us observe now the geometric interpretation of the fact that two blending TP_3 bases $\boldsymbol{u}, \boldsymbol{v}$ for which (1.2) holds satisfy $\boldsymbol{u} \preceq_{TP_3} \boldsymbol{v}$. This means that $\boldsymbol{v} = \boldsymbol{u} A$, where A is a stochastic TP_3 matrix. From Proposition 3.4 of [2] it can be deduced that A transforms convex polygons into convex polygons. If $U_0 \cdots U_n$ and $V_0 \cdots V_n$ are the control polygons of a planar curve γ with respect to $\boldsymbol{u}$ and $\boldsymbol{v}$, we have that $(U_0, \ldots, U_n)^T = A(V_0, \ldots, V_n)^T$. In consequence, if the control polygon $V_0 \cdots V_n$ is convex then the control polygon $U_0 \cdots U_n$ is also convex: $\boldsymbol{u}$ is "more convexity preserving" than $\boldsymbol{v}$. So, a blending TP_3 basis minimal for $\preceq_{TP_3}$ satisfying (1.2) will have optimal convexity preserving properties. By Corollary 2.7 a B-basis is minimal for $\preceq_{TP_3}$ and therefore a blending B-basis satisfying (1.2) has optimal convexity preserving properties.

In Theorem 2.1 we give a generalization of the concept of B-basis for spaces $\mathcal{U}$ possessing nonnegative bases. In Corollary 2.6, we shall provide new characterizations of a B-basis.

§2. Main Results

In the next theorem we construct nonnegative bases $\boldsymbol{u} = (u_0, \ldots, u_n)$ with

$$\inf\left\{\frac{u_j(t)}{u_{j-1}(t)} \,\middle|\, t \in \Omega, u_{j-1}(t) \neq 0\right\} = 0, \quad j = 1, \ldots, n, \tag{2.1}$$

$$\inf\left\{\frac{u_j(t)}{u_{j+1}(t)} \,\middle|\, t \in \Omega, u_{j+1}(t) \neq 0\right\} = 0, \quad j = 0, \ldots, n-1. \tag{2.2}$$

Given a totally positive basis $\boldsymbol{v}$, it was shown in [5] that there exists a B-basis $\boldsymbol{u}$ (which satisfies (1.1) and, in particular, (2.1) and (2.2)) such that

$$\boldsymbol{v} = \boldsymbol{u}A, \quad A \text{ is a totally positive matrix.} \tag{2.3}$$

Now we generalize the concept of B-basis in the following sense: if $\boldsymbol{v}$ is a nonnegative basis, we can always obtain a nonnegative basis $\boldsymbol{u}$ satisfying (2.1), (2.2) and (2.3).

Theorem 2.1. *Let $\mathcal{U}$ be a vector space of functions defined on $\Omega \subseteq \mathbb{R}^s$ and let $\boldsymbol{v} = (v_0, \ldots, v_n)$ be a nonnegative basis of $\mathcal{U}$. Then there exists a nonnegative basis $\boldsymbol{u} = (u_0, \ldots, u_n)$ of $\mathcal{U}$ satisfying (2.1), (2.2) and (2.3). Furthermore, if the functions in $\boldsymbol{v}$ do not satisfy the corresponding conditions (2.1) or (2.2), then the matrix A in (2.3) is not a diagonal matrix.*

Proof: If the functions in $\boldsymbol{v}$ satisfy (2.1) and (2.2) then the result holds for $\boldsymbol{u} = \boldsymbol{v}$ and $A = I$ (the identity matrix).

If (2.1) does not hold for $\boldsymbol{v}$, let i_1 be the least index such that

$$\inf\left\{\frac{v_i(t)}{v_{i-1}(t)} \,\middle|\, t \in \Omega, v_{i-1}(t) \neq 0\right\} = \varepsilon > 0. \tag{2.4}$$

Let $\boldsymbol{u}^{i_1} = (u_0^{(i_1)}, \ldots, u_n^{(i_1)})$ be a new basis of $\mathcal{U}$ defined by

$$u_h^{(i_1)} := \begin{cases} v_h, & \text{if } h \neq i_1, \\ v_{i_1} - \varepsilon v_{i_1-1}, & \text{if } h = i_1. \end{cases} \tag{2.5}$$

It is easy to see that $\boldsymbol{u}^{i_1}$ is a nonnegative basis for which (2.1) holds for all $j \leq i_1$. On the other hand, $\boldsymbol{v} = \boldsymbol{u}^{i_1} B^{(i_1)}$, where $B^{(i_1)} = (b_{hl}^{(i_1)})_{h,l=0}^n$ is the totally positive matrix given by

$$b_{hl}^{(i_1)} := \begin{cases} 1 & \text{if } h = l, \\ \varepsilon & \text{if } h = i_1 - 1, \, l = i_1, \\ 0 & \text{otherwise.} \end{cases} \tag{2.6}$$

Now let $i_2 (> i_1)$ be the least index such that (2.1) does not hold for $\boldsymbol{u}^{i_1}$, if it exists. Then, analogously to the previous case, we can construct a new nonnegative basis $\boldsymbol{u}^{i_2}$ such that (2.1) holds for all $j \leq i_2$ satisfying that $\boldsymbol{u}^{i_1} = \boldsymbol{u}^{i_2} B^{(i_2)}$, where $B^{(i_2)}$ is a totally positive matrix. Iterating the previous procedure, after a finite number of steps we obtain a nonnegative basis $\boldsymbol{u^{i_p}}$ such that (2.1) holds and $\boldsymbol{v} = \boldsymbol{u^{i_p}} B^{(i_p)} B^{(i_{p-1})} \cdots B^{(i_1)}$, where $B^{(i_p)}, \ldots, B^{(i_1)}$ are totally positive matrices.

Now let us denote with $\boldsymbol{w} = (w_0, \ldots, w_n)$ the basis $\boldsymbol{u^{i_p}}$: $w_j = u_j^{(i_p)}$, $j = 0, \ldots, n$. If (2.2) holds for $\boldsymbol{w}$ the result follows. Otherwise, let j_1 be the greatest index such that

$$\inf \left\{ \frac{w_{j_1}(t)}{w_{j_1+1}(t)} \,\middle|\, t \in \Omega, w_{j_1+1}(t) \neq 0 \right\} = \varepsilon > 0. \tag{2.7}$$

Let $\boldsymbol{w^{j_1}} = (w_0^{(j_1)}, \ldots, w_n^{(j_1)})$ be a new basis of $\mathcal{U}$ defined by

$$w_h^{(j_1)} := \begin{cases} w_h, & \text{if } h \neq j_1, \\ w_{j_1} - \varepsilon w_{j_1+1}, & \text{if } h = j_1. \end{cases} \tag{2.8}$$

Obviously, $w_h^{(j_1)} \geq 0$ if $h \neq j_1$ and (2.2) holds for $\boldsymbol{w^{j_1}}$ for all $j > j_1$. Using (2.7) and (2.8) it can be deduced that $w_{j_1}^{(j_1)} \geq 0$ and that (2.2) holds for $j = j_1$. In order to check that $\boldsymbol{w^{j_1}}$ also satisfies (2.1) it remains to see that

$$\inf \left\{ \frac{w_{j_1}^{(j_1)}(t)}{w_{j_1-1}^{(j_1)}(t)} \,\middle|\, t \in \Omega, w_{j_1-1}^{(j_1)}(t) \neq 0 \right\} = 0 \tag{2.9}$$

and that

$$\inf \left\{ \frac{w_{j_1+1}^{(j_1)}(t)}{w_{j_1}^{(j_1)}(t)} \,\middle|\, t \in \Omega, w_{j_1}^{(j_1)}(t) \neq 0 \right\} = 0. \tag{2.10}$$

Since $w_{j_1}^{(j_1)} = w_{j_1} - \varepsilon w_{j_1+1} \leq w_{j_1}$ and (2.1) holds for $\boldsymbol{w}$ for $j = j_1$, we can deduce (2.9). On the other hand, if $w_{j_1}(t) \neq 0$ and $w_{j_1}(t) \neq \varepsilon w_{j_1+1}(t)$, one has that

$$\frac{w_{j_1+1}^{(j_1)}(t)}{w_{j_1}^{(j_1)}(t)} = \frac{(w_{j_1+1}(t)/w_{j_1}(t))}{1 - \varepsilon(w_{j_1+1}(t)/w_{j_1}(t))}$$

and, using that (2.1) holds for $\boldsymbol{w}$ for $j = j_1 + 1$, it is easy to deduce (2.10).

In conclusion, $\boldsymbol{w^{j_1}}$ is a nonnegative basis of $\mathcal{U}$ such that (2.1) holds for all j and (2.2) holds for all $j \geq j_1$. Besides, $\boldsymbol{w} = \boldsymbol{w^{j_1}} C^{(j_1)}$, where $C^{(j_1)} = (c_{hl}^{(j_1)})_{h,l=0}^n$ is the totally positive matrix given by

$$c_{hl}^{(j_1)} := \begin{cases} 1 & \text{if } h = l, \\ \varepsilon & \text{if } h = j_1 - 1, l = j_1, \\ 0 & \text{otherwise.} \end{cases} \tag{2.11}$$

Continuing analogously if it is necessary, we can finally obtain (after a finite number of steps) a nonnegative basis $\boldsymbol{w^{j_m}}$ such that (2.1) and (2.2) hold and

$$\boldsymbol{w} = \boldsymbol{w^{j_m}} C^{(j_m)} C^{(j_{m-1})} \cdots C^{(j_1)},$$

where $C^{(j_m)}, \ldots, C^{(j_1)}$ are totally positive matrices. Now the basis $\boldsymbol{u} = (u_0, \ldots, u_n)$ with $u_j := w_j^{(j_m)}$, $j = 0, \ldots, n$, proves the first part of theorem with the matrix $A = C^{(j_m)} C^{(j_{m-1})} \cdots C^{(j_1)} B^{(i_p)} B^{(i_{p-1})} \cdots B^{(i_1)}$, which is totally positive by Theorem 3.1 of [1]. Let us observe also that the second part of the theorem follows from the form of the previous matrices (see (2.6) and (2.11)). ∎

Corollary 2.2. *If $\boldsymbol{u} = (u_0, \ldots, u_n)$ is a nonnegative basis of a vector space of functions $\mathcal{U}$ which is minimal for $\preceq_{TP}$ then $\boldsymbol{u}$ satisfies (2.1) and (2.2).*

Proof: Let us assume that $\boldsymbol{u}$ is minimal for $\preceq_{TP}$ but it does not satisfy (2.1) or (2.2). By the previous theorem, there exists a nonnegative basis $\boldsymbol{b}$ if $\mathcal{U}$ satisfying (2.1) and (2.2) such that $\boldsymbol{u} = \boldsymbol{b}A$, where A is a nonsingular totally positive matrix which is not diagonal. Since $\boldsymbol{u}$ is minimal for $\preceq_{TP}$, by Lemma 3.2 (ii) of [7] A must be a generalized permutation matrix and, since it is not diagonal, there exists $i \in \{0, \ldots, n\}$ such that $a_{ii} = 0$, in contradiction with Corollary 3.8 of [1]. ∎

Remark 2.3. *With the same proof of Lemma 2.1 of [5], it can be proved that if $(u_0, \ldots, u_n)$ is a TP_2 system defined on $I \subseteq \mathbb{R}$ and $D_l = \{t \in I | u_l(t) \neq 0\}$ $(l = 0, \ldots, n)$, then the functions $u_i(t)/u_j(t)$ defined on D_j are increasing (resp., decreasing) for $j < i$ (resp., for $j > i$).*

The next theorem shows that condition (1.1) can be weakened by (2.1) and (2.2).

Theorem 2.4. *Let $\mathcal{U}$ be a vector space of functions defined on $I \subseteq \mathbb{R}$ and let $(b_0, \ldots, b_n)$ be a TP_2 basis of $\mathcal{U}$. If $(b_0, \ldots, b_n)$ satisfies (2.1) and (2.2), then (1.1) holds.*

Proof: Let us assume that $j < i$ in (1.1). Let $D_l = \{t \in I | b_l(t) \neq 0\}$. In order to prove that $\inf\left\{\frac{b_i(t)}{b_j(t)} \,\middle|\, t \in D_j\right\} = 0$, we can suppose without loss of generality that $D_j \subseteq D_i$. Let us prove that $D_j \subseteq D_{i-1}$. Let us assume that there exists $t_0 \in D_j (\subseteq D_i)$ such that $b_{i-1}(t_0) = 0$ and we shall obtain a contradiction. Since $(b_0, \ldots, b_n)$ is TP_2 we have for any $t \geq t_0$ that

$$0 \leq \begin{vmatrix} b_{i-1}(t_0) & b_i(t_0) \\ b_{i-1}(t) & b_i(t) \end{vmatrix} = -b_i(t_0)b_{i-1}(t).$$

Since $t_0 \in D_i$ we obtain that $b_{i-1}(t) = 0, \forall t \geq t_0$. Since b_{i-1} is a nonnegative and nonzero function there exists $t_1 < t_0$ such that $b_{i-1}(t_1) > 0$, and, using again that $(b_0, \ldots, b_n)$ is TP_2, we have that

$$0 \leq \begin{vmatrix} b_j(t_1) & b_{i-1}(t_1) \\ b_j(t_0) & b_{i-1}(t_0) \end{vmatrix} = -b_j(t_0)b_{i-1}(t_1).$$

Since $t_0 \in D_j$ we obtain that $b_{i-1}(t_1) = 0$, which is a contradiction. Thus $D_j \subseteq D_{i-1}$.

Let $x \in D_j$ and let $M := (b_j/b_{i-1})(x)(> 0)$. Since $\inf_{t \in D_{i-1}} b_i/b_{i-1} = 0$, given $\varepsilon > 0$ there exists $t \in D_{i-1}$ such that $(b_i/b_{i-1})(t) < \varepsilon M$. If $x < t$, since b_i/b_{i-1} is an increasing function by Remark 2.3, we have that $(b_i/b_{i-1})(x) < \varepsilon M$ and so $(b_i/b_j)(x) = ((b_i/b_{i-1})/(b_j/b_{i-1}))(x) < \varepsilon$. If $t \leq x$, since b_j/b_{i-1} is a decreasing function by Remark 2.3, we have that $0 < (b_j/b_{i-1})(x) \leq (b_j/b_{i-1})(t)$ and therefore $t \in D_j$ and $(b_i/b_j)(t) = (b_i/b_{i-1})(t)/(b_j/b_{i-1})(t) \leq (b_i/b_{i-1})(t)/(b_j/b_{i-1})(x) < \varepsilon$. So $\inf_{t \in D_j} b_i(t)/b_j(t) = 0$. If $j > i$ we could reason analogously to the previous case: now, if we assume that $D_j \subseteq D_i$ we would obtain that $D_j \subseteq D_{i+1}$. ∎

The next example shows that in the previous theorem we cannot remove the hypothesis of $(b_0, \ldots, b_n)$ is TP_2.

Example 2.5. *Let $\boldsymbol{u} = (u_0, u_1, u_2)$ be a basis of nonnegative functions given by*

$$u_0(t) := \begin{cases} 1 & \text{if } t \in [0,1) \cup [2,3] \\ 0 & \text{if } t \in [1,2), \end{cases} \qquad u_1(t) := \begin{cases} 0 & \text{if } t \in [0,1) \\ 1 & \text{if } t \in [1,3], \end{cases}$$

$$u_2(t) := \begin{cases} 1 & \text{if } t \in [0,1/2] \cup [2,3] \\ 0 & \text{if } t \in (1/2,2). \end{cases}$$

Then one can check that $\boldsymbol{u}$ satisfies (2.1) and (2.2) but it does not satisfy (1.1) because $\inf\{u_0(t)/u_2(t)|t \in [0,1/2] \cup [2,3]\} = 1$.

As a consequence of the two previous theorems and other results we have

Corollary 2.6. *Let $\mathcal{U}$ be a vector space of functions defined on $I \subseteq \mathbb{R}$ with a totally positive basis. If $\boldsymbol{b} = (b_0, \ldots, b_n)$ is a TP_2 basis of $\mathcal{U}$, then the following conditions are equivalent:*
  (i) *$(b_0, \ldots, b_n)$ satisfies (1.1),*
 (ii) *$\boldsymbol{b}$ is minimal for $\preceq$,*
(iii) *$\boldsymbol{b}$ is a B-basis,*
 (iv) *$(b_0, \ldots, b_n)$ satisfies (2.1) and (2.2),*
  (v) *$\boldsymbol{b}$ is minimal for $\preceq_{TP_k}$, for $k \geq 1$,*
 (vi) *$\boldsymbol{b}$ is minimal for $\preceq_{TP}$.*

Proof: (i) $\Longrightarrow$ (iii) is a consequence of Remark 2.3 and of Theorem 3.4 of [6].
(iii) $\Longrightarrow$ (iv) is a consequence of Proposition 3.12 of [5].
(iv) $\Longrightarrow$ (i) is a consequence of Theorem 2.4.
(iii) $\Longrightarrow$ (ii) is a consequence of Theorem 3.9 of [7].
(ii) $\Longrightarrow$ (v) is obvious.
(v) $\Longrightarrow$ (vi) is obvious.
(vi) $\Longrightarrow$ (iv) is a consequence of Corollary 2.2. ∎

Let us observe that the equivalence of (iv) with (ii) and (i) improves Theorem 3.9 of [7].

In Theorem 4.2 of [5] it was proved that a space with a blending totally positive basis has a unique blending B-basis. As a consequence of this fact and the previous theorem, we can deduce the following corollary.

Corollary 2.7. *If k is a positive integer and U is a finite dimensional vector space of functions defined on an interval I ⊆ ℝ with a blending totally positive basis, then the blending B-basis is the unique blending TP_2 basis of U which is minimal for $\preceq_{TP_k}$.*

In the case $k = 3$, the previous corollary proves the optimal convexity preserving properties of a B-basis satisfying (1.2), as was noted in Section 1.

Acknowledgments. Both authors were partially supported by DGICYT PB93-0310 and by the EU project CHRX-CT94-0522.

References

1. Ando, T., Totally positive matrices, Linear Algebra Appl. **90** (1987), 165–219.

2. Carnicer, J. M., M. García-Esnaola, and J. M. Peña, Generalized convexity preserving transformations, Comput. Aided Geom. Design **13** (1996), 179–197.

3. Carnicer, J. M., M. García-Esnaola, and J. M. Peña, Convexity of rational curves and total positivity, Journal of Computational and Applied Mathematics **71** (1996), 365–382.

4. Carnicer, J. M. and J. M. Peña, Shape preserving representations and optimality of the Bernstein basis, Advances in Computational Mathematics **1** (1993), 173–196.

5. Carnicer, J. M. and J. M. Peña, Totally positive bases for shape preserving curve design and optimality of B-splines, Comput. Aided Geom. Design **11** (1994), 635–656.

6. Carnicer, J. M. and J. M. Peña, Characterizations of the optimal Descartes' rules of signs, to appear in Math. Nachr.

7. Carnicer, J. M. and J. M. Peña, Total positivity and optimal bases, in *Total Positivity and its Applications*, M. Gasca and C.A. Micchelli (eds.), Kluwer Academic Press, 1996, 133–155.

8. Farouki, R. T. and T. N. T. Goodman, On the optimal stability of the Bernstein basis, to appear in Math. Comp.

9. Goodman, T. N. T., Shape preserving representations, in *Mathematical Methods in Computer Aided Geometric Design*, T. Lyche and L.L. Schumaker (eds.), Academic Press, Boston, 1989, 333–351.

M. García-Esnaola and J. M. Peña
Departamento de Matemática Aplicada.
Universidad de Zaragoza.
50009 Zaragoza, SPAIN
mgesnaola@mcps.unizar.es, jmpena@posta.unizar.es

Rational Interpolation of the Unit Circle

Th. Gensane

Abstract. We consider interpolating data on the unit circle, and look for the rational parameterizations of the circle which interpolate them. We transform this non-linear problem into a linear one and generate all the solutions of our problem. We also tackle the search for solutions of lower degree.

§1. Introduction

The problems relating to the circle are inherently non-linear. Nevertheless, using rational parameterizations for circular arcs is important in CAGD, and so two ways are generally considered in doing so: either accept to solve non linear equations, or approximate the circle, for instance by pieces of polynomial curves. We propose here a complex formulation of all *the rational parameterizations of the unit circle*; we denote by $\mathcal{C}$ the set of such curves: $\mathcal{C} = \{Z(t) = (X(t), Y(t))^T : X(t) \text{ and } Y(t) \text{ are rational and } X^2(t) + Y^2(t) = 1\}$. This formulation enables us to transform some non-linear problems into a linear one. This paper does not deal with offset curves although the formulation we give is connected to the one of Pythagorean-hodograph curves, see [4,5,6,7,11,12], the Gauss map of these curves, $t \to n(t)$ being rational. The reader can find a projective approach to the sphere and other quadrics in [2,3].

After setting forth the interpolation problem (Section 2), we generate all solutions (Section 3). We show (Section 4), that the search for a solution of lower degree will require a long analysis in order to find good interpolation schemes.

§2. The Problem

The problem to be solved is:

Given $m \in \mathbb{N}$, $(\gamma_0, \cdots, \gamma_m) \in \mathbb{N}^{m+1}$, real numbers $t_0 < t_1 < \cdots < t_m$ and complex numbers η_i^j, find interpolating curves $Z(t) \in \mathcal{C}$ so that

$$Z^{(j)}(t_i) = \eta_i^j \text{ for } i = 0, \cdots, m, \text{ and } j = 0, \cdots, \gamma_i. \tag{1}$$

Curves and Surfaces with Applications in CAGD
A. Le Méhauté, C. Rabut, and L. L. Schumaker (eds.), pp. 127–134.
Copyright © 1997 by Vanderbilt University Press, Nashville, TN.
ISBN 0-8265-1293-3.
All rights of reproduction in any form reserved.

The equations (1) will be denoted in the following by $(Z^{(j)}(t_i)) = (\eta_i^j)$, and the set of the couples (i, j) by Γ.

A necessary condition for the existence of such a curve is given by Leibniz's formula applied to the map $t \to X^2(t) + Y^2(t)$; after setting $\eta_i^j = \alpha_i^j + i\beta_i^j$ ($i^2 = -1$), we obtain at each parameter t_i:

$$
\text{(H)} \begin{cases}
(\alpha_i^0)^2 + (\beta_i^0)^2 = 1 \\
\alpha_i^0 \alpha_i^1 + \beta_i^0 \beta_i^1 = 0 \\
\quad \vdots \\
\displaystyle\sum_{q=0}^{\gamma_i - 1} \binom{\gamma_i - 1}{q} (\alpha_i^q \alpha_i^{\gamma_i - q} + \beta_i^q \beta_i^{\gamma_i - q}) = 0.
\end{cases}
$$

It will be shown in the Proposition 3 that the conditions (H), are also sufficient for existence of a rational interpolating curve $Z(t) \in \mathcal{C}$ of the η_i^j.

§3. Determination of all the Solutions

We now give the representation of the curves of $\mathcal{C}$ that we are using.

Proposition 1. *A rational curve $Z(t)$ belongs to $\mathcal{C}$ iff there exist two polynomials M and N such that $Z(t) = \frac{z(t)}{\bar{z}(t)}$, where $z(t) = M(t) + iN(t)$ and $\bar{z}(t)$ is its conjugate.*

Proof: Setting $Z(t) = \frac{P(t)}{R(t)} + i\frac{Q(t)}{R(t)}$, the polynomial identity $P^2(t) + Q^2(t) = R^2(t)$ implies the existence of M, N, K such that $P = K(M^2 - N^2)$ and $Q = 2KMN$ (see for instance [4,6]). The result follows. ∎

It now appears natural to transform our interpolation problem on $Z(t)$ into one on the polynomial curve $z(t)$; it consists in determining the admissible derivatives of $z(t)$ which give a solution $Z(t)$. Let us remark that the argument of $z(t)$ is half of the one of $Z(t)$.

Definition 2. *The data (b_i^j) is called admissible iff $b_i^0 \neq 0$ for all $i = 0, \cdots, m$ and*

$$(z^{(j)}(t_i)) = (b_i^j) \implies (Z^{(j)}(t_i)) = (\eta_i^j).$$

We now adopt some notations used in the proofs of Propositions 3 and 4: Let $Z(t) = \frac{z(t)}{\bar{z}(t)}$ be a curve of $\mathcal{C}$. Setting $Z(t) = X(t) + iY(t)$ and $z(t) = x(t) + iy(t)$ we obtain the matrix relation $A(t) \cdot z(t) = 0$, where

$$A(t) = \begin{pmatrix} X(t) - 1 & Y(t) \\ -Y(t) & X(t) + 1 \end{pmatrix} \quad \text{and} \quad z(t) = \begin{pmatrix} x(t) \\ y(t) \end{pmatrix}. \tag{2}$$

Considering the j-th order derivatives of each component of $A(t)$, we get the matrices denoted by $A^{(j)}(t)$. In relation to the interpolation data $\eta_i^j = \alpha_i^j + i\beta_i^j$, we denote by A_i^j the matrices

$$A_i^0 = \begin{pmatrix} \alpha_i^0 - 1 & \beta_i^0 \\ -\beta_i^0 & \alpha_i^0 + 1 \end{pmatrix} \quad \text{and} \quad A_i^j = \begin{pmatrix} \alpha_i^j & \beta_i^j \\ -\beta_i^j & \alpha_i^j \end{pmatrix} \quad \text{for } j > 0.$$

Proposition 3. *Let* (η_i^j) *satisfy the conditions* (H).
(i) *The data* (b_i^j) *is admissible iff for all* $i \in \{0, \cdots, m\}$, *the complex number* b_i^0 *is not zero, and*

$$
\begin{cases}
A_i^0 \cdot b_i^0 = 0 \\
A_i^0 \cdot b_i^1 = -A_i^1 \cdot b_i^0 \\
\quad \vdots \\
A_i^0 \cdot b_i^{\gamma_i} = -\displaystyle\sum_{q=1}^{\gamma_i} \binom{\gamma_i}{q} \cdot (A_i^q \cdot b_i^{\gamma_i - q}).
\end{cases}
\tag{3}
$$

(ii) *The previous system admits of solutions, and thus, the necessary conditions* (H) *are also sufficient.*

Proof: (i) Let us remark that the interpolation condition $Z^{(j)}(t_i) = \eta_i^j$ is equivalent to the matrix identity $A^{(j)}(t_i) = A_i^j$. If (b_i^j) is admissible, we consider $z(t)$ so that $(z^{(j)}(t_i)) = (b_i^j)$. We have $A(t_i) \cdot z(t_i) = 0$ and $A_i^0 \cdot b_i^0 = 0$. Leibniz's formula applied to $A(t) \cdot z(t) = 0$ at $t = t_i$, γ_i times gives the necessary condition.

Conversely, we prove by induction that the condition is sufficient. Let us consider $b_i^0 \neq 0, b_i^1, \cdots, b_i^{\gamma_i}$ satisfying (3) and an interpolating curve $z(t)$ of these values. We have $A_i^0 \cdot b_i^0 = 0$ and $A(t_i) \cdot b_i^0 = 0$. The determinant of the matrix $A_i^0 - A(t_i)$ is equal to $((\alpha_i^0 - X(t_i))^2 + (\beta_i^0 - Y(t_i))^2$, which is equal to zero since b_i^0 is not. Hence $Z(t_i) = \eta_i^0$.

If $Z^{(j)}(t_i) = A_i^j$ for all $j \in \{0, \cdots, l - 1\}$ and $l \leq \gamma_i$, we apply l times the Leibniz formula to $A(t) \cdot z(t) = 0$ at $t = t_i$ and substitute b_i^j for $z^{(j)}(t_i)$, we get $(A^{(l)}(t_i) - A_i^l) \cdot b_i^0 = 0$. A similar argument on the determinant gives $Z^{(l)}(t_i) = \eta_i^l$. We have finally shown that (b_i^j) is admissible.

(ii) We denote by E_j the $j + 1$-th equation of (3). The set of solutions of E_0 is a space vector of dimension 1. We suppose that $E_0, \cdots, E_j$ has solutions. We are showing the compatibility of the right-hand member of E_{j+1}. Setting $b_i^j = x_i^j + iy_i^j$, the equation E_{j+1} becomes

$$
\begin{cases}
(\alpha_i^0 - 1)x_i^{j+1} + \beta_i^0 y_i^{j+1} = -\displaystyle\sum_{q=1}^{j+1} \binom{j+1}{q}(\alpha_i^q x_i^{j+1-q} + \beta_i^q y_i^{j+1-q}) \\
-\beta_i^0 x_i^{j+1} + (\alpha_i^0 + 1)y_i^{j+1} = -\displaystyle\sum_{q=1}^{j+1} \binom{j+1}{q}(-\beta_i^q x_i^{j+1-q} + \alpha_i^q y_i^{j+1-q}).
\end{cases}
$$

We suppose that $\beta_i^0 \neq 0$; in the contrary case, it is sufficient to consider a rotation such as the points η_i^0 are all different from 1 and -1. Multiplying the first equation by $-\beta_i^0$ and the second by $\alpha_i^0 - 1$ and substracting them, after considering the two relations between $\alpha_i^0, \beta_i^0, x_i^0$ and y_i^0 given by E_0, we obtain

$$2y_i^0\alpha_i^0\alpha_i^{j+1} + 2y_i^0\beta_i^0\beta_i^{j+1} = -\beta_i^0 \sum_{q=1}^{j} \binom{j+1}{q}(\alpha_i^q x_i^{j+1-q} + \beta_i^q y_i^{j+1-q})$$

$$- (\alpha_i^0 - 1)\sum_{q=1}^{j} \binom{j+1}{q}(-\beta_i^q x_i^{j+1-q} + \alpha_i^q y_i^{j+1-q}).$$

(4)

So, the equation E_{j+1} has solutions iff the point $(\alpha_i^{j+1}, \beta_i^{j+1})$ belongs to the straight line

$$\alpha_i^0 U + \beta_i^0 V = s_{j+1},$$

(5)

where s_{j+1} is the right member of (4) divided by $2y_i^0$. Let us remark that the $j + 2$-th line of the assumption (H) implies that the point $(\alpha_i^{j+1}, \beta_i^{j+1})$ belongs to the straight line

$$\alpha_i^0 U + \beta_i^0 V = -\sum_{k=1}^{j} \binom{j}{k}(\alpha_i^k \alpha_i^{j+1-k} + \beta_i^k \beta_i^{j+1-k}).$$

(6)

In order to conclude the proof, we show that the two straight lines (5) and (6) are equal. We consider a point (A, B) of the straight line (5). The point (i) applied with $\gamma_i = j+1$ implies there exists a curve $Z(t) = X(t) + iY(t) \in \mathcal{C}$ such that $Z^{(l)}(t_i) = X^{(l)}(t_i) + iY^{(l)}(t_i) = \alpha_i^l + i\beta_i^l$ for all $l \in \{0, \cdots, j\}$ and $Z^{(j+1)}(t_i) = A + iB$ (the complex numbers $b_i^0, \cdots, b_i^j$ being solutions to equations $E_o, \cdots, E_j$ and $A + iB$ being a solution to E_{j+1}, these complex numbers constitute admissible data). We consider the $j + 1$-order derivatives of $X^2(t) + Y^2(t) = 1$ at $t = t_i$ and get by Leibniz's formula:

$$X(t_i)X^{(j+1)}(t_i) + Y(t_i)Y^{(j+1)}(t_i) =$$

$$- \sum_{k=1}^{j} \binom{j}{k}(X^{(k)}(t_i)X^{(j+1-k)}(t_i) + Y^{(k)}(t_i)Y^{(j+1-k)}(t_i),$$

which shows that (A, B) belongs to the straight line (6), and thus the equality of the two straight lines. Finally, the compatibility is obtained, up to $j = \gamma_i - 1$. ∎

We have at our disposal all the admissible data and thereby all the interpolating curves $Z(t)$ of (η_i^j). But, if we want to determine them explicitly, we need to choose one of them and apply the following proposition.

Proposition 4. *Let (b_i^j) be admissible, another data (a_i^j) is admissible iff there exist reals λ_i^j, $\lambda_i^0 \neq 0$ such that*

$$\begin{cases} a_i^0 = \lambda_i^0 b_i^0 \\ a_i^1 = \lambda_i^0 b_i^1 + \lambda_i^1 b_i^0 \\ \quad\vdots \\ a_i^{\gamma_i} = \lambda_i^0 b_i^{\gamma_i} + \cdots + \binom{\gamma_i}{k}\lambda_i^k b_i^{\gamma_i-k} + \cdots + \lambda_i^{\gamma_i} b_i^0. \end{cases}$$

(7)

Proof: Let (b_i^j) be admissible and $z_1(t)$ be a polynomial curve such that $(z_1^{(j)}(t_i)) = (b_i^j)$. Let us remark that if P is a real polynomial, the data $((P \cdot z_1)^{(j)}(t_i))$ is admissible (it suffices to consider the derivatives of $\frac{P \cdot z_1}{P \cdot z_1}$). Then, if the data (a_i^j) satisfies (7) for all $i \in \{0, \cdots, m\}$, we have $a_i^j = (P \cdot z_1)^{(j)}(t_i)$ if $P^{(k)}(t_i) = \lambda_i^k$ for all $k \in \{0, \cdots, j\}$, and so the data (a_i^j) is admissible.

Conversely, let us consider (a_i^j) and (b_i^j) as being two admissible data, and two curves $z(t)$, $z_1(t)$ which interpolate them respectively. We show that the condition is necessary by induction. The arguments of a_i^0 and b_i^0 are both equal to half of that of η_i^0, and we get $a_i^0 = \lambda_i^0 b_i^0$. We adopt notations similar to those in (2) and consider the two matrices $A(t)$ and $A_1(t)$ relative to $z(t)$ and $z_1(t)$. We have $A^{(j)}(t_i) = A_1^{(j)}(t_i)$. We assume there exist $\lambda_i^0, \cdots, \lambda_i^j \in \mathbb{R}$ such that $a_i^j = \sum_{k=0}^{j} \binom{j}{k} \lambda_i^k b_i^{j-k}$ for all $j \in \{0, \cdots, l-1\}$ and $l \leq \gamma_i$. Applying the Leibniz formula to $A(t) \cdot z(t)$ at $t = t_i$, we obtain

$$A(t_i) \cdot z^{(l)}(t_i) = -\sum_{p=1}^{l} \left(\binom{l}{p} A^{(p)}(t_i) \cdot \left(\sum_{q=0}^{l-p} \binom{l-p}{q} \lambda_i^q z_1^{(l-p-q)}(t_i) \right) \right). \quad (8)$$

Since $\sum_{p=0}^{l-q} \binom{l-q}{p} A^{(p)}(t_i) z_1^{(l-p-q)}(t_i) = (A_1 \cdot z_1)^{(l-q)}(t_i) = 0$ and $\binom{l}{p}\binom{l-p}{q} = \binom{l}{q}\binom{l-q}{p}$, we get

$$A(t_i) \cdot z^{(l)}(t_i) = \sum_{q=0}^{l-1} \binom{l}{q} A(t_i) \cdot z_1^{(l-q)}(t_i).$$

So, since the rank of the matrix $A(t_i)$ is one, the identity

$$A(t_i) \cdot \left(z^{(l)}(t_i) - \sum_{q=0}^{l-1} \binom{l}{q} z_1^{(l-q)}(t_i) \right) = 0$$

implies that there exists λ_i^l such that

$$z^{(l)}(t_i) = \sum_{q=0}^{l-1} \binom{l}{q} z_1^{(l-q)}(t_i) + \lambda_i^l z_1(t_i)$$

which concludes the proof. ∎

§4. Solutions of Lower Degree

We are now able to construct the set of all the good derivatives of $z(t)$ giving a solution $Z(t)$ of our problem. A natural question is how to determine the solutions of lower degree. We do not have any theoretical result on this problem. But, we can choose a specific problem leading to the resolution of a square linear system and find the solution if the matrix is regular. Such an interpolation scheme is given in Example 5. Unfortunately, the search for lower degree can lead to a non-linear problem, but we have avoided these difficulties using the form $Z(t) = \frac{z(t)}{\bar{z}(t)}$. Yet, on the whole, does the true nature of the problem come back? For instance, the way we proceed in Example 6 in order to find some smooth parametrizations of the full circle leads to a problem of decreasing the rank of a matrix whose coefficients depend on arbitrary reals.

Example 5. Here we consider $m = 2$ and $\gamma = (k, 0, k)$, i.e. the interpolation of two points and their first derivatives and a middle point. We are looking for a Bézier curve $z(t) = \sum_{i=0}^{k+1} P_i B_i^{k+1}(t)$ such that

- $\sum_{i=0}^{k+1} P_i B_i^{k+1}(t_1) = \lambda_1^0 b_1^0$ (interpolation at the middle point),

- $\frac{(k+1)!}{(k+1-j)!} \Delta^j P_0 = \sum_{l=0}^{j} \binom{j}{l} \lambda_0^l b_0^{j-l}$ (interpolation of the j-th derivative at the first point), for all $j \in \{0, \cdots, k\}$,

- $\frac{(k+1)!}{(k+1-j)!} \Delta^j P_{k+1-j} = \sum_{l=0}^{j} \binom{j}{l} \lambda_2^l b_2^{j-l}$ (interpolation of the j-th derivative at the end point), for all $j \in \{0, \cdots, k\}$.

We obtain a linear system of $4k + 6$ real equations in $4k + 7$ unknowns. Let us remark that if we consider $\lambda_1^0(0, \cdots, 0, b_1^0)$ as the right-hand side, the solutions of the square system we get, if any, are homothetic to the ones we obtain with $\lambda_1^0 = 1$. So, that is also the case of the control polygon of $z(t)$. Hence, if the matrix is regular, the curves $Z(t) = \frac{z(t)}{\bar{z}(t)}$ generated by the values of λ_1^0 are all equal.

We have experimented with this method to approximate non-rational offsets $r(t) + dn(t)$ of a rational curve $r(t)$ by a curve $r(t) + d\,Z(t)$ (see [10] for a similar method, but not with a curve $Z(t) \in \mathcal{C}$). The good results we have obtained and the empirical existence of a solution $Z(t)$ of degree $2k + 2$ allowed us to be optimistic. In fact, it is possible to construct an example of two points and two derivatives on the circle for which all the curves of degree 4 go through the same point at a prescribed parameter t_1. In order to find them, we consider an admissible data $b_0^0 = \mathrm{i}$, $b_0^1 = c + \mathrm{i}d$, $b_2^0 = 1 + \mathrm{i}x$, $b_2^1 = e + \mathrm{i}f$ and $b_1^0 \in \mathbb{C}$. A curve $z(t) = \sum_{i=0}^{2} P_i B_i^2(t)$ interpolates the values $a_0^0 = b_0^0$, $a_0^1 = b_0^1 + \alpha b_0^0$, $a_2^0 = \beta b_2^0$ and $a_2^1 = \beta b_2^1 + \gamma b_2^0$ iff $P_0 = \mathrm{i}$, $P_1 = \frac{c}{2} + \mathrm{i}(1 + \frac{d+\alpha}{2})$ and $P_2 = \frac{-cx+d+2+\alpha}{ex-f}(1 + \mathrm{i}x)$. We use the real number α for the interpolation of the third point λb_1^0, where $\lambda \in \mathbb{R}$. We just need to remark that $z(t_1) = A + \alpha B$, where

$$A = \frac{a_1 c}{2} + a_2 \frac{-cx+d+2}{ex-f} + \mathrm{i}(a_0 + a_1(1 + \frac{d}{2}) + a_2 \frac{x(-cx+d+2)}{ex-f})$$

$$B = -\frac{a_2}{ex-f} + \mathrm{i}(\frac{a_1}{2} - \frac{a_2 x}{ex-f})$$

with $a_i = B_i^2(t_1)$. If t_1, x, c, d, e, f are fixed such that $A = 0$ and $B \neq 0$, all the curves $z(t)$ determined by the choice of α, give a curve $Z(t)$ which goes through the point $B/\overline{B}$ at $t = t_1$. So, the interpolation of another middle point is impossible with a curve $Z(t)$ of degree 4.

Example 6. We construct parameterizations of the full circle on $[0, 1]$. We want to connect as smoothly as possible, i.e. to find a curve $Z(t) = \frac{z(t)}{\overline{z}(t)}$ such that $Z^{(j)}(0) = Z^{(j)}(1)$ for all $j \in \{0, \cdots, k\}$. These conditions become

$$z^{(j)}(1) = \sum_{l=0}^{j} \binom{j}{l} \alpha_l z^{(j-l)}(0) \quad \text{for all } j = 0, \cdots, k,$$

where $\alpha_0, \cdots, \alpha_k \in \mathbb{R}$. If we are looking for a curve $z(t)$ of degree k, i.e. $z(t) = \sum_{i=0}^{k} P_i B_i^k(t)$, its derivatives at the first point satisfy $z^{(j)}(0) = \frac{k!}{(k-j)!} \Delta^j P_0$, and at the end point point $z^{(j)}(1) = \frac{k!}{(k-j)!} \Delta^j P_{k-j}$ for all $j \in \{0, \cdots, k\}$. Thus, its Bézier polygon $(P_0, \cdots, P_k)$ satisfies

$$D \cdot \begin{pmatrix} \Delta^0 P_k \\ \vdots \\ \Delta^k P_0 \end{pmatrix} = \begin{pmatrix} \alpha_0 & 0 & & 0 \\ \alpha_1 & k\alpha_0 & & \\ \vdots & & \ddots & \\ \alpha_k & \cdots & & k!\alpha_0 \end{pmatrix} \cdot \begin{pmatrix} \Delta^0 P_0 \\ \vdots \\ \Delta^k P_0 \end{pmatrix},$$

where D is a diagonal matrix. So, the vector of the progressive differences $(\Delta^0 P_0, \cdots, \Delta^k P_0)$ is solution of a linear system $(L - D \cdot U) \cdot X = 0$, where L is the above lower triangular matrix and U the triangular matrix given by $(\Delta^0 P_0, \cdots, \Delta^k P_0) = U \cdot (\Delta^0 P_k, \cdots, \Delta^k P_0)$. The search for the points $(\Delta^0 P_0, \cdots, \Delta^k P_0)$ not all on a same straight line through zero (and then of $(P_0, \cdots, , P_k)$), are equivalent to the search for $\alpha_0, \cdots, \alpha_k$ such that the rank of $L - D \cdot U$ is less than $k - 1$. This approach gives us a constructive way to generate $z(t)$ if any. The curves we have found, using the software *Mathematica*, are the same as the ones given in [8], where the authors use some changes of parameter in the parameterization of degree 2 of the circle. They obtain continuity C^3 with a parameterization of degree 6, and C^5 with degree 8 or 10.

We can also represent the curves $Z(t) = \frac{z(t)}{\overline{z}(t)}$ using the BR-form, see [6,9]. If $z(t) = \sum_{i=0}^{n} P_i B_i^n(t)$, we get

$$Z(t) = \frac{\sum_{i \in I} \beta_i R_i B_i^{2n}(t)}{\sum_{i=0}^{2n} \beta_i B_i^{2n}(t)} + \frac{\sum_{i \in \bar{I}} \vec{U}_i B_i^{2n}(t)}{\sum_{i=0}^{2n} \beta_i B_i^{2n}(t)}$$

with $\beta_k = \sum_{i=0}^{k} \lambda_n^{i,k-i} P_i \cdot \bar{P}_{k-i}$, $\lambda_n^{i,j} = \binom{n}{i}\binom{n}{j}/\binom{2n}{i+j}$, $I = \{i/\beta_i \neq 0\}$, $\bar{I} = \{i/\beta_i = 0\}$, $R_k = (\sum_{i=max(0,k-n)}^{min(k,n)} \lambda_n^{i,k-i} P_i \cdot P_{k-i})/(\sum_{i=max(0,k-n)}^{min(k,n)} \lambda_n^{i,k-i} P_i \cdot \bar{P}_{k-i})$, for $k \in I$ and $\vec{U}_k = \sum_{i=max(0,k-n)}^{min(k,n)} \lambda_n^{i,k-i} P_i \cdot P_{k-i}$ for $k \in \bar{I}$.

The weights of a rational curve $Z(t) \in \mathcal{C}$ can be studied geometrically. For instance, we prove easily that the weigths of any quartic parameterization of the full circle cannot be all strictly positive. For an analytic proof of this result, see [1].

References

1. Chou, Jin J., Higher order Bézier circles, Computer-Aided Design **27** (1995), 303–309.

2. Dietz, R., Hoshek, J., Jüttler, B., An algebraic approach to curves and surfaces on the sphere and on other quadrics, Comput. Aided Geom. Design **10** (1993), 211–229.

3. Dietz, R., Hoshek, J., Jüttler, B., Rational patches on quadric surfaces, Computer-Aided Design **27** (1995), 27–40.

4. Farouki, R.T., Sakkalis, T., Pythagorean hodographs, IBM J. Res. Develop. **34** (1990), 736–752.

5. Farouki, R.T., The conformal map $z \to z^2$ of the hodograph plane, Comput. Aided Geom. Design **11** (1994), 363–390.

6. Gensane, Th., Courbes Rationnelles à Parallèles Rationnelles et Approximation de Courbes Parallèles, Thesis of the University of Lille, 1993.

7. Fiorot, J.C., Gensane, Th., Characterization of the rational curves with rational offsets, in *Curves and Surfaces in Geometric Design*, Laurent, P.-J., A. Le Méhauté, and L. L. Schumaker (eds.), A. K. Peters, Wellesley, 1994, 153–160.

8. Fiorot, J. C., Cattiaux-Huillard, I., Uniform point distribution on a circle, in *Curves and Surfaces with Applications in CAGD*, A. Le Méhauté, C. Rabut, and L. L. Schumaker (eds.), Vanderbilt Univ. Press, Nashville, 1997, 103–110.

9. Fiorot, J.C., Jeannin, P., *Rational Curves and Surfaces, Applications to CAD*, J.Wiley and Sons, Chichester, 1992.

10. Lee, In-K, Kim, M-S, Elber, G., Planar curve offset based on circle approximation, Computer-Aided Design **28**, 8 (1996), 617–630.

11. Pottmann, H., Rational curves and surfaces with rational offsets, Comput. Aided Geom. Design **12** (1995), 175–192.

12. Pottmann, H., Curves design with rational Pythagorean-hodograph curves, Adv. Comp. Math. **3** (1995), 147–170.

Thierry Gensane
Université du Littoral, L.M.A., bât. H. Poincaré
50, rue F. Buisson B.P. 699
62228 Calais Cedex FRANCE
gensane@lma.univ-littoral.fr

Designing Nonlinear Models for Flexible Curves

S. Girard, B. Chalmond and J. M. Dinten

Abstract. This paper deals with the modeling of a flexible curve using a set of examples. The problem is addressed from the data analysis point of view. Each curve is interpretated as a point in a high-dimensional space. Auto-Associative Composite Models (AACM) are proposed to approximate this set of points. Those new models are interesting for two main reasons. On the one hand, they take advantage of efficient Projection Pursuit Regression algorithms adapted to the auto-associative case. On the other hand, they extend the Principal Component Analysis (PCA) theoretical properties to the non-linear case and offer better approximation and generalization results.

§1. Problem Statement

The goal of this study is to build a model of a flexible curve: when several samples of this curve are observed, differences can appear. Figure 1a shows an example of such curve observations when the maximum location is variable. We want to build a curve model able to take into account such deformations.

A training set of several samples of the flexible curve is used as the starting point of the modeling [11]. Let us call N the number of samples. The curve numbered j is described by the sampled values of a function g^j on n sites $\{s_1 \ldots s_n\}$, and therefore is represented by the n-dimensional vector

$$x^j =^t \left(g^j(s_1) \ldots g^j(s_n) \right).$$

The training set of curves is interpreted as a cloud $\mathcal{X}$ of N-vectors, each vector of the cloud being in $\mathbb{R}^n$. Our approach consists in designing a model of this vector set by considering the implicit equation

$$G(\beta, x) = 0,$$

where $G(\beta, .)$ is a function from $\mathbb{R}^n$ to $\mathbb{R}^n$, β is a parameter set that will be described later, and x is an n-dimensional vector standing for the generic curve. This model is said to be *auto-associative* as opposed to regression-like

Curves and Surfaces with Applications in CAGD 135
A. Le Méhauté, C. Rabut, and L. L. Schumaker (eds.), pp. 135–142.
Copyright © 1997 by Vanderbilt University Press, Nashville, TN.
ISBN 0-8265-1293-3.
All rights of reproduction in any form reserved.

models $y = f(\beta, x) + \varepsilon$. We focus on models of which equation defines a manifold. In this case, we call the *dimension of the model* the dimension of the underlying manifold. The model is required to have two properties :

Approximation property.

On average, the training set points x^j have to be close to the manifold points $\{x | G(\beta, x) = 0\}$. This can be controlled using the residual error

$$\varepsilon^2(G, \mathcal{X}) = \sum_{j=1}^{N} \|G(\beta, x^j)\|^2.$$

We shall write $r^j(d) = G(\beta, x^j)$ for the residual error after approximating x^j by the d-dimensional model.

Generalization property.

This is the "dual" property. Every point x generated by the model has to be close to at least one point of the training set. The cross-validation criterion is a good measure of this property :

$$\mathcal{V}(G, \mathcal{X}) = \sum_{j=1}^{N} \|G^{[j]}(\beta, x^j)\|^2,$$

$G^{[j]}$ being the model estimated on the training set $\mathcal{X} \setminus \{x^j\}$. Let us note that the error $\varepsilon^2(G, \mathcal{X})$ is globally decreasing when the model dimension, d, is increasing, whereas for high d values the model has poor generalization properties. In this way, choosing the model dimension is a compromise. To make the choice easier, we propose in the next section a composite model which is built iteratively. At each iteration, the model dimension is incremented until the model has both good approximation and generalization properties.

§2. Auto-Associative Composite Models

The considered models are defined by a function $G(\beta, .)$ given by the *composition* of d functions

$$G(\beta, .) = \prod_{k=d}^{1} G^k(\beta^k, .).$$

Note that the composition is accomplished by first applying G^1, then G^2, etc. This explains the order of the composition product. It will be shown that, under some hypothesis, the model dimension is exactly d, the number of terms entering in the composition product. Each of these terms $G^k(\beta^k, .)$ is a function from $\mathbb{R}^n$ to $\mathbb{R}^n$. Because of the "curse of dimensionality" [10], the design of such a function remains difficult: in high dimensional spaces, the number of data is usually not sufficient to build multivariate functions. To overcome this problem, the vector set is modeled only in some directions. The direction's choice is then an important problem known as Projection Pursuit [6,10].

A *projection index* $I(\mathcal{X}, a)$ is defined to measure the quality of any direction a. Its maximization provides the best direction. Two supplementary operators are introduced :

- P^a projection on the a axis, $P^a : \mathbb{R}^n \to \mathbb{R}$,
- S^α smoothing function with parameters α, $S^\alpha : \mathbb{R} \to \mathbb{R}^n$.

A linear projection is chosen $P^a(x) = \langle x, a \rangle$ which allows to derive theoretical results on the model properties (see [8] for an example of non-linear projection). The smoothing function choice is more difficult as it determines the model's class (linear/non-linear ...) and will be described in Section 4. Each term of the composition product is defined by

$$G^k(\beta^k, .) = Id_{\mathbb{R}^n} - S^{\alpha^k} P^{a^k}.$$

Associating the P^{a^k} and S^{α^k} functions creates a bottleneck in the same way as in the perceptrons design [9]. This technique allows to compress a n-dimensional vector x^j into a scalar value $\langle a^k, x^j \rangle$, and so to solve the curse of dimensionality problem. Indeed, S^{α^k} is a univariate function, also called *restoration function*, because of its ability to rebuild a $\mathbb{R}^n$ vector from a compressed one. For instance, in the PCA case, the restoration function is a retro-projection. Finally, $G^k(\beta^k, x^j)$ computes the difference between a vector x^j and its projection-restoration $S^{\alpha^k} P^{a^k}(x^j)$.

We can now give the Auto-Associative Composite Model definition.

Definition 1. *A model is called* composite *if G can be written as*

$$G(\beta, x) = \left(\prod_{k=d}^{1} (Id_{\mathbb{R}^n} - S^{\alpha^k} P^{a^k}) \right)(x),$$

with $\forall k \in \{1 \ldots d\}$:

- P^{a^k} *projection on the axis $a^k \in \mathbb{R}^n$;*
- S^{α^k} *a restoration function with parameters α^k.*

The set of parameters $\beta = \{(\alpha^k, a^k), k = 1 \ldots d\}$ contains both the projection axis and the restoration function parameters.

We shall define an iterative algorithm to build the model. Let us consider first the 1-dimensional model which will initialize the algorithm : $x - S^{\alpha^1} P^{a^1} x = 0$.

Lemma 1. *Provided the projection of the restoration function is the identity function*

$$P^{a^1} S^{\alpha^1} = Id_{\mathbb{R}}, \tag{1}$$

the following results hold:

- *the model represents a 1-dimensional manifold ;*
- *each point residual is orthogonal to the projection axis ;*

$$\langle a^1, r^j(1) \rangle = 0, \quad \forall j \in \{1 \ldots N\}.$$

Proof: The first result's proof is technical, see [7] for details. The second one is easily verified by scalar products:

$$\langle a^1, r^j(1) \rangle = \langle a^1, x^j \rangle - \langle a^1, S^{\alpha^1} P^{a^1} x^j \rangle = \left(Id_{\mathbf{R}} - P^{a^1} S^{\alpha^1} \right) P^{a^1} x^j = 0,$$

using property (1). ∎

Assuming condition (1) means that the projection has to be the left-inverse of the restoration : a scalar restoration-projection should be the same scalar value. This natural constraint allows to iterate the model design. Once the 1-dimensional model is built $x - S^{\alpha^1} P^{a^1} x = 0$, the second step is to model the residuals $r^j(1) = G(\beta^1, x^j)$ with the equation $r(1) - S^{\alpha^2} P^{a^2} r(1) = 0$. According to Lemma 1, those residuals are located in the a^1 orthogonal space, S^{α^2} and a^2 can then also be chosen in this subspace. The algorithm is the following :

Algorithm.
  1) *Initialization.*

    $d \leftarrow 0$
    $r^j(0) \leftarrow x^j$ *for* $j = 1 \ldots N$
  2) *Computation of the $d + 1$th parameters :*
  - *Computation of the axis a^{d+1}.*

$$\begin{cases} a^{d+1} = arg \max_a I(r(d), a) \\ \langle a^{d+1}, a^k \rangle = 0 \; for \; 1 \leq k \leq d \end{cases}$$

  - *Computation of the smoothing parameters α^{d+1}.*

$$\begin{cases} \alpha^{d+1} = arg \min_{\alpha} \sum_{j=1}^{N} \left\| r^j(d) - S^{\alpha} P^{a^{d+1}} r^j(d) \right\|^2 \\ P^{a^k} S^{\alpha^{d+1}} = 0 \; for \; 1 \leq k \leq d \\ P^{a^{d+1}} S^{\alpha^{d+1}} = Id_{\mathbf{R}} \end{cases}$$

  3) *Residuals update.*
    $r^j(d + 1) \leftarrow r^j(d) - S^{\alpha^{d+1}} P^{a^{d+1}} r^j(d)$ *for* $j = 1 \ldots N$
    $d \leftarrow d + 1$
  4) *Go to 2) "if necessary" (see theorem 2)*

Theorem 1. *The d-terms model built by the algorithm can be written as*

$$\left(\prod_{k=d}^{1} (Id_{\mathbf{R}^n} - S^{\alpha^k} P^{a^k}) \right) (x) = 0,$$

with the following conditions :

  - $\forall k \in \{1 \ldots d\} \; P^{a^k} S^{\alpha^k} = Id_{\mathbf{R}}$;
  - $\forall k \in \{1 \ldots d\} \; P^{a^l} S^{\alpha^k} = 0$ for $l < k$;
  - $\forall k, l \in \{1 \ldots d\} \; \langle a^k, a^l \rangle = \delta_{kl}$.

§3. Model Properties

First we give three approximation properties of the model, and then we present results for comparing this model to classical data analysis approaches.

Theorem 2. *Auto-Associative Composite Models have the following properties :*

1) *The d-terms model represents a d-dimensional manifold.*
2) *The residual error is decreasing when d is increasing.*
3) *With $d = n$, the model is exact.*

In other words, the approximation properties are improved at each iteration (assertion 2) by building a higher dimensional manifold (assertion 1). Moreover, the algorithm provides the exact solution, that is to say all the residuals are zero, in a finite number of steps (assertion 3). Proofs can be found in [7]. It is interesting to note that the AACM manifold approximation properties generalizes the PCA linear subspace approximation properties.

Auto-Associative Composite Models can be compared to the regression additives models as well. For instance in Projection Pursuit Regression (PPR) the following regression model is considered [3]: $y = \sum_{k=1}^{d} f^k \left(\langle a^k, x \rangle \right) + \varepsilon$. The regression function is described by a superimposition of univariate functions thanks to a similar algorithm [5]. Let us give the Auto-Associative Additive Model definition.

Definition 2. *A model is called additive if G can be written as*

$$G(\beta, x) = \left(Id_{\mathbf{R}^n} - \sum_{k=1}^{d} S^{\alpha^k} P^{a^k} \right) (x).$$

It is quite easy to show that additive models are a particular case of composite models. The converse proposition is more interesting: the only composite model which is additive is also linear.

Theorem 3. *An additive model built with the previous algorithm is necessarily linear and (assuming μ is the mass center of the vector cloud) is written as*

$$G(\beta, x) = x - \mu - \sum_{k=1}^{d} \langle a^k, x - \mu \rangle a^k.$$

This is the model built by a PCA.

It appears that using additive PPR models in auto-associative mode demands to choose linear restoration functions, and so to use the well-known PCA. Yet, the PCA linear model is not adapted to represent the wide class of vector set resulting from curves deformation. An example will be shown in Section 5. We propose to give up the additive property and to use a composite model instead. Such a model combines the PCA-like approximation properties with an efficient implementation similar to PPR algorithms. The implementation details are discussed in the next section.

§4. Model Implementation

To implement the algorithm, we need to specify its second step by describing the restoration function and projection index choices.

The kind of restoration functions determines the model nature. For instance, it has been seen with Theorem 3 that choosing linear restoration functions leads to a PCA. We propose to use spline restoration functions which have shown their performances in regression frameworks [4]. The number of nodes to use is determined by cross-validation [12] so as to obtain a good balance between approximation and generalization. Let us emphasize that even if the obtained model is nonlinear, the estimation of the spline parameters α in the algorithm's second step is done by solving a linear system.

This estimation can also be interpreted as a way to build a restoration function S^α such that $S^\alpha P^a \simeq Id_{\mathbf{R}}$. This can be related to the previous condition $P^a S^\alpha = Id_{\mathbf{R}^n}$, and shows the link between the model approximation performances and the projection characteristics. The best case happens when P^a is injective because this allows us to find smoothing parameters satisfying exactly $S^\alpha P^a = Id_{\mathbf{R}}$. In most cases, it is not possible to find an axis a leading to an injective projection P^a. The goal of the projection index $I(a, \mathcal{X})$ is then to furnish axes minimizing the number of superimpositions while projecting the training set. Superimpositions are detected by looking for points in the training set whose closest neighbour is not preserved after projection :

$$I(\mathcal{X}, a) = \sum_{i=1}^{N} \sum_{j \neq i} \left[\left(x^j \ closest \ to \ x^i \right) \Rightarrow \left(P^a(x^j) \ closest \ to \ P^a(x^i) \right) \right].$$

See [7] for its analytical expression. The maximization of the index is difficult. The non-continuity with respect to a prevents the use of steepest descent technics. Only a Simulated Annealing algorithm [1] ensures reaching the global maximum of the index.

§5. Results

The modeling method is applied to the set of curves with variable maximum location (Figure 1a.). Each curve represents a function $s \rightarrow g_t(s)$ sampled on a fixed subdivision $(s_i)_{1 \leq i \leq 50}$. A training set of $N = 100$ curves is obtained by generating t^j values uniformly distributed. This training set is interpretated as a 100 vector cloud in a 50-dimensional space :

$$\mathcal{X} = \left\{ x^j \in \mathbb{R}^n, x_i^j = g_{t^j}(s_i) \right\}.$$

By considering the projection on the principal plane, computed by PCA, we can see in Figure 1d that the points are distributed on a 1-dimensional manifold. This is not surprising as long as the vector set is generated using a 1-parameter deformation, the maximum location.

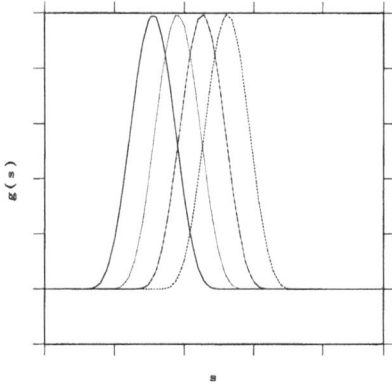

(a) A part of the training set

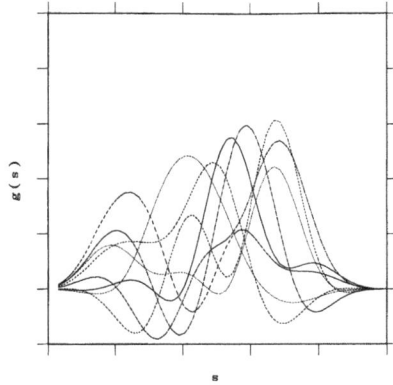

(b) PCA simulations

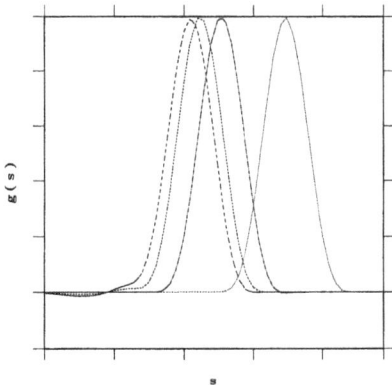

(c) AACM simulations

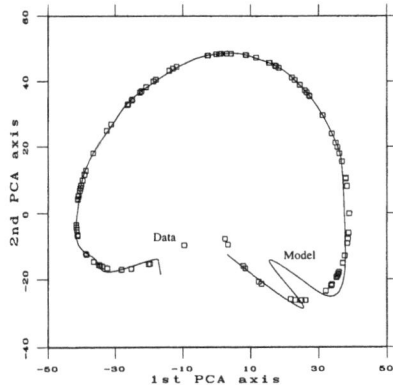

(c) Principal plane projections

Fig. 1. Flexible curves modeling.

PCA models are ill-adapted: the vector set has not a linear structure, and so, to get small residuals a 5-dimensional model has to be considered. The approximation property is then obtained at the expense of the generalization property. As the model dimension is larger than the data intrinsic dimension, unrealistic simulations are obtained (Figure 1b.). Thus, the Auto-Associative Composite Model is used to overcome the linear model limitations. Using a 1-dimensional model with 36 nodes allows us to obtain both interesting approximation and generalization properties. This can be seen in Figure 1d. where the training set points are close to the manifold, and in Figure 1c. where the model simulations look like the training set curves.

References

1. Azencott, R., Simulated annealing, Séminaire BOURBAKI 40ème année **697** (1988), 1–15.

2. Chalmond, B., S. Girard, and J. M. Dinten, Parametric Non-linear Auto-associative Models for Data Representation by a Manifold, submitted.

3. Diaconis, P., and M. ShahShahani, On nonlinear functions of linear combinations, *SIAM Journal of Scientifical Stat. Comput.* **5(1)** (1984), 175–191.

4. Eubank, R. L., *Spline Smoothing and Non-Parametric Regression*, Decker, 1990.

5. Friedman, J. H., W. Stuetzle, Projection Pursuit Regression, *JASA* **76(376)** (1981), 817–823.

6. Friedman, J. H., Exploratory Projection Pursuit, *JASA* **82(397)** (1987), 249–266.

7. Girard, S., *Design and Statistical Learning of Non linear Auto-Associative Models*, PhD thesis (in french), Cergy-Pontoise University, France, 1996.

8. Hastie, T., and W. Stuetzle, Principal Curves, *JASA* **84(406)** (1989), 502–516.

9. Hertz, J., A. Krogh, and K. Palmer, *The Theory of Neural Computation*, Addison Wesley, 1991.

10. Hubert, P. J., Projection Pursuit, *The Annals of Statistics* **13(2)** (1985), 435–525.

11. Rice, J. A., and B. W. Silverman, Estimating the mean and covariance structure nonparametrically when the data are curves, *JRSS B* **53(1)** (1991), 233–243.

12. Stone, M., Cross-validatory choice and assessment of statistical prediction, *JRSS B* **36** (1974), 111–147.

S. Girard, J.M. Dinten
LETI (CEA - Technologies Avancées) DSYS - CEN/G
17 avenue des Martyrs, 38054 Grenoble Cedex 9, FRANCE
girard@dsys.ceng.cea.fr

B. Chalmond
Cergy-Pontoise University and ENS Cachan (DIAM-CMLA)
2, av. A. Chauvin, 95302 Cergy Cedex, FRANCE
Bernard.Chalmond@cmla.ens-cachan.fr

Sk–spline Interpolation on the Torus using Number Theoretic Knots

Sonia M. Gomes, Alexander K. Kushpel,
Jeremy Levesley, and David L. Ragozin

Abstract. For a fixed, continuous, periodic kernel k, an sk–spline is a function of the form $sk(x) = c_0 + \sum_{j=1}^{n} c_j K(x - x_j)$, where $\sum_{j=1}^{n} c_j = 0$. In this paper we examine the convergence of sk–spline interpolation on the d–dimensional torus for $d \geq 2$, where K comes from the associated Sobolev class. An important feature of the interpolation process described here is the number theoretic ideas underpinning the choice of interpolation knots.

§1. Introduction

In this paper we continue to develop the methods of Kushpel [8] for analysing the error in interpolation when using sk–splines, which were extended in [10] to study interpolation on compact abelian groups using sk–splines. We also apply the results to the special case of the d–dimensional torus $\mathbb{T}^d$. Given a kernel $K \in \mathbf{C}(\mathbb{T}^d)$, and a set of n distinct knots $\mathbf{w}_1, \ldots, \mathbf{w}_n$, an *sk-spline* is a function of the form

$$sk(\mathbf{x}) = c_0 + \sum_{i=1}^{n} c_i K(\mathbf{x} - \mathbf{w}_i),$$

where the coefficients $c_1, \ldots, c_n$ satisfy the condition

$$\sum_{i=1}^{n} c_i = 0.$$

Such functions are natural generalisations of periodic polynomial splines for which the kernel K is an appropriate Bernoulli monospline. They have proved useful in interpolation and in computing n–widths of sets of smooth functions;

Curves and Surfaces with Applications in CAGD
A. Le Méhauté, C. Rabut, and L. L. Schumaker (eds.), pp. 143–149.
Copyright © 1997 by Vanderbilt University Press, Nashville, TN.
ISBN 0-8265-1293-3.
All rights of reproduction in any form reserved.

see [6,7], wherein the reader can find more information on univariate sk–splines. An overview of results concerning sk–splines may be found in [9].

The desire here and in [10] is to approximate functions in the Sobolev class

$$W_p^{\mathbf{r}} = \left\{ \phi : \|\phi^{(\mathbf{r})}\|_p = \left(\int_{\mathbb{T}^d} |\phi^{(\mathbf{r})}|^p \right)^{\frac{1}{p}} \leq 1 \right\},$$

where $\mathbf{r} = (r_1, r_2, \ldots, r_d) \in \mathbb{R}^d$, $1 < \alpha = r_1 = \cdots = r_\nu < r_{\nu+1} \leq \cdots \leq r_d$, is a multiindex. We will use the fact that

$$W_p^{\mathbf{r}} = \mathbb{R} \oplus K * U_p = \{c + K * \phi, \ c \in \mathbb{R}, \ \phi \in U_p\},$$

U_p being the unit ball of L_p,

$$K(\mathbf{x}) = \prod_{k=1}^d \sum_{z_k \in \mathbb{Z}\backslash\{0\}} |z_k|^{-r_k} e^{-iz_k x_k} = \sum_{\mathbf{z} \in \mathbb{Z}^d} k_{\mathbf{z}} e^{-i\mathbf{z}\mathbf{x}},$$

and

$$k_{\mathbf{z}} = \begin{cases} |z_1|^{-r_1} \ldots |z_d|^{-r_d}, & \text{if } z_i \neq 0 \text{ for any } 1 \leq i \leq d, \\ 0, & \text{otherwise.} \end{cases}$$

In this article we wish to adopt a different construction of interpolants to that in [10], in which cardinal interpolants are generated by taking a tensor product of univariate cardinal interpolants. If the sk–spline interpolant to a function f on an n–point knot set Δ_n is written $\mathrm{sk}(f, \Delta_n)$ then the approach of [10] yielded, in the special case of Sobolev classes, the result

$$\sup_{f \in W_p^{\mathbf{r}}} \|f - \mathrm{sk}(f, \Delta_n)\|_q \ll n^{(-\alpha + p^{-1} - q^{-1})/d},$$

for $1 \leq p \leq 2 \leq q \leq \infty$ with $p^{-1} - q^{-1} \geq 1/2$.

However, from the results of, for example [1], we might expect an error estimate more of the form

$$\sup_{f \in W_p^{\mathbf{r}}} \|f - \mathrm{sk}(f, \Delta_n)\|_q \ll n^{-\alpha + (p^{-1} - q^{-1})_+} (\log n)^\beta, \tag{1}$$

for some $\beta \in \mathbb{R}$, for an interpolant at n points.

In this paper we describe an interpolant which employs the same kernel K, as before, but in this case interpolates at knots which have almost minimal discrepency in the sense described by Kuipers and Niederreiter [5].

The paper will be organised as follows: in Section 2 we will describe the interpolation points we will use, and give an exact representation of the cardinal sk–splines for this knot set. In Section 3 we prove that the sk–spline at this knot set satisfies an error estimate of the form given in (1) above.

§2. Cardinal Sk–splines

Let $\mathbb{T}^d$ be $[0, 2\pi)^d$, and for a fixed prime number Q, let $\mathbf{g} = (g_1, g_2, \ldots, g_d) \in [0, 2\pi)^d$, with $g_k = 2\pi m_k/Q$, for some $1 \leq m_k \leq Q - 1$, $1 \leq k \leq d$. Then, for every such $\mathbf{g}$, $Q\mathbf{g} = \mathbf{0}$. Let $\Delta_Q^{\mathbf{g}} = \{0, \mathbf{g}, 2\mathbf{g}, \ldots, (Q-1)\mathbf{g}\}$. Using Theorem 11 of [10], we know that $\mathrm{sk}(f, \Delta_Q^{\mathbf{g}})$, the sk–spline interpolant to f with kernel K exists and is unique, and in particular, we infer the existence of the cardinal sk–spline

$$\widetilde{\mathrm{sk}}_Q^{\mathbf{g}}(\mathbf{x}) = \begin{cases} 1, & \mathbf{x} = \mathbf{0}, \\ 0, & \mathbf{x} = m\mathbf{g}, \ 1 \leq m \leq Q - 1. \end{cases}$$

Proposition 1. *Let*

$$\rho_j(\mathbf{x}) = \sum_{l=1}^{Q} K(\mathbf{x} + l\mathbf{g}) e^{-\frac{2\pi i l j}{Q}}.$$

Then

$$\widetilde{\mathrm{sk}}_Q^{\mathbf{g}}(\mathbf{x}) = \frac{1}{Q} \left\{ 1 + \sum_{j=1}^{Q-1} \frac{\rho_j(\mathbf{x})}{\rho_j(\mathbf{0})} \right\}.$$

Remark 2. *The cardinal sk–spline, which is 1 at $l\mathbf{g}$ and 0 at the other knots is simply the shift $\widetilde{\mathrm{sk}}_Q^{\mathbf{g}}(\cdot - l\mathbf{g})$, $0 \leq l \leq Q - 1$.*

For the analysis of Section 3 we require the following lemma, which bounds the error when interpolating complex exponentials using sk–splines. A proof of this may be found in [2].

Lemma 3. *Let $\bar{k}_1 = \bar{k}_1^1 \ldots \bar{k}_1^d$, $\bar{k}_1^s = \max\{1, |k_1^s|\}$, where all equivalences in this paper are modulo Q. † Then*

$$\left| e^{i\mathbf{z}\mathbf{x}} - \sum_{l=1}^{Q} e^{il\mathbf{z}\mathbf{g}} \widetilde{\mathrm{sk}}_Q^{\mathbf{g}}(\mathbf{x} - l\mathbf{g}) \right| \leq 2 \min \left\{ 1, \left(\sum_{\mathbf{m}\mathbf{g} \equiv \mathbf{0}}^{*} \bar{k}_{\mathbf{z}+\mathbf{m}} \right) / \bar{k}_{\mathbf{z}} \right\}.$$

Proposition 1, Remark 2, and Lemma 3 can be easily proved by direct calculation.

† Here and in what follows we shall use the symbol $\sum^{*}$ to indicate a sum that does not include $\mathbf{0}$.

§3. Error Estimates.

Theorem 4. *There is a* $\mathbf{g}^*$, *as decribed above, such that*

$$\delta_Q(W_p^{\mathbf{r}}, L_\infty) = \sup_{f \in W_p^{\mathbf{r}}} \|f - \mathrm{sk}(f, \Delta_Q^{\mathbf{g}^*})\|_\infty \ll Q^{-\alpha + \frac{1}{p}} (\log Q)^{\alpha\nu + \frac{d-1}{p}}; \quad Q \to \infty.$$

Proof: For fixed $\mathbf{x} \in \mathbb{T}^d$, and any fixed $\mathbf{g}$,

$$|f(\mathbf{x}) - [\mathrm{sk}(f, \Delta_Q^{\mathbf{g}})](\mathbf{x})| = \left| c + (K * \phi)(\mathbf{x}) - \sum_{j=1}^{Q} [c + (K * \phi)(j\mathbf{g})] \widetilde{\mathrm{sk}}_Q^{\mathbf{g}}(\mathbf{x} - j\mathbf{g}) \right|$$

$$= \left| (K * \phi)(\mathbf{x}) - \sum_{j=1}^{Q} (K * \phi)(j\mathbf{g})] \widetilde{\mathrm{sk}}_Q^{\mathbf{g}}(\mathbf{x} - j\mathbf{g}) \right|,$$

for some $\phi \in U_p$, where we have used the fact that $\sum_{j=1}^{Q} \widetilde{\mathrm{sk}}_Q^{\mathbf{g}}(\mathbf{x} - j\mathbf{g}) = 1$ for every $\mathbf{x} \in \mathbb{T}^d$. Using Hölder's inequality we can bound

$$\left| K * \phi(\mathbf{x}) - \sum_{j=1}^{Q} [K * \phi(j\mathbf{g})] \widetilde{\mathrm{sk}}_Q^{\mathbf{g}}(\mathbf{x} - j\mathbf{g}) \right|$$

$$\leq \|\phi\|_p \left\| K(\cdot - x) - \sum_{j=1}^{Q} K(\cdot - j\mathbf{g}) \widetilde{\mathrm{sk}}_Q^{\mathbf{g}}(\mathbf{x} - j\mathbf{g}) \right\|_{p'} \leq \left(\sum_{\mathbf{z} \in \mathbb{Z}^d} c_{\mathbf{z}}^p(\mathbf{x}) \right)^{\frac{1}{p}},$$

$$\tag{2}$$

using the Haudorff–Young inequality for $1 \leq p \leq 2$ (see Zygmund [11]), where

$$c_{\mathbf{z}}(\mathbf{x}) = (2\pi)^{-d} \int_{\mathbb{T}^d} \left\{ K(\mathbf{y} - \mathbf{x}) - \sum_{j=1}^{Q} K(\mathbf{y} - j\mathbf{g}) \widetilde{\mathrm{sk}}_Q^{\mathbf{g}}(\mathbf{x} - j\mathbf{g}) \right\} e^{-i\mathbf{z}\mathbf{y}} \, d\mathbf{y}$$

$$= k_{\mathbf{z}} \left\{ e^{-i\mathbf{z}\mathbf{x}} - \sum_{j=1}^{Q} e^{-i\mathbf{z}j\mathbf{g}} \widetilde{\mathrm{sk}}_Q^{\mathbf{g}}(\mathbf{x} - j\mathbf{g}) \right\}.$$

$$\tag{3}$$

To estimate the right hand side of (2), we divide $\mathbb{Z}^d$ into two regions, $\Gamma = \{\mathbf{z} : \bar{z}_1 \ldots \bar{z}_d \leq [Q/4]\}$ and $\mathbb{Z}^d \setminus \Gamma$. We then bound

$$\left(\sum_{\mathbf{z} \in \mathbb{Z}^d \setminus \Gamma} c_{\mathbf{z}}^p(\mathbf{x}) \right)^{\frac{1}{p}} \leq \left(\sum_{\mathbf{z} \in \mathbb{Z}^d \setminus \Gamma} k_{\mathbf{z}}^p \right)^{\frac{1}{p}}$$

$$\leq 2 \left(\sum_{\mathbf{z} \in \mathbb{Z}^d \setminus \Gamma} \frac{1}{|z_1 z_2 \ldots z_d|^{\alpha p}} \right)^{\frac{1}{p}} \tag{4}$$

$$\ll Q^{-\alpha + \frac{1}{p}} (\log Q)^{\frac{d-1}{p}},$$

using Lemma 3 and Lemma 7.7 of [3]. Using Lemma 3 again we have, for any $\mathbf{x} \in T^d$,

$$\left(\sum_{\mathbf{z} \in \Gamma} c_{\mathbf{z}}^p(\mathbf{x}) \right)^{1/p} \leq 2 \left(\sum_{\mathbf{z} \in \Gamma} \left(\sum_{\mathbf{mg} \equiv 0}^{*} \overline{k}_{\mathbf{z}+\mathbf{m}} \right)^p \right)^{1/p}$$

$$\leq 2 (\operatorname{card} \Gamma)^{1/p} \max_{\mathbf{z} \in \Gamma} \left(\sum_{\mathbf{mg} \equiv 0}^{*} \overline{k}_{\mathbf{z}+\mathbf{m}} \right),$$

and since by Lemma 7.7 of [3] $\operatorname{card} \Gamma \ll Q (\log Q)^{d-1}$, we need only show that for any $\mathbf{z} \in \Gamma$

$$\sigma_{\mathbf{z}} := \sum_{\mathbf{mg} \equiv 0}^{*} \overline{k}_{\mathbf{z}+\mathbf{m}} \ll Q^{-\alpha} (\log Q)^{\alpha\nu}. \tag{5}$$

Let us put $\mathbf{m} = \mathbf{n}Q + \mathbf{m}'$, where $\mathbf{m}' = (m_1', ..., m_d') \in \Psi_Q := \{\mathbf{m}' \in Z^d : -[(Q-1)/2] \leq m_s' \leq [Q/2], 1 \leq s \leq d\}$. Then, it is straightforward to show that, for some constant C,

$$\overline{k}_{\mathbf{z}+\mathbf{m}} \leq C \begin{cases} \overline{k}_{\mathbf{n}} \overline{k}_{\mathbf{m}'}, & \mathbf{n} \text{ and } \mathbf{m}' \neq \mathbf{0}, \\ \overline{k}_{[Q\mathbf{n}]/2}, & \mathbf{m}' = \mathbf{0}. \end{cases}$$

Since $\alpha > 1$,

$$\sigma_{\mathbf{z}} \ll \sum^{*} \overline{k}_{\mathbf{n}} \sum_{\mathbf{m}'\mathbf{g} \equiv 0, \, \mathbf{m}' \in \Psi_Q,}^{*} \overline{k}_{\mathbf{m}'} + \sum^{*} \overline{k}_{[Q\mathbf{n}]/2}, \tag{6}$$

and direct calculation shows that the second sum on the right is bounded by $CQ^{-\alpha}$ for some constant C. After summation over $\mathbf{n} \in \mathbb{Z}^d$ we see that

$$\sum^{*} \overline{k}_{\mathbf{n}} \sum_{\mathbf{m}'\mathbf{g} \equiv 0, \, \mathbf{m}' \in \Psi_Q,}^{*} \overline{k}_{\mathbf{m}'} \leq \sum_{\mathbf{m}'\mathbf{g} \equiv 0, \mathbf{m}' \in \Psi_Q}^{*} \overline{k}_{\mathbf{m}'}$$

$$\leq \left(\sum_{\mathbf{m}'\mathbf{g} \equiv 0, \mathbf{m}' \in \Psi_Q}^{*} (\overline{k}_{\mathbf{m}'})^{1/\alpha} \right)^{\alpha}, \tag{7}$$

due to Minkowski's inequality. Since $\mathbf{m}' \neq \mathbf{0}$ (see [5]) then for the average

$$S := \frac{1}{(Q-1)^d} \sum_{\mathbf{g} \in G} \sum_{\mathbf{m}'\mathbf{g} \equiv 0, \mathbf{m}' \in \Psi_Q}^{*} (\overline{k}_{\mathbf{m}'})^{1/\alpha},$$

where $G := \{\mathbf{g} \in \mathbb{Z}^d : 1 \leq g_s \leq Q - 1, \ 1 \leq s \leq d\}$, we have an estimate

$$S \leq Q^{-1} \sum_{\mathbf{m}' \in \Psi_Q}^{*} (\overline{k}_{\mathbf{m}'})^{1/\alpha},$$

because of Q is prime and , consequently, the maximum number of solutions of equation $\mathbf{m}'\mathbf{g} \equiv 0$ for any fixed $\mathbf{m}' \in \Psi_Q$ is $(Q-1)^{d-1}$ (see e.g. [5]). Thus, there is a $\mathbf{g}^* \in G$ such that

$$
\left(\sideset{}{^*}\sum_{\mathbf{m}'\mathbf{g}^* \equiv 0, \mathbf{m}' \in \Psi_Q} (\overline{k}_{\mathbf{m}'})^{1/\alpha} \right)^{\alpha} \leq \left(Q^{-1} \sideset{}{^*}\sum_{\mathbf{m}' \in \Psi_Q} (\overline{k}_{\mathbf{m}'})^{1/\alpha} \right)^{\alpha}
$$

$$
\leq \left(Q^{-1} \sum_{\mathbf{m}' \in \Psi_Q} \prod_{s=1}^{\nu} \overline{m'_s}^{-1} \prod_{s=\nu+1}^{d} \overline{m'_s}^{-r_s/\alpha} \right)^{\alpha} \ll Q^{-\alpha}(\log Q)^{\alpha\nu}. \qquad (8)
$$

Comparison of (6), (7) and (8) gives us (5). ∎

Using a duality argument we can show that also

$$
\sup_{f \in W_1^r} \|f - \mathrm{sk}(f, \Delta_Q^{\mathbf{g}^*})\|_{p'} \ll Q^{-\alpha + \frac{1}{p}}(\log Q)^{\alpha\nu + \frac{d-1}{p}}; \quad Q \to \infty.
$$

Application of the Riesz–Thorin interpolation theorem (see [11]) yields the following result:

Corollary 5. *Let $1 \leq p \leq 2 \leq q \leq \infty$, with $p^{-1} - q^{-1} \geq 1/2$. Then, for any $f \in W_p^r$,*

$$
\sup_{f \in W_p^r} \|f - \mathrm{sk}(f, \Delta_Q^{\mathbf{g}^*})\|_q \ll Q^{-\alpha + p^{-1} - q^{-1}}(\log Q)^{\alpha\nu + (d-1)(\frac{1}{p} - \frac{1}{q})}; \quad Q \to \infty.
$$

Remark 6.

1) *The order of the logarithmic term in the estimate of Theorem 5, and hence Corollary 6, can be decreased by a different choice of the region Γ in the proof of Theorem 5. A description of a more appropriate Γ is given in [2].*

2) *The vector $\mathbf{g}^*$ can be constructed in practice using Fibonacci numbers; see Korobov [4].*

References

1. E. M. Galeev, Approximation fo Fourier sums of sets of functions with bounded derivatives, Mat. Zametki **22** (1978), 197–211.

2. S. M. Gomes, A. K. Kushpel, J. Levesley, and D. L. Ragozin, sk-spline interpolation using grids with almost minimal discrepency, in preparation.

3. H. L. Keng, and W. Yuan, *Applications of Number Theory to Numerical Analysis*, Springer Verlag, Beijing, 1981.

4. N. M. Korobov, *Trigonometric Sums and Their Applications*, Nauka, Moscow, 1989.

5. L. Kuipers and H. Niederreiter, *Uniform Distribution of Sequences*, Wiley Interscience, New York, 1974.

6. A. K. Kushpel, Sharp estimates of widths of convolution classes, Izvestia Acad. Nauk SSSR **52** (1988), 1315–1332.

7. A. K. Kushpel, sk–splines and exact estimates of n–widths of functional classes in the space $C_{2\pi}$, Preprint 85.51, Inst. Math. Acad. Sci. of Ukraine SSR (1985).

8. A. K. Kushpel, Rate of convergence of then sk–spline interpolant on classes of convolutions, Inst. Math. Acad. Nauk Ukrain SSR (1987), 50–58.

9. A. K. Kushpel, J. Levesley, and W. A. Light, Approximation of smooth functions by sk–splines, Technical Report 1996/27 of the Department of Mathematics and Computer Science, University of Leicester.

10. Levesley, J. and Kushpel, A. K., Generalised sk–spline interpolation on compact abelian groups, Technical Report 1995/1, Department of Mathematics and Computer Science, University of Leicester.

11. A. Zygmund, *Trigonometric Series II*, CUP, 1959.

David L. Ragozin
University of Washington
Department of Mathematics
Seattle, WA 98195-4350
rag@math.washington.edu

Jeremy Levesley and Alexander K. Kushpel
University of Leicester
Department of Mathematics and Computer Science
Leicester LE1 7RH, ENGLAND
jl1@mcs.le.ac.uk
ak99@mcs.le.ac.uk

Sonia M. Gomes
Department of Applied Mathematics
State University of Campinas
Caixa Postal 6065
13081-970 Campinas SP, BRAZIL
soniag@ime.unicamp.br

Shape Preserving Interpolation
by G^2 Curves in Three Dimensions

T. N. T. Goodman and B. H. Ong

Abstract. In [4] we have considered some shape preserving criteria for curve interpolation in space with convexity, inflections, torsion, collinearity and coplanarity, and have derived a local scheme for generating shape preserving curves interpolating given data in space. However, this previous scheme does not produce interpolating curves with continuous variation of their osculating planes across the data points. An almost identical set of shape preserving criteria is considered here, and a local scheme which generates G^2 shape preserving interpolating curves in space is described. In particular, the osculating planes of these interpolating curves vary continuously along the curves.

§1. Introduction

In [4] we have considered some criteria for shape preserving curve interpolation in space, similar to those considered in [6], and constructed a local interpolating scheme based upon these criteria which generates curves which preserve the convexity, inflections, signs of torsion, coplanarity and collinearity displayed by the given data. The curves are unit tangent continuous, and although they are also continuous in curvature magnitude when no collinear data are involved, their osculating planes in general do not vary continuously across the data points. Here we consider an almost identical set of criteria for shape preserving curve interpolation, and construct with the piecewise rational cubics a local scheme which satisfies these criteria. The curve generated is G^2 when no collinear data are involved; in particular, its osculating plane varies continuously along the curve.

The shape preserving criteria are listed in the next section. In Section 3 we define the vector M_i which is normal to the osculating plane at data point I_i and depends on two positive parameters ℓ_i and r_i. The general construction of an arbitrary segment of the interpolating curve to a data set which does not contain four or more consecutive coplanar data points is also given in

Curves and Surfaces with Applications in CAGD
A. Le Méhauté, C. Rabut, and L. L. Schumaker (eds.), pp. 151–158.
Copyright © 1997 by Vanderbilt University Press, Nashville, TN.
ISBN 0-8265-1293-3.
All rights of reproduction in any form reserved.

the same section. The conditions on the positive parameters ℓ_i and r_i to ensure that the criteria for torsion, convexity and inflections are satisfied are discussed in Section 4, and the values of ℓ_i and r_i are determined locally and numerically in Section 5. When there are four or more consecutive coplanar data points, planar segment will be constructed. We shall see in Section 6 that this causes the non-planar segment which is just before or just after it to be given a different construction which replaces a single rational cubic segment with two. We conclude with some numerical examples. As space is limited, we shall skip details, omit the treatment for the end points of open curves, and refer the reader to [5] for more details and examples.

§2. Shape Preserving Criteria

$R(u)$ denotes the curve interpolating data $\{I_i \in \mathbb{R}^3 : 1 \leq i \leq n\}$ with $R(u_i) = I_i$, where $u_1 < u_2 < \ldots < u_n$. Let $\tau(u)$ be the torsion of $R(u)$, $\kappa(u) = \frac{R'(u) \times R''(u)}{|R'(u)|^{3/2}}$, $L_i = I_{i+1} - I_i$, $N_i = L_{i-1} \times L_i$ and $\Delta_i = [L_{i-1}, L_i, L_{i+1}]$, where $[\cdot, \cdot, \cdot]$ is a scalar triple product. A plane normal to vector N is denoted by S_N and the orthogonal projection onto S_N by P_N. For completeness, the shape preserving criteria are listed in full below.

(i) If $N_i \cdot N_{i+1} > 0$, then $\forall N = \lambda N_i + \mu N_{i+1}$, with $\lambda \geq 0$ and $\mu \geq 0$ and $N \neq 0$, $P_N R(u)$, $u \in [u_i, u_{i+1}]$, is convex and $\kappa(u) \cdot N > 0$.

(ii) If $N_i \cdot N_{i+1} < 0$, then $\kappa(u_i) \cdot N_i > 0$, $\kappa(u_{i+1}) \cdot N_{i+1} > 0$ and the number of sign changes of $\kappa(u) \cdot N$ in $[u_i, u_{i+1}]$ is one, $\forall N = \lambda N_i + \mu N_{i+1}$ where $\lambda, \mu \in \mathbb{R}$, with $\lambda\mu \leq 0$ and $N \neq 0$.

(iii) If $N_i \cdot N_{i+1} = 0$, then $\kappa(u) = 0$, $\forall u \in (u_i, u_{i+1})$.

(iv) If $\Delta_j \neq 0$ for $i - 1 \leq j \leq i + 1$, then $\tau(u)\Delta_i > 0$, $\forall u \in (u_i, u_{i+1})$.

(v) If $\Delta_i = 0$, then $\tau(u) = 0$, $\forall u \in (u_i, u_{i+1})$.

(vi) If $|N_i| = 0$, then $\kappa(u) = 0$, $\forall u \in (u_{i-1}, u_{i+1})$.

§3. Generation of the Piecewise Rational Cubic Curve

We shall first consider a given set of data which does not contain four or more consecutive coplanar data points. The scheme is local, so it suffices just to describe the generation of one curve segment. Each pair of consecutive data points I_i and I_{i+1} are joined by a rational cubic of the form

$$R_i(t) = \frac{I_i\alpha_i(1-t)^3 + B_i t(1-t)^2 + C_i t^2(1-t) + I_{i+1}\beta_i t^3}{\alpha_i(1-t)^3 + t(1-t)^2 + t^2(1-t) + \beta_i t^3}, \quad 0 \leq t \leq 1,$$

$$(3.1)$$

where the weights $\alpha_i, \beta_i > 0$ and $B_i, C_i \in \mathbb{R}^3$. When three or more consecutive data points are collinear, they are joined by straight line segments instead of rational cubics, thus fulfilling criterion (vi). However, the curvature is then discontinuous at the knot where a linear segment joins a non-linear segment.

The osculating plane at I_i is defined by the vector M_i which is normal to the osculating plane and given by

$$M_i = N_i - \ell_i L_{i-1} \times L_{i+1} - r_i L_{i-2} \times L_i, \quad \text{for some } \ell_i, r_i > 0.$$

The ℓ_i and r_i are the parameters used to fulfill the convexity, inflections and torsion criteria. The conditions imposed upon them are discussed in the next section. The definition for M_i is motivated by the observation that $M_i \cdot L_i$ and $M_{i+1} \cdot L_i$ have the same sign as Δ_i, that is $M_i \cdot L_i = \ell_i \Delta_i$ and $M_{i+1} \cdot L_i = r_{i+1} \Delta_i$. In [4] the tangent direction at each I_i is given by the vector T_i^* defined on the plane S_{N_i} as

$$T_i^* = a_i L_{i-1} + b_i L_i, \tag{3.2}$$

where $a_i = |N_i \cdot \kappa_{i+1}^*||L_i|^2$, $b_i = |N_i \cdot \kappa_{i-1}^*||L_i|^2$, $\kappa_i^* = \frac{2 N_i}{|L_{i-1}||L_i||I_{i+1} - I_{i-1}|}$. The tangent direction T_i to the curve at I_i is defined as the vector which when projected onto the plane S_{N_i} yields T_i^*. Hence $T_i = T_i^* - k_i N_i$, where

$$k_i = \frac{T_i^* \cdot M_i}{N_i \cdot M_i} = \frac{a_i r_i \Delta_{i-1} + b_i \ell_i \Delta_i}{|N_i|^2 - \ell_i[N_i, L_{i-1}, L_{i+1}] - r_i[N_i, L_{i-2}, L_i]}. \tag{3.3}$$

The curvature $|\kappa(u_i)|$ at I_i is defined as $|\kappa_i^*|$ which is the curvature of the circle passing through the data points I_{i-1}, I_i and I_{i+1}.

By G^2 continuity we obtain

$$B_i = I_i + s_i T_i, \quad C_i = I_{i+1} - t_i T_{i+1}, \tag{3.4}$$

where $s_i = (L_i \cdot M_{i+1})/(T_i \cdot M_{i+1})$, $t_i = (L_i \cdot M_i)/(T_{i+1} \cdot M_i)$. The weights α_i and β_i are then fixed by the curvature magnitudes. The torsion of a rational cubic segment has a constant sign which is the same as that of its Bézier polygon (see [1]). Thus the torsion of curve segment i has the same sign as that of τ_i, where

$$\begin{aligned} \tau_i &= [B_i - I_i, \ L_i, \ I_{i+1} - C_i] \\ &= ((a_i b_{i+1} - k_i k_{i+1}|L_i|^2)\Delta_i - (a_i k_{i+1} + b_{i+1} k_i)N_i \cdot N_{i+1}). \end{aligned} \tag{3.5}$$

When the torsion criterion (iv) is satisfied, the signs of $[B_i - I_i, C_i - B_i, L_i]$ and $M_i \cdot L_i$ are equal and as I_i, B_i and C_i lie on plane S_{M_i}, hence M_i is in the direction of $(B_i - I_i) \times (C_i - B_i)$ which is the direction of the binormal at I_i. Then $T_i \cdot M_{i+1}$ and $T_{i+1} \cdot M_i$ are positive multiples of Δ_i, thus ensuring s_i and t_i in (3.4) are positive.

§4. Conditions for Torsion, Convexity and Inflections

To satisfy the torsion criterion (iv), the parameters ℓ_j and r_j for $j = i, i + 1$ should be chosen so that the sign of τ_i in (3.5) is the same as that of Δ_i. Moreover, they also have to satisfy the conditions from the convexity and inflections criteria below.

Suppose that $\Delta_i \neq 0$. Then the ith rational cubic curve segment is not planar. Consider the case when $N_i \cdot N_{i+1} \neq 0$. We require the binormals and the discrete binormals at the same data point to be less than an angle of $\frac{\pi}{2}$ apart, that is

$$M_i \cdot N_i > 0, \quad M_{i+1} \cdot N_{i+1} > 0. \tag{4.1}$$

In addition, if $N_i \cdot N_{i+1} > 0$, we would like to have

$$M_i \cdot N_{i+1} > 0, \quad M_{i+1} \cdot N_i > 0, \tag{4.2}$$

and if $N_i \cdot N_{i+1} < 0$, then we would require

$$M_i \cdot N_{i+1} < 0, \quad M_{i+1} \cdot N_i < 0. \tag{4.3}$$

When $N_i \cdot N_{i+1} > 0$, then $\forall N = \lambda N_i + \mu N_{i+1}$, with $\lambda \geq 0$ and $\mu \geq 0$ and $N \neq 0$, the torsion criterion and the conditions in (4.1) and (4.2) are sufficient to ensure that the curve segment between I_i and I_{i+1}, when projected onto the plane S_N, is convex.

When $N_i \cdot N_{i+1} < 0$, by the torsion criterion and conditions (4.1) and (4.3) the projection of $I_i B_i C_i I_{i+1}$ onto S_N, $N = \lambda N_i + \mu N_{i+1}, \lambda\mu \leq 0, N \neq 0$, has one inflection. Hence the projection of the curve segment onto S_N also has one inflection (see [3]) and criterion (ii) is satisfied.

If $N_i \cdot N_{i+1} = 0$, the scheme generates a linear segment between I_i and I_{i+1}. This is precisely the limiting curve as $N_i \cdot N_{i+1} \to 0$. As $N_i \cdot N_{i+1} \to 0$, by the choice of ℓ_j and $r_j, j = i, i+1$ as described in the next section, $\ell_j \to 0$ and $r_j \to 0$ and thus $M_i \to N_i$ and $M_{i+1} \to N_{i+1}$ which in turn cause s_i and t_i in (3.4) to tend to zero, thus the curve segment tends to a linear segment.

§5. Effect and Determination of Parameters ℓ_i and r_i

When $\ell_j, r_j, j = i, i+1$, are very small, the Bézier polygonal arcs $B_i - I_i$ and $I_{i+1} - C_i$ are very short and thus the curve segment is tight. As ℓ_j and r_j increase, the "tightness" diminishes and the curve becomes "loose". But when the ℓ_j and r_j are increased further, the curve begins to "tighten up" again. This is so because when ℓ_j and r_j are large, the weights α_i and β_i of the curve segment are large and thus $R_i(t)$ in (3.1) tends to a linear segment.

We shall now show how the criteria in (4.1), (4.2) and (4.3) are satisfied by some numerical approach on the choice of $\ell_j, r_j, j = i, i+1$. ℓ_i and r_i are found in the expressions $N_j \cdot M_i = N_j \cdot N_i - \ell_i[N_j, L_{i-1}, L_{i+1}] - r_i[N_j, L_{i-2}, L_i], j = i-1, i, i+1$. To fulfill (4.1)–(4.3), it suffices to have $\ell_i < \frac{N_j \cdot N_i}{2[N_j, L_{i-1}, L_{i+1}]}$ and $r_i < \frac{N_j \cdot N_i}{2[N_j, L_{i-2}, L_i]}, j = i-1, i, i+1$, where if the expression on the right of the inequality is negative, that inequality is considered void. If at least one of the $\frac{N_j \cdot N_i}{2[N_j, L_{i-1}, L_{i+1}]}, j = i-1, i, i+1$, is positive, we define ℓ_{i1} to be the minimum of the positive ones from the three; if otherwise ℓ_{i1} would be left undefined since ℓ_i is unconstrained. The value r_{i1} for r_i is defined similarly. When the values ℓ_{i1} and r_{i1} are defined, let $\lambda_i = \frac{r_{i1}}{\ell_{i1}}$, otherwise let $\lambda_i = 1$.

There is a range of possible values for ℓ_i and r_i which satisfy the torsion, convexity and inflections conditions. We suggest using the values obtained in the following way as an estimate to the values to be chosen.

We use as a guide the lengths derived in our previous scheme [4] to be the desirable lengths for the projections of the Bézier polygonal arcs $B_i - I_i$ and $I_{i+1} - C_i$. That is, the desirable length ℓ_{ab} for $|B_i^* - I_i|$, B_i^* being the projection of B_i onto S_{N_i}, is

$$\frac{2\sin b\,|L_i|}{2\gamma k\sin b|\cos\psi| + (1-\gamma)k|L_i||\kappa_{i+1}^*||\cos\psi| + 2k\left|\frac{\sin a\cos b}{\cos\psi}\right| + 2|\cos a\sin b|} \tag{5.1}$$

where $0 < \gamma < 1$, a and b are the smaller angles between the tangent T_i and L_i, L_i and tangent T_{i+1} respectively, and ψ the angle between N_i and N_{i+1}. For $N_i \cdot N_{i+1} > 0$, $k = 1$, but when $N_i \cdot N_{i+1} < 0, k \geq 1$ so that the length suggested by (5.1) is shorter as we would like the curve segment for the inflection case to be less "round". We shall use ℓ_{ab} to obtain the approximate corresponding value of ℓ_{i+1} and hence also r_{i+1}, by using the ratio λ_{i+1}. By $(3.4),(3.3)$ and approximating T_i as T_i^*, we obtain

$$|B_i^* - I_i| \approx \left|\frac{\ell_{i+1}\lambda_{i+1}}{a_i + \ell_{i+1}p_i}\right|, \tag{5.2}$$

where $p_i = b_i\lambda_{i+1} - a_i\frac{[L_{i-1},L_i,L_{i+2}]}{\Delta_i}$.

For $|B_i^* - I_i| = \ell_{ab}$, the corresponding value h_{i+1} of ℓ_{i+1} is approximated from (5.2) above. By using an expression, similar to (5.1) and denoted by ℓ_{cd}, as the desired length for $I_{i+1} - C_i^*$, where C_i^* is the projection of C_i onto plane $S_{N_{i+1}}$, the approximate value h_i of ℓ_i is determined similarly. At each data point I_i, there are two values of h_i suggested by the above procedure, one from each of the segment having I_i as an end point. The average of these two suggested values is denoted by ℓ_{i2} and r_{i2} is defined as $\ell_{i2} \times \lambda_i$.

Now we shall take the starting values for the parameter ℓ_i to be the minimum of ℓ_{i1} and ℓ_{i2} and for r_i the minimum of r_{i1} and r_{i2}. Recall that we have earlier mentioned that ℓ_{i1} and r_{i1} might have been left undefined. When this happens, the starting values for ℓ_i and r_i are just ℓ_{i2} and r_{i2}.

We shall need to test for each segment i whether $\ell_j, r_j, j = i, i+1$ satisfy the torsion criterion. If the torsion criterion is not satisfied, then their values are decreased but the ratio λ_j is always kept constant. The test function used for the torsion criterion will ensure that, when for some values of these four parameters the torsion criterion for segment i is satisfied, then decreasing these parameters further while keeping the ratio λ_j fixed, the decreased values also satisfy the torsion criterion. This property is necessary since at the common end point I_i of two adjacent segments, ℓ_i and r_i have to satisfy the torsion criterion for both segments. Thus after each segment has been processed to satisfy the torsion criterion, the final values for ℓ_i and r_i at I_i will each be the smaller of the two corresponding values.

The sign of the torsion for the *ith* curve segment is the same as that of τ_i in (3.5). The torsion criterion is tested by a test function which is modified from τ_i. This is done as follows. With the initial values of $\ell_j, r_j, j = i, i+1$, the signs of k_i and k_{i+1} can be computed so that if any of the last three terms in (3.5), namely, $-|L_i|^2 k_i k_{i+1}\Delta_i$, $-a_i k_{i+1} N_i \cdot N_{i+1}$ and $-b_{i+1} k_i N_i \cdot N_{i+1}$, is of the same sign as Δ_i, then that term is removed from (3.5) and we shall refer to the resulting expression consisting of the remaining terms as τ_i^*. Observe that if ℓ_j and r_j decrease but with the ratio λ_j fixed, then k_j decreases in magnitude. Hence, when the initial values of ℓ_j and r_j are substituted into τ_i^*, if τ_i^* has the same sign as Δ_i, then the torsion criterion is satisfied, otherwise ℓ_j and r_j are reduced till τ_i^* has the same sign as Δ_i. With the initial values of ℓ_j and r_j as upper bounds and zeros as lower bounds, the process of reduction is carried out by using the method of bisection. We observe that once ℓ_j and r_j satisfy the torsion criterion, then when these values are decreased but with their ratio fixed the new values will still satisfy the torsion criterion.

§6. Planar, Pre-planar and Post-planar Segments

When four consecutive data I_{i-1}, I_i, I_{i+1} and I_{i+2} are coplanar, $\Delta_i = 0$ and so a planar segment is constructed between I_i and I_{i+1} to fulfill criterion (v). We may assume that $N_i \neq 0$ and $N_{i+1} \neq 0$ for otherwise the segment between I_i and I_{i+1} is just linear. This planar segment lies in the plane of the four planar data which is S_{N_i} or $S_{N_{i+1}}$, thus M_i and M_{i+1} at I_i and I_{i+1} have to be defined as N_i and N_{i+1} respectively. It is generated by using the two dimensional rational cubic scheme in [2] or the scheme in [4].

The above coincidence of the binormal with the discrete binormal causes problem to the non-planar segments just before or just after the planar segment which, for convenience, shall be referred to respectively as pre-planar and post-planar segments. If the construction of these segments proceeds as described in Section 3, then it is clear from (3.4) that each of them will have one of its inner Bézier points coincide with a data point. To avoid this undesirable situation, an alternative construction is used whereby two rational cubics instead of one are generated between the two consecutive data points concerned. We shall just describe the construction of a post-planar segment between two consecutive data points I_i and I_{i+1} as a similar construction holds for a pre-planar segment. There are two cases, $N_i \cdot N_{i+1} > 0$ and $N_i \cdot N_{i+1} < 0$ and here we only describe briefly the former.

The construction is motivated by the de Casteljau's subdivision though it is only a very rough imitation of it. The polygon $AB^*\overline{C}G$ (Fig. 1) is generated where $A = I_i$ and $G = I_{i+1}$, B^* lies on plane S_{N_i} with $B^* - I_i$ in the direction of T_i^* in (3.2) and $\overline{C}$ on $S_{M_{i+1}}$ with $I_{i+1} - \overline{C}$ in the direction of T_{i+1} in (3.3). The lengths $|B^* - I_i|$ and $|I_{i+1} - \overline{C}|$ are, whenever possible, given by (5.1) with N_{i+1} being replaced by M_{i+1}. An approximate "subdivision" is applied to this polygon to yield two sub Bézier polygons $ABCD$ and $DEFG$. Let P be the point of intersection of the line through A and B^* with plane $S_{M_{i+1}}$. The line Γ through P and G lies on planes S_{N_i} and $S_{M_{i+1}}$. Let Q be the

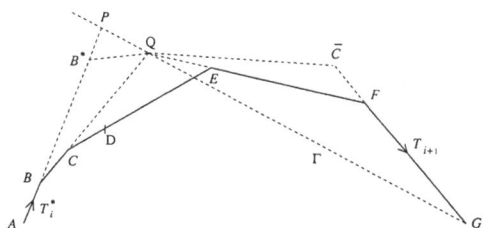

Fig. 1. Construction of post-planar segment when $N_i \cdot N_{i+1} > 0$.

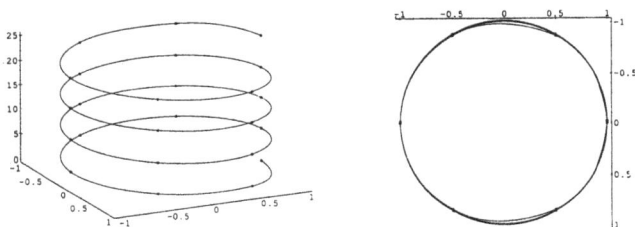

Fig. 2. Two views of the interpolating curve for Example 1.

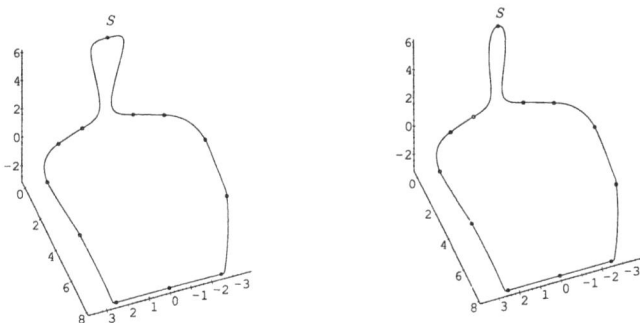

Fig. 3. (a) Interpolating curve, (b) Modified interpolating curve for Example 2.

average of the points of intersection of segment PG with the projections of segment $B^*\overline{C}$ onto S_{N_i} and $S_{M_{i+1}}$ respectively. Q approximately "subdivides" $B^*\overline{C}$ in the ratio $s : 1 - s$, where $s = |Q - B^*|/(|Q - B^*| + |Q - \overline{C}|)$. By completing the subdivision process, points B, F, C, E and D are obtained which divide respectively the line segments $AB^*, \overline{C}G, BQ, QF$ and CE in the ratio $s : 1 - s$, thus producing two sub control polygons $ABCD$ and $DEFG$. Let the curvature magnitude at D be $(1 - s)|\kappa_i^*| + s|\kappa_{i+1}^*|$, then the two sets of weights for the two rational cubic sub segments with $ABCD$ and $DEFG$ as their Bézier polygons are determined.

This construction, which is always feasible by taking ℓ_{i+1} sufficiently small, ensures G^2 continuity. The torsion of the sub segment which has $DEFG$ as its Bézier polygon has the same sign as Δ_i. The convexity of the projection of the composite segment formed by the two rational cubics can be achieved when ℓ_{i+1} and r_{i+1} are sufficiently small.

§7. Numerical Results

The data of Example 1 (Fig. 2) come from a helix $r(\theta) = (\sin\theta, \cos\theta, \theta)$ sampled at an interval of $\frac{\pi}{3}$ in θ. In this example, in order to produce a curve which is close to the initial helix (except for the first two and last two segments), we scale down the reference values ℓ_{ab} and ℓ_{cd} suggested by (5.1) by a factor of 0.85. Though the scheme does not guarantee continuity of the torsion across the data points, the torsion of this interpolating curve is continuous along the curve except at the first two and the last two data points.

Example 2 is a closed curve with 12 data points obtained from [6]. There are two subsets of coplanar data points, namely $\{I_i \,|\, 2 \leq i \leq 5\}$ and $\{I_i \,|\, 9 \leq i \leq 12\}$. Thus segments 3 and 10 are planar while segments 2 and 8, 4 and 11 are respectively pre-planar and post- planar, each with two pieces of rational cubics. The interpolating curve preserves the symmetry of the data. At S the initial default curvature magnitude assigned is very small and the interpolating curve appears "flat" in the neighbourhood of S (Fig. 3a). But when this curvature magnitude is increased by a factor of 20, the curve turns more sharply at S (Fig. 3b). Thus the curvature assigned at a data point could be used as a parameter to alter the shape of the curve locally. Segments 6 and 7 are linear which agree with the collinearity of data points $I_j, 6 \leq j \leq 8$.

References

1. M. S. Floater, Derivatives of rational Bézier curves, Comput. Aided Geom. Design **9** (1992), 161–174.

2. Goodman, T. N. T., Shape preserving interpolation by parametric rational cubic splines, in *International Series of Numerical Mathematics* **86**, Birkhäuser Verlag Basel, 1988, 149–158.

3. Goodman, T. N. T., Inflections on curves in two and three dimensions, Comput. Aided Geom. Design **8** (1991), 37–50.

4. Goodman, T. N. T. and Ong, B. H., Shape preserving interpolation by space curves, submitted for publication and also available as Report AA/963, Univ. of Dundee, 1996.

5. Goodman, T. N. T. and Ong, B. H., Shape preserving interpolation by G^2 space curves, Report AA/964, Univ. of Dundee, 1996.

6. Kaklis, P. D. and Karavelas, M. T., Shape preserving interpolation in $\mathbb{R}^3$, preprint, 1995.

Fairing Bicubic B-Spline Surfaces using Simulated Annealing

Stefanie Hahmann and Stefan Konz

Abstract. In this paper we present an automatic fairing algorithm for bicubic B-spline surfaces. The fairing method consists of a knot removal and knot reinsertion step which locally smoothes the surface. The simulated-annealing search strategy is used to search for the global minimum of the fairing measure.

§1. Introduction

Free form surfaces are an indispensable part of powerful CAD-systems. Bicubic tensor product B-splines, which are the subject of this paper, are often used in geometric modelling due to their well known advantages which result basically from the local support of the basis splines and their dependence on the knot vector. However, the designer is not always satisfied with the fairness or smoothness of the resulting surface obtained from interpolation or approximation of some data sets. A fair surface can be obtained by two different ways:

- modelling surfaces with constraints (see e.g. [10,2,16,7,18,8]),
- post-processing surface fairing (see e.g. [4,14,9]).

In the present paper we introduce a new *post-processing* fairing method for bicubic tensor product B-spline surfaces based on the knot-removal reinsertion step [6] combined with a search strategy, called *simulated annealing*. This method will be local in contrast to Kjellander's method for bicubic surfaces [14]. The principal reason for incorporating a search strategy into the fairing process is to overcome the drawback of other iterative algorithms which stop when the fairing measure achieves its first local minimum.

Section 2 will recall an algorithm of Farin/Sapidis for curves [20]. A fairing algorithm for B-spline surfaces based on a special search strategy is presented in Section 3. The heuristic *simulated-annealing algorithm*, depending on a probability function and some other parameters searches for a *global minimum* of the fairness measure. It results in an optimal surface with respect to its fairness. Some examples are shown at the end.

Curves and Surfaces with Applications in CAGD   159
A. Le Méhauté, C. Rabut, and L. L. Schumaker (eds.), pp. 159–168.
Copyright © 1997 by Vanderbilt University Press, Nashville, TN.
ISBN 0-8265-1293-3.
All rights of reproduction in any form reserved.

§2. Knot Removal for B-spline Curves and Surfaces

Given the positive integers n, m and k, l and $\boldsymbol{u} = (u_i)_{i=0}^{n+k}$, $\boldsymbol{v} = (v_j)_{j=0}^{m+l}$ two sequences of real numbers with $u_i < u_{i+k}$ and $v_j < v_{j+l}$ $(i, j = 0, \ldots, n, m)$, the B-spline basis functions associated with the *knot vector* $\boldsymbol{u}$ are denoted by $(N_{i,k,u})_{i=0}^{n}$ (or simply by $N_{i,k}$) and are assumed to be normalized to sum to one. The same holds for $(N_{j,l,v})_{j=0}^{m}$. A parametric *tensor product B-spline surface* $\boldsymbol{X}$ in $\mathbb{R}^3$ of order (k, l) is then defined by

$$\boldsymbol{X}(u, v) = \sum_{i=0}^{n} \sum_{j=0}^{m} \boldsymbol{d}_{ij} N_{i,k}(u) N_{j,l}(v), \quad (u, v) \in [u_{k-1}, u_{n+1}] \times [v_{l-1}, v_{m+1}]$$

(1)

where the coefficients $\boldsymbol{d}_{ij} \in \mathbb{R}^3$ form the control net. Note that the B-spline $\boldsymbol{X}(t)$ is only defined over $[u_{k-1}, u_{n+1}] \times [v_{l-1}, v_{m+1}]$. $u_k, \ldots, u_n, v_l, \ldots, v_m$ are called *inner knots*. For more details about B-splines, see [3,21].

The knot removal step consists of approximating a given spline $x = \sum_0^n d_i N_{i,k,t}$ by a spline $\tilde{x} = \sum_0^{n-1} \tilde{d}_i N_{i,k,\tilde{t}}$ on a knot vector $\tilde{t}$ which is a subsequence of t with one knot less. While *knot insertion* doesn't change the shape of the curve, the "inverse" process of knot removal cannot be carried out in general without changing the shape, except the case where the curve is C^{k-1} at t_j, i.e. its continuity order is higher than it should be according to its multiplicity). In the future we will use knot removal for bicubic surfaces, therefore we restrict the following considerations to the cubic case.

Different possibilities to determine approximated solutions of the *knot removal problem* exist: From the approximation point of view, knot removal can be solved by $\min \|\tilde{x} - x\| : \{\tilde{x} \in S_{k,\tilde{t}}\}$ with respect to some appropriate norm $\|\cdot\|$. For more details see [17]. In general, all control points are involved in that knot removal procedure, which is therefore a global one.

We want our fairing step to work as local as possible. This means, that after reinsertion of the removed knot t_j, a minimum number of control points $\tilde{d}_i$ of the curve $\tilde{x}$ should differ from the original control points d_i. The knot removal problem can locally be solved by calculating an approximate solution of an over-determined (3,2)-system of linear equations. Detailed description of these knot removal algorithms can be found in [6,5].

Farin/Sapidis' fairing algorithm for cubic splines is based on the fact, that after a knot-removal-reinsertion step of the knot t_j the new curve $\tilde{x}(t)$ is C^3 at t_j. Hence, the *local fairness measure* $z_j = |\kappa'(t_j^+) - \kappa'(t_j^-)|$ decreases to zero. By repeating this procedure several times always at the most offending knot (i.e. t_k with $z_k = \max(z_i) : \{i \in (4, \ldots, n)\}$), they could expect, that the sum of the local measures

$$\xi = \sum_{i=4}^{n} z_i$$

(2)

decreases also. The algorithm stops if there is no more improvement of the fairness measure ξ. This is an algorithm according to the following fairness criterion:

(C) *A C^2-cubic B-spline curve is fair if the curvature function is continuous, has the appropriate sign, and is as close as possible to a piecewise monotone function with as few as possible monotone pieces [20].*

We want to close this section with the conclusion that a knot-removal-reinsertion step has a local fairing effect, and can therefore be used in a fairing method for the entire curve, or tensor product surface. The main problem is to find a sequence of knots such that repeated knot-removal-reinsertion at these knots leads to a (global) minimum of the global fairness measure ξ.

§3. Surface Fairing Method

The curve fairing method can easily be extended to tensor product surfaces. Of course, one can simply apply the curve scheme to each row and each column of control points of the surface control net. This procedure has two main disadvantages: the fairness measures are not adapted to the surface geometry and therefore don't justify the selection of a knot as the most offending one. And the "stop"-criterion for the algorithm doesn't guarantee to find the global minimum of the fairness measure ξ (2). The algorithm stops when the first local minimum of that function is reached. And this is probably not the lowest one. Now a surface fairing algorithm is presented which is based on knot-removal-reinsertion (which is a very fast fairing step) and which uses simulated annealing to *optimize the fairing process.*

When searching for a global minimum, one can do it by spanning a so called *search tree* (see Figure 1).

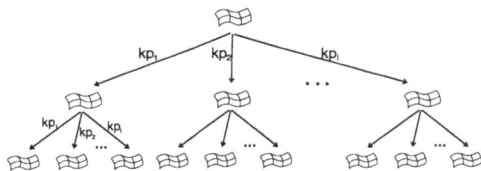

Figure 1. Search tree of the iterative fairing method.

The search tree has as its root the given surface to be faired. Each vertex of the tree represents a B-spline surface that results from a fairing step (knot removal and reinsertion) at one knot of the parent vertex. The edges of the tree are marked with the knot pair kp_i which is treated by the knot operations. The search tree lists systematically all possible steps of the fairing algorithm. Each path from the root to a leaf of the tree describes the history of one algorithm call.

At least one of the leaves represents a surface with a global minimum of the fairing measure. The global minimum (after k iterations) can only be found when performing an exponential number $O((n \cdot m)^k)$ of fairing steps. The aim of a search strategy is to find this path without constructing the whole tree because of the exponential number of vertices. The five main characteristics of our fairing method are now presented:

3.1. Knot operation

We use *Knot-Removal-Reinsertion* (KRR) as a local fairing step for the surface. To do so, the knot operations (Section 2) have to be adapted to the tensor product surface description. It's an inherent property of tensor products that a lot of algorithms on surfaces (e.g. de Boor algorithm, degree elevation, etc.) can be reduced to the univariate curve algorithms with respect to the two parameters u and v. Different strategies can be adopted for KRR of a B-spline surface:

- KKR in u-direction: *for $j = 4$ to m do KKR of u_i at (u_i, v_j),*
- KKR in v-direction: *for $i = 4$ to n do KKR of v_j at (u_i, v_j),*
- KRR in both directions, first in u and then in v direction, or the other way round.

Notice, that there is no symmetry in the third KRR-step. Changing u and v leads to different results, because the second step depends on the first.

3.2. Fairing measure

The *local measure* $(z_{ij}^u + z_{ij}^v)$ of a knot pair (u_i, v_j) from a surface $\boldsymbol{X}(u, v)$ which evaluates the fairness of the knot pair is defined by the partial derivation of a curvature function $g(u, v)$ (e.g. Gaussian curvature $\kappa_{min} \cdot \kappa_{max}$, or $\kappa_{min}^2 + \kappa_{max}^2$, etc.) along one of the two parameters of the surface:

$$z_{ij}^u = \frac{\left| \frac{\partial g}{\partial u}(u_i, v_j) \right|}{\left\| \frac{\partial X(u,v)}{\partial u} \right\|}, \qquad z_{ij}^v = \frac{\left| \frac{\partial g}{\partial v}(u_i, v_j) \right|}{\left\| \frac{\partial X(u,v)}{\partial v} \right\|}. \tag{3}$$

According to the selected curvature function the automatic fairing algorithm will evaluate a knot pair in a special manner.

Finally the sum of the local measures of all knot pairs will define the *global fairness measure* ξ which will be minimized by the fairing algorithm:

$$\xi = \sum_{i,j} z_{i,j}^u + z_{ij}^v. \tag{4}$$

Remark: One can look for the local measures z_{ij}^u, z_{ij}^v independent of each other, and apply KRR only in u or v-direction.

3.3. Visualization utility

An appropriate technique for emphasizing visually the fairing effect is a light reflection method. The fairing method reduces surface irregularities and provides a *more pleasing shape*. Those aesthetic aspects are well captured by isophotes, reflection line or highlight lines [19,15,1,11], because they are very sensitive to changes in the surface normals. We chose the isophote method.

3.4. Scheduler and termination condition

The selection of the knot pair and the search for the global minimum of the fairing measure can be done in the following way.

Simulated Annealing
Simulated annealing is a heuristic method to solve large optimization problems. It ensures *"with high probability"* to find a global minimum of the function of interest, where iterative methods often get stuck at a local minimum. In 1983, IBM researchers [12,13] found an interesting analogy in material physics, where the heuristic process of an optimal annealing of metal is a quit difficult task. That's where the algorithm got it's name.

Simulated annealing is a stepwise algorithm which is allowed to make *good steps* (i.e. steps which improve some measure of quality) and to make *bad steps* (those that do not). Those bad steps are necessary in order to leave a valley of a local minimum. The probability that bad steps are accepted decreases while the algorithm proceeds.

This algorithm can be compared with the *hill climbing* model, see Figure 2. Searching in this case the maximum of a function one can imagine that a hill climber is searching for the highest top. Depending of the departure he has to walk on tops (good steps) and valleys (bad steps) in order to reach the highest top. Therefore he must not stop walking when he arrives at the first top. In this case he would only find a local maximum. Sometimes the climber ought to take a descending path. This helps to evade from a local maximum. In general the topography of the problem isn't known in advance. Hence it is natural to use a probability for the acceptance of bad steps and in order to ensure that the algorithm stops.

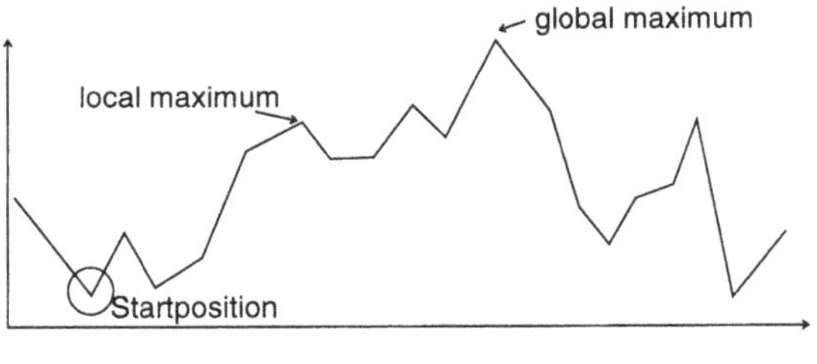

Figure 2. Topography of an optimization process.

In our fairing context, the measure of quality is the global fairness measure ξ (4), which has to be minimized by a sequence of fairing steps, the knot-removal-reinsertion (KRR)-steps. At each stage of the algorithm a knot pair (u_i, v_j) of the current surface is chosen either randomly or by a ranking list with respect to the local fairness measure $z_{ij}^u + z_{ij}^v$ (3) at each knot pair. While the KRR-step increases ξ, the steps are accepted as good steps. Otherwise, bad steps are accepted with a certain probability. As the probability to make bad steps gets smaller during the algorithm it gets less and less possible to leave the current valley of the fairness measure and the algorithm stops there.

Let *max_kp* be the number of inner knot pairs. If the number of stages of the algorithm is controlled by the integer *max_step*, the probability function P at each stage t_k $(k = 1, \ldots, max\_step)$ should be a monotonously decreasing function. Some examples are shown in Figure 3.

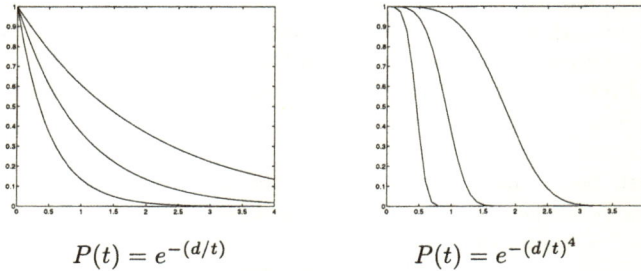

$$P(t) = e^{-(d/t)} \qquad\qquad\qquad P(t) = e^{-(d/t)^4}$$

Figure 3. Probability of accepting bad steps (d = 0.5, 1, 2).

<u>Algorithm:</u> *(Simulated annealing for fairing surfaces)*

for $k = 1 \ldots, end\_temp$ **do**
    do
        for $r = 1 \ldots, max\_step$ **do**
            choose randomly a knot pair *kp*
            Let $\tilde{\boldsymbol{X}} := KRR(\boldsymbol{X})$ be the faired surface after a KRR-step
            if $\xi_X > \xi_{\tilde{X}}$
            then $\boldsymbol{X} := \tilde{\boldsymbol{X}}$; b_count = 1   /* good step */
            else choose a random number $0 \le \nu \le 1$
                if $\nu < P(t_k)$
                then $\boldsymbol{X} := \tilde{\boldsymbol{X}}$; b_count = 1   /* bad step */
                else b_count += 1   /* bad step refused */
        end for
      until (b_count $\ge$ *blimit*)   /* no more improvement */
end for

The decreasing sequence of real numbers $t_1 > t_2 > \cdots > t_{max\_step}$ is called *annealing schedule* and controls the probability that bad steps will be done. The t_k correspond to temperatures in the annealing of metal. More they get smaller, less a bad step will be probable. Once the annealing schedule is fixed, for example as a linear schedule $t_k = \frac{max\_kp}{k}$, an appropriate factor d in the probability function P has to be chosen (see Figure 3).

At each stage k the number of good steps is limited by *max_step*, in order to avoid to make too many good steps at the beginning and to get stuck at a local minimum in a false valley. A general suggestion we can make for the choice of the parameter *max_step* is to take it half or less of *max_kp*.

There are two possibilities for the algorithm to stop. In general the algorithm should stop when the making of a bad step was refused *blimit* times in succession (due to the decreasing probability of making bad steps). In this case there is no more way out of the current valley. The actual minimum is accepted as the optimal one. When the algorithm seems to be in an infinite loop, it stops after a maximal number of stages *end_temp* (final temperature). This should be an emergency stop, and *end_temp* has to be chosen big enough to allow a regular stop.

The simulated annealing algorithm doesn't guarantee to find a global minimum, but it's well known that it's an effective way to search for one. This algorithm has already been used in CAGD by L. Schumaker [22] for computing optimal triangulations.

§4. Results

One fairing step consists of KRR applied at one knot pair. Under all KRR algorithms (Section 2) we favorite the

minimal region algorithm (mr-KRR),

where the distance of the old control net from the new one can be relatively high, because only one control point is changed to ensure the C^3-continuity (in u or v-direction) at the selected knot. This leads to big changes in the fairness measure but also in the shape of the surface, and the

least squares algorithm (ls-KRR),

which minimizes the squared distance between the old and new control net by moving three control points. Therefore the shape of the surface doesn't change very hard but the fairing effect isn't as high as with the first knot removal method.

The user has to decide which aspect he prefers: preservation of the surface shape or maximal fairing effect.

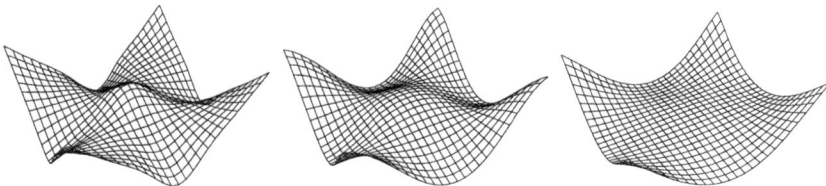

Figure 4. Fairing effect of the two different KRR-steps:
(a) *unfair surface* (b) *least-squares KRR* (c) *minimal-region KRR*

The first example illustrate the different fairing effects of these two knot-removal algorithms for surfaces, see Figure 4. The unfair surface (Fig. 4a) has only 25 inner knots. One expects from a fairing method that the high variations of curvature, which occur in the middle of the patch and on the boundary, become smaller or disappear. We then run the simulated annealing fairing (parameters: $g(u,v) := \kappa_{max}^2 + \kappa_{min}^2$ (3), $max\_step = blimit = 10$) once with the least-squares KRR (32 steps) (Figure 4b) and once with the minimal-region KRR (71 steps) (Figure 4c). As it was assumed above, the

ls-KRR is better adapted for shape preserving fairing, while a more flattening
fairing effect is achieved with the mr-KRR.

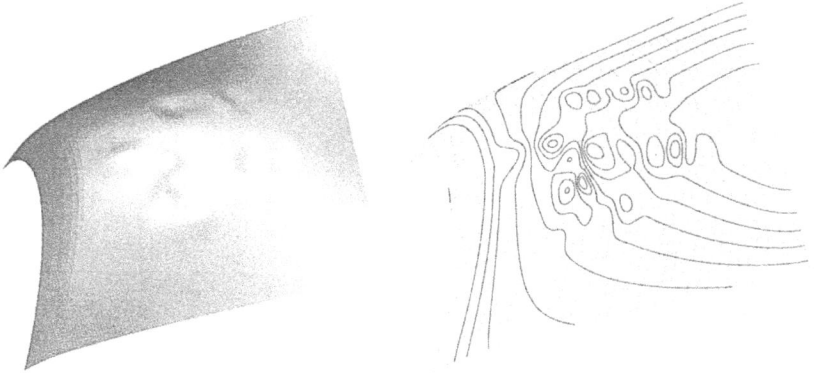

Figure 5. *Unfair surface with isophote analysis.*

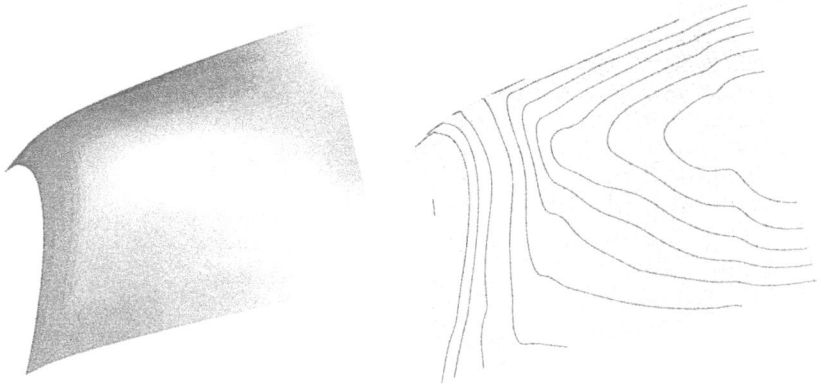

Figure 6. *Faired surface with isophote analysis.*

If all inner knots are involved in the fairing algorithm, only the four surface
corner control points are not modified. But it is also possible to apply the
fairing method to a subset of knots. This can be necessary if the boundary
curves and tangent planes are not allowed to move, because the surface needs
to fit into a whole patchwork of surfaces. Figures 5 and 6 show an example of
locally fairing a surface (i.e. the boundary curves will not be affected). The
given surface has 256 inner knots. The control points near the border are
kept fixed. The simulated annealing with least-squares fairing (ls-KRR) is
then applied to only 121 knot in the area, where the surfaces needs to be
faired (see the isophote analysis in Figure 5). It can be observed that several
runs of the simulated annealing algorithm always stops after $\sim$400 steps with

nearly the same result. i.e. a faired surface, as in Figure 6.

For comparison, notice that a systematic search for the global minimum of the fairness measure (after only 10 KRR-steps+ 10 search tree levels) would have needed 121^{10}!!! KRR-steps.

References

1. Beier, K.-P., Highlight-line algorithm for realtime surface quality assessment, Computer Aided Design **26** (1994), 268–277.

2. Bonneau, G.-P., and H. Hagen, Variational design of rational Bézier curves and surfaces, in *Curves and Surfaces in Geometric Design*, P.-J. Laurent, A. Le Méhauté, and L. L. Schumaker (eds.), A. K. Peters, Wellesley MA, 1994, 51–58.

3. De Boor, C., *A Practical Guide to Splines*, Springer, New York, 1978.

4. Brunet, P., Increasing the smoothness of bicubic spline surfaces, Computer Aided Geometric Design **2** (1985), 157–164.

5. Eck, M., and J. Hadenfeld, Knot removal for B-spline curves, Computer Aided Geometric Design **12** (1995), 259–282.

6. Farin, G., G. Rein, N. Sapidis, and A. J. Worsey, Fairing cubic B-spline curves, Computer Aided Geometric Design **4** (1987), 91–103.

7. Ferguson, D. R., P. D. Frank, and K. Jones, Surface shape control using constrained optimization on the B-spline representation, Computer Aided Geometric Design **5** (1988), 87–103.

8. Greiner, G., Variational design and fairing of spline surfaces, Proc. Eurographics 1994, 143–154.

9. Hadenfeld, J., Local energy fairing of B-spline surfaces, in *Mathematical Methods for Curves and Surfaces*, Morten Dæhlen, Tom Lyche, Larry L. Schumaker (eds.), Vanderbilt University Press, Nashville & London, 1995, 203–212.

10. Hagen, H., and P. Santarelli, Variational design of smooth B-spline surfaces, in *Topics in Surface Modelling*, H. Hagen (ed.), SIAM, 1992, 85–92.

11. Hahmann, S., Visualization techniques for surface analysis, to be published in *Data Visualization Techniques*, C. Bajaj (ed.), John Wiley, 1996.

12. Kirkpatrick S., C. D. Gelatt, M. P. Vecchi, Science **220** (1983), 671–677.

13. Kirkpatrick S., Optimization by simulated annealing, J. Stat. Phys. **34** (5/6), (1984), 975–987.

14. Kjellander, J. A., Smoothing of bicubic parametric surfaces, Computer Aided Design **15** (1983), 289–293.

15. Klass, R., Correction of local irregularities using reflection lines, Computer Aided Design **12** (1980), 73–77.

16. Lott, N. J., and D. I. Pullin, Method for fairing B-spline surfaces, Computer Aided Design **20** (1988), 597–604.

17. Lyche, T., and K. Morken, Knot removal for parametric B-spline curves and surfaces, Computer Aided Geometric Design **4** (1987), 217–230.

18. Moreton, H. P., and C. H. Séquin, Functional optimization for fair surface design, Comp. Graph. **26** (1992), 167–176.

19. Poeschl, T., Detecting surface irregularities using isophotes, Computer Aided Geometric Design **1** (1984), 163–168.

20. Sapidis, N., and G. Farin, Automatic fairing algorithm for B-spline curves, Computer Aided Design **22** (1990), 121–129.

21. Schumaker, L. L., *Spline Functions: Basic Theory*, Wiley, New York, 1981.

22. Schumaker, L. L., Computing optimal triangulations using simulated annealing, Computer Aided Geometric Design **10** (1993), 329–345.

Stefanie Hahmann
Laboratoire LMC-IMAG
Université INP de Grenoble
F-38041 Grenoble cedex 9, FRANCE
Stefanie.Hahmann@imag.fr

Stefan Konz
Universität Kaiserslautern
Fachbereich Informatik
D-67653 Kaiserslautern, GERMANY
ko@nemetschek.de

Positivity and Convexity Criteria for Bernstein-Bézier Polynomials Over Simplices

Tian-Xiao He

Abstract. This paper discusses linear criteria for convexity and positivity of Bernstein-Bézier polynomials over simplices.

§1. Introduction

We discuss linear criteria for convexity and positivity of Bernstein-Bézier polynomials of several variables over simplices. This section will give notations and lemmas. The main results on the convexity and positivity criteria of Bernstein-Bézier polynomials defined on an arbitrary simplex will be presented in Section 2. First, we give some notations and some properties about the basis functions of Bernstein-Bézier polynomials.

Let $\mathbf{x}^0, \ldots \mathbf{x}^s \in \mathbf{R}^s$, $s \geq 1$, $\mathbf{x}^i = (x_1^i, \ldots, x_s^i)$ and consider the convex hull $T_s := \langle \mathbf{x}^0, \ldots, \mathbf{x}^s \rangle = \{\sum_{i=0}^s \alpha_i \mathbf{x}^i : \sum_{i=0}^s \alpha_i = 1, \alpha_i \geq 0\}$. This convex hull is called an *s-simplex* if its signed volume is nonzero. Suppose that $\langle \mathbf{x}^0, \ldots, \mathbf{x}^s \rangle$ is an s-simplex. Then any $\mathbf{x} \in \mathbf{R}^s$ can be identified by an $(s+1)$-tuple $\lambda = (\lambda_0, \ldots, \lambda_s)$, the *barycentric coordinates of* $\mathbf{x}$ *relative to the s-simplex* $\langle \mathbf{x}^0, \ldots, \mathbf{x}^s \rangle$. Thus each $\lambda_i = \lambda_i(\mathbf{x})$ is a linear polynomial in $\mathbf{x}$ with $\sum_{i=0}^s \lambda_i = 1$, and if $\mathbf{x} \in \langle \mathbf{x}^0, \ldots, \mathbf{x}^s \rangle$, then $\lambda_i \geq 0$.

For any $\beta = (\beta_0, \ldots, \beta_s) \in \mathbf{Z}_+^{s+1}$, and $n \in \mathbf{Z}_+$, we will use the usual multivariate notation $\lambda^\beta = \lambda_0^{\beta_0} \cdots \lambda_s^{\beta_s}$, $\beta! = \beta_0! \cdots \beta_s!$, $|\beta| = \beta_0 + \cdots + \beta_s$. Hence,

$$\phi_\beta^n(\lambda) := \frac{n!}{\beta!} \lambda^\beta \qquad (1)$$

is a polynomial in $\pi_{|\beta|}^s$, the space of all polynomials in one variable of order $|\beta| + 1$, or degree at most $|\beta|$. With any set $\{a_\beta^n\} = \{a_\beta^n\}_{\beta \in \mathbf{Z}_+^{s+1}, |\beta|=n} \subset \mathbf{R}$ one may associate the polynomial

$$p_n(\mathbf{x}) = B_n[\{a_\beta^n\}; \lambda] = \sum_{|\beta|=n} a_\beta^n \phi_\beta^n(\lambda), \qquad (2)$$

Curves and Surfaces with Applications in CAGD
A. Le Méhauté, C. Rabut, and L. L. Schumaker (eds.), pp. 169–176.

Copyright ℗ 1997 by Vanderbilt University Press, Nashville, TN.
ISBN 0-8265-1293-3.
All rights of reproduction in any form reserved.

which is called a *Bernstein-Bézier polynomial of total degree n relative to the s-simplex* $\langle \mathbf{x}^0, \ldots, \mathbf{x}^s \rangle$. In addition, $\{a_\beta^n : |\beta| = n\}$ in (2) is called the set of Bézier coefficients of the polynomial p_n.

In the following, we will use the notation $D_{\mathbf{y}} = \sum_{i=1}^s y_i \frac{\partial}{\partial x_i}$, where $\mathbf{x} = (x_1, \ldots, x_s)$ and $\mathbf{y} = (y_1, \ldots, y_s)$. For $\mathbf{y} = \mathbf{x}^i - \mathbf{x}^j$, we denote $D_{ij} = D_{\mathbf{y}} = D_{\mathbf{x}^i - \mathbf{x}^j}$, $i \neq j$. By using the barycentric coordinates $\{\lambda_\ell\}_{\ell=0}^s$ of $\mathbf{x} \in \mathbf{R}^s$ relative to an s-simplex $T_s = \langle \mathbf{x}^0, \ldots, \mathbf{x}^s \rangle$, we can write $\mathbf{x} = \sum_{\ell=0}^s \lambda_\ell \mathbf{x}^\ell$. If we define that $E_i a_\alpha := a_{\alpha + e^i}$ and $\triangle_{ij} a_\alpha^n = E_i a_\alpha^n - E_j a_\alpha^n$, where $\mathbf{e}^i = (\delta_{ij})_{j=0}^s$ denotes the ith coordinate vector in $\mathbf{R}^{s+1}$, we have

$$D_{ij} p_n = n \sum_{|\alpha|=n-1} (E_i - E_j) a_\alpha^n \phi_\alpha^{n-1}(\lambda) = n \sum_{|\alpha|=n-1} \triangle_{ij} a_\alpha^n \phi_\alpha^{n-1}(\triangle). \qquad (3)$$

For any direction V, there exists a vector $\mathbf{c}_V = (c_1, c_2, \ldots, c_s)$ such that $V = \sum_{i=1}^s c_i(\mathbf{x}^i - \mathbf{x}^0)$. Thus, from (3), we have $D_V p_n(\mathbf{x}) = n \sum_{|\alpha|=n-1} \mathbf{c}_V^T \mathbf{b}_\alpha \phi_\alpha^{n-1}(\lambda)$ and

$$D_V^2 p_n(\mathbf{x}) = n(n-1) \sum_{|\alpha|=n-2} q_{\alpha,s}(\mathbf{c}_V) \phi_\alpha^{n-2}(\lambda), \qquad (4)$$

where $c_V = (c_1, \ldots, c_s)^T$, $\mathbf{b}_\alpha = (\triangle_{i0} a_\alpha^n)_{i=1}^s$, $q_{\alpha,s}(\mathbf{c}_V) = \mathbf{c}_V^T Q_{\alpha,T_s} \mathbf{c}_V$, and

$$Q_{\alpha,T_s} := (\triangle_{i0} \triangle_{j0} a_\alpha^n)_{i,j=1}^{s,s}, \qquad (5)$$

for $|\alpha| = n - 2$. Obviously, $p_n(\mathbf{x}) = B_n[\{a_\beta^n\}; \lambda]$ is convex on T_s if and only if $D_V^2 p_n(\mathbf{x}) \geq 0$ for any directional vector V and at any point $\mathbf{x} \in T_s$.

In order to discuss the convexity of Bernstein-Bézier polynomials over simplex T_s, we need the following lemmas.

Lemma 1. *For functions $\phi_\beta^n(\lambda)$ defined by (1), the following two inequalities hold:*

$$0 \leq \phi_\beta^n(\lambda) \leq \frac{(n-1)!}{\beta! n^{n-1}} (\sum_{i=0}^s \beta_i \lambda_i)^n, \qquad (6)$$

$$0 \leq \phi_\beta^n(\lambda) \leq \frac{(n-1)!}{\beta!} \sum_{i=0}^s \beta_i \lambda_i^n. \qquad (7)$$

Proof: (6) and (7) can be deduced from the mean inequalities

$$\prod_{k=1}^m a_k^{u_k} \leq (\sum_{k=1}^m u_k a_k / \sum_{k=1}^m u_k)^{\sum_{k=1}^m u_k} \qquad (8)$$

and

$$\prod_{k=1}^m a_k^{v_k} \leq \sum_{k=1}^m a_k v_k, \qquad (9)$$

respectively, where m is any positive integer, $a_k \geq 0$, $u_k > 0$, $v_k > 0$, and $\sum_{k=1}^{m} v_k = 1$ (cf. [6]). In fact, inequality (7) has been proved by Wang and Liu in [11] by using mean inequality (9). We now prove inequality (6) by using mean inequality (8), which is more fundamental than inequality (9). Noting that $\sum_{i=0}^{s} \beta_i = n$, from inequality (8), we obtain

$$\phi_\beta^n(\lambda) = \frac{n!}{\beta!} \prod_{i=0}^{s} \lambda_i^{\beta_i} \leq \frac{n!}{\beta!} \left(\frac{\sum_{i=0}^{s} \lambda_i \beta_i}{\sum_{i=0}^{s}} \right)^{\sum_{i=0}^{s} \beta_i}$$

$$= \frac{n!}{\beta!} \left(\frac{\sum_{i=0}^{s} \lambda_i \beta_i}{n} \right)^n = \frac{(n-1)!}{\beta! n^{n-1}} \left(\sum_{i=0}^{s} \lambda_i \beta_i \right)^n. \blacksquare$$

Remark 1. Obviously, the equal signs in (6) and (7) hold if and only if all the $\{\lambda_k\}_{k=0}^{s}$ are equal.

The second lemma is about the product of two Bernstein-Bézier polynomials, and can be proved by direct computation.

Lemma 2. *We have*

$$B_n[\{a_\alpha\}; \lambda] \cdot B_m[\{b_\beta\}; \lambda] = B_{n+m}[\{h_\gamma\}; \lambda], \tag{10}$$

where

$$h_\gamma = \sum_{|\alpha|=n} (a_\alpha b_{\gamma-\alpha}) \binom{\gamma}{\alpha} / \binom{n+m}{n},$$

$|\gamma| = n + m$, *and* $b_{\gamma-\alpha} = 0$ *for* $|\gamma - \alpha| > m$.

§2. Positivity and Convexity Criteria

The following positive criterion will also be used to give some convexity criteria later. Define $K = \{\mathbf{k} : |\mathbf{k}| = n, \mathbf{k} \neq n\mathbf{e}^0, \ldots, n\mathbf{e}^s, a_{\mathbf{k}} < 0\}$.

Theorem 1. *Let* $B_n[\{a_{\mathbf{k}}\}; \lambda]$ *be a Bernstein-Bézier polynomial defined on* T. *If its Bézier coefficients satisfy either*

$$a_{n\mathbf{e}^i} + (n-1)! \sum_{\mathbf{k} \in K} \frac{k_i}{\mathbf{k}!} a_{\mathbf{k}} \geq 0, \tag{11}$$

or

$$a_{n\mathbf{e}^i} + (n-1)! \sum_{\mathbf{k} \in K} (\frac{s+1}{n})^{n-1} \frac{k_i^n}{\mathbf{k}!} a_{\mathbf{k}} \geq 0, \tag{12}$$

for $i = 0, 1, \cdots, s$, *where* $\mathbf{e}^i$ *is the ith coordinate vector, then* $B_n[\{a_{\mathbf{k}}\}; \lambda] \geq 0$.

Proof: Condition (11) was shown by Dahmen in [5]. Its special case of dimension two was given earlier by Wang and Liu in [11]. From (7), we have

$$a_{\mathbf{k}} \phi_{\mathbf{k}}^n(\lambda) \geq a_{\mathbf{k}} \frac{(n-1)!}{\mathbf{k}!} \sum_{i=0}^{s} k_i \lambda_i^n$$

for any $a_\mathbf{k} < 0$. Thus,

$$B_n[\{a_\mathbf{k}\}; \lambda] = \sum_{|\mathbf{k}|=n} a_\mathbf{k} \phi_\mathbf{k}^n(\lambda) \geq \sum_{i=0}^{s} a_{n\mathbf{e}^i} \lambda_i^n + \sum_{\mathbf{k} \in K} a_\mathbf{k} \phi_\mathbf{k}^n(\lambda)$$

$$\geq \sum_{i=0}^{s} \left(a_{n\mathbf{e}^i} + \sum_{\mathbf{k} \in K} \frac{(n-1)! k_i}{\mathbf{k}!} a_\mathbf{k} \right) \lambda_i^n.$$

Hence, if (11) holds, then $B_n[\{a_\mathbf{k}\}; \lambda] \geq 0$.

We will now prove condition (12). From (6), we have

$$a_\mathbf{k} \phi_\mathbf{k}^n(\lambda) \geq a_\mathbf{k} \frac{(n-1)!}{\mathbf{k}! n^{n-1}} \left(\sum_{j=0}^{s} k_j \lambda_j \right)^n$$

for any $a_\mathbf{k} < 0$. By using the above inequality and inequality (11), we have

$$B_n[\{a_\mathbf{k}\}; \lambda] = \sum_{|\mathbf{k}|=n} a_\mathbf{k} \phi_\mathbf{k}^n(\lambda) \geq \sum_{i=0}^{s} a_{n\mathbf{e}^i} \lambda_i^n + \sum_{\mathbf{k} \in K} a_\mathbf{k} \phi_\mathbf{k}^n(\lambda)$$

$$\geq \sum_{i=0}^{s} a_{n\mathbf{e}^i} \lambda_i^n + \sum_{\mathbf{k} \in K} a_\mathbf{k} \frac{(n-1)!}{\mathbf{k}! n^{n-1}} \left(\sum_{i=0}^{s} \lambda_i k_i \right)^n$$

$$\geq \sum_{i=0}^{s} a_{n\mathbf{e}^i} \lambda_i^n + \sum_{\mathbf{k} \in K} a_\mathbf{k} \frac{(n-1)!}{\mathbf{k}! n^{n-1}} \sum_{|\beta|=n} \frac{n!}{\beta!} \prod_{i=0}^{s} (\lambda_i k_i)^{\beta_i}$$

$$\geq \sum_{i=0}^{s} a_{n\mathbf{e}^i} \lambda_i^n + \sum_{\mathbf{k} \in K} a_\mathbf{k} \frac{(n-1)!}{\mathbf{k}! n^{n-1}} \sum_{|\beta|=n} \frac{(n-1)!}{\beta!} \sum_{i=0}^{s} \beta_i (\lambda_i k_i)^n$$

$$\geq \sum_{i=0}^{s} \left(a_{n\mathbf{e}^i} \lambda_i^n + \sum_{\mathbf{k} \in K} a_\mathbf{k} \frac{(n-1)! k_i^n}{\mathbf{k}! n^{n-1}} \lambda_i^n \left(\sum_{|\beta|=n} \frac{(n-1)!}{\beta!} \beta_i \right) \right)$$

$$\geq \sum_{i=0}^{s} \left(a_{n\mathbf{e}^i} + \sum_{\mathbf{k} \in K} \frac{(n-1)! k_i^n}{\mathbf{k}!} \left(\frac{s+1}{n} \right)^{n-1} a_\mathbf{k} \right) \lambda_i^n.$$

Hence, if (12) holds, then $B_n[\{a_\mathbf{k}\}; \lambda] \geq 0$.

Finally, we point out that conditions (11) and (12) do not cover each other. In fact, if $k_i \leq n/(s+1)$; i.e., $k_i \geq k_i^n (\frac{s+1}{n})^{n-1}$, then condition (12) is weaker than condition (11); otherwise, condition (11) is weaker than (12). ∎

We will now give some convexity criteria of Bernstein-Bézier polynomials over simplices. We first define a function $\mathbf{c}_V$ associated with $q_{\alpha,s}(\mathbf{c}_V)$ as follows:

$$w_\alpha(\mathbf{c}_V) = \begin{cases} 0, & \text{if } q_{\alpha,s}(\mathbf{c}_V) \geq 0, \\ 1, & \text{if } q_{\alpha,s}(\mathbf{c}_V) < 0, \end{cases}$$

where $|\alpha| = n-2$ and $\alpha_k \neq n-2$, $k = 1, 2, \ldots, s$. If $|\alpha| = n-2$ and $\alpha_k = n-2$, $k = 1, 2, \ldots, s$, then $w_\alpha(\mathbf{c}_V) = 1$. Thus, we have the following result.

Theorem 2. *Let $r_\alpha \in \{0,1\}$ for $|\alpha| = n$ and $\alpha \neq n\mathbf{e}^k$, $k = 0, 1, \ldots, s$; $r_\alpha = 1$ for $\alpha = n\mathbf{e}^k$, $k = 0, 1, \ldots, s$. The Bernstein-Bézier polynomial $p_n(v) = B_n[\{\alpha_k^n\}; \lambda]$ is convex on T_s if for any fixed $u \in \{0, 1, \ldots, s\}$ and $v = 0, 1, \ldots, s$, $v \neq u$, its Bézier coefficients satisfy either*

$$\sum_{|\alpha|=n-2} \left(\sum_{k=0}^s \alpha_k^2\right)^{\frac{n}{2}} \frac{r_\alpha}{\alpha!} \triangle_{vu} \triangle_{vu} a_\alpha \geq$$

$$\sum_{\substack{w=0,1,\cdots,s \\ w \neq u,v}} \left| \sum_{|\alpha|=n-2} \left(\sum_{k=0}^s \alpha_k^2\right)^{\frac{n}{2}} \frac{r_\alpha}{\alpha!} \triangle_{vu} \triangle_{wu} a_\alpha \right| \quad (13)$$

or

$$\sum_{|\alpha|=n-2} \frac{\alpha_k}{\alpha!} r_\alpha \triangle_{vu} \triangle_{vu} a_\alpha \geq \sum_{\substack{w=0,1,\cdots,s \\ w \neq u,v}} \left| \sum_{|\alpha|=n-2} \frac{\alpha_k}{\alpha!} r_\alpha \triangle_{vu} \triangle_{wu} a_\alpha \right| (14)$$

Theorem 3. *Let $r_\alpha \in \{0,1\}$ for $|\alpha| = n$ and $\alpha \neq n\mathbf{e}^k$, $k = 0, 1, \ldots, s$. and $r_\alpha = 1$ for $\alpha = n\mathbf{e}^k$, $k = 0, 1, \ldots, s$. The Bernstein-Bézier polynomial $p_n(v) = B_n[\{\alpha_k^n\}; \lambda]$ is convex on T_s if for $k = 0, 1, \cdots, s$, any fixed $u \in \{0, 1, \ldots, s\}$, and $v = 0, 1, \ldots, s$, $v \neq u$, its Bézier coefficients satisfy either*

$$\sum_{|\alpha|=n-2} \left(\sum_{k=0}^s \alpha_k^2\right)^{\frac{n}{2}} \frac{r_\alpha}{\alpha!} \left[\triangle_{vu} \triangle_{vu} a_\alpha - \sum_{\substack{w=0,1,\cdots,s \\ w \neq u,v}} |\triangle_{vu} \triangle_{wu} a_\alpha|\right] \geq 0. \quad (15)$$

or

$$\sum_{|\alpha|=n-2} \frac{\alpha_k}{\alpha!} r_\alpha \left(\triangle_{vu} \triangle_{vu} a_\alpha - \sum_{\substack{w=0,1,\cdots,s \\ w \neq u,v}} |\triangle_{vu} \triangle_{wu} a_\alpha|\right) \geq 0. \quad (16)$$

Proof: It is sufficient to prove Theorem 2 since Theorem 3 is implied by Theorem 2. Because condition (14) has been given in [5], we will only need to prove condition (13). Without loss of generality, we can assume $u = 0$. Noting inequality (6) and the Cauchy-Schwarz inequality, from equation (4), we have

$$D_V^2 p_n = n(n-1) \sum_{|\alpha|=n-2} q_{\alpha,s}(\mathbf{c}_V) \phi_\alpha^{n-2}(\lambda)$$

$$\geq n(n-1) \sum_{|\alpha|=n-2} q_{\alpha,s}(\mathbf{c}_V) w_\alpha \phi_\alpha^{n-2}(\lambda)$$

$$\geq n(n-1) \sum_{|\alpha|=n-2} q_{\alpha,s}(\mathbf{c}_V) w_\alpha \frac{(n-1)!}{\alpha! n^{n-1}} \left(\sum_{i=0}^s \alpha_i \lambda_i\right)^n$$

$$\geq n(n-1) \sum_{|\alpha|=n-2} q_{\alpha,s}(\mathbf{c}_V) w_\alpha \frac{(n-1)!}{\alpha! n^{n-1}} \left(\sum_{k=0}^s \alpha_k^2\right)^{\frac{n}{2}} \left(\sum_{k=0}^s \lambda_k^2\right)^{\frac{n}{2}}$$

$$= \frac{(n-1)(n-1)!}{n^{n-2}} \left(\sum_{k=0}^{s} \lambda_k^2 \right)^{\frac{n}{2}} \mathbf{c}_V^T \left[\sum_{|\alpha|=n-2} \left(\sum_{k=0}^{s} \alpha_k^2 \right)^{\frac{n}{2}} Q_{\alpha,T_s} \frac{w_\alpha}{\alpha!} \right] \mathbf{c}_V$$

$$= \frac{(n-1)(n-1)!}{n^{n-2}} \left(\sum_{k=0}^{s} \lambda_k^2 \right)^{\frac{n}{2}} \mathbf{c}_V^T \left[\sum_{|\alpha|=n-2} \left(\sum_{k=0}^{s} \alpha_k^2 \right)^{\frac{n}{2}} \frac{w_\alpha}{\alpha!} \triangle_{i0}\triangle_{j0}a_\alpha \right]_{i,j=1}^{s} \mathbf{c}_V.$$

If the symmetric matrix

$$A = \left[\sum_{|\alpha|=n-2} \left(\sum_{k=0}^{s} \alpha_k^2 \right)^{\frac{n}{2}} \frac{r_\alpha}{\alpha!} \triangle_{i0}\triangle_{j0}a_\alpha \right]_{i,j=1}^{s}$$

is strongly diagonally dominant; i.e.,

$$\sum_{|\alpha|=n-2} \left(\sum_{k=0}^{s} \alpha_k^2 \right)^{\frac{n}{2}} \frac{r_\alpha}{\alpha!} \triangle_{i0}\triangle_{i0}a_\alpha \geq$$

$$\sum_{\substack{j=1 \\ j \neq i}}^{s} \left| \sum_{|\alpha|=n-2} \left(\sum_{k=0}^{s} \alpha_k^2 \right)^{\frac{n}{2}} \frac{r_\alpha}{\alpha!} \triangle_{i0}\triangle_{j0}a_\alpha \right|,$$

then matrix A and matrix

$$B = \left[\sum_{|\alpha|=n-2} \left(\sum_{k=0}^{s} \alpha_k^2 \right)^{\frac{n}{2}} \frac{w_\alpha}{\alpha!} \triangle_{i0}\triangle_{j0}a_\alpha \right]_{i,j=1}^{s}$$

are both positive semi-definite. Hence, $D_V^2 p_n \geq 0$, and p_n is convex on T_s.

Similar to the argument about conditions (11) and (12) at the end of the proof of Theorem 1, we claim that conditions (13) and (14) are independent. Thus, Theorem 2 is proved. ∎

Remark 2. A stronger convexity condition is implied by inequalities (13) and (14) and it is as follows: $\triangle_{vu}\triangle_{vu}a_\alpha \geq \sum_{\substack{w=0,1,\cdots,s \\ w \neq u,v}} | \triangle_{vu}\triangle_{wu}a_\alpha |$ (cf. [5] and [7]). Here, $u \in \{0,1,\ldots,s\}$, $v = 0,1,\ldots,s$, and $v \neq u$. Chang and Davis's condition [1] is also a special case of the results of Theorems 2 and 3.

Finally, we will discuss another approach for finding the convexity criteria from the positivity criteria shown in Theorem 1. From equation (5), we have $D_V^2 p_n = n(n-1) \sum_{|\alpha|=n-2} \mathbf{c}_V^T Q_{\alpha,T_s} \mathbf{c}_V \phi_\alpha^{n-2}(\lambda)$. If $s = 2$, Q_{α,T_2} is equal to

$$Q_{\alpha,T_2} = \left[\sum_{|\alpha|=n-2} \triangle_{i0}\triangle_{j0}a_\alpha \phi_\alpha^{n-2}(\lambda) \right]_{i,j=1}^{2}$$

$$= \left[\begin{matrix} \sum_{|\alpha|=n-2} \triangle_{10}\triangle_{10}a_\alpha \phi_\alpha^{n-2}(\lambda) & \sum_{|\alpha|=n-2} \triangle_{10}\triangle_{20}a_\alpha \phi_\alpha^{n-2}(\lambda) \\ \sum_{|\alpha|=n-2} \triangle_{20}\triangle_{10}a_\alpha \phi_\alpha^{n-2}(\lambda) & \sum_{|\alpha|=n-2} \triangle_{20}\triangle_{20}a_\alpha \phi_\alpha^{n-2}(\lambda) \end{matrix} \right].$$

Using Lemma 2, we obtain the determinant of the matrix Q_{α,T_2} as

$$\det Q_{\alpha,T_2} = \sum_{|\gamma|=2n-4} \sum_{|\alpha|=n-2} \left[(\triangle_{10}\triangle_{10}a_\alpha)(\triangle_{20}\triangle_{20}a_{\gamma-\alpha}) - \right.$$
$$\left. - (\triangle_{10}\triangle_{20}a_\alpha)(\triangle_{10}\triangle_{20}a_{\gamma-\alpha}) \right] \phi_\gamma^{2n-4}(\lambda).$$

Hence, if the determinant is non-negative and $\sum_{|\alpha|=n-2} \triangle_{10}\triangle_{10}a_\alpha \phi_\alpha^{n-2}(\lambda) \geq 0$,
then $D_V^2 p_n \geq 0$ and p_n is convex. Therefore, from the positivity criteria shown in Theorem 1, we can obtain another type of convexity criteria by using the following notation:

$$\triangle_\alpha^{(u)} = \triangle_{vu}\triangle_{vu}a_\alpha \tag{17}$$

and

$$\nabla_\gamma^{(u)} = \sum_{|\alpha|=n-2} \left[(\triangle_{vu}\triangle_{vu}a_\alpha)(\triangle_{wu}\triangle_{wu}a_{\gamma-\alpha}) \right.$$
$$\left. - (\triangle_{vu}\triangle_{wu}a_\alpha)(\triangle_{vu}\triangle_{wu}a_{\gamma-\alpha}) \right], \tag{18}$$

where $|\alpha| = n - 2$, $|\gamma| = 2n - 4$, $u \in 0,1,2$, $v, w = 0,1,2$, $v \neq u$, $w \neq u$, and $w \neq v$. Hence, from respective the inequality (11) and (12), we immediately have the following theorems. Let $A := \{\alpha : |\alpha| = n - 2, \alpha \neq (n - 2)e^0, (n - 2)e^1, (n - 2)e^2, \triangle_\alpha^{(u)} < 0\}$ and $\Gamma := \{\gamma : |\gamma| = 2n - 4, \gamma \neq (2n - 4)e^0, (2n - 4)e^1, (2n - 4)e^2, \nabla_\gamma^{(u)} < 0\}$.

Theorem 4. *Let $B_n[\{a_k\}; \lambda]$ be a Bernstein-Bézier polynomial defined on T. If the Bézier coefficients of $B_n[\{a_k\}; \lambda]$ satisfy either (i) or (ii) for $i = 0,1,2$, then it is convex.*

$$(i) \quad \triangle_{(n-2)e^i}^{(u)} + (n - 3)! \sum_{\alpha \in A} \frac{\alpha_i}{\alpha!} \triangle_\alpha^{(u)} \geq 0$$

and

$$\nabla_{(2n-4)e^i}^{(u)} + (2n - 5)! \sum_{\gamma \in \Gamma} \frac{\gamma_i}{\gamma!} \nabla_\gamma^{(u)} \geq 0,$$

$$(ii) \quad \triangle_{(n-2)e^i}^{(u)} + \frac{2^{n-3}(n - 3)!}{(n - 2)^{n-3}} \sum_{\alpha \in A} \frac{\alpha_i^{n-2}}{\alpha!} \triangle_\alpha^{(u)} \geq 0$$

and

$$\nabla_{(2n-4)e^i}^{(u)} + \frac{2^{2n-5}(2n - 5)!}{(2n - 4)^{2n-5}} \sum_{\gamma \in \Gamma} \frac{\gamma_i^{2n-4}}{\gamma!} \nabla_\gamma^{(u)} \geq 0,$$

where $i = 0,1,2$, $\triangle_\alpha^{(u)}$ and $\nabla_\gamma^{(u)}$ are defined in (17) and (18) respectively, and e^i is the ith coordinate vector.

Remark 3. Chang and Feng's condition [2] is a special case of Theorem 4.

Acknowledgments. Many thanks to Larry L. Schumaker for suggesting improvements.

References

1. Chang, G. and P.J. Davis, The convexity of Bernstein polynomials over triangles, J. Approx. Th., **40** (1984), 11-28.

2. Chang, G. and Y. Feng, An improved condition for the convexity of Bernstein-Bézier surfaces over triangles, Comp. Aided Geom. Des. **1** (1984), 279-283.

3. Chui, C.K., H.C. Chui, and T.X. He, Shape-preserving interpolation, in *Workshop on Computational Geometry*, A. Conte and V. Demichelis (eds.), World Scientific Singapore, 1993, 21-75.

4. Dahmen, W. and C.A. Micchelli, Convexity of multivariate Bernstein polynomials and box spline surfaces, Studia Scientiarum Mathematicarum Hungarica **23** (1988), 265-287.

5. Dahmen, W. Convexity and Bernstein-Bézier polynomials, *Curves and Surfaces*, P. J. Laurent, A. Le Méhauté, and L. L. Schumaker (eds.), 1991, 107-134.

6. Hardy, G.H., J.E. Littlewood, and G. Pólya, *Inequalities*, Cambridge Univ. Press, London/New York, 1934.

7. He, T.X., Shape criteria of Bernstein-Bézier polynomials over simplexes, Comp. Math. Appl. **30**(1995), 317-333.

8. He, T.X. and L. L. Schumaker, Convexity of spherical Bernstein-Bézier polynomials and circular Bernstein-Bézier curves, 1996.

9. Micchelli, C.A. and A. Pinkus, Some remarks on nonnegative polynomials on polyhedra, in: T.W. Anderson, K.B. Athreya, and D.L. Iglehart, Eds., *Probability, Statistics, and Mathematics, Papers in Honor of Samuel Karlin*, Academic Press, New York, 1989, 163-186.

10. Nadler, E., Nonnegativity of bivariate quadratic functions on a triangle, Comp. Aided Geom. Des. **9** (1992), 195-205.

11. Wang, Z. and Q. Liu, An improved condition for the convexity and positivity of Bernstein-Bézier surfaces over triangles, Comp. Aided Geom. Des. **5** (1988), 269-275.

12. Zhou, C.Z., On the convexity of parametric Bézier triangular surfaces, Comp. Aided Geom. Des. **7** (1990), 459-463.

Tian-Xiao He
Department of Mathematics
Illinois Wesleyan University
Bloomington, IL 61702-2900
the@sun.iwu.edu

Fitting Uncertain Data with NURBS

Wolfgang Heidrich, Richard Bartels, and George Labahn

Abstract. Fitting of uncertain data, that is, fitting of data points that are subject to some error, has important applications for example in statistics and for the evaluation of results from physical experiments. Fitting in these problem domains is usually achieved with polynomial approximation, which involves the minimization of an error at discrete data points. Norms typically used for this minimization include the l_1, l_2 and l_∞ norms, which are chosen depending on the problem domain and the expected type of error on the data points. In this paper we describe how the l_1 and l_∞ norms can be applied to integral and rational B-spline fitting as a linear programming problem. This allows for the use of B-splines and NURBS for the fitting of uncertain data.

§1. Introduction

An important field of applications for approximation techniques is the fitting of uncertain data that occurs, for example, in statistics or as a result of measurements in experiments. Approximation techniques in this context usually involve fitting a polynomial through a set of data points that are subject to some error. The type of error depends on the specific application domain, and could for example be uniformly distributed over the data points, or concentrated in a few outliers.

Depending on the application domain and the expected type of error, different norms are selected for the minimization process. In particular, the l_2 norm is typically used for data sets in which every data point is subject to a small, normally distributed error. The l_1 norm is well-suited for removing a small set of outliers from a set of data points with otherwise high precision. Finally, the l_∞ norm is appropriate if every single data point is very precise, and therefore the approximation error should be evenly distributed amongst the data points.

In the context of B-spline approximation, the l_2 norm has traditionally been used almost exclusively. Consequently, the form of B-spline approximation described in most of the literature is not well suited for handling, for example, uncertain data with outliers.

Curves and Surfaces with Applications in CAGD
A. Le Méhauté, C. Rabut, and L. L. Schumaker (eds.), pp. 177–184.
Copyright ☉ 1997 by Vanderbilt University Press, Nashville, TN.
ISBN 0-8265-1293-3.

All rights of reproduction in any form reserved.

In the following we will describe how the l_1 and l_∞ norms can be applied to integral and rational spline approximation problems using a linear programming approach. This allows for the efficient combination of B-splines and NURBS with standard techniques for fitting uncertain data.

§2. Preliminaries

The task of B-spline approximation is fitting a B-spline curve with N control points $\mathbf{c} = [c_1, \ldots, c_N]^T$ through a set of $M > N$ (typically $M \gg N$) data points $\mathbf{d} = [d_1, \ldots, d_M]^T$ at given parameter values $u_1, \ldots, u_M$. Such an approximation problem leads to an over-determined system of a linear equations

$$\underbrace{\begin{bmatrix} B_1(u_1) & \cdots & B_N(u_1) \\ B_1(u_2) & \cdots & B_N(u_2) \\ \vdots & \ddots & \vdots \\ B_1(u_M) & \cdots & B_N(u_M) \end{bmatrix}}_{\mathbf{B}} \cdot \begin{bmatrix} c_1 \\ \vdots \\ c_N \end{bmatrix} = \begin{bmatrix} d_1 \\ d_2 \\ \vdots \\ d_M \end{bmatrix}, \qquad (1)$$

where $B_i(u)$ is the ith B-spline basis function.

Since this system of equations is over-determined, in general an exact solution does not exist. Therefore, instead of directly solving (1), it is necessary to find an approximate solution by minimizing the error $\|\mathbf{B} \cdot \mathbf{c} - \mathbf{d}\|$ with respect to some norm $\|.\|$.

As we have stated above, the typical choice for this norm in the B-spline literature is the least-squares (l_2) norm. It can be shown that a least-squares solution of (1) can be obtained as the solution of the linear equation system

$$\mathbf{B}^T\mathbf{B} \cdot \mathbf{c} = \mathbf{B}^T \cdot \mathbf{d}. \qquad (2)$$

This square system of size $N \times N$ can be solved efficiently, which is one reason for the success of this approach.

In contrast to the least-squares norm, the l_1 and l_∞ minimizations cannot be obtained as the solutions of a linear equation system. Fortunately, it is possible to formulate the minimization using these two norms as a linear programming problem, which also allows for an efficient implementation.

A *linear programming problem* is a problem of the form

Minimize

$$\mathbf{q}^T \cdot \mathbf{x} - q_0$$

subject to

$$\mathbf{A} \cdot \mathbf{x} = \mathbf{b}.$$

In addition, some of the variables x_i may be restricted to non-negative values: $x_i \geq 0$.

The expression $\mathbf{q}^T \cdot \mathbf{x} - q_0$ which is to be minimized is called the *objective function*, while $\mathbf{A} \cdot \mathbf{x} = \mathbf{b}$ is a set of *constraints* that are to be fulfilled. The name "linear programming problem" reflects the fact that both the objective function and the constraints are linear in the unknowns $\mathbf{x}$.

Problems of this general form are usually solved using the *simplex method* [3], or one of its descendants. It is important to note that optimized versions of the simplex method exist for various problems of more specific forms.

§3. Integral B-spline Fitting with the l_1 and l_∞ Norms

The l_1 B-spline approximation $\|\mathbf{B} \cdot \mathbf{c} - \mathbf{d}\|_1$ of a data set can be reduced to a linear programming problem. Introducing $\mathbf{b}_i$ as a shorthand notation for the ith row-vector of the B-spline matrix $\mathbf{B}$, the approximation using the l_1 norm can be written as

$$\min_{\mathbf{c}} \sum_{i=1}^{M} |\mathbf{b}_i^T \mathbf{c} - d_i|.$$

The first step of transforming this to a linear programming problem is to make the function linear by removing the absolute value function. This can be achieved by introducing two vectors $\mathbf{p}$ and $\mathbf{n}$ of *slack variables* p_i and n_i. Using these two vectors of variables, we can write

$$\mathbf{b}_i^T \mathbf{c} - d_i = p_i - n_i \quad ; \quad p_i \geq 0, \ n_i \geq 0. \tag{3}$$

The intention is to have

$$p_i = \begin{cases} \mathbf{b}_i^T \mathbf{c} - d_i & ; \mathbf{b}_i^T \mathbf{c} - d_i \geq 0 \\ 0 & ; \text{otherwise} \end{cases} \tag{4}$$

and

$$n_i = \begin{cases} 0 & ; \mathbf{b}_i^T \mathbf{c} - d_i \geq 0 \\ -\mathbf{b}_i^T \mathbf{c} + d_i & ; \text{otherwise} \end{cases}. \tag{5}$$

Using these equations, the expression

$$\sum_{i=1}^{M} p_i + n_i$$

becomes the objective function and (3) becomes the constraint. We define $\mathbf{0}_N$ to be a vector of N zeroes, and $\mathbf{1}_N$ to be a vector of N ones. The linear programming problem for the l_1 approximation can now be written as

Minimize

$$\begin{bmatrix} \mathbf{0}_N^T & \mathbf{1}_M^T & \mathbf{1}_M^T \end{bmatrix} \cdot \begin{bmatrix} \mathbf{c} \\ \mathbf{p} \\ \mathbf{n} \end{bmatrix} \tag{6}$$

subject to

$$\begin{bmatrix} \mathbf{B} & -\mathrm{Id}_M & \mathrm{Id}_M \end{bmatrix} \cdot \begin{bmatrix} \mathbf{c} \\ \mathbf{p} \\ \mathbf{n} \end{bmatrix} = \mathbf{d} \quad ; \quad p_i \geq 0, \ n_i \geq 0.$$

The constraints assure that (3) actually holds, while the minimization of the objective function guarantees (4) and (5).

This linear programming problem can be solved using the normal simplex method. However, it is also possible to use an optimized version of the simplex method that makes use of the special structure of this problem [3].

The minimization with respect to the l_∞ norm can be expressed as a linear programming problem in a similar fashion. The l_∞ fit to some data points $\mathbf{d}$ is given as

$$\min_{\mathbf{c}} \left(\max_{i=1...M} \left| \mathbf{b}_i^T \cdot \mathbf{c} - d_i \right| \right).$$

By defining $c_0 := \max_{i=0...M} \left| \mathbf{b}_i^T \cdot \mathbf{c} - d_i \right|$, the approximation process can be rewritten as a linear programming problem. The expression c_0 becomes the objective function, while the constraints are given as

$$
\begin{aligned}
c_0 &\geq 0 \\
c_0 &\geq -\mathbf{b}_i^T \cdot \mathbf{c} + d_i \Leftrightarrow \quad \mathbf{b}_i^T \cdot \mathbf{c} + c_0 \geq d_i \\
c_0 &\geq \quad \mathbf{b}_i^T \cdot \mathbf{c} - d_i \Leftrightarrow -\mathbf{b}_i^T \cdot \mathbf{c} + c_0 \geq -d_i.
\end{aligned}
\tag{7}
$$

In matrix form this results in

Minimize

$$[\, \mathbf{0}_N^T \quad 1 \,] \cdot \begin{bmatrix} \mathbf{c} \\ c_0 \end{bmatrix} \tag{8}$$

subject to

$$\begin{bmatrix} \mathbf{B} & 1 \\ -\mathbf{B} & 1 \end{bmatrix} \cdot \begin{bmatrix} \mathbf{c} \\ c_0 \end{bmatrix} \geq \begin{bmatrix} \mathbf{d} \\ -\mathbf{d} \end{bmatrix} \quad ; \quad c_0 \geq 0.$$

It should be noted that this is a more general type of linear programming problem than stated above, since the constraints in this case are inequalities. By the introduction of additional slack variables, however, this problem can be transformed so that only equalities are used.

The resulting linear programming problem can again be solved either using the standard simplex algorithm or a version that is especially optimized for the structure of this specific problem [3].

§4. Rational B-spline Fitting

While approximations with integral B-splines are often sufficient, it is sometimes desirable to use rational B-splines (NURBS). In order for the solution to be useful, it is then essential to ensure that the weights of the rational B-spline curve remain positive. A clever scheme for rational B-spline fitting with positive weights has recently been proposed by Ma and Kruth [2].

The basic idea of this approach is to separate the calculation of the weights from the actual fitting process. The weights are obtained as the solution of a homogeneous linear equation system. In order to ensure that the weights remain positive, Ma and Kruth use quadratic programming with constraints to compute this solution. Once the weights have been obtained, they are substituted into the linear equation system for rational approximation, and then the usual integral approximation process is applied to the system.

In the following we shall briefly review this approach before we describe how quadratic programming can be replaced by more efficient linear programming.

The approximation of M data points d_i with a NURBS curve with N control points c_i and weights w_i yields the following set of equations:

$$d_j = \frac{\sum_{i=1}^{N} w_i c_i B_i(u_j)}{\sum_{i=1}^{N} w_i B_i(u_j)} \quad ; \quad j = 1 \dots M. \tag{9}$$

Recall that each of these expressions actually consists of e equations, where e is the dimension of the space. Following the setting in [2], we will in the following assume that the space is three-dimensional, that is, $e = 3$. This is only to make the notation more readable and comprehensible, and does not constitute a limitation of the algorithm itself.

Rewriting these linear equations in matrix form, and then performing some matrix manipulations (see [1] and [2] for details), yields

$$\begin{bmatrix} \mathbf{B}^T\mathbf{B} & & & -\mathbf{B}^T\mathbf{D}_x\mathbf{B} \\ & \mathbf{B}^T\mathbf{B} & & -\mathbf{B}^T\mathbf{D}_y\mathbf{B} \\ & & \mathbf{B}^T\mathbf{B} & -\mathbf{B}^T\mathbf{D}_z\mathbf{B} \\ & & & \mathbf{M} \end{bmatrix} \cdot \begin{bmatrix} \mathbf{c}_x \\ \mathbf{c}_y \\ \mathbf{c}_z \\ \mathbf{w} \end{bmatrix} = \begin{bmatrix} \mathbf{0}_N \\ \mathbf{0}_N \\ \mathbf{0}_N \\ \mathbf{0}_N \end{bmatrix}, \tag{10}$$

where

$$\begin{aligned} \mathbf{M} = \ & \mathbf{B}^T\mathbf{D}_x^2\mathbf{B} + \mathbf{B}^T\mathbf{D}_y^2\mathbf{B} + \mathbf{B}^T\mathbf{D}_z^2\mathbf{B} \\ & -(\mathbf{B}^T\mathbf{D}_x\mathbf{B})(\mathbf{B}^T\mathbf{B})^{-1}(\mathbf{B}^T\mathbf{D}_x\mathbf{B}) \\ & -(\mathbf{B}^T\mathbf{D}_y\mathbf{B})(\mathbf{B}^T\mathbf{B})^{-1}(\mathbf{B}^T\mathbf{D}_y\mathbf{B}) \\ & -(\mathbf{B}^T\mathbf{D}_z\mathbf{B})(\mathbf{B}^T\mathbf{B})^{-1}(\mathbf{B}^T\mathbf{D}_z\mathbf{B}). \end{aligned} \tag{11}$$

In this equation, $\mathbf{D}_x$, $\mathbf{D}_y$ and $\mathbf{D}_z$ are diagonal matrices holding the components of the data points. $\mathbf{c}_x = [c_1^x w_1, \dots, c_N^x w_N]^T$ is the vector of the x-components of all control points in homogeneous form. Similarly, $\mathbf{c}_y$ and $\mathbf{c}_z$ are the vectors of the y- and z-components of the control points, respectively, and $\mathbf{w}$ is the vector of weights. $\mathbf{B}$ is the B-spline matrix from Sections 2 and 3.

Thus, the weights have been separated from the control points, as they can be determined from the homogeneous linear equation system

$$\mathbf{M} \cdot \mathbf{w} = \mathbf{0}_N. \tag{12}$$

These weights are then substituted into (9), and finally Ma and Kruth propose to apply a standard least-squares approximation, as described in Section 2, to each component of the control points separately.

Note that the null space of $\mathbf{M}$ might have a dimension larger than 1, in which case more than one solution of (12) exists, and a specific one has to be chosen. On the other hand, the matrix $\mathbf{M}$ may also have full rank, in which case an exact solution of (11) does not exist, so that a minimization process becomes necessary in order to obtain an approximation.

The authors of [2] suggest the use of the least-squares norm for determining the weights. This minimization process yields an approximation in cases where $\mathbf{M}$ is non-singular, and selects a specific set of weights in cases where multiple solutions exist. A least-squares approximation of (12) can be obtained from the singular value decomposition (SVD, see [1] for details).

However, it is usually desirable to have only positive weights, because otherwise singularities in the the resulting curve may arise. Since it is not clear how a set of positive weights can be obtained from the SVD in an automated fashion, Ma and Kruth also propose another approach based on solving the quadratic programming problem

Minimize

$$\mathbf{w}^T \cdot \mathbf{M}^T \mathbf{M} \cdot \mathbf{w}$$

subject to

$$w_i > 0.$$

While this approach works and yields the desired results, it relies on quadratic programming, which is relatively inefficient. On the other hand, since $\mathbf{M}$ is not geometrically meaningful, it is not clear why the least-squares norm should be preferred over other norms for this minimization.

We therefore propose the use of the l_1 or l_∞ norm, which can be implemented more efficiently as a linear programming problem, similar to the implementation of the integral fitting process. A modified version of (6) for determining a set of positive weights with the l_1 norm is given by

Minimize

$$[\,\mathbf{0}_N^T \quad \mathbf{1}_N^T \quad \mathbf{1}_N^T \quad \mathbf{1}_N^T\,] \cdot \begin{bmatrix} \mathbf{w} \\ \mathbf{p} \\ \mathbf{n} \\ \mathbf{s} \end{bmatrix} \tag{13}$$

subject to

$$\begin{bmatrix} \mathbf{M} & -\mathbf{Id}_N & \mathbf{Id}_N & \\ \mathbf{Id}_N & & & -\mathbf{Id}_N \end{bmatrix} \cdot \begin{bmatrix} \mathbf{w} \\ \mathbf{p} \\ \mathbf{n} \\ \mathbf{s} \end{bmatrix} = \begin{bmatrix} \mathbf{0}_N \\ \mathbf{1}_N \end{bmatrix} \quad ; \quad p_i \geq 0,\ n_i \geq 0,\ s_i \geq 0.$$

The additional vector $\mathbf{s}$ of slack variables is used to ensure that the resulting weights obey the condition $\mathbf{w} = \mathbf{1}_N + \mathbf{s}$, with $s_i \geq 0$. Therefore the smallest resulting weight will be 1. Since all weights can be scaled by a constant factor without changing the resulting curve, this is not a restriction for the algorithm.

The linear programming Problem (8) for the l_∞ norm can be modified in a similar fashion, and also allows for the computation of positive weights.

Having computed the weights with one of these norms, it is then possible to determine the control points of the approximating NURBS curve by applying one of the l_1, l_2 or l_∞ norms to (9), similarly to the integral case.

§5. Results

There are several differences between the scenario described in this paper and traditional fitting of uncertain data. First of all, splines as described in this paper define *parametric* instead of *functional* curves. However, since the B-spline approximation treats every component of the vector space separately, the parametric approximation can be seen as a composition of several independent functional approximations.

Another difference is that splines have different approximation properties than polynomials. Because of the local control property of B-splines, it is to be expected that a spline approximation locally changes to meet outliers better, and that the removal of outliers is therefore not as good as in the polynomial case. On the other hand, with splines it is possible to create good approximations to data sets with a more complex shape. The negative effects of the local control property can partly be compensated by using the error-specific norms.

Figure 1 shows two examples for integral quadratic B-spline approximation using different norms. From top to bottom the figure contains the original function, the l_1, l_2 and l_∞ fit.

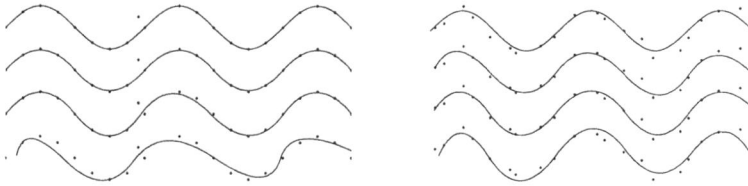

Fig. 1. Fitting of uncertain data with integral B-splines..

Both data sets consists of 21 points that have been originally sampled from a sine curve. The image on the left contains a single outlier that is being removed nicely by the l_1 fit. The data set on the right has been generated by jittering the original points with a normally distributed error. In this situation, the least-squares fit produces the best results.

The left side of Figure 2 shows another example for an integral quadratic fit with different norms (from top to bottom: l_1, l_2 and l_∞).

Fig. 2. Fitting of precise data (left) and reproduction of a quarter circle (right).

Under the assumption that the positions of the data points are *exact*, the l_∞ fit produces the best result, since it manages to distribute the error evenly across the parameter domain, while the l_1 norm treats the second, fifth and eighth point as outliers, and is therefore not able to correctly represent the shape of the data set. The quality of the l_2 norm is somewhere inbetween these two results.

We have also implemented rational fitting using the l_1 and l_∞ norms to determine the weights. In our experiments, we were not able to find any differences in the quality of the weights obtained using these two norms compared to the least-squares approach proposed by the authors of [1].

In particular, experiments show that both the l_1 and l_∞ norms are capable of reproducing conic sections and other sampled NURBS curves, as long as enough data points are provided together with the exact parameterization. An example for such an approximation is shown on the right side of Figure 2.

References

1. Wolfgang Heidrich, Spline Extensions for the MAPLE Plot System. Master Thesis, Department of Computer Science, University of Waterloo, 1995.

2. Weiyin Ma and Jean-Pierre Kruth, Mathematical Modeling of Free-Form Curves and Surfaces from Discrete Points with NURBS, in *Curves and Surfaces in Geometric Design*, P.-J. Laurent, A. Le Méhauté, and L. L. Schumaker (eds.), A. K. Peters, Wellesley MA, 1994, 319–326.

3. G. A. Watson, *Approximation Theory and Numerical Methods*. John Wiley & Sons, 1980.

Wolfgang Heidrich
Universität Erlangen
Graphische Datenverarbeitung
Am Weichselgarten 9
D-91058 Erlangen, GERMANY
Heidrich@informatik.uni-erlangen.de

Richard Bartels, George Labahn
Department of Computer Science
University of Waterloo
Waterloo, Ontario, CANADA
N2L 3G1
rhbartel@cgl.uwaterloo.ca, glabahn@daisy.uwaterloo.ca

Interpolation and Approximation
with Developable Surfaces

Josef Hoschek and Mike Schneider

Abstract. Developable surfaces (ruled surfaces of Gaussian curvature $k \equiv 0$) can be described with help of dual B–Spline curves in $\mathbb{R}^3$. The developable surfaces are envelopes of a one parameter set of planes. In this paper we will develop algorithms for interpolation and approximation of lines, points and planes with developable surfaces and discuss applications such as feeder and blankholder constructions.

§1. Introduction

Developable surfaces can be obtained by twisting and bending of a flat surface such as sheet metal or paper, leather, plastic or glass without stretching, folding or creasing. Those surfaces occur in many applications such as windshield design, binder (blankholder) surfaces for sheet metal forming processes, aircraft skins, ship hulls, shoes and clothing, ductwork and feeder and many other areas. Developable surfaces are a special subclass of ruled surfaces (surfaces generated by a straight line moving in space), and can be constructed by combination of cylinders, cones and tangent surfaces of space curves. Recently, geometric methods for constructing developable surfaces in CAD environments have been developed in various papers: For a planar curve and two points in a parallel plane Aumann [1] constructed a connecting developable surface between a cubic Bézier curve and a suitable quartic Bézier curve. A projective geometric approach to developable surfaces has been developed by Ravani and co–authors [3,4]. Lang, Röschel [15] have constructed rational $(1, n)$–Bézier representations of developable surfaces. The concept of dual Bézier– and B–Spline surfaces proposed in [11], has been efficiently used for description, interpolation and approximation with developable surfaces [13,17] additionally, geometric properties of developable surfaces could be developed [18,20]. The report [8] gives a comprehensive overview on applications and extends Aumann's approach [1]. Referring to [2,5,6] a special application

Curves and Surfaces with Applications in CAGD
A. Le Méhauté, C. Rabut, and L. L. Schumaker (eds.), pp. 185–202.
Copyright ⊚ 1997 by Vanderbilt University Press, Nashville, TN.
ISBN 0-8265-1293-3.

185

All rights of reproduction in any form reserved.

(shoulder surface for a feeder) is developed by mainly using differential geometric techniques. In the present paper we will use the dual description of developable surfaces for solving different interpolation problems (lines, points, planes, planes with points and lines etc.) [19] and different approximation problems (lines, set of scattered data points) [19,21]. The feeder construction and strategies for blankholder constructions will be discussed as applications.

§2. Developable and Dual B–Spline Surfaces

The usual B–Spline representation of a rational B–Spline curve in Euclidian 3–space E^3 has the parametric representation

$$\mathbf{X}(t) = \sum_{i=0}^{n} \mathbf{D}_i N_{ik}(t),$$

where the $\mathbf{D}_i$ are the control points in homogeneous coordinates and $N_{ik}(t)$ are the normalized B–Spline functions of order k. If the $\mathbf{D}_i$ are not at infinity, we can write $\mathbf{D}_i = (\omega_i, \omega_i x_i, \omega_i y_i, \omega_i z_i)$ with the weights $\omega_i \neq 0$ and $\mathbf{d}_i = (x_i, y_i, z_i)$ as Cartesian coordinate vector of the control points. We will only consider open rational B–Spline curves. Thus we choose a knot vector $\mathbf{T} := (v_0 = v_1 = \cdots = v_{k-1}, v_k, \cdots, v_n, v_{n+1} = \cdots = v_{n+k})$ with a monotone sequence v_i. If the knot sequence has only the values $\mathbf{T} = (v_0 = v_1 = \cdots = v_{k-1}, v_k = v_{k+1} = \cdots = v_{2k-1})$ the B–Spline description changes to the Bézier representation.

If we use the *principal of duality* from projective geometry we have to exchange

$$\textbf{points} \quad \Longleftrightarrow \quad \textbf{planes}$$

and obtain as a representation of a dual B–Spline curve

$$\mathbf{Y}(t) = \sum_{i=0}^{n} \mathbf{U}_i N_{ik}(t) =: (y_0(t), y_1(t), y_2(t), y_3(t)), \tag{1}$$

where the vectors $\mathbf{U}_i$ are the homogeneous plane coordinate vectors of the control planes. Formula (1) represents a one parameter set of planes with the explicit equation in E^3

$$y_0(t) + y_1(t)x + y_2(t)y + y_3(t)z = 0. \tag{2}$$

As an envelope, a one parameter set of planes has a developable surface, and its edge of regression is the locus of the singularities. In general we want to create strips of developable B-Spline-surfaces that are regular, i.e. containing no singular points. With the well-known properties of B-Spline-curves [12] follows that $\mathbf{U}_0$ and $\mathbf{U}_n$ are tangent planes at the boundaries of the developable surface over the given knot vector, the lines $\mathbf{U}_0 \cap \mathbf{U}_1$ and $\mathbf{U}_n \cap \mathbf{U}_{n-1}$ determine the boundary generators of the developable surface and $\mathbf{U}_0 \cap \mathbf{U}_1 \cap \mathbf{U}_2$ and $\mathbf{U}_n \cap \mathbf{U}_{n-1} \cap \mathbf{U}_{n-2}$ determine the points of regression at the

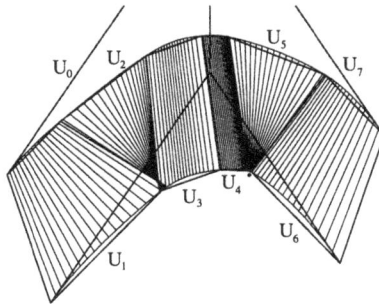

Fig. 1. Developable B–Spline surface of degree 2 (set of cones with corners) and its control planes U_i.

boundary generators. Further, it can be shown that a dual B–Spline curve is formed of developable Bézier surfaces pieced together along the rulings to the knots $t = v_l$ of multiplicity μ_l with the continuity $C^{k-\mu_l-1}$ [17].

Fig. 1 contains an example of a developable B–Spline surface with order $k = 3$ and $n = 7$. The control planes are denoted by thick lines, the surface consists of parts of cylinders and cones, some vertices of the cones are marked additionally.

The generators of the developable surface follow as intersection of the plane $\mathbf{Y}(t)$ and its derivatives $\dot{\mathbf{Y}}(t)$, and have the direction vectors

$$\mathbf{a}(t) = \mathbf{N}(t) \wedge \dot{\mathbf{N}}(t) \tag{3}$$

with $\mathbf{N}(t)$ as normal vector of the plane (2). The edge of regression or cuspoidal edge is obtained as the intersection $\mathbf{Y}(t) \cap \dot{\mathbf{Y}}(t) \cap \ddot{\mathbf{Y}}(t)$. It is a Bézier- or B–Spline curve of degree $3k - 9$ and possesses $C^{k-\mu_l-3}$ continuity at $t = v_l$. Its explicit representation can be described by

$$\mathbf{C}(t) = \mathbf{Y}(t) \wedge \dot{\mathbf{Y}}(t) \wedge \ddot{\mathbf{Y}}(t). \tag{4}$$

An explicit formula of the edge of regression for dual Bézier curves in the Bernstein basis see in [20].

Algorithms for converting the dual representation of developable surfaces to standard tensor product form have been given in [17]. They are particularly simple if the patch is to be confined by planar boundary curves. This is due to the fact that the intersection of a developable NURBS surface $\mathbf{Y}$ with a plane $\mathbf{U}$ is a NURBS curve; its control lines (of its dual form) are simply the intersections of the control planes of $\mathbf{Y}$ with the plane $\mathbf{U}$.

§3. Interpolation with Developable Surfaces

Now we will deal with two types of problems for interpolation with developable surfaces:

Problem 1. *Lines (generators), points, tangent planes with suitable parameter values are given and should be interpolated by a developable surface.*

Problem 2. *A curve $X_1(t)$ and some points P_i with parameter values t_i are given, required is a developable surface interpolating $X_1(t)$, P_i and the generators $g_i = P_i \cup X_1(t_i)$. The region of interest is the domain between $X_1(t)$ and a curve $X_2(t)$ through the points P_i.*

To retain linearity for the interpolation process, we will use a special representation of lines: if two planes with the plane vectors V_1 and V_2 intersect in a line $g = V_1 \cap V_2$, then they determine a *pencil of planes* through g. Each plane V of the pencil can be described with the parameters λ_1, λ_2 by

$$V = \lambda_1 V_1 + \lambda_2 V_2. \tag{5}$$

If we consider Problem 1, we will determine a developable B–Spline surface $Y(t)$ interpolating at described parameter values $t_{\alpha j}$

a) given tangent planes $T_j (j = 1, ..., M_1)$
b) given generators $g_j (j = 1, ..., M_2)$
c) given tangent planes and generators $\tilde{T}_j, \tilde{g}_j (j = 1, ..., M_3)$ with $\tilde{g}_j \in \tilde{T}_j$
d) given points $P_j (j = 1, ..., M_4)$
e) given points and tangent planes $P_j^*, T_j^* (j = 1, ..., M_5)$ with $P_j^* \in T_j^*$.

The case a) can be described by the vector valued equation

$$\tau_{1j} T_j = \sum_{i=0}^{n} U_i N_{ik}(t_{1j}), \tag{6a}$$

where the real factors τ_{1j} and the components of U_i are the unknowns. (6a) contains four scalar equations.

In case b), we describe the lines according to (5) by the intersection of two (suitable chosen) planes H_j, E_j. Thus we have the conditions

$$\alpha_{2j} H_j + \beta_{2j} E_j = \sum_{i=0}^{n} U_i N_{ik}(t_{2j}), \quad \gamma_{2j} H_j + \delta_{2j} E_j = \sum_{i=0}^{n} U_i \dot{N}_{ik}(t_{2j}), \tag{6b}$$

with the real factors $\alpha, \beta, \gamma, \delta$ and the components of U_i as unknowns. (6b) contains 8 equations.

For case c), we describe the generators $\tilde{g}_j$ as intersections of $\tilde{T}_j$ and suitable chosen auxiliary planes $\tilde{H}_j$ and obtain

$$\tau_{3j} \tilde{T}_j = \sum_{i=0}^{n} U_i N_{ik}(t_{3j}), \quad \lambda_{3j} \tilde{T}_j + \mu_{3j} \tilde{H}_j = \sum_{i=0}^{n} U_i \dot{N}_{ik}(t_{3j}), \tag{6c}$$

with the real factors τ, λ, μ and the components $\mathbf{U}_i$ as unknowns. (6c) leads to 8 equations.

Case d) can be described by the scalar equations (equations of planes)

$$\left(\sum_{i=0}^{n}\mathbf{U}_iN_{ik}(t_{4j})\right)\cdot\mathbf{P}_j = 0, \quad \left(\sum_{i=0}^{n}\mathbf{U}_i\dot{N}_{ik}(t_{4j})\right)\cdot\mathbf{P}_j = 0, \qquad (6d)$$

with points $\mathbf{P}_j$ in homogeneous coordinates and the unknown control planes $\mathbf{U}_i$ (2 equations).

The last case e) can be described by the equations

$$\left(\sum_{i=0}^{n}\mathbf{U}_iN_{ik}(t_{5j})\right)\cdot\mathbf{P}_j^* = 0, \quad \left(\sum_{i=0}^{n}\mathbf{U}_i\dot{N}_{ik}(t_{5j})\right)\cdot\mathbf{P}_j^* = 0, \qquad (6e)$$

$$\tau_{5j}\mathbf{T}_j^* = \sum_{i=0}^{n}\mathbf{U}_iN_{ik}(t_{5j}).$$

Now we have 6 equations, the real unknowns τ_{5j} and the unknown components of the planes $\mathbf{U}_i$.

Additionally, we can prescribe boundary conditions: The properties of the dual B–Spline curves imply for case c) the following conditions on the control planes $\mathbf{U}_0$, $\mathbf{U}_1$, $\mathbf{U}_{n-1}$, $\mathbf{U}_n$:

$$\mathbf{U}_0 = \tau_{31}\tilde{\mathbf{T}}_1, \quad \mathbf{U}_1 = \lambda_{31}\tilde{\mathbf{T}}_1 + \mu_{31}\tilde{\mathbf{H}}_1 \qquad (7)$$

$$\mathbf{U}_n = \tau_{3M_3}\tilde{\mathbf{T}}_{M_3}, \quad \mathbf{U}_{n-1} = \lambda_{3M_3}\tilde{\mathbf{T}}_{M_3} + \mu_{3M_3}\tilde{\mathbf{H}}_{M_3}$$

with the unknown real factors τ, λ, μ. In (7) one of the real factors can be arbitrary chosen: we set $\tau_{31} \equiv 1$; thus (7) leads additionally to 5 unknowns. If we sum up all the numbers of equations we obtain

$$4M_1 + 8M_2 + 8(M_3 - 2) + 2M_4 + 6M_5,$$

while from the equations (6a)–(6e) and (7), the number of unknowns is

$$M_1 + 4M_2 + 3(M_3 - 2) + M_5 + 5$$

(real factors in (6a)–(6e), with respect to (7)). With (7) we have as unknown control planes $\mathbf{U}_2, ..., \mathbf{U}_{n-2}$ with $4(n - 3)$ unknown components. The total number of the unknowns and the number of equations must be equal! Thus we get the necessary balance condition for a solvable linear system [19]

$$3M_1 + 4M_2 + 5M_3 + 2M_4 + 5M_5 - 4n - 3 = 0.$$

This Diophantine equation must be fulfilled to get a solvable system. Additionally, the parameters $t_{\alpha j}$ of the given elements have to satisfy some interlacing conditions on the knots v_i to get a nonsingular linear system. The shape of the interpolating surface depends significantly on the chosen parametrization. It can happen that singularities (the edge of regression) appear in the region of interest of the considered surface, but in general it is possible to change the parameter values $t_{\alpha j}$ with help of optimization techniques to force out the singularities [13]. Fig. 2 gives an example of a well shaped strip of a developable surface interpolating 3 generators and tangent planes and 4 points ($k = 6$, $n = 5$). The developable strip of interest should be determined by two curves (thick lines in Fig. 2). We have chosen the points and the generators with help of these curves. One can observe that the curve with the points (front curve) is much better approximated than the second curve.

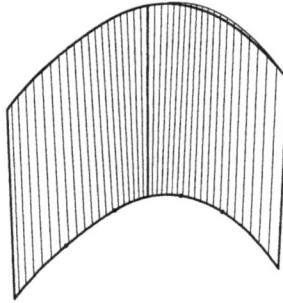

Fig. 2. A developable surface of order $k = 6$ interpolating 4 points and 3 generators and tangent planes.

Now we discuss Problem 2: Aumann [1] has developed a special solution for a given planar cubic Bézier curve $\mathbf{X}_1(t)$, where $\mathbf{X}_2(t)$ is constructed as a planar Bézier curve of degree 4 in a plane parallel to the osculating plane of $\mathbf{X}_1(t)$, while the two boundary generators (and the boundary points P_1, P_2 of $\mathbf{X}_2(t)$) of the developable surface patch are, additionally, given. We will extend this approach to arbitrary (nonplanar) curves $\mathbf{X}_1(t)$ and to arbitrary points $\mathbf{P}_i$ with parameter values t_i (may be as points of a second curve $\mathbf{X}_2(t)$). The points $\mathbf{X}_1(t_i)$ and $\mathbf{P}_i$ determine the generators of the required developable surface.

Because a developable surface can be interpreted as an envelope of a one parameter set of planes, the normal vector $\mathbf{N}(t)$ of the required developable surface D must fulfill the condition

$$\mathbf{N}(t) \cdot \dot{\mathbf{X}}_1(t) = 0, \tag{8}$$

while the tangent planes $\mathbf{T}$ of D have the equation

$$\mathbf{T}: \ \mathbf{N}(t) \cdot \mathbf{Y} - \mathbf{N}(t) \cdot \mathbf{X}_1(t) = 0 \tag{9}$$

with $\mathbf{Y} = (x, y, z)^T$. The rulings of a developable surface are determined by $\mathbf{T} \cap \dot{\mathbf{T}}$, and thus we obtain with (8)

$$\dot{\mathbf{T}}: \ \dot{\mathbf{N}}(t) \cdot \mathbf{Y} - \dot{\mathbf{N}}(t) \cdot \mathbf{X}_1(t) = 0. \tag{10}$$

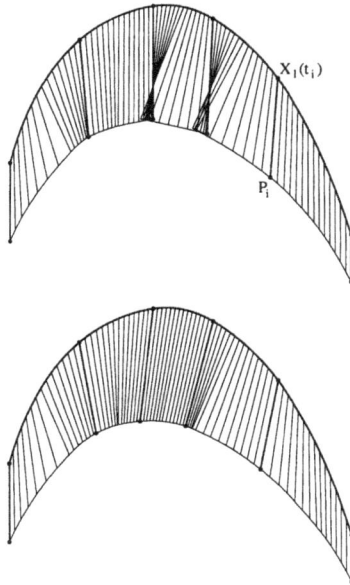

Fig. 3a, 3b. Interpolation of a curve $\mathbf{X}_1(t)$ and lines $\mathbf{X}_1(t_i) \cup \mathbf{P}_i$ where six points $\mathbf{P}_i$ are chosen on a second curve $\mathbf{X}_2(t)$. One can observe the effect of a good and a bad parametrization.

We assume that the given B–Spline curve $\mathbf{X}_1(t)$ has the components $\mathbf{X}_1(t) = (x(t), y(t), z(t))^T$, thus the vectors $\mathbf{N}_1 = (-\dot{y}(t), \dot{x}(t) + \dot{z}(t), -\dot{y}(t))^T$ and $\mathbf{N}_2 = (\dot{y}(t) + \dot{z}(t), -\dot{x}(t), -\dot{x}(t))^T$ are perpendicular to the tangent vector $\dot{\mathbf{X}}_1$ of $\mathbf{X}_1$ and in general linearly independent. A suitable normal vector $\mathbf{N}$ of the required developable surface can be described by a linear combination of $\mathbf{N}_1$ and $\mathbf{N}_2$ by

$$\mathbf{N}(t) = (1 - \lambda(t))\mathbf{N}_1(t) + \lambda(t)\mathbf{N}_2(t) \tag{11}$$

with $\lambda(t) := \sum_{j=0}^{s} \alpha_j N_{jk}(t)$ as B–Spline function, where α_j are real unknowns.

Now we choose m points $(\mathbf{P}_i, t_i)$ (may be from a curve $\mathbf{X}_2(t)$) and insert these points $\mathbf{P}_i$ and the normal vector (11) in the equations (9) and (10). We obtain a linear system of $2m$ equations for the $s+1$ unknowns α_j. Therefore, we have the following balance condition for the existence of a solution of Problem 2:

$$2m = s + 1.$$

Figs. 3a,b contain examples to this interpolation problem: Two non-planar B–Spline curves $\mathbf{X}_1(t), \mathbf{X}_2(t)$ of order 4 are given over $t \in [0,1]$. In Fig. 3a six generators $\mathbf{X}_1(t_i) \cup \mathbf{P}_i$ are determined by the same (equidistant) parameter values on $\mathbf{X}_1$ and on $\mathbf{X}_2$ ($\mathbf{P}_i = \mathbf{X}_2(t_i)$) - one can observe that this parametrization leads to some singularities in the area of interest between $\mathbf{X}_1(t)$ and $\mathbf{X}_2(t)$. In Fig. 3b corresponding points $\mathbf{X}_1(t_i)$ and $(\mathbf{P}_i, t_i^*)$ from $\mathbf{X}_2(t)$ are determined by a "rolling plane" on both curves: Since a developable surface is an envelope of an one parameter set of planes, the tangent planes of the required surface must be tangential to both curves. Thus the points of contact of a rolling plane $\mathbf{X}_1(t_i)$ and $\mathbf{X}_2(t_i^*)$ determine suitable generators of the required developable surface. One can observe in Fig. 3b that this generic parametrization don't lead to singularities.

Singularities cannot appear in the area between the two curves, as long as the contact points of the rolling plane induce a monotonic map between the parameters t of $\mathbf{X}_1(t)$ (or only a part of the parameter interval) and the parameter function $t^*(t)$ of $\mathbf{X}_2(t^*(t))$. As soon as monotonicity of the contact points is disturbed, singularities will appear in the region between the curves $\mathbf{X}_1$ and $\mathbf{X}_2$.

In Fig. 3a the t_i are equidistant on $\mathbf{X}_1(t)$ and $\mathbf{X}_2(t)$, in Fig. 3b correspond to the t_i from $\mathbf{X}_1(t)$ suitable t_i^* on $\mathbf{X}_2(t)$ are determined by a rolling plane. The developable surface has order 8 (as $\lambda(t)$ in (11) has order 3).

§4. Approximation with Developable Surfaces

In this section we will consider the following problems:

Problem A. *Given a set of lines G_j with parameter values t_j and two sets of points $(\mathbf{P}_j^1, \mathbf{P}_j^2)$ with $\mathbf{P}_j^1, \mathbf{P}_j^2 \in G_j$. Required is a developable surface whose generators g_j are as close as possible to the given lines, and the approximation should only be evaluated in the strip between $\mathbf{P}_j^1$ and $\mathbf{P}_j^2$.*

Problem B. *Given a set of scattered data points $\mathbf{P}_j$ (points from a laser scan). Required is a developable surface whose generators g_j are as close as possible to the given points.*

The crucial point of approximation with developable surfaces is the choice of an appropriate error measurement. For Problem A we will measure the error distance with help of the shortest distances from the points $(\mathbf{P}_j^1, \mathbf{P}_j^2)$ to the generators g_j with the parameter values t_j on the required surface $\mathbf{Y}(t)$ (see Fig. 4).

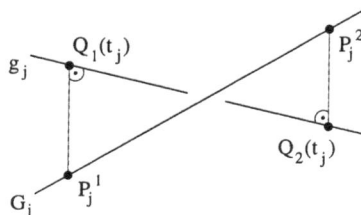

Fig. 4. Error measurement for approximation of given lines G_j.

Therefore we have to minimize the error function

$$\sum_{j}\left\{\alpha\left(\|\mathbf{Q}_1(t_j)-\mathbf{P}_j^1\|^2+\|\mathbf{Q}_2(t_j)-\mathbf{P}_j^2\|^2\right)+\beta\,\frac{\|\mathbf{a}(t_j)\times\mathbf{Z}_j\|^2}{\|\mathbf{a}(t_j)\|^2\|\mathbf{Z}_j\|^2}\right\}\longrightarrow\text{min.}\quad(12)$$

with $\mathbf{Q}_i(t_j)$ as feet of the perpendiculars of $\mathbf{P}_j^1,\mathbf{P}_j^2$ to g_j, $\mathbf{a}(t_j):=\mathbf{Y}(t_j)\cap\dot{\mathbf{Y}}(t_j)$ as direction vector of the generators g_j of the approximation surface and with $\mathbf{Z}_j:=\mathbf{P}_j^1-\mathbf{P}_j^2$.

For Problem B, we will minimize the distance of the given points $\mathbf{P}_i(i=1,...,r)$ with suitable parameter values t_i to the generators of the required developable surface $\mathbf{Y}(t_i)\cap\dot{\mathbf{Y}}(t_i)$. We reduce the distance of the point $\mathbf{P}_i$ with the parameter value t_i from the tangent plane $\mathbf{Y}(t_i)$ and from the plane of the first derivative $\dot{\mathbf{Y}}(t_i)$ as a substitute for $g_i=\mathbf{Y}(t_i)\cap\dot{\mathbf{Y}}(t_i)$.

$$F(u)=\sum_{i=1}^{r}\left[\left(\mathbf{P}_i\cdot\frac{\mathbf{Y}(t_i)}{\|\mathbf{Y}(t_i)\|}\right)^2+\left(\mathbf{P}_i\cdot\frac{\dot{\mathbf{Y}}(t_i)}{\|\dot{\mathbf{Y}}(t_i)\|}\right)^2\right]\longrightarrow\text{min.}\quad(13)$$

Unfortunately, both objective functions lead to a nonlinear optimization problem. Therefore, we have to carefully select

- an initial guess of the required surface,

- a suitable initial parametrization of the given lines or points

and introduce parameter correction to improve iteratively the initial parametrization.

In **case a)** we start with a quasi interpolating developable surface as an initial guess: the control planes $\mathbf{W}_i^0$ ($i=1,...,n-1$) of the initial surface may be determined by the planes through the given lines $G_i=\mathbf{P}_i^1\cup\mathbf{P}_i^2$ and the previous points $\mathbf{P}_{i-1}^1$. At the boundary lines the tangent planes $\mathbf{W}_0^0$ and $\mathbf{W}_n^0$ may be chosen suitably. As initial guess of the parametrization we use a chordal parametrization of the images of the normal vectors $\mathbf{N}_i$ ($\|\mathbf{N}_i\|=1$) of the control planes $\mathbf{W}_i^0$. Now we consider the components of the control planes $\mathbf{W}_1^0,...,\mathbf{W}_{n-1}^0$ as unknowns and minimize (12) with help of a gradient algorithm. With these new control planes $\mathbf{W}_i^1$ we start the parameter correction: we fix the control planes $\mathbf{W}_i^1$ and change the parameter values $t_i\longrightarrow t_j^1$ while minimizing (12) again. With these new parameter values t_i^1 we restart the minimization of (12) with the components of the control planes $\mathbf{W}_i^1$ and get new control planes $\mathbf{W}_i^2$. We stop this iterative optimization if the reduction of the error equation (12) is less then a suitable chosen ε. Fig. 5 contains an approximation of 17 generators (15 pairs of points and the boundary lines) by a quadratic B–Spline surface with 21 control planes. It was chosen $\alpha=1000$, $\beta=0$. After 20 iterations we obtain as maximal distance error $\delta_{max}=0.119\,mm$ by an extension of $170\,mm\times300\,mm$ of the whole area of interest.

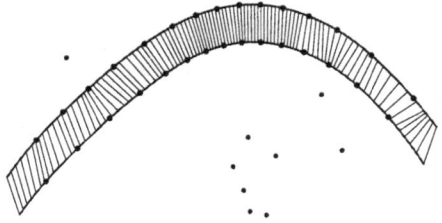

Fig. 5. Approximation of 17 given lines (15 pairs of points and two boundary lines) by a quadratic B–Spline surface. The approximation surface consists of cones and cylinders. The corners of the cones (singularities) are also plotted.

Case b) comes from the automotive industry: a surface generated by bending of a metal sheet is given, and we require a representation of the surface in a CAD–system. The surface will be digitized with a laser scanner by a large number of points $\mathbf{P}_i$ (Fig. 6 contains an example with 5286 points; part of a car door). The first step of our approach is to generate an initial guess of suitable rulings of the required developable surfaces: As basis of the generation of the generators we construct an approximating triangulation of the given set of points following an algorithm of Hoppe [10], where the facets of the triangulation approximate the set of points within an error tolerance δ. Now we choose a triangle Δ in the interior (near to the center of gravity of the points $\mathbf{P}_i$) and rotate a plane ε_i normal to Δ around the normal vector $\mathbf{N}_i$ of Δ in the center of gravity of Δ. ε_i intersects the triangulation in polygonal lines and we determine with a least square approximation that polygonal line which is closest to a line and choose this lines $\mathbf{G}_i^0$ as first guess of the generators of the required surface. Within a given distance tolerance we move the center of rotation to a new triangle in the neighbourhood of the starting triangle Δ and start the rotation process again. At the end of that step we have a sequence of lines (and pairs of points $\mathbf{P}_i^1, \mathbf{P}_i^2$ on these lines at the boundaries of the region of interest) as an initial guess of the generators for the required surface. The method may fail if planar regions appear in the triangulation. In this case the approximation surface may consist of developable patches connected by parts of planes.

Now we can proceed similarly to Problem A. While in Problem A the lines to be approximated are fixed, in Problem B we only have a guess of the generating lines. Therefore, we choose as normal vectors of the control planes $\mathbf{W}_i^0$ for the initial developable surface the mean values $\mathbf{N}_i$ of both planes determined by the lines $\mathbf{P}_i^1 \cup \mathbf{P}_i^2$ and containing $\mathbf{P}_{i-1}^1$ or $\mathbf{P}_{i+1}^1$. Additionally $\mathbf{W}_i^0$ should pass through the point $\mathbf{P}_i^1$.

To minimize (13) further we need a parametrization of all given points $\mathbf{P}_i$. First we parametrize the lines $\mathbf{G}_i^0$ with help of the chordal distances of the normal image $\mathbf{N}_i$ ($\|\mathbf{N}_i\| = 1$) of the control planes $\mathbf{W}_i^0$, then we connect two lines $\mathbf{G}_i^0$ and $\mathbf{G}_{i+1}^0$ by a bilinear Coons patch, project the points $\mathbf{P}_i$ between

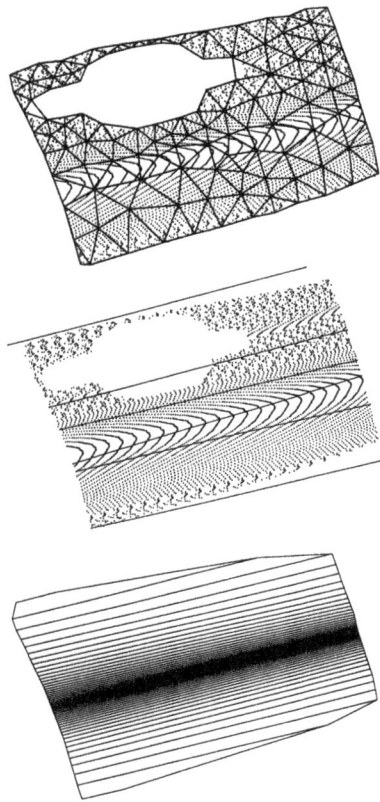

Fig. 6. Set of scattered points from a car door, the corresponding triangulation (first figure), the initial guess of the generators (second figure) and the approximating developable surface (third figure).

$\mathbf{G}_i^0$ and $\mathbf{G}_{i+1}^0$ perpendicular on this patch and get parameter values of these points with respect to the parameter values t_i and t_{i+1} of the lines $\mathbf{G}_i^0$ and $\mathbf{G}_{i+1}^0$.

After this preparation we minimize the objective function (13) with a suitable gradient method, the components of the control planes $\mathbf{W}_1^0, ..., \mathbf{W}_{n-1}^0$ are the unknowns. Then we change the parameter $t_i \longrightarrow \tilde{t}_i$ of each point $\mathbf{P}_i$ by minimizing $\delta = \sum_{i=1}^r \|\mathbf{P}_i - g_i\|$, where g_i are the generators for $t = t_i$ on the developable surface. After this parameter correction we start again minimizing (13) with the components of the control planes as unknowns. Fig. 6 demonstrates the approximation of a set of scattered points: In the first figure we have the triangulation; in the second figure the initial guess of the generators of the required surface; the third figure shows a set of generators of a developable surface with a prescribed error tolerance.

§5. Applications

As applications of developable surfaces we will discuss two examples:

Example I. *Construct the shoulder surface of a packing machine.*

Example II. *Construct the blankholder surface for a sheet metal forming process.*

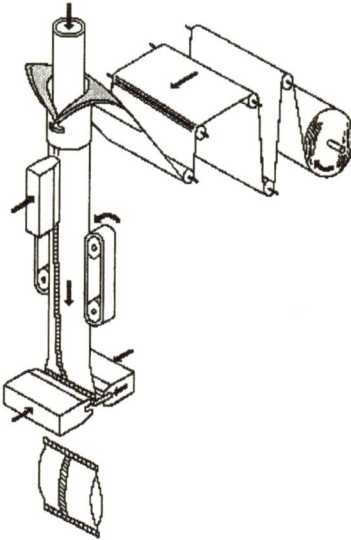

Fig. 7. Functionality of a feeder (grey) (see [6]).

The shoulder surface of a packing machine (*feeder*) should guide the packing material (paper or plastic sheet) from a horizontal roll into a vertical circular cylinder where it is folded against the inner wall (see Fig. 7). Inside the cylinder the sheet is sealed at the bottom and at the front side to form a pack. The pack is filled from above by dropping the product to be packed (powder, candy).

The shoulder surface S must be developable [2], [5], [6] andintersects the circular cylinder C in a space curve (*shoulder curve*) $\mathbf{X}(t)$. Because the cylinder C and the shoulder surface S have to be isometrically bended into the plane a common shoulder curve $\mathbf{X}(t) = S \cap C$ on both surfaces must have the same geodesic curvature κ_g [7], [14] at common points.

We first consider the shoulder curve $\mathbf{X}(t)$ as a curve on a circular cylinder of radius r with the parametric representation

$$\mathbf{X}(t) = (r\cos t, r\sin t, z(t)) \tag{14a}$$

The tangent planes of the shoulder surface S must contain the tangents

$$\dot{\mathbf{X}}(t) = (-r\sin t, r\cos t, \dot{z}(t)) \tag{14b}$$

of the shoulder curve (14a). Therefore, the normal vector $\mathbf{N}$ of these tangent planes can be described as a linear combination of normal vectors $\mathbf{N}_1$ and $\mathbf{N}_2$ of the tangent vector (14b). Two suitable normal vectors may be

$$\mathbf{N}_1(t) = (\cos t, \sin t, 0), \quad \mathbf{N}_2(t) = (0, -\dot{z}(t), r\cos t),$$

and the normal vector $\mathbf{N}$ to the tangent planes of S is

$$\mathbf{N}(t) = \frac{(1 - \lambda(t))\mathbf{N}_1(t) + \lambda(t)\mathbf{N}_2(t)}{\|(1 - \lambda(t))\mathbf{N}_1(t) + \lambda(t)\mathbf{N}_2(t)\|}, \tag{15}$$

where $\lambda(t)$ is a suitable function. The shoulder curve $\mathbf{X}(t)$ has the same geodesic curvature on both surfaces iff

$$\det(\dot{\mathbf{X}}, \ddot{\mathbf{X}}, \mathbf{N}_z) = \det(\dot{\mathbf{X}}, \ddot{\mathbf{X}}, \mathbf{N}) \qquad (16)$$

with $\mathbf{N}_z$ as normal vector of the cylinder. (16) is a quadratic equation for λ and has the solution $\lambda_1 = 0$ and $\lambda_2 = \lambda(t)$ (according to the well known theorem [7] that there exist only two developable surfaces through $\mathbf{X}(t)$, which have the same common curve after their isometric bending into the plane). From (16), it also follows that the generators from S and C intersect the common shoulder curve $\mathbf{X}(t)$ with the same angle γ. (16) can be solved easily with a formula manipulation system. In Figure 8 we have chosen (see [2]) $z(t) = \alpha \cosh(\beta t)$ with real α, β. The quantities r, α, β can be used to prescribe the angle of the generators of S in the highest point $t = 0$ of $\mathbf{X}(t)$ with the cylinder.

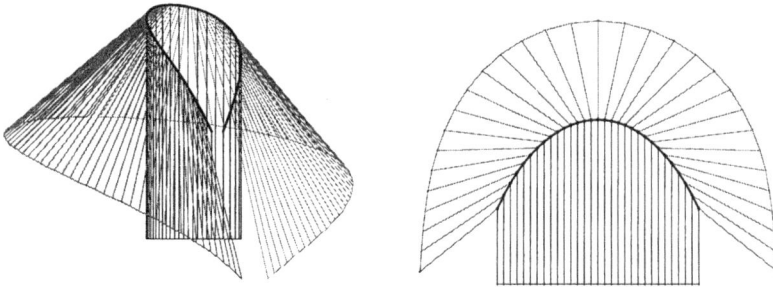

Fig. 8. An example of a feeder and its isometric bending in the plane.

Now we will move to Example II. In a stamping die a binder (*blankholder*) surface is needed, to fix the metal sheet during the metal forming process. This binder surface should be a developable surface to avoid stretching of the metal sheet before the forming process starts. There are different strategies used for blankholder construction during a die–face design:

Strategy I. *A closed curve (punch line) is given as one boundary curve of the blankholder. The punch line is the rim between the binder surface and that part of the metal sheet which will be deformed during the metal forming process to the desired part of a car-body.*

Strategy II. *The desired die–face is intersected by parallel planes. These intersection curves will be approximated by suitable convex curves, whose convex hull should simulate the initial shape of a metal sheet when starting the metal forming process.*

If we discuss **Strategy I** we may assume that the punch line $\mathbf{C}(t)$ is a given closed curve with some restrictions to the slope of the rulings of the

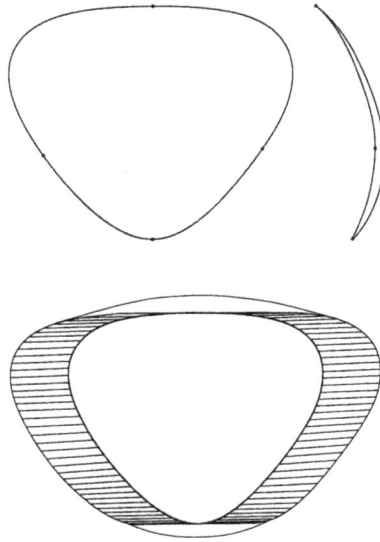

Fig. 9. A punch line with handle points and a suitable binder surface.

required binder surface. The binder surface must be not to steep with respect to the direction of the die's moving to avoid gliding of the metal sheet within the tool [16]. The usual value for the slope is less than 15 degrees. To get potential rulings of the binder surface, we have to simulate the behaviour of the metal sheet during the beginning of the punching process: if the metal sheet is moved into the punch tool it will contact the punch line in (at least) two points $\mathbf{P}_i$ and $\tilde{\mathbf{P}}_i$, i.e., the metal sheet is the common tangent plane of the punch line in these points. The line connecting these two points could be one ruling of the required binder surface. These lines are often called *common torsal lines*. If we roll the metal sheet along the punch line we obtain as boundary positions of the metal sheet the points $\mathbf{a}_i$ were the contact points $\mathbf{P}_i$ and $\tilde{\mathbf{P}}_i$ coincide. At these points the osculating plane of $\mathbf{C}(t)$ can be used as continuation of the required binder surface. These boundary points are called *handle points* and are determined by vanishing torsion of $\mathbf{C}(t)$, i.e.

$$\det\left(\mathbf{C}(t_0), \dot{\mathbf{C}}(t_0), \ddot{\mathbf{C}}(t_0)\right) = 0 \qquad (17)$$

with t_0 as parameter value of the handle point. Vice versa we can start at the handle points constructing a family of torsal lines with a pursuit algorithm: the parameter values of the contact points $\mathbf{P}_i$ and $\tilde{\mathbf{P}}_i$ may be u and v, the tangents in these points are $\dot{\mathbf{x}}(u) = \dot{\mathbf{C}}(u)$, $\mathbf{y}'(v) = \mathbf{C}'(v)$, the position vectors of $\mathbf{P}_i$ and $\tilde{\mathbf{P}}_i$ are $\mathbf{x}(u) = \mathbf{C}(u)$ and $\mathbf{y}(v) = \mathbf{C}(v)$. A common tangent plane of the punch line is determined by

$$\det\left(\dot{\mathbf{x}}(u), \mathbf{y}'(v), \mathbf{y}(v) - \mathbf{x}(u)\right) = 0. \qquad (18)$$

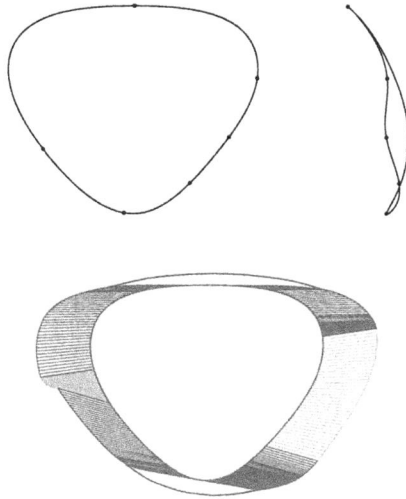

Fig. 10. Binder surface for a twisted punch line.

To get a solution of (18), we start at a handle point $t = t_0$ (as zero of (17)) of the punch line $\mathbf{C}(t)$, move the tangent plane to $t = u$ and calculate v as zero of (18) with a Newton iteration. In general there exists more than one handle point on the closed punch line. If we start at different handle points, we get different families of (discrete) torsal lines. If one of these families of common torsal lines g_i covers the whole punch line without self intersections we can construct the required binder surface: We also choose between the torsal lines g_i further points $\mathbf{Q}_i$ on the punch line and construct a suitable parametrization of the lines g_i and the points $\mathbf{Q}_i$. These elements are now interpolated or approximated by a developable surface according to the methods from Sections 3 and 4. Fig. 9 (top) contains a punch line with handle points and a cross projection to show that the curve is not a cylindric one. Fig. 9 (bottom) shows a suitable binder surface to this punch line.

If the punch line is very twisted, the rolling plane has in some positions more than two points of contact. In this situation we start with our rolling plane as in Fig. 9 at a suitable chosen handle point and stop the algorithm as soon as the common torsal lines begin to intersect in the strip of interest (the expected binder surface, see Fig. 10). The gap between the common torsal lines will be filled by a ruled surface. Because of the continuity conditions, it may happen that it is impossible to fill the gap with a developable surface, then the Gaussian curvature of the filling ruled surface should be as small as possible. In the plot at the top of Fig. 10 two projections show the twisted punch line and the bottom figure gives a suitable binder surface. The filling surface (grey) has Gaussian curvature $-0.010 \leq k \leq 0$.

Now we move to **Strategy II** of binder construction: we assume that some intersection curves of the required die–tool are given. Fig. 11 (top figure)

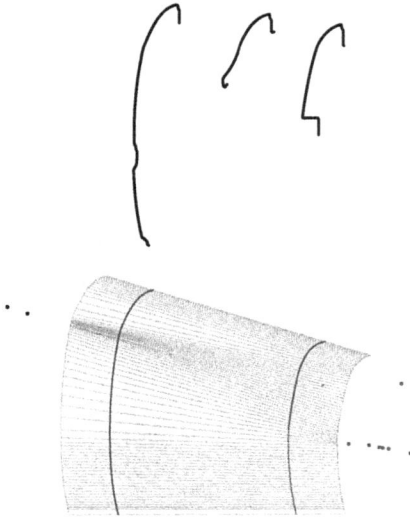

Fig. 11. Construction of the initial shape of the metal sheet based on intersection curves (top figure) of a die–tool. The first and the third curve are approximated by convex curves, the convex approximation of the first curve is suitably extended (grey curve), the second curve is omitted. The common developable surface has order $k = 3$, the corners of the cones are plotted additionally.

shows an example with three given intersection curves of a die for a car-body's wing. It is necessary to construct the convex hull of these intersection curves. Therefore we cancel the middle curve and approximate both other curves by convex curves. To get unique points of contact for a rolling plane we extend the curve at the right hand side suitably by a piece of a convex curve (plotted grey in Fig. 11). Now we role a plane along these convex curves and get an initial guess of the metal sheet lying in the die–tool (see Fig. 11). Some CAD–systems use this technique of approximating convex curves, but in general they don't use a rolling plane, they only connect points on both curves with the same parameter values. In general, this straight forward construction of the ruled surface lead to a nondevelopable surface. In our case the common torsal lines are now interpolated or approximated by a developable surface thus we can guarantee that the initial shape of the metal plate is really developable. After the construction of this initial shape, the designer of the die–tool can choose a suitable closed curve on this initial guess to get the punch line or the rim of the binder surface. The rest of the initial guess can be used as part of the binder surface, some unnecessary parts can be cut off.

§6. Conclusion

We have developed some solutions for application problems with developable surfaces but a lot of questions are still open:

- how to get algorithms for controlling the edge of regression. The strip of interest of a developable surface must be free of points of regression!
- how to get an approximation of a given surface by developable surfaces (for example construct a rotational container with optimal pieces of developable surfaces).
- how to construct the isometric bending in a plane for a developable surface in our dual description.

We hope to find solutions of these problems and other applications for developable surfaces in our further research.

References

1. Aumann, G., Interpolation with developable Bézier patches, Computer Aided Geometric Design **8** (1991), 409–420.

2. Culpin, D., A Metal–Bending Problem, Math. Scientist **5** (1980), 121–127.

3. Bodduluri, R. M. C., and B. Ravani, Geometric design and fabrication of developable surfaces, ASME Adv. Design Autom. **2** (1992), 243–250.

4. Bodduluri, R. M. C., and B. Ravani, Design of developable surfaces using duality between plane and point geometries, Computer–Aided Design **10** (1993), 621–632.

5. Boersma, J., and J. Molenaar, Case study from industry, SIAM Review **3** (1995), 406–422.

6. Dietz, G., and H. Günther, CAD/CAM von Formschultern für Schlauch-beutel-Form-Füll-Verschließ-Maschinen, Wiss. Z. Tech. Univ. Dresden **34** (1985).

7. Forsyth, A. R., *Lectures on the Differential Geometry of Curves and Surfaces*, Cambridge University Press, Cambridge, 1912.

8. Frey, W. H., and D. Bindschadler, Computer aided design of a class of developable Bézier surfaces, General Motors R&D Publication 8057 (1993).

9. Hanen, M., Conception des surfaces serre–flan et étude de l'évolution de la surface d'impact, Direction des Méthodes Carrosserie R.N.U.R. Boulogne–Billancourt, 1977.

10. Hoppe, H., T. DeRose, T. Duchamp, J. McDonald and W. Stuetzle, Surface reconstruction from unorganized points. Computer Graphics (SIGGRAPH'92 Proceedings) 26(2) (1992), 71–78.

11. Hoschek, J., Dual Bézier curves and surfaces, in *Surfaces in Computer Aided Geometric Design*, R.E. Barnhill & W. Boehm, eds., North Holland 1983, 147–156.

12. Hoschek, J., and D. Lasser, *Fundamentals of Computer Aided Geometric Design*, AK Peters, Wellesley, Massachusetts, 1993.

13. Hoschek, J., and H. Pottmann, Interpolation and Approximation with Developable B–Spline Surfaces. in *Mathematical Methods for Curves and Surfaces*, M. Dæhlen, T. Lyche, L.L. Schumaker (eds.), Vanderbilt University Press 1995, 255–264.

14. Kreyszig, E., Differential Geometry, Dover, New York, 1991.

15. Lang, J., and O. Röschel, Developable $(1, n)$–Bézier surfaces, Computer Aided Geometric Design **9** (1992), 291–298.

16. Lange, R., Umformtechnik Band **3**, Springer (1990), 413–425.

17. Pottmann, H., and G. Farin, Developable rational Bézier and B–Spline surfaces, Computer Aided Geometric Design **12** (1995), 513–531.

18. Pottmann, H., Rational curves and surfaces with rational offsets, Computer Aided Geometric Design **12** (1995), 175–192

19. Schneider, M., Approximation von Blechhalterflächen mit Torsen in B–Spline–Darstellung. Diplomarbeit, TH Darmstadt, 1995.

20. Schwanecke, U., Untersuchung geometrischer Eigenschaften von dualen Bézier– und B–Spline–Kurven mit Anwendungen auf Interpolationsprobleme. Diplomarbeit, TH Darmstadt, 1996.

21. Vatter, R., Approximation von Datenpunkten mit Torsen in B–Spline–Darstellung. Diplomarbeit, TH Darmstadt, 1996.

22. Vogt, C.D., Analyse und Werkzeuge zur Unterstützung des Auslegungsprozesses von Ziehstufen großer freiformflächiger Blechbauteile, PhD–Thesis, ETH Zürich, 1994.

Josef Hoschek
Fachbereich Mathematik
Technische Hochschule Darmstadt
Schloßgartenstraße 7
D–64289 Darmstadt, GERMANY
hoschek@mathematik.th--darmstadt.de

Mike Schneider
Fachbereich Mathematik
Technische Hochschule Darmstadt
Schloßgartenstraße 7
D–64289 Darmstadt, GERMANY
mschneider@mathematik.th--darmstadt.de

Sectional Curvature–Preserving Interpolation of Contour Lines

Bert Jüttler

Abstract. A sequence of given contour curves is interpolated by a surface composed of tensor–product B–spline patches. The interpolation scheme preserves the signs of the sectional curvature of the contours. Based on an appropriate linearization of the shape constraints we formulate this task as a quadratic programming problem which is solved with the help of an active set strategy.

§1. Introduction

Methods for the generation of surfaces from given contour line data are required in several applications, ranging from the reconstruction of bone surfaces from medical images to the construction of ship hulls. For an overview over related literature the reader is referred to the survey articles by Schumaker [7] and Unsworth [8]. It is desirable that the generated surfaces preserve (at least approximately) the shape of the given contour data. In order to achieve this property, Kaklis and Ginnis [6] proposed to use interpolation by polynomial splines of non–uniform degree whereby the degrees of the spline segments act as tension parameters. If the degree of the spline segments is chosen high enough, then the surface preserves the shape of the contours. In the present paper we will outline a different approach; the shape–preserving property is guaranteed with the help of additional linear constraints to the control points.

We assume that B–spline representations of some contour curves of the surface are already known. They can be found with the help of a method for shape preserving least–square approximation by polynomial parametric spline curves which has been developed in [4]. We interpolate these curves by a surface which preserves their shape, i.e., all segments of level curves interpolating between two convex segments of contour lines are convex (sectional curvature–preserving interpolation). Based on an appropriate linearization of the shape constraints we are able to formulate this task as a quadratic programming (QP) problem. We then construct an initial solution which is very close to the optimum and solve the QP problem using an active set strategy. The method is illustrated by an example.

Curves and Surfaces with Applications in CAGD
A. Le Méhauté, C. Rabut, and L. L. Schumaker (eds.), pp. 203–210.
Copyright © 1997 by Vanderbilt University Press, Nashville, TN.
ISBN 0-8265-1293-3.
All rights of reproduction in any form reserved.

§2. Sectional Curvature–Preserving Interpolation

The $C+1$ given contour curves, represented by open B–spline curves (see [3])

$$\mathbf{x}_i(t) = \sum_{j=0}^{D_i} \mathbf{d}_{i,j} N_{i,j}^d(t) \quad t \in [0,1], \quad i = 0, ..., C \tag{1}$$

of degree d, with the associated *heights* $(z_i)_{i=0,...,C}$ (i.e., $x_{i,3}(t) \equiv z_i$), are to be interpolated by a C^l ($l = 1, 2$) surface $\mathbf{y}(z,t)$ with the parameter domain $(z,t) \in [z_0, z_C] \times [0,1]$. The contour curves are defined over possibly different knot sequences $\mathcal{T}_i$ with $d+1$–fold boundary knots 0 and 1 whereby all inner knots have multiplicity $d-l$. We assume that points with the same parameter t on adjacent contours (1) correspond to each other. The third coordinate function of the interpolating surface will simply be equal to the z–coordinate, $y_3(z,t) \equiv z$. In addition to the interpolation of the given contours, $\mathbf{y}(z_i,t) \equiv \mathbf{x}_i(t)$ for $i = 0, ..., C$, the surface $\mathbf{y}(z,t)$ is to preserve the sectional curvature of the given contours. This notion has been introduced by Kaklis and Ginnis [6]:

Definition 1. *The surface $\mathbf{y}(z,t)$ is said to be a sectional curvature–preserving (sc–p) interpolant if it possesses the following property for any pair $\mathbf{x}_{i-1}(t)$, $\mathbf{x}_i(t)$ of adjacent contour curves $(i = 1, ..., C)$: if both contours possess non–positive (resp. non–negative) curvatures at a point $t = t_0$, then also the curvatures of all interpolating contours $\mathbf{y}(z_0,t)$ with $z_0 \in [z_{i-1}, z_i]$, constant, are non–positive (resp. non–negative) at this point.*

The interpolating surface $\mathbf{y}(z,t)$ is an sc–p interpolant if the two inequalities $[\dot{\mathbf{x}}_{i-1}(t_0), \ddot{\mathbf{x}}_{i-1}(t_0)] \geq 0$ (resp. ≤ 0) and $[\dot{\mathbf{x}}_i(t_0), \ddot{\mathbf{x}}_i(t_0)] \geq 0$ (resp. ≤ 0) imply $[\dot{\mathbf{y}}(z,t_0), \ddot{\mathbf{y}}(z,t_0)] \geq 0$ (resp. ≤ 0) for all $(z,t) \in [z_{i-1}, z_i] \times [0,1]$ with $1 \leq i \leq C$, whereby " $\dot{}$ " denotes the differentiation $\frac{\partial}{\partial t}$ with respect to t. The abbreviation $[\vec{p}, \vec{q}] = p_1 \cdot q_2 - p_2 \cdot q_1$ means the third component of the cross product of the vectors $\vec{p}, \vec{q} \in \mathbb{R}^3$. Moreover we denote by $\sphericalangle(\vec{p}, \vec{q}) \in (-\pi, \pi]$ the oriented angle between the top views (orthogonal projections onto the plane $z = 0$) of the two vectors $\vec{p}, \vec{q}$, whereas $\|\vec{p}\| = \sqrt{(p_1^2 + p_2^2)}$ is the length of the top view of the vector $\vec{p}$.

The construction of the interpolating surface is based on the following two additional assumptions.

(i) The contour curves and their control polygons have coinciding shape. More precisely, if $\mathbf{x}_i(t)$ has non–negative (resp. non–positive) curvature for some of its polynomial spline segments, then the angles between adjacent legs of the corresponding control polygon $(\mathbf{d}_{i,j})_{j=p,...,q}$ are also non–negative (resp. non–positive) but less than $\frac{\pi}{d-1}$ (greater then $-\frac{\pi}{d-1}$),

$$0 \leq \sphericalangle(\Delta_{[2]}\mathbf{d}_{i,j}, \Delta_{[2]}\mathbf{d}_{i,j+1}) < \frac{\pi}{d-1}$$
$$(\text{ resp. } -\frac{\pi}{d-1} < \sphericalangle(\Delta_{[2]}\mathbf{d}_{i,j}, \Delta_{[2]}\mathbf{d}_{i,j+1}) \leq 0) \tag{2}$$

for $j=p, ..., q$ with $\Delta_{[2]}\mathbf{d}_{i,j} = \mathbf{d}_{i,j+1} - \mathbf{d}_{i,j}$. (Note that the lower index of the difference operator always refers to the number of the index where it applies to.)

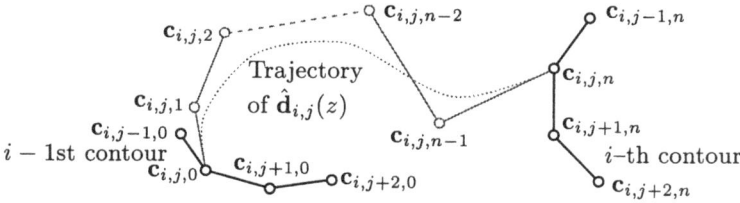

Fig. 1. The definition of the interpolating surface.

(*ii*) Inflection or flat points of the contours $\mathbf{x}_i(t)$ occur only at knots. In the case of C^2 spline curves, the presence of an inflection or flat point causes the three neighbouring control points to be collinear and we even assume to have $\ddot{\mathbf{x}}_i(t_{\mathrm{infl}}) = \vec{0}$ at this point.

The first assumption can always be made true by inserting additional knots into the knot vectors $\mathcal{T}_i$ of the contour curves. The upper (resp. lower) bounds for the angles are due to a sufficient convexity criterion by Goodman [2] which will be used in order to guarantee the property of sc–p interpolation. The second assumption is automatically satisfied if the contour curves have been constructed with the help of the algorithm for shape preserving least–square approximation presented in [4].

§3. Definition of the Surface and Continuity Constraints

The interpolating surface $\mathbf{y}(z,t)$ is defined as a composition of C tensor–product B–spline surfaces $(\mathbf{y}_i(z,t))_{i=1..C}$. The degree of the parameter lines $t = \mathrm{const}$ is equal to $n \geq 2\,l+2$, whereas the contour curves $z = \mathrm{const}$ are of degree d. The i–th surface patch $\mathbf{y}_i(z,t)$ is defined over the parameter domain $(z,t) \in [z_{i-1}, z_i] \times [0,1]$ and it possesses the parametric representation

$$\mathbf{y}_i(z,t) = \sum_{j=0}^{\hat{D}_i} \hat{\mathbf{d}}_{i,j}(z) \cdot \hat{N}_{i,j}^d(t) \qquad (i = 1, ..., C) \tag{3}$$

with the $\hat{D}_i + 1$ *contour control points*

$$\hat{\mathbf{d}}_{i,j}(z) = \sum_{k=0}^{n} \mathbf{c}_{i,j,k} \cdot B_k^n\left(\frac{z - z_{i-1}}{\Delta_{[1]} z_{i-1}}\right), \quad j = 0, ..., \hat{D}_i; \quad \Delta_{[1]} z_i = z_{i+1} - z_i, \tag{4}$$

running on Bézier curves with control points $(\mathbf{c}_{i,j,k})_{k=0,...,n}$, cf. Figure 1. The B–spline basis functions $(\hat{N}_{i,j}^d(t))_{j=0,...\hat{D}_i}$ in (3) are defined over the union $\hat{\mathcal{T}}_i$ of the knot vectors $\mathcal{T}_{i-1}$ and $\mathcal{T}_i$ of the adjacent contours. The blending functions $B_k^n(u) = \binom{n}{k} u^k (1-u)^{n-k}$ are the Bernstein polynomials of degree n. The third components of the control points $\mathbf{c}_{i,j,k} \in \mathbb{R}^3$ of the surface $\mathbf{y}_i(z,t)$ are chosen according to $c_{i,j,k,3} = (1 - \frac{k}{n}) \cdot z_{i-1} + \frac{k}{n} \cdot z_i$ which implies $y_{i,3}(z,t) \equiv z$. Moreover, the first and last control points $\mathbf{c}_{i,j,0}, \mathbf{c}_{i,j,n}$ $(j = 0, ..., \hat{D}_i)$ of the trajectories of the contour control points (4) result immediately from the interpolation conditions. They are obtained by representing the adjacent contour curves $\mathbf{x}_{i-1}(t)$ and $\mathbf{x}_i(t)$ as B-spline curves over the knot vector $\hat{T}_i$ with the help of the knot insertion algorithm, cf. [3].

The first and second components of the remaining control points are unknown yet. They will be computed by solving an appropriate optimization problem. Due to the required order $l = 1, 2$ of differentiability they are subject to the *continuity constraints*

$$\left(\frac{\partial}{\partial z}\right)^\lambda \mathbf{y}_i(z,t)\bigg|_{z=z_i} \equiv \left(\frac{\partial}{\partial z}\right)^\lambda \mathbf{y}_{i+1}(z,t)\bigg|_{z=z_i} \tag{5}$$

for $\lambda = 1, ..., l$ and $i = 1, ..., C-1$. Note that the B–spline basis functions of adjacent surface patches are defined over the possibly different knot vectors $\hat{T}_i$ and $\hat{T}_{i+1}$. After representing both sides of (5) over the union of these knot vectors we obtain a set of linear equations for the control points $\mathbf{c}_{i,j,k}$ by comparing the coefficients. The set of linear equations obtained from (5) is denoted by CC_i. Due to the different knot vectors of adjacent surface patches, it includes certain not–a–knot–type conditions for the unknown control points. Resulting from the choice of the polynomial degree $n \geq 2l+2$ of the parameter lines t=const, each control point $\mathbf{c}_{i,j,k}$ is subject to one set CC_i of continuity constraints at most.

§4. Shape Constraints

Now we consider the conditions on one segment $\mathbf{y}_i(z,t)$ of the interpolating surface which are implied by the desired shape of the contour curves. According to the assumptions made in §2, the shape of the given contours $\mathbf{x}_{i-1}(t) = \mathbf{y}_i(z_{i-1}, t)$ and $\mathbf{x}_i(t) = \mathbf{y}_i(z_i, t)$ coincides with the shape of the control polygons $(\hat{\mathbf{d}}_{i,j}(z_{i-1}))_{i=0,...,\hat{D}_i}$ and $(\hat{\mathbf{d}}_{i,j}(z_i))_{i=0,...,\hat{D}_i}$. We denote by

$$\Delta_{[2]}\hat{\mathbf{d}}_{i,j}(z) = \sum_{k=0}^{n} B_k^n\left(\frac{z-z_i}{\Delta_{[1]}z_{i-1}}\right) \cdot \Delta_{[2]}\mathbf{c}_{i,j,k} \quad (j = 0, ..., \hat{D}_i - 1) \tag{6}$$

the difference vectors $\hat{\mathbf{d}}_{i,j+1} - \hat{\mathbf{d}}_{i,j}$ of adjacent contour control points. The difference vectors at $z=z_{i-1}$ and $z=z_i$ are already known from the interpolation conditions. The following conditions are sufficient for the desired property of sectional curvature–preserving interpolation.

1.) If for two adjacent difference vectors of contour control points the inequality

$$0 \leq \sphericalangle\left(\Delta_{[2]}\hat{\mathbf{d}}_{i,j}(z), \Delta_{[2]}\hat{\mathbf{d}}_{i,j+1}(z)\right) < \frac{\pi}{d-1}$$
$$(\text{resp. } 0 \geq \sphericalangle\left(\Delta_{[2]}\hat{\mathbf{d}}_{i,j}(z), \Delta_{[2]}\hat{\mathbf{d}}_{i,j+1}(z)\right) > -\frac{\pi}{d-1}) \tag{7}$$

holds for both boundaries $z=z_i$ and $z=z_{i-1}$ whereby the angle $\sphericalangle(...)$ vanishes once at most, then we ensure that it is even true for *all* $z \in [z_{i-1}, z_i]$.

2.) If the angle in (7) vanishes for $z=z_{i-1}$ *and* $z=z_i$ then the control points $\hat{\mathbf{d}}_{i,j+1}(z_{i-1})$ and $\hat{\mathbf{d}}_{i,j+1}(z_i)$ are an affine combination of their neighbours,

$$\hat{\mathbf{d}}_{i,j+1}(z_{i-1}) = (1 - \rho) \cdot \hat{\mathbf{d}}_{i,j}(z_{i-1}) + \rho \cdot \hat{\mathbf{d}}_{i,j+2}(z_{i-1})$$
$$\hat{\mathbf{d}}_{i,j+1}(z_i) = (1 - \sigma) \cdot \hat{\mathbf{d}}_{i,j}(z_i) + \sigma \cdot \hat{\mathbf{d}}_{i,j+2}(z_i)$$
(8)

with some constants $\rho, \sigma \in \mathbb{R}$. Resulting from the assumption (ii) made in §2, these numbers are equal, $\sigma = \rho$. So we can add the linear equations

$$\mathbf{c}_{i,j+1,k} = (1 - \rho) \cdot \mathbf{c}_{i,j,k} + \rho \cdot \mathbf{c}_{i,j+2,k} \text{ for } k = 0, ..., n \qquad (9)$$

to the set of shape constraints. Note that these equations are compatible with the continuity conditions obtained from (5).

The second case may happen for C^2 surfaces if both contour curves have an inflection with the same parameter value t_{infl}. The two sets of constraints obtained from 1.) and 2.) guarantee the following property: for each sub–polygon $(\hat{\mathbf{d}}_{i,j}(z))_{j=p,...,q}$ $(0 \leq p < q \leq \hat{D}_i)$ of the contour control polygons the angles between adjacent legs are always non–negative (resp. non–positive) for all $z \in [z_{i-1}, z_i]$ and smaller than $\frac{\pi}{d-1}$ (resp. greater than $-\frac{\pi}{d-1}$), provided that this is true for the boundaries $z=z_{i-1}$ and $z=z_i$. So it is possible to apply Goodman's sufficient convexity criterion [2]. Therefore the constraints imply that the interpolating contour curves preserve the curvature signs of the given contours.

The conditions obtained from 1.) are guaranteed with the help of linear inequalities which are constructed using the following observation.

Lemma 2. *Let a constant* $\lambda \in \mathbb{R}$ *and four vectors* $\vec{u}_0, \vec{u}_1, \vec{v}_0, \vec{v}_1 \in \mathbb{R}^3$ *satisfying* $\|\vec{u}_0\|=\|\vec{u}_1\|=\|\vec{v}_0\|=\|\vec{v}_1\|=1$ *and* $0 \leq \sphericalangle (\vec{u}_0, \vec{u}_1) < \pi$, $0 \leq \sphericalangle (\vec{v}_0, \vec{v}_1) < \frac{\pi}{d-1}$ *be given. If the control points fulfill the linear inequalities*

$$[-\vec{u}_1, \Delta_{[2]}(\mathbf{c}_{i,j,k} - \lambda\mathbf{c}_{i,j+1,k})] \geq 0, \quad [\vec{u}_0, \Delta_{[2]}\mathbf{c}_{i,j+1,k}] \geq 0,$$
$$[\Delta_{[2]}(\mathbf{c}_{i,j,k} - \lambda\mathbf{c}_{i,j+1,k}), \vec{u}_0] \geq 0, \quad [\Delta_{[2]}\mathbf{c}_{i,j+1,k}, \vec{u}_1] \geq 0,$$
(10)

and

$$[\vec{v}_0, \Delta_{[2]}\mathbf{c}_{i,j,k}] \geq 0, \quad [\Delta_{[2]}\mathbf{c}_{i,j+1,k}, \vec{v}_1] \geq 0 \qquad (11)$$

for $k = 0, ..., n$, *then the trajectories of the contour control points satisfy the relation in the first line of (7) for all* $z \in [z_{i-1}, z_i]$.

Proof: The inequalities (10) and (11) imply that the difference vectors of the control points $\mathbf{c}_{i,j,k}$ can be separated as shown in Figures 2a and b. Due to the convex hull property, this is also true for the difference vectors $\Delta_{[2]}\mathbf{d}_{i,j}(z)$ of the contour control points for $z \in [z_{i-1}]$.

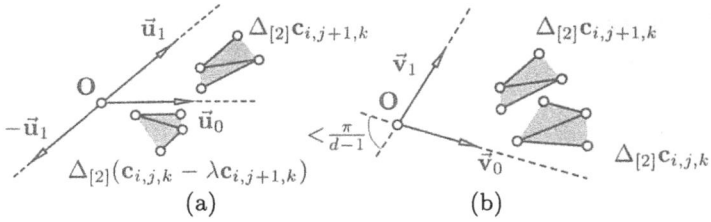

Fig. 2. Linearization of the shape constraints (top view).

As evident from Figure 2a the inequalities (10) lead to

$$[\Delta_{[2]}(\mathbf{d}_{i,j}(z)-\lambda\mathbf{d}_{i,j+1}(z)), \Delta_{[2]}\mathbf{d}_{i,j+1}(z)]=[\Delta_{[2]}\mathbf{d}_{i,j}(z), \Delta_{[2]}\mathbf{d}_{i,j+1}(z)] \geq 0,$$
$$(12)$$

thus they guarantee the left–hand side of (7). Similarly the inequalities (11) imply the right–hand side of (7). ∎

We introduced the constant λ (based on the identity (12)) in order to keep the number of required inequalities as small as possible. Similarly one can also modify the second argument of the bracket product $[.,.]$; this yields a mirrored version of Lemma 2.

For generating the linear inequalities which ensure the constraints obtained from 1.) one has to choose a couple of constants and bounding vectors. This is done automatically with the help of algorithms described in the report [5]. Due to space limitations we are not able to describe these algorithms in more detail. They are based on so–called *reference curves* $(\Delta_{i,j}^{\mathrm{sprl}}(z))_{j=0,...,\hat{D}_i-1}$ which represent the expected turns of the difference vectors of the contour control points. For example, the reference curves can be chosen as segments of Archimedean spirals which interpolate the difference vectors $\Delta_{[2]}\mathbf{d}_{i,j}(z_{i-1})$, $\Delta_{[2]}\mathbf{d}_{i,j}(z_i)$ of the control points of the given contour curves.

Lemma 2 can also be applied to subsegments $z \in [z_a, z_e] \subseteq [z_{i-1}, z_i]$ of the trajectories $\hat{\mathbf{d}}_{i,j}(z)$ of the contour control points. We then have to replace the control points $(\mathbf{c}_{i,j,k})_{k=0,...,n}$ by those of the subsegments which result from the de Casteljau scheme. Sometimes it is necessary to use this idea in order to obtain the linearized constraints, see [5].

§5. Computing the Control Points

The unknown components of the control points $\mathbf{c}_{i,j,k}$ of the interpolating spline surface are found by minimizing an appropriate objective function subject to the linear shape and continuity constraints. The objective function is chosen such that the transition of the control polygons of adjacent contours $(\mathbf{x}_i(t))_{i=0,...,C}$ becomes as smooth as possible. For this we take sample points from the reference curves and minimize the corresponding least–square sum. Moreover we add a "tension term" for the trajectories of the first and the last contour control points, e.g. the sum of the squared lengths $\|\Delta_{[3]}\mathbf{c}_{i,0,k}\|^2$ and

Given contour curves:

$z_3 = 32$

$z_4 = 40$

$z_1 = 10$

$z_2 = 17$

$z_0 = 0$

Interpolating contours:

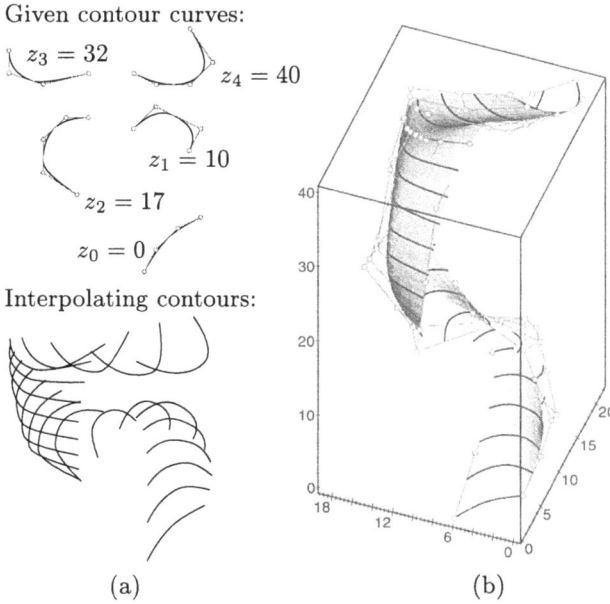

(a)   (b)

Fig. 3. A spline surface which interpolates five given contours.

$\|\Delta_{[3]}\mathbf{c}_{i,\hat{D}_i,k}\|^2$ of their control polygons ($i=1,...,n-1$). This guarantees the uniqueness of the solution, see [5].

We obtain a quadratic function of the control points $\mathbf{c}_{i,j,k}$ ($i=1,...,C$; $j=0,...,\hat{D}_i$; $k=1,...,n-1$). The minimization of the objective function under the linear equality and inequality constraints ensuring the desired shape and continuity properties therefore leads to a *quadratic programming problem* which is solved with the help of an active set strategy as described in the textbook by Fletcher [1]. This strategy requires an initial solution which has to be constructed first with the help of linear programming, i.e., with the simplex algorithm. We choose the initial solution as close as possible to the optimal one. This is achieved by choosing the objective function of the auxiliary linear programming problem as the l^1 norm (taken in the linear space of the unknown components of the control points) of the difference to the solution of the unconstrained problem. The latter one is obtained by minimizing the quadratic objective function under equality constraints (which arise from the continuity and interpolation conditions) only. It can be computed using Lagrangian multipliers leading to a system of linear equations for the unknown components of the control points.

Note that the existence of solutions is not automatically guaranteed. In our examples, the feasible region of the LP problem was always non–empty. It can be shown that solutions exist, provided that the polynomial degree n of the parameter lines $t = $ const has been chosen high enough [5].

As an example we show a spline surface which has been obtained by interpolating five contour curves. The given contours are described by B–spline curves of order three which are defined over different knot vectors. The interpolating C^1 spline surface of degree $(2, 3)$ preserves the signs of the sectional curvature. Figure 3a shows the top view of the contour curves, whereas the resulting spline surface and its control points have been drawn in Figure 3b. The construction of the surface led to a quadratic programming problem with 92 unknowns, 48 equality constraints and 201 inequality constraints. In our implementation we use the equality constraints for eliminating a part of the unknowns from the problem; this yielded a QP problem with only 44 unknowns. Only four inequalities are active for the final solution.

Acknowledgments. The author thanks Professor Tim Goodman for his interest in the presented work. The financial support by the European Union through the network FAIRSHAPE is also gratefully acknowledged.

References

1. Fletcher, R., *Practical Methods of Optimization*, Wiley, Chichester 1991.

2. Goodman, T. N. T., Inflections on curves in two and three dimensions, Comput. Aided Geom. Design **8** (1991), 37–50.

3. Hoschek, J., and D. Lasser, *Fundamentals of Computer Aided Geometric Design*, A. K. Peters, Wellesley MA, 1993.

4. Jüttler, B., Shape preserving least–square approximation by polynomial parametric spline curves, University of Dundee, Applied Analysis Report 965, 1996, to appear in Comput. Aided Geom. Design.

5. Jüttler, B., Sectional Curvature Preserving Approximation of Contour Lines, University of Dundee, Applied Analysis Report 966, 1996.

6. Kaklis, P. D. and A. I. Ginnis, Sectional–Curvature Preserving Skinning Surfaces, National Technical University of Athens, Ship–Design Laboratory Technical Report, to appear in Comput. Aided Geom. Design.

7. Schumaker, L. L., Reconstructing 3D objects from cross–sections, in *Computation of Curves and Surfaces*, W. Dahmen, M. Gasca and C. A. Micchelli (eds.), Kluwer, Dordrecht, 1990, 275–309.

8. Unsworth, K., Recent developments in surface reconstruction from planar cross–sections, in *Computer Aided Geometric Design*, C. A. Micchelli and H. B. Said (eds.), Annals of Numerical Mathematics **3** (1996), 401–422.

Bert Jüttler
Technische Hochschule Darmstadt
Fachbereich Mathematik, AG 3
Schloßgartenstraße 7
64289 Darmstadt, GERMANY
juettler@mathematik.th-darmstadt.de

Analysis of Curvature-Related
Surface Shape Properties

Johannes Kaasa and Geir Westgaard

Abstract. This is a description of two new measures we use in surface shape analysis, developed by our colleague Even Mehlum. The first is a second order curvature measure, and the other is a third order measure for variation of curvature. We analyse the difference in complexity between curvature and curvature variation, which serves as a motivation for the definition of the third order measure. The presentation is concluded with an example showing the usefulness of the new measures.

§1. Introduction

A major challenge in Computer Aided Geometric Design is to develop methods for surface modelling and manipulation which give the designer sufficient control over the shape of the surface. Properties related to surface curvature are important tools in analysis of surface shapes, see for instance [2,3,4,5]. We discuss experiments with two new measures for curvature and variation of curvature, introduced and calculated by our colleague Even Mehlum.

In Section 2 we describe the new measures. We present them without proofs and we skip the basics of differential geometry (a differential geometry background can be found for instance in [1]). In Section 3 we compare the complexity of curvature and variation of curvature. Section 4 contains an example.

§2. Two New Measures for Curvature and Variation of Curvature

Traditional curvature measures are combinations of the normal curvature in the principal directions:

Gaussian curvature K is a product of the principal curvatures, *mean curvature* H is a mean of the principal curvatures, *absolute curvature* A is a sum of the absolute values of the principal curvatures and *total curvature* T is a sum of their squares.

Curves and Surfaces with Applications in CAGD
A. Le Méhauté, C. Rabut, and L. L. Schumaker (eds.), pp. 211–215.
Copyright © 1997 by Vanderbilt University Press, Nashville, TN.
ISBN 0-8265-1293-3.
All rights of reproduction in any form reserved.

To take into consideration the complete behaviour of the normal curvature, Mehlum has introduced the *Mehlum second order curvature measure*, where the square of the normal curvature k_n is integrated in all tangent directions θ around the surface point:

$$M_2 = \frac{1}{\pi}\int_0^\pi k_n^2(\theta)d\theta = \frac{3}{2}H^2 - \frac{1}{2}K. \tag{1}$$

A property equally important as curvature is variation of curvature. Mehlum has introduced the *Mehlum third order curvature variation measure* by integrating the square of the arc length derivative of the normal curvature in all tangent directions around the surface point:

$$M_3 = \frac{1}{\pi}\int_0^\pi (\frac{d}{ds}k_n(\theta))^2 d\theta. \tag{2}$$

To calculate this expression, it is necessary to first find $\frac{d}{ds}k_n(\theta)$ as a function of θ (the angle in the tangent plane relative to the maximum principal direction). Not surprisingly, this is a rather complex expression. If $S(u,v): U \subset \mathbb{R}^2 \to \mathbb{R}^3$ is the surface in question, N the unit surface normal, E, F, G the first fundamental form coefficients, and e, f, g the coefficients of the second fundamental form, we introduce some utility variables to make things more tractable:

$$\begin{aligned}
&\sigma = \sqrt{EG - F^2}, \\
&\alpha = \langle S_{uu}, S_u \rangle/\sigma^2, \beta = \langle S_{uu}, S_v \rangle/\sigma^2, \gamma = \langle S_{uv}, S_u \rangle/\sigma^2, \\
&\delta = \langle S_{uv}, S_v \rangle/\sigma^2, \epsilon = \langle S_{vv}, S_u \rangle/\sigma^2, \mu = \langle S_{vv}, S_v \rangle/\sigma^2, \\
&a = Ff - Ge, b = Fe - Ef, c = Fg - Gf, d = Ff - Eg, \\
&P = \langle N, S_{uuu} \rangle + 3(a\alpha + b\beta), \\
&Q = \langle N, S_{uuv} \rangle + c\alpha + d\beta + 2a\gamma + 2b\delta, \\
&S = \langle N, S_{uvv} \rangle + 2c\gamma + 2d\delta + a\epsilon + b\mu, \\
&T = \langle N, S_{vvv} \rangle + 3(c\epsilon + d\mu), \\
&D = \sqrt{H^2 - K}, \\
&\varphi = \arccos(\frac{g - HG}{GD}), \varphi = \arcsin(\frac{fG - gF}{\sigma GD}), \\
&\psi = \arccos(\frac{\sigma}{\sqrt{EG}}), \psi = \arcsin(\frac{F}{\sqrt{EG}}).
\end{aligned} \tag{3}$$

The derivative of the normal curvature is then

$$\begin{aligned}
\frac{d}{ds}k_n(\theta) = \frac{1}{\sigma^3}[&PG^{\frac{3}{2}}\sin^3(\theta + \frac{\varphi}{2}) + 3QGE^{\frac{1}{2}}\cos(\theta + \frac{\varphi}{2} + \psi)\sin^2(\theta + \frac{\varphi}{2}) \\
+ &3SG^{\frac{1}{2}}E\cos^2(\theta + \frac{\varphi}{2} + \psi)\sin(\theta + \frac{\varphi}{2}) + TE^{\frac{3}{2}}\cos^3(\theta + \frac{\varphi}{2} + \psi)].
\end{aligned} \tag{4}$$

This can be viewed as a third order version of the Euler formula, which expresses the connection between normal curvature and tangential rotation angle.

Integrating the square of this we get Mehlum's third order curvature variation:

$$
\begin{aligned}
M_3 = [&5G^3P^2 + (EG + 4F^2)(9GQ^2 + 9ES^2 + 6GPS + 6EQT) \\
&+ 5E^3T^2 - 2F(3EG + 2F^2)(PT + 9QS) - 30F(G^2PQ + E^2ST)] \quad (5) \\
&/[16(EG - F^2)^3].
\end{aligned}
$$

Roughly speaking, we can say that M_3 reveals more about the shape than M_2, since it is based on one higher derivative.

§3. Remarks on the Complexity of Curvature Variation

In this section we plot an example of the normal curvature and its derivative, as a function of the rotational angle in the surface tangent plane, see Figure 1. The motive is to illustrate the difference in complexity between these measures, and thereby the difference in complexity between curvature and variation of curvature.

The test is performed on a water turbine blade. The abscissa of the plot represents the rotational angles, with regard to the maximum principal direction. This means that we have principal directions at 0, $\frac{\pi}{2} \approx 1.6$, $\pi \approx 3.1$ and $\frac{3\pi}{2} \approx 4.7$.

We see from the graph on the left that the normal curvature has a trigonometric behaviour between the principal directions, as can be expected from the Euler formula. It is therefore quite acceptable to express curvature as a combination of the principal curvatures.

Variation of curvature is another matter. The equation (4) and the graph on the right show that the principal curvatures do not give many clues about the shape of the function. We must therefore conclude that a definition of curvature variation should utilise information from more directions than the principal ones, as is achieved in (5).

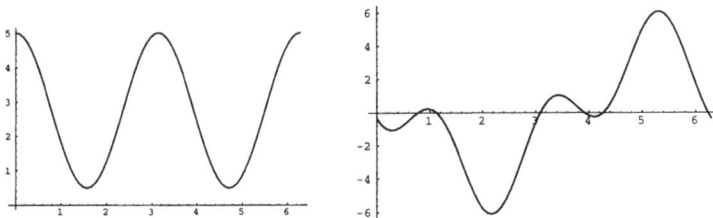

Fig. 1. The normal curvature and its derivative.

§4. An Example with Mehlum Curvature Measures

To get a better understanding of the nature of the different curvature measures, in this section we give an example.

In Figure 2 we show a spherelike NURBS surface. The order is 3 in both parameter directions, and the number of coefficients is 5 by 3. At a first glance it looks as if it has a smooth curvature and variation of curvature. However, the nature of the surface is revealed by the second and third order Mehlum curvature, especially the variation of curvature in Figure 4 has profound features. One has to take into account that the squaring in (1) and (2) contributes to this: if the normal curvature (or its variation) is less than 1 it is shrunk, if it is greater than 1 it is blown up.

The example illustrates how well-suited the third order Mehlum curvature is for shape analysis.

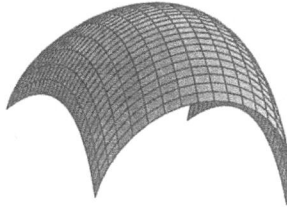

Fig. 2. A NURBS surface with a spherelike shape.

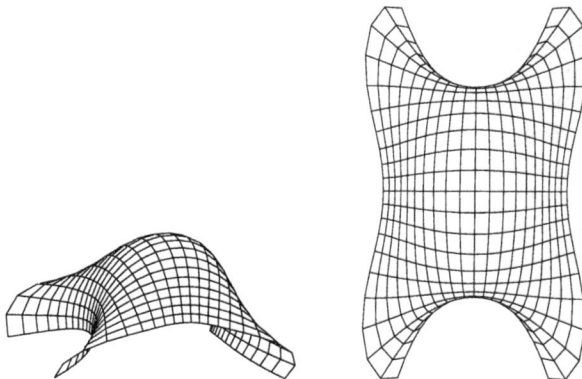

Fig. 3. The second order Mehlum curvature of the surface as an offset grid.

Fig. 4. The third order Mehlum curvature of the surface as an offset grid.

References

1. Do Carmo, M. P., *Differential Geometry of Curves and Surfaces*, Prentice-Hall, New Jersey, 1976.

2. Greiner, G., A. Kolb, R. Pfeifle, H.-P. Seidel, P. Slusallek, M. Encarnação, and R. Klein, A platform for visualizing curves and surfaces, Computer-Aided Design **27** (1995), 559–566.

3. Hagen, H., S. Hahmann, and T. Schreiber, Visualization and computation of curvature behaviour of freeform curves and surfaces, Computer-Aided Design **27** (1995), 545–552.

4. Maekawa, T., F.-E. Wolter, and N. M. Patrikalakis, Umbilics and lines of curvature for shape interrogation, Comput. Aided Geom. Design **13** (1996), 133–161.

5. Moreton, H. P., Simplified curve and surface interrogation via mathematical packages and graphics libraries and hardware, Computer-Aided Design **27** (1995), 523–543.

Johannes Kaasa and Geir Westgaard
SINTEF, Institute for Applied Mathematics
P.O. BOX 124, Blindern
N-0314 Oslo, NORWAY
Johannes.Kasa@si.sintef.no
Geir.Westgaard@si.sintef.no

On an Almost–Convex–Hull Property

Daniela P. Kacsó and Hans–Jörg Wenz

Abstract. Some operators in CAGD do not have the convex hull property since they are not positive. An example is the Overhauser spline which has other desirable properties. However, it is intuitively clear that the graph of the Overhauser spline is not too far from the convex hull of its control points. The present paper introduces a generalization of the convex hull property in order to quantify this intuitive feeling.

§1. Introduction

The convex hull property is well understood and it is shared by many operators in the field of CAGD, for instance by the Bernstein–Bézier operator and the Schoenberg operator. In $\mathbb{R}^2$, the convex hull property reads as follows: Given a sequence of points f_i, $i = 0, \ldots, n$, the operator $\mathcal{A} : (\mathbb{R}^2)^{n+1} \to (span\{b_0, \ldots, b_n\})^2$, $\mathcal{A}(f_0, \ldots, f_n; \cdot) = \sum_{i=0}^n f_i b_i$ with the real valued basis functions b_i, $i = 0, \ldots, n$, is said to have the *convex hull property* whenever $\mathcal{A}(f_0, \ldots, f_n; t)$ lies in the convex hull $[f_0, \ldots, f_n]$ of control points for any $t \in I \subseteq \mathbb{R}$. Here, I is the parameter interval.

As Farin [4], p. 35, points out, the importance of the classical convex hull property lies in interference checking. Instead of checking whether two given curves have a collision one checks whether the bounding boxes of the respective convex hulls overlap.

It is well known that the operator $\mathcal{A}$ has the convex hull property if and only if the corresponding functions b_i, $i = 0, \ldots, n$, form a partition of unity and are all nonnegative in the parameter interval I (and thus $\mathcal{A}$ is a nonnegative operator). However, there are operators in CAGD and approximation theory which possess a lot of desirable properties, but are not nonnegative, and thus do not have the convex hull property. Among these are the interpolating cubic splines with all kinds of boundary conditions, the Overhauser spline [3,4,6] as well as a family of modified Bernstein operators of Cao and Gonska [2].

The paper is organized as follows: In Section 2 we present some preliminary considerations in the univariate case which help to understand our generalization of the convex hull property in Section 3. Finally, we include some examples in Section 4.

Curves and Surfaces with Applications in CAGD
A. Le Méhauté, C. Rabut, and L. L. Schumaker (eds.), pp. 217–222.
Copyright. © 1997 by Vanderbilt University Press, Nashville, TN.
ISBN 0-8265-1293-3.

All rights of reproduction in any form reserved.

Fig. 1. Center of gravity for $q_x \cdot q_y \geq 0$.

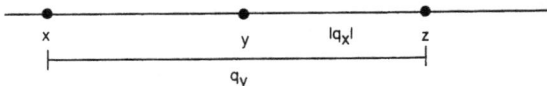

Fig. 2. Center of gravity for $q_x > 0$, $q_y < 0$.

Fig. 3. Blown up convex hull of two points.

§2. Preliminary Considerations

First we look at the following problem: Given two points $x < y$ on the real axis with charges q_x and q_y, respectively, the center of charges (in analogy to the center of gravity) is given by

$$z = \frac{xq_x + yq_y}{q_x + q_y} \ , \tag{1}$$

$q_x \neq -q_y$. Now our task is to find a possibly small interval in which the center of charge is to be found, once a certain *a priori* knowledge concerning the charges q_x and q_y is available. For $q_x \cdot q_y \geq 0$, we have the situation $x \leq z \leq y$ and $z - x = |q_y|$, $y - z = |q_x|$, see Fig. 1. For $q_x \cdot q_y < 0$, one possible situation is $q_x > 0$, $q_y < 0$. Then we have $x < y \leq z$ and $z - x = q_y$, $y - z = q_x$, see Fig. 2.

Let us assume without loss of generality $q_x + q_y = 1$ and $-m \leq q_x, q_y \leq 1 + m$ with $0 \leq m$. Thus, from (1) we conclude

$$x(1 + m) + y(-m) \leq z = xq_x + yq_y \leq x(-m) + y(1 + m) \ ,$$

or, equivalently

$$x - m(y - x) \leq z \leq y + m(y - x) \ ,$$

which means that the center of charge, z, is to be found in an interval $[a, b]$ that is obtained by "blowing up" the convex hull $[x, y]$ by a factor $1 + 2m$, see Fig. 3.

§3. The Almost–Convex–Hull Property

To understand our general concept, it is enough to consider the case of plane curves. From this, the case of curves and surfaces in higher dimensions should be well understood.

Suppose we are given parameter values $t_0 < \ldots < t_n$ as well as the respective control points $\boldsymbol{f}_i = (x(t_i), y(t_i))^T$, $i = 0, \ldots, n$. Moreover, let the operator $\mathcal{A}$ be given by

$$\mathcal{A}(\boldsymbol{f}_0, \ldots, \boldsymbol{f}_n; \cdot) = \sum_{i=0}^{n} \boldsymbol{f}_i b_i \qquad (2)$$

with real–valued functions $b_i : I := [t_0, t_n] \to \mathbb{R}$, $i = 0, \ldots, n$, satisfying $\sum_{i=0}^{n} b_i(t) = 1$, $t \in I$. With the notation $g^+ := \max\{g, 0\}$ and $g^- := \max\{-g, 0\}$ (pointwise definition) for the positive and negative part of a real–valued function, we can write $\mathcal{A}$ as

$$\mathcal{A}(\boldsymbol{f}_0, \ldots, \boldsymbol{f}_n; \cdot) = \sum_{i=0}^{n} \boldsymbol{f}_i b_i = \sum_{i=0}^{n} \boldsymbol{f}_i (b_i^+ - b_i^-) = \sum_{i=0}^{n} \boldsymbol{f}_i b_i^+ - \sum_{i=0}^{n} \boldsymbol{f}_i b_i^- . \qquad (3)$$

We claim here that $\sum_{i=0}^{n} b_i^+ > 0$ holds on I. Assume to the contrary that for some parameter τ we have $\sum_{i=0}^{n} b_i^+ = 0$. Then we conclude

$$1 = \sum_{i=0}^{n} b_i(\tau) = -\sum_{i=0}^{n} b_i^-(\tau) \le 0 ,$$

which is a contradiction. Thus, from (3) we can now conclude

$$\mathcal{A}(\boldsymbol{f}_0, \ldots, \boldsymbol{f}_n; t) = \sum_{i=0}^{n} \boldsymbol{f}_i b_i^+(t) \quad \text{for} \quad \sum_{i=0}^{n} b_i^-(t) = 0 \qquad \text{(Case I)}$$

and

$$\mathcal{A}(\boldsymbol{f}_0, \ldots, \boldsymbol{f}_n; t) = \left(\sum_{j=0}^{n} b_j^+(t) \right) \cdot \sum_{i=0}^{n} \boldsymbol{f}_i \frac{b_i^+(t)}{\sum_{j=0}^{n} b_j^+(t)}$$

$$+ \left(-\sum_{j=0}^{n} b_j^-(t) \right) \cdot \sum_{i=0}^{n} \boldsymbol{f}_i \frac{b_i^-(t)}{\sum_{j=0}^{n} b_j^-(t)}$$

$$\text{for} \quad \sum_{i=0}^{n} b_i^-(t) \ne 0. \qquad \text{(Case II)}$$

In Case I we know that $\mathcal{A}(\boldsymbol{f}_0, \ldots, \boldsymbol{f}_n; t)$ lies in the convex hull of the $\boldsymbol{f}_i$, $i = 0, \ldots, n$. Observe that in case II the nonnegative functions $\dfrac{b_i^+}{\sum_{j=0}^{n} b_j^+}$ and

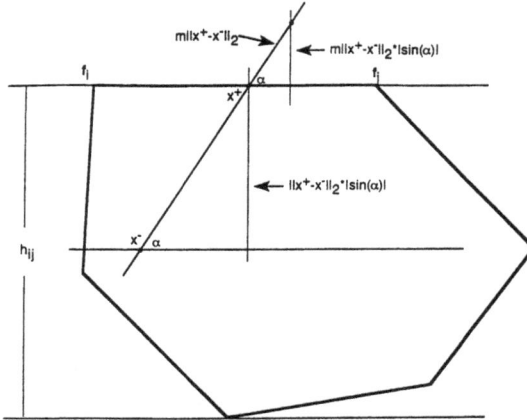

Fig. 4. Construction of an almost–convex–hull in two dimensions.

$\frac{b_i^-}{\sum_{j=0}^n b_j^-}$, $i = 0, \ldots, n$, respectively, form a partition of unity. Therefore, the centers of positive and negative charge

$$x^+(t) = \sum_{i=0}^n \boldsymbol{f}_i \frac{b_i^+(t)}{\sum_{j=0}^n b_j^+(t)} \quad \text{and} \quad x^-(t) = \sum_{i=0}^n \boldsymbol{f}_i \frac{b_i^-(t)}{\sum_{j=0}^n b_j^-(t)}$$

are also situated within the classical convex hull of the control points $\boldsymbol{f}_i$, $i = 0, \ldots, n$.

With $q^+ := \sum_{i=0}^n b_i^+$ and $q^- := \sum_{i=0}^n b_i^-$ we can now write

$$\mathcal{A}(\boldsymbol{f}_0, \ldots, \boldsymbol{f}_n; \cdot) = q^+ x^+ - q^- x^- . \tag{4}$$

Here, the point $x^-(t)$ is defined to equal $x^+(t)$ in case $q^-(t) = 0$.

We remark here that our function q^- is closely related to the so–called Lebesgue function $L = \sum_{i=0}^n |b_i|$ which is useful for studying interpolation processes (see de Boor [1], p. 25 and p. 234):

$$L = \sum_{i=0}^n |b_i| = \sum_{i=0}^n b_i^+ + b_i^- = \sum_{i=0}^n b_i^+ - b_i^- + 2b_i^- = \sum_{i=0}^n b_i + 2\sum_{i=0}^n b_i^- = 1 + 2q^- .$$

Now observe that in Case II, the points $x^+(t)$ and $x^-(t)$ lie in the convex hull of the control points $\boldsymbol{f}_i$, $i = 0, \ldots, n$, and that $\mathcal{A}(\boldsymbol{f}_0, \ldots, \boldsymbol{f}_n; t)$ is situated on the line ℓ through $x^+(t)$ and $x^-(t)$. This leads us back to our considerations in the previous section. According to our remarks there, and with $m := \max_{t \in I} \sum_{i=0}^n b_i^-$ and the fact $q^+(t) + q^-(t) = 1$, $t \in I$, the point $\mathcal{A}(\boldsymbol{f}_0, \ldots, \boldsymbol{f}_n; t)$ lies on ℓ, at most $m \cdot \|x^+(t) - x^-(t)\|_2$ apart from the segment $[x^+(t), x^-(t)]$.

If ℓ intersects some segment $[\boldsymbol{f}_i, \boldsymbol{f}_j]$ of the boundary of $[\boldsymbol{f}_0, \ldots, \boldsymbol{f}_n]$ and $\mathcal{A}(\boldsymbol{f}_0, \ldots, \boldsymbol{f}_n; t)$ is not in $[\boldsymbol{f}_0, \ldots, \boldsymbol{f}_n]$, then an inspection of Figure 4 shows that the Euclidean distance of $\mathcal{A}(\mathbf{f}_0, \ldots, \mathbf{f}_n; t)$ from the line $\ell_{i,j}$ through $[\mathbf{f}_i, \mathbf{f}_j]$ is at most $m \cdot h_{ij}$ with

$$h_{ij} := \max_{k \in \{0,\ldots,n\}} \min_{\mathbf{x} \in \ell_{ij}} \|\mathbf{f}_k - \mathbf{x}\|_2 . \tag{5}$$

To summarize the discussions above, we state the following result:

Theorem 1. *Suppose we are given control points* $\mathbf{f}_i$, $i = 0, \ldots, n$, *together with the respective parameters* $t_0 < \ldots < t_n$ *and some operator* $\mathcal{A}(\mathbf{f}_0, \ldots, \mathbf{f}_n; \cdot)$ *as in (2). Moreover, let* $m := \max_{t \in I} \sum_{i=0}^{n} b_i^-(t)$, *let* $H_1, \ldots, H_p$ *be the unique distinct (not disjoint) closed halfplanes with*

$$[\mathbf{f}_0, \ldots, \mathbf{f}_n] = \bigcap_{j=1}^{p} H_j ,$$

let ℓ_j *be the straight line defining* H_j, $j = 1, \ldots, p$, *let*

$$h_j := \max_{k \in \{0,\ldots,n\}} \min_{\mathbf{x} \in \ell_j} \|\mathbf{f}_k - \mathbf{x}\|_2 , \quad j = 1, \ldots, p ,$$

and let finally the closed halfplane $\tilde{H}_j \supseteq H_j$ *be defined by a parallel to* ℓ_j *in Euclidean distance* $m \cdot h_j$, $j = 1, \ldots, p$. *Then we have*

$$\mathcal{A}(\mathbf{f}_0, \ldots, \mathbf{f}_n; t) \in G := \bigcap_{j=1}^{p} \tilde{H}_j, \quad t \in I,$$

where the set on the right-hand side is the generalized convex hull of the control points $\mathbf{f}_i$, $i = 0, \ldots, n$.

Remarks.

1) By construction, G is convex and invariant with respect to affine transformations.
2) For $m = 0$ (i.e., for positive operators), $G = [\mathbf{f}_0, \ldots, \mathbf{f}_n]$.
3) If the basis functions b_i in (2) have local support, a local version of the generalized convex hull property above is easily derived.
4) For a translation of the theorem to the case of curves and surfaces in $\mathbb{R}^d$, $d > 2$, read *halfplane* as *halfspace* and *straight line* as *hyperplane*.

§4. Examples

As special examples of non–positive operators we consider here the interpolating Overhauser spline (piecewise cubic spline interpolation with Bessel-conditions at the knots) [3,4,6] with 6 equidistant points as well as a six control point operator of Cao and Gonska [2,5]. Following [2] and with the notations therein, we considered a convolution operator with Jackson kernel of order 3 and a composite trapezoidal rule with respect to the $N_1 + 1 = 6$

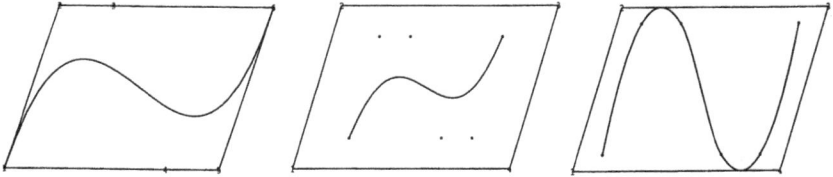

Fig. 5. Comparison of the Bernstein, Cao-Gonska, and Overhauser operators.

knots $y_j = \cos j\pi/5$, $j = 0, \ldots, 5$, corrected by means of a Boolean sum in order to obtain reproduction of linear functions for the resulting operator. The characteristic magnitudes are $m = 1/8$ for Overhauser and $m \leq 0.31$ for Cao and Gonska.

Figure 5 shows the control points, the corresponding functions, and their almost–convex–hulls when applying the operators of Bernstein, Cao-Gonska, and Overhauser to the data $(1,1), (2,2), (3,2), (4,1), (5,1), (6,2)$.

References

1. Boor, C. de, *A Practical Guide to Splines*, Springer Verlag, New York, 1978.

2. Cao, J.–d. and H. H. Gonska, Computation of DeVore–Gopengauz–type approximants, in *Approximation Theory VI*, C. Chui, L. Schumaker, and J. Ward (eds), Academic Press, New York, 1989, 117–120.

3. Cao, J.–d., H. H. Gonska, and D. P. Kacsó, On some polynomial curves derived from trigonometric kernels, in *Curves and Surfaces in Geometric Design*, A. Le Méhauté, C. Rabut, and L. L. Schumaker (eds), Vanderbilt University Press, Nashville TN, 1997, 53–60.

4. Degen, W. L. F., The shape of the Overhauser spline, Computing Supplement **10** (1995), 117–128.

5. Farin, G., *Curves and Surfaces for Computer Aided Geometric Design*, 3rd. ed., Academic Press, NY, 1993.

6. Overhauser, A. W., Analytic definition of curves and surfaces by parabolic blending, Ford Motor Company Technical Report SL 68–40, 1968.

Daniela P. Kacsó, Hans–Jörg Wenz
Dept. of Mathematics
University of Duisburg
D–47057 Duisburg, GERMANY
kacso@informatik.uni-duisburg.de
wenz@informatik.uni-duisburg.de

Developable Surfaces with Creases

Y. L. Kergosien

Abstract. We give examples of creased developable surfaces and a criterion for the necessity of using creases when fitting an applicable surface to a given boundary curve in space. Using developable surfaces with creases, we give conditions for the existence of a family of isometric deformations of a cylindrical box in the symmetric case and show how to construct it. Applications to the computer synthesis of realistic images of creased paper and some post-buckling shapes of shells are presented.

§1. Developable Surfaces and Creases

Developable surfaces got their name from their property of being locally isometric to pieces of planes (the applicability property). Developables have a practical importance in the metal industry for modelling the surfaces obtained from the isometric deformations of thin plane parts. The isometry requirement comes from the approximate unstretchability of the material in such conditions. Some results on the approximation of developable surfaces are available [11,8,1,7].

As a definition [9], we may take one of the geometric differential conditions on the surface which imply applicability under some regularity assumption. Namely a *developable* is either: (1) a regular surface with an everywhere vanishing Gaussian curvature, (2) the envelope of a 1-parameter family of planes, (3) a ruled surface such that along any of its generators the tangent plane is constant. Definition (2) permits considering also singular developables; definition (1) allows for surfaces combining pieces of planes and several pieces of developables defined as in (2).

We shall repeatedly refer to the *generators* of a developable as the generators of the ruled surface of definition (3). Note that starting from a 1-family of planes as in (2), one builds the surface first finding the stationary points of the family of planes (the characteristic points), which constitute a line for each plane of the family (the characteristic line, along which the common tangent plane touches the surface). The envelope is generated as a ruled surface

Curves and Surfaces with Applications in CAGD
A. Le Méhauté, C. Rabut, and L. L. Schumaker (eds.), pp. 223–230.
Copyright © 1997 by Vanderbilt University Press, Nashville, TN.
ISBN 0-8265-1293-3.
All rights of reproduction in any form reserved.

by these characteristic lines, which are classically called the generators of the developable.

Planes, cylinders, and cones are developables, as well as the surfaces generated by the tangents to a given curve in space, which can have irregular points constituting edges of regression. However, these irregular points are still very tame when compared to the cases studied by Lebesgue [6]. In this paper we shall be interested in simpler cases of irregularity to model the creases which may appear on surfaces when crushing thin sheets of paper or metal, either with mechanically free edges or within boxes such as cans:

Definition. *A regular developable patch is a developable homeomorph to a disk with a C^2 interior and a piecewise C^2 boundary. A developable surface with creases is a locally applicable surface which is the union of a finite number of regular developable patches, and which is C^1 except at a finite number of points (the crease points) or a finite number of piecewise regular curves (the crease curves).*

Irregular points on developables to model such phenomena as paper creasing have been mentioned and studied early [4]. With industrial motivation, [2] analyzes the geometry of a piece of plane when creased along a curve and progressively bent isometrically.

§2. The Need for Creases

Let us study why and when it is necessary to introduce creases on an inextensible surface (e.g., modeling a sheet of paper) when the shape of its boundary is altered. We start with the preliminary problem of finding a non-singular developable surface through a given boundary curve [5,3]. For a solution, we admit either a regular developable or a developable surface with creases such that at any point M of the boundary, $l(r)$ being the length of the set of points on the surface at a distance r from M, the limit of $l(r)/r$ equals π. For the intended modeling, the surface should have no self-intersection; here we shall only care about local self-intersections, since detecting multi-local ones requires entirely different methods. A curve like $(x, y, z) = (\cos\theta, \sin\theta, \cos 2\theta)$ which bounds two different regular developables (with generators parallel either to the x-axis or the y-axis) proves that unicity does not hold in general.

2.1. Regular Case

We start with a boundary for which a regular solution is known, and we look for a family of regular developables fitting the family of boundaries. Notice that in general, the boundaries to be fitted are not those of the isometric images of a fixed applicable surface. Some latitude in choosing the path allows for the use of generic methods. Starting with a boundary which leads to a regular developable as in Fig. 1, the developable is the envelope of a family of planes which are bitangent to the curve. After a small change in the curve, one can compute a new family of bitangent planes and its envelope, giving the solution for the new boundary. Practically, one directly finds

Fig. 1. Swallowtail appearing in the envelope of bitangent planes and removed.

the generators looking for couples of points (M, M') on the curve such that $det(M' - M, t_M, t_{M'}) = 0$, where t_M, $t_{M'}$ are non-zero vectors tangent to the curve at M and M', respectively. Proper choice of the initial condition for that search is important to stay in the same continuous family of solutions.

2.2. Singularities, their Generic Type and Detecting Them

Along some paths in the space of boundary curves, it may happen that the envelopes become singular. Generically for envelopes of planes, this occurs within a local singularity known as the *swallowtail* [10], and slightly modifying the path permits considering only those generic singularities. The self-intersection and the two edges of regression of the swallow-tail make the surface now inappropriate as a solution to our problem. In view of the next application, it is useful to be able to detect singularities inside a developable patch. Given two regular segments of boundary curve B_1, B_2, and a regular parametrization s of them both with $B_1 = \{M(s) \in \mathbb{R}^3, \ s_1 \leq s \leq s_2\}$, we suppose that a differentiable function ϕ defined on $[s_1, s_2]$ exists such that the lines $(M(s), M(\phi(s)))$ remain transversal to B_1 and B_2, with

$$\forall s, \ s_1 \leq s \leq s_2, \quad M(\phi(s)) \in B_2, \quad det(M(\phi(s)) - M(s), t(s), t(\phi(s))) = 0,$$

where $t(s) = M'(s)$ and $t(\phi(s)) = M'(\phi(s))$ are vectors tangent to the curve at $M(s)$ and $M(\phi(s))$ respectively. Let $n(s)$ be a unit vector normal to $M(\phi(s)) - M(s)$ and coplanar with $t(s)$ and $t(\phi(s))$; we suppose that $(t(s).n).(t(\phi(s)).n) < 0$ holds in $[s_1, s_2]$. The following result is then easily proved:

Proposition. *A sufficient condition for the developable patch generated by the line segments $(M(s), M(\phi(s))$ $(s_1 \leq s \leq s_2)$ to be regular is that $\phi' < 0$. The patch has a singularity in its interior if there exists s with $\phi'(s) > 0$.*

2.4. Replacing a Singular Patch by a Conic Patch

It is possible to remove the singularity replacing part of the surface by a conic patch so as to achieve at the same time C^1 junction of the patch to the rest of the surface, applicability (2π angle for the cone), and required conditions at the boundary curve. One should remove at least the part of the surface (call it the bad patch) between the two generators whose tangent plane is the tritangent plane. For a couple of generators (one on each side of the bad patch) one chooses a point on the intersection of the two corresponding tangent planes and one builds the cone from it to the boundary of the removed patch. This achieves C^1 junction of the patch. Among the cones constructed in that way, at least one has total angle more than 2π: The cone vertex being on the boundary where the two generators which bound the bad patch meet, the limit in the boundary condition is more than π. This makes that point inappropriate and shows that vertices taken close to it and in the tritangent plane between the two limit generators lead to cone angles more than 2π. One can also find cone vertices with an angle less than 2π; hence, by continuity a cone can be found with an angle equal to 2π.

§3. Simulating Creased Paper

In [5] the former results were used to model the evolution of a rectangular sheet of paper under some forces exerted on its boundary. Deformations were required to be isometric. The internal forces were restricted to bending moments with no moments at the crease, external forces being interactively applied on points of the boundary. Starting with a regular developable, a polyhedral approximation of it was defined, together with a mapping of the nodes on the unbent plane sheet. Among all the infinitesimal movements of the nodes and their images on the unbent sheet, those respecting the constraint of isometry constitute a linear subspace. Notice that the permitted movements include displacement of the nodes within the sheet. The simulation numerically integrated the differential equation obtained by projecting the vector of forces on that hyperplane. After each step, ϕ as above is computed for each patch, and the need for introducing creases is checked. When needed, a cone crease modification was performed before resuming the evolution with a new polygonal representation. Curve creases were also used (polygonal curves), both choices leading to satisfactory realism.

§4. Crushing Boxes: the Case of a Cylindrical Box

We call *(creased) developable boxes* the surfaces which are homeomorphic to orientable compact 2-manifolds and obtained by gluing (creased) developable patches along *rim curves* where the local applicability property is not necessarily true. To find global isometric deformations of these objects within the set of creased developable boxes raises a problem of existence. We now study the special case of a cylindrical box, under a constraint of symmetry, and prove in this case that the isometries exist. The family of isometric deformations will however be only 1-dimensional, a computationally interesting property contrasting to the much larger families of deformations we met in the case of free ended developable sheets.

The box S to be creased is made of three parts: two disks $\{(x, y, z) \in \mathbb{R}^3 \mid x^2 + y^2 < R^2,\ z = h\}$ and $\{(x, y, z) \in \mathbb{R}^3 \mid x^2 + y^2 < R^2,\ z = -h\}$ glued to a segment of cylinder $\{(x, y, z) \in \mathbb{R}^3 \mid x^2 + y^2 = R^2,\ -h \leq z \leq h\}$ along two rim circles. We shall look for a family of creased developable boxes which are isometric to it, symmetric across the planes $P : (y = 0)$ and $Q : (z = 0)$, and with the patch structure of Fig. 2, where the directions of the generators are specified (only dashed patches get bent): C, C' are cone vertices, with CC' a crease line. All the points on one side of a plane parallel to the axis of the cylinder and to the crease segment will be left invariant.

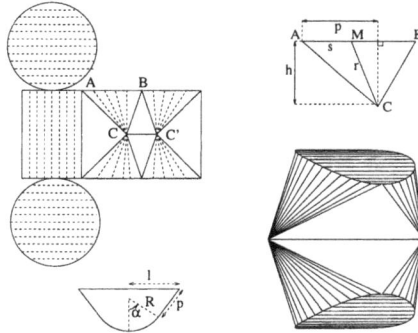

Fig. 2. Development, horizontal cross section, and creasing of the cylinder.

4.1. Existence of the Global Isometry

Theorem. *Given positive $h, R, \alpha \in \mathbb{R}$, and defining $f(s) = d(s) + r(s) - l$, with $d(s) = R\sin(\alpha + s/R)$, $r(s) = \sqrt{h^2 + (p - s)^2}$, $l = R(\pi - \alpha) - p$, and $p = R(\pi - \alpha - \sin\alpha)/(1 + \cos\alpha)$, a sufficient condition for the cylindrical box of height $2h$ and radius R to be isometric to a creased developable box with symmetric patch decomposition as in Fig. 2, where $2\alpha R$ is the length of the unbent rim, is that $\forall s \in \mathbb{R}$, $0 < s < (\pi - \alpha)R$, $f(s) > 0$. A necessary condition is that $\forall s \in \mathbb{R}$, $0 < s < (\pi - \alpha)R$, $f(s) \geq 0$.*

Proof: For each α, let i_α be the isometry to be found from S to the creased developable box $i_\alpha(S)$ where $(x > R\cos\alpha)$ is the part of S left invariant. On the upper rim of S we measure angles from $(R, 0, h)$ and arc length s from $(R\cos\alpha, R\sin\alpha, h)$, also using $\theta = \alpha + s/R$.

For any α, assuming the existence of the isometry, $i_\alpha(S) \cap Q$ is a curve of constant length $2\pi R$. From the tangency condition at T (Fig. 2), one checks that $\|C - T\| = p$ and $\|C' - C\| = 2l$ as defined above, determining the coordinates of C. Let $M(s) = (x(s), y(s), z(s))$ be a point on the upper rim of S. One can locate $i_\alpha^{-1}(C)$ on S and compute on it the geodesic distance $dist_S(i_\alpha^{-1}(C), M(s))$ from $i_\alpha^{-1}(C)$ to $M(s)$, found to be $r(s)$ defined above. Since in the conic patch generators go from C to the rim, this distance is equal to $dist_{\mathbb{R}^3}(C, i_\alpha(M(s))$, i.e., the Euclidean distance in $\mathbb{R}^3$ from C to $i_\alpha(M(s))$, where s is also equal to the arc length at $i_\alpha(M(s))$ on the bent rim. In a similar way, going back to S one can check that $dist_{\mathbb{R}^3}(i_\alpha(M(s)), P) = y(s) = d(s)$, because the generators on the bent upper disk are lines normal to P. To build the required isometry, it is enough to find a regular rim curve

$N(s) = (x(s), y(s), z(s))$ parameterized by its arc length s, such that, with the formerly computed d, r, and C:

$$\forall s,\ 0 < s < (\pi - \alpha)R \qquad y(s) = d(s), \qquad \|C - N(s)\| = r(s).$$

For a given α, let us consider the locus of the points located in $\mathbb{R}^3$ at a distance $r(s)$ from C and at a distance $\mathrm{d}(s)$ from P, for all s between αR and πR: It is a surface of revolution Σ_α around an axis normal to P through C, with a curve m_α as a meridian.

That points on Σ_α exist for all s follows from the inequalities which secure the intersection of the sphere with midpoint C and radius $r(s)$ and the plane $y = d(s)$): $d(s) > l - r(s)$ (1), and $d(s) < l + r(s)$ (2), where l is the Euclidean distance from C to P. Inequality (2) holds, so that only (1) has to be checked, as required in the hypothesis of the theorem. We now show that it is a sufficient condition.

On Σ_α, s defines a function which is constant on the parallel circles. To find the required rim curve for $i_\alpha(S)$, we find a curve on Σ_α such that, calling a its arc length and defining $s(a)$ by restriction of the function s to the curve, $s(a) = a$ along the curve. If the curve is regular enough, this is equivalent to the differential condition $\dot{s} = \dot{a}$.

Let us show that Σ_α, or equivalently m_α, is regular. We parameterize m_α as $(y(s), v(s))$ with $y(s) = d(s)$, $v(s) = \sqrt{(r(s))^2 - (l - d(s))^2}$. Now $dy/ds = 0$ only at the point where y is maximal, i.e., for $\theta = \pi/2$. There, $dv/ds = (\partial v/\partial r)(\partial r/\partial s)$ and $\partial v/\partial r \neq 0$. But $dr/ds = 0$ only if on the triangle which develops the conic patch the cone vertex projects orthogonally on the point of the rim where $\theta = \pi/2$. This happens only when $\alpha = 0$, a case excluded by the hypotheses (for $\alpha = 0$, the creased box may exist but is non-rigid and thus non unique).

We now go back to the rim curve to be found as a curve on Σ_α, and call β the angle of the tangent to that curve with the gradient of s on Σ_α: $|ds/da| = \|grad\ s\|.|cos\beta|$. But $\|grad\ s\| \geq 1$ along m_α, so everywhere on Σ_α the angle β can be found so as to get $ds/da = 1$. The rim can thus be constructed as a curve along which s varies from 0 to $(\pi - \alpha)R$, crossing all the iso-s lines. At its end, the rim reaches P and its tangent is perpendicular to P since $\|grad\ s\| = 1$ there. This achieves C^1 junction of the conic patch.

The second inequality of the theorem is obviously necessary since the rim has to scan all possible values of s from 0 to $(\pi - \alpha)R$, with $M(s)$ at a distance $d(s)$ from P and a distance $r(s)$ from the cone vertex for all these s. If $f(s) < 0$ for some s, P is too far from the cone vertex for $M(s)$ to exist. ∎

Remark. The inequality required for the existence of the isometric creasing does not care about possible self-intersection of the cylinder (the rim crossing Q) which happens for small values of α (i.e., extreme bendings) when the ratio R/h is large. For the ratio small enough, existence and physical realizability hold for $\alpha \in [0, \pi]$.

4.2. Numerical Implementation

Instead of integrating the differential equation that would give the rim in the former setting, but still relying on the former conditions for existence, we use a discrete scheme and compute the rim of a cylindrical box built on a polyhedral regular polyhedron inscribed in the disk.

At each vertex of the new polygonal rim we approximate the distance to the cone vertex by the distance computed on the original circular cylinder using the triangular patch as before. We know the distance to the former vertex together with the angle of the rim segment with the former generator, which leads to the distance δ to P. The vertex thus lies in the plane R parallel to P and at a distance δ from it; the solution is found as one of the intersections of two circles in that plane. Computation time for the rim is very short. Building and rendering the facets of the whole crushed can is a standard task for specialized hardware.

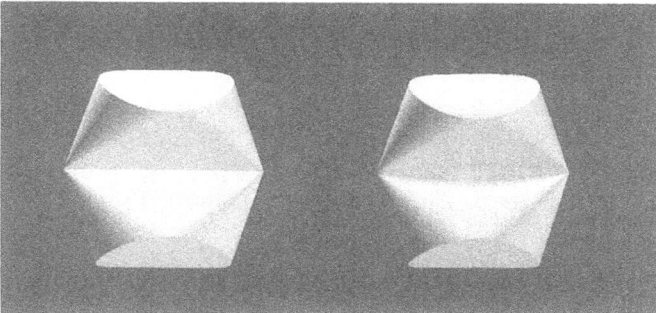

§5. Partially Elastic Models

Physical creasing of thin metallic shells involves some departures from the unstretchability assumption. Experimentally creased cylindrical cans show a satisfactory agreement to the patch organization of our model with approximately triangular parts, conic parts, and high curvature regions appearing at about the same position as the cone vertices of the model. However, a creased segment is not observed between these cone vertices, a definite departure from the model. Realistic images are obtained (Fig. 3), by first computing an isometric developable box creasing as before, then replacing the two triangles across the crease segment by a single elastic plate bent to join the boundary of the removed patch [5]. The energy used allows some stretching besides bending. Restricting the elastic modeling to that part keeps the computational load to low levels.

Acknowledgments. Part of this work was done in collaboration with H. Gotoda and T. L. Kunii at University of Tokyo and University of Aizu, Japan.

References

1. Aumann, G., Interpolation with developable Bezier patches, Computer Aided Geometric Design, **8** (1991), 409–420.

2. Duncan, J. P., and J. L. Duncan, Folded developables, Proc. R. Soc. Lond. **383** (1982), 191–205.

3. Hoschek, J., and M. Schneider, Interpolation and approximation with developable surfaces, in *Curves and Surfaces with Applications in CAGD*, A. Le Méhauté, C. Rabut, and L. L. Schumaker (eds.), Vanderbilt Univ. Press, Nashville, 1997, 185–202.

4. Huffman, D. A., Curvature and creases: A primer on paper, IEEE Trans. Computers, Vol. C–25 **10** (1976), 1010–1019.

5. Kergosien, Y. L., H.Gotoda, and T. L. Kunii, Bending and creasing virtual paper, IEEE CG&A **14**, No 1 (Jan. 1994), 40–48.

6. Lebesgue, H., Integrale, longueur, aire, Ann. di Matem. Pura et Applic. Serie 3, **7** (1902), 231–361.

7. Pottmann, H., and G. Farin, Developable Rational Bézier and B-spline surfaces, Comp. Aided Geom. Design 12 (1995), 513-531.

8. Redont, P., Representation and deformation of developable surfaces, Computer Aided Design **21** (1) (Jan/Feb 1989), 13–20.

9. Stoker, J. J., *Differential Geometry*, Wiley, New York, 1969.

10. Thom, R., Sur la théorie des enveloppes, J. Mathématique, **41** (1962), 177–192.

11. Weiss, G., and P. Furtner, Computer-aided treatment of developable surfaces, Computers and Graphics **12**, **1** (1988), 39–51.

Y. L. Kergosien
Université de Cergy-Pontoise
Informatique, 2, Av. Adolphe Chauvin
95302 Cergy-Pontoise Cedex, FRANCE
kergos@u-cergy.fr

Universal Parameterizations
of Some Rational Surfaces

Rimvydas Krasauskas

Abstract. The universal parameterizations of quadric surfaces and two cases of torus surfaces are introduced. These general constructions are applied to the representation problem of a triangular (resp. quadrangular) surface patch as a Bézier triangle (resp. tensor product patch) of the minimal degree with given boundary rational Bézier curves.

§1. Introduction

Quadric surfaces and Dupin cyclides are important for geometric design as low implicit degree basic surface types. Since powerful Bézier parametric constructions now are dominating, the parameterization problem of these surfaces becomes actual.

In this paper we introduce the universal parameterization concept which is a common generalization of several parameterization constructions earlier proposed for quadrics [3,4], torus [7] and ring cyclide surfaces [9]. The investigation is concentrated on the main problem: how to find a rational Bézier triangular or tensor product patch of minimal degree on the surface with fixed boundary curves.

We formulate the main problem in Section 2. Then we define the universal parameterization and sketch an algorithm of the solution in Section 3. Section 4 is devoted to the examples of three cases of quadrics and two cases of torus surfaces. Most proofs can be found in my preprint [7].

§2. Formulation of the Main Problem

Recall that a real n-dimensional projective space is defined as a space of all lines in $\mathbb{R}^{n+1}$ going through the origin $\mathbf{0} = (0, \ldots, 0)$. Note that a group $\mathbb{R}^* = \mathbb{R} \setminus \{0\}$ acts on the space $\mathbb{R}^{n+1} \setminus \{\mathbf{0}\}$ by the formula $r * (x_0, \ldots, x_n) = (rx_0, \ldots, rx_n)$ (elementary facts about group actions can be found in the book [1]). Then we can define $\mathbb{R}\,P^n$ equivalently as the space of orbits

Curves and Surfaces with Applications in CAGD
A. Le Méhauté, C. Rabut, and L. L. Schumaker (eds.), pp. 231–238.
Copyright © 1997 by Vanderbilt University Press, Nashville, TN.
ISBN 0-8265-1293-3.
All rights of reproduction in any form reserved.

$(\mathbb{R}^{n+1} \setminus \{0\})/\mathbb{R}^*$. There is a natural projection $\Pi \colon \mathbb{R}^{n+1} \setminus \{0\} \to \mathbb{R}\,\mathrm{P}^n$, which maps $(x_0, \ldots, x_n)$ to its orbit denoted by $[x_0, \ldots, x_n]$. These are actually homogeneous coordinates of a corresponding point in $\mathbb{R}\,\mathrm{P}^n$.

Similarly *complex* projective spaces $\mathbb{C}\mathrm{P}^n$ can be defined: just change all $\mathbb{R}$ to $\mathbb{C}$ in the previous definition. Also we will need the *weighted* real projective plane $\mathbb{R}\,\mathrm{P}(1,1,2) = (\mathbb{R}^3 \setminus \{0\})/\mathbb{R}^*$. Here the group $\mathbb{R}^*$ acts with "weights": $r * (x_0, x_1, x_2) = (rx_0, rx_1, r^2x_2)$ (see [6], p. 128).

Let D be one of three standard affine domains: an interval $I = [0,1]$, a triangle or a square. Define a *rational* Bézier curve σ (resp. triangular, tensor product surface) in $\mathbb{R}\,\mathrm{P}^3$ as a composition

$$\sigma \colon D \xrightarrow{\ \gamma\ } \mathbb{R}^4 \setminus \{0\} \xrightarrow{\ \Pi\ } \mathbb{R}\,\mathrm{P}^3, \quad D = I \ (\text{resp. triangle, square}),$$

where γ is a polynomial Bézier curve (resp. triangular, tensor product surface) called *lifting* of σ. Note that it is a *non-standard* definition of rational Bézier curves and surfaces. This is exactly the Fiorot and Jeannin [5] approach after the restriction onto some affine part $\mathbb{R}^3 \subset \mathbb{R}\,\mathrm{P}^3$.

A $(k+1)$-*chain* of curves $\langle c_0, \ldots, c_k \rangle$ in some space X is a sequence $c_0, c_1, \ldots, c_k \colon I \to X$ such that $c_i(1) = c_{i+1}(0)$, $i = 0, \ldots, k-1$. Call the chain *closed* iff $c_0(0) = c_k(1)$. We will consider mostly 3- and 4-chains of curves. They naturally appear as boundaries of Bézier triangular and tensor product patches. For a surface patch σ denote by $\partial\sigma$ its boundary chain.

Now we are ready to formulate the main problem.

Main Problem. *Find a rational Bézier triangular patch (resp. a Bézier tensor product patch) s on an algebraic surface $V \subset \mathbb{R}\,\mathrm{P}^3$ of minimal degree with a boundary chain of curves ∂s equal or similar to a given 3-chain (resp. 4-chain) of curves on V.*

Here the word *equal* means that we look for an extension of a given rational map on the boundary to the interior of the patch. Often in practice we need more delicate conditions. Call two Bézier curves $c, c' \colon I \to \mathbb{R}\,\mathrm{P}^3$ *similar* iff they have the same endpoints and images: $c(I) = c'(I)$. Let a polynomial Bézier curve γ with control points $p_0, \ldots, p_d$ be a lifting of c in $\mathbb{R}^4 \setminus \{0\}$. Then (under some mild conditions) there exists $\rho > 0$ such that $c' = \Pi \circ \Phi_\rho^d \gamma$ where a curve $\Phi_\rho^d \gamma$ has control points $p_0, \rho p_1, \ldots, \rho^d p_d$ (see [7,10]).

§3. Definition of Universal Parameterization

First let us consider relations between arbitrary rational c/s (abbreviation for curves/surfaces) in $\mathbb{R}\,\mathrm{P}^3$ and their polynomial liftings. Call a polynomial map $\gamma \colon D \to \mathbb{R}^4 \setminus \{0\}$ *irreducible* iff all coordinate polynomials γ_i are mutually prime. Using factorization of polynomials it is easy to prove:

Lemma 1. *(i) Every rational c/s has some irreducible lifting.*

(ii) If two polynomials c/s $\gamma, \gamma' \colon D \to \mathbb{R}^4 \setminus \{0\}$ have the same projections $\Pi \circ \gamma = \Pi \circ \gamma'$ and γ' is irreducible, then $\gamma = r * \gamma'$ for some polynomial map $r \colon D \to \mathbb{R}^*$; γ is also irreducible iff $r = \mathrm{const}$.

The group $\mathbb{R}^*$ naturally acts on a set of all polynomial maps to $\mathbb{R}^4 \setminus \{0\}$.

Corollary 2. *There is 1–1 correspondence* $\gamma \mapsto \Pi \circ \gamma$ *between the set of orbits* {*all irreducible polynomial c/s* $D \to \mathbb{R}^4 \setminus \{0\}$ } $/ \mathbb{R}^*$ *and the set of all rational c/s* $D \to \mathbb{R} \mathrm{P}^3$.

Consider rational c/s $\sigma : D \to V$ on an algebraic surface $V \subset \mathbb{R} \mathrm{P}^3$. According to Corollary 2 they correspond to irreducible polynomial maps $\gamma : D \to \Pi^{-1}(V)$ which can be extended to the whole affine line/plane $\gamma : \mathbb{R}^m \to \Pi^{-1}(V) \cup \{0\}$, $m = 1, 2$. When $m = 2$ we can regard $\Pi \circ \gamma$ as a rational parameterization of V. Note that its natural domain is $\mathbb{R}^2 \setminus \gamma^{-1}(0)$. Here $\gamma^{-1}(0)$ is a set of base points of the parameterization . It is useful to consider cases when $m > 2$. A domain of such m-dimensional parameterization will be a Zariski open $\mathbb{R}^k \setminus \{(m-2)$-dimensional algebraic subset} subset of an affine space $\mathbb{R}^m$.

Definition 3. *A universal parameterization of the rational surface V consists of (i) a m-dimensional rational parameterization* $\Pi_V : U_V \to V$, *(ii) an action of an algebraic group* G_V *on* U_V, *(iii) a distinguished set of V-irreducible polynomial maps to* U_V *closed relatively to* G_V*-action; provided that*

(U1) *fibers* $\Pi_V^{-1}(x)$ *over all non-singular points* $x \in V$ *coincide with orbits of the action;*

(U2) *any irreducible lifting* $u_V : U_V \to \Pi^{-1}(V)$ *of* Π_V, *defines 1–1 correspondence* $\delta \mapsto u_V \circ \delta$ *between the sets of orbits* {*all V-irreducible polynomial c/s in* U_V } $/ G_V$ *and* {*all irreducible polynomial c/s in* $\Pi^{-1}(V)$ } $/ \mathbb{R}^*$.

The condition (U1) enables to define a division operation between any two points in U_V from the fiber over non-singular point: $y/z \in G_V$ is a unique element such that $y = (y/z) * z$.

The condition (U2) has a reformulation more convenient for applications:

Lemma 4. *The condition (U2) is equivalent to the following two statements:*

(i) *for every irreducible c/s* $\gamma : D \to \Pi^{-1}(V)$ *there is some V-irreducible c/s* $\delta : D \to U_V$ *such that* $\gamma = u_V \circ \delta$.

(ii) *if two polynomial c/s* δ *and* δ' *in* U_V *have the same projections* $\Pi_V \circ \delta = \Pi_V \circ \delta'$ *and* δ' *is V-irreducible, then* $\delta = g * \delta'$ *for some rational c/s in* G_V; δ *is V-irreducible iff* $g = \mathrm{const}$.

We call a c/s $g : D \to G_V$ from Lemma 4(ii) *admissible* to a curve δ'.

Suppose that an algebraic surface $V \subset \mathbb{R} \mathrm{P}^3$ has a universal parameterization $\Pi_V : U_V \to V$ with a fixed irreducible lifting $u_V : U_V \to \Pi^{-1}(V)$. Consider now curves and chains of curves on V. Fix a non-singular point $x \in V$ and its lifting $y \in U_V$. For any rational curve $c : I \to V$, $c(0) = x$, one can find the unique V-irreducible lifting $\tilde{c} : I \to U_V$, $y = \tilde{c}(0)$. Indeed using (U2) we find *some* V-irreducible lifting δ and notice that a point $\delta(0)$ is from the same fiber as y. Using (U1) and the division put $\tilde{c}(t) = y/\delta(0) * \delta(t)$. Uniqueness follows from Lemma 4(ii).

Obviously, step by step we can lift any chain of curves to some V-irreducible chain in U_V. Suppose our chain $\langle c_0, \ldots, c_k \rangle$ in V was closed. Then there is no reason why its V-irreducible lifting $\langle \tilde{c}_0, \ldots, \tilde{c}_k \rangle$ should be closed. Nevertheless, points $\tilde{c}_0(0)$ and $\tilde{c}_k(1)$ are in the same fiber over x. Hence an element of the group G_V is defined $\mathcal{O}(\langle c_0, \ldots, c_k \rangle) = \tilde{c}_0(0)/\tilde{c}_k(1)$. It is easy to check that this element does not depend on y. Now suppose some closed lifting (not necessary V-irreducible) exists. Then apply Lemma 5(ii) and obtain some chain of admissible curves in the group G_V. Hence we come to the following proposition:

Proposition 5. *Let* $\xi = \langle c_0, \ldots, c_k \rangle$ *be a closed chain of rational curves on the surface* V. *Then*

(i) ξ *has closed* V-*irreducible lifting iff* $\mathcal{O}(\xi) = 1 \in G_V$;

(ii) *the set of all closed liftings of* ξ *in* U_V *is in 1–1 correspondence with the set of admissible (to the* V-*irreducible lifting) chains of curves in the group* G_V *connecting the unit 1 with* $\mathcal{O}(\xi)$.

If we gradually change the parameterization of some curve in a chain ξ moving ρ (see the definition of Φ_ρ^d, Section 2) from 1 to a different value, then $\mathcal{O}(\xi)$ traces some non-constant *reparameterization* curve in the group G_V.

Algorithm (Solution of Main Problem).

Input: a closed chain of curves $\xi = \langle c_0, \ldots, c_k \rangle$, $k = 2, 3$, *on the surface* V.

1) *Find a* V-*irreducible lifting* $\langle \tilde{c}_i \rangle$ *and calculate* $\mathcal{O}(\xi) = \tilde{c}_0(0)/\tilde{c}_k(1)$.

2) *Reparameterize some curves* $c_i \mapsto c_i'$ *in order to simplify* $\mathcal{O}(\xi)$.

3) *IF it is possible THEN find the optimal admissible chain* $\langle g_i \rangle$ *connecting* 1 *and* $\mathcal{O}(\xi)$ *in* G_V *and calculate the closed lifting* $\chi = \langle g_i * \tilde{c}_i \rangle$ *ELSE stop.*

4) *IF it is possible THEN fill the closed chain* χ *with a surface patch* δ *in* U_V *and calculate the projection* $s = \Pi \circ \delta$ *to the* V *ELSE stop.*

Output: a surface patch s *with boundary* $\partial s = \xi$.

Note that possibilities of Steps 3 and 4 of this algorithm are mostly related to topological possibilities to fill "holes". Here we skip all details, and suppose in the following section that we deal with situations when the answer is positive.

§4. Examples

Here all examples of universal parameterization are described according to the same template: notation of the surface in $\mathbb{R}P^3$, equation in homogeneous coordinates x_0, x_1, x_2, x_3, singular points (if any), definition of u_V, group G_V, its action on U_V, definition of V-irreducible c/s via coordinates in U_V. In the item *Results* the results of application of our algorithm are presented. The following notations will be used: degree of a chain of curves $\deg(\xi) = \langle \deg c_0, \ldots, \deg c_k \rangle$, bidegree of a tensor product patch $\deg s = \binom{d}{d'}$.

It will be convenient to use complex numbers. We use letters x_i, y_i, r, s for real variables and z_i, λ, μ for complex ones. We identify $\mathbb{C} \cong \mathbb{R}^2$ and use the notations Rez, Imz, $|z|$ and $\bar{z}$ for the real part, imaginary part, modulus, and complex conjugate of a complex number z, respectively.

Sphere

S: $x_1^2 + x_2^2 + x_3^2 = x_0^2$.

$$u_S\colon \mathbb{C}^2 \longrightarrow \mathbb{R} \times \mathbb{C} \times \mathbb{R}, \quad (z_0, z_1) \mapsto (|z_0|^2 + |z_1|^2, 2\, z_0\bar{z}_1, |z_0|^2 - |z_1|^2).$$

It is easy to check that this is exactly the formula of the generalized stereographic projection [3]. Just put $z_0 = p_2 + p_1 i$ and $z_1 = p_0 + p_3 i$.

The group $G_S = \mathbb{C}^* = \mathbb{C}\backslash\{0\}$ acts on $U_S = \mathbb{C}^2\backslash\{0\}$: $\lambda*(z_0, z_1) = (\lambda\cdot z_0, \lambda\cdot z_1)$. (z_0, z_1) is S-irreducible iff $\gcd(z_0, z_1) = 1$. The projection $\Pi\colon U_S \to S$ naturally factors:

$$\mathbb{C}^2 \setminus \{0\} \xrightarrow{\Pi'} \mathbb{C}P^1 \xrightarrow{\iota_S} S$$

In fact the mapping ι_S is an isomorphism to the sphere S in the sense of real algebraic geometry. It is also closely related to the classical Riemann sphere and Hopf fibration constructions (see [1], Section 4.3).

Results

Input: $\deg \xi = \langle 2d, 2d, 2d \rangle$. Output: triangular patch $\deg s = 2d + 2$.

Input: $\deg \xi = \langle 2d, 2d', 2d, 2d' \rangle$. Output: tensor pr. patch $\deg s = \binom{2d}{2d'+2}$.

Comments. When $d = 1$ this is the result of [3]. Here we use a trivial fact: any element $\mathcal{O} \in \mathbb{C}^*$ can be joined with 1 via some linear 2-chain.

Hyperbolic Paraboloid

H: $x_0 x_3 = x_1 x_2$.

$$u_H\colon \mathbb{R}^4 \to \mathbb{R}^4, \quad (y_0, y_1, y_2, y_3) \mapsto (y_0 y_3, y_1 y_3, y_0 y_2, y_1 y_2).$$

The group $G_H = \mathbb{R}^* \times \mathbb{R}^*$ acts on $U_H = (\mathbb{R}^2\backslash\{0\}) \times (\mathbb{R}^2\backslash\{0\})$:

$$(r, s) * (y_0, y_1, y_2, y_3) = (r \cdot y_0, r \cdot y_1, s \cdot y_2, s \cdot y_3)$$

(y_0, y_1, y_2, y_3) is H-irreducible iff $\gcd(y_0, y_1) = \gcd(y_2, y_3) = 1$.

Formula $(y_0, y_1, y_2, y_3) \mapsto ([y_0, y_1], [y_2, y_3])$ defines a factorization

$$(\mathbb{R}^2\backslash\{0\}) \times (\mathbb{R}^2\backslash\{0\}) \xrightarrow{\Pi \times \Pi} \mathbb{R}P^1 \times \mathbb{R}P^1 \xrightarrow{\iota_H} H.$$

The map ι_H is well-known in algebraic geometry as the Segre map and is an isomorphism. Note that every c/s of degree d on H corresponds to a pair of c/s on $\mathbb{R}P^1$ of degrees d_0, d_1: $d_0 + d_1 = d$. We denote such "bidegrees" (d_0, d_1) in results bellow. One can follow DeRose [2] and use the model $(\mathbb{R}P^1)^2$ for arbitrary tensor product surfaces.

Results

Input: $\deg \xi = \langle (d_0, d_1), (d_0, d_1), (d_0, d_1) \rangle$.

Output: triangular patch $\deg s = d_0 + d_1 + 1$.

Input: $\deg \xi = \langle (d_0, d_1), (d_0', d_1'), (d_0, d_1), (d_0', d_1') \rangle$.

Output: tensor product patch $\deg s = \binom{d_0+d_1}{d_0'+d_1'+1}$ if $d_0 d_1' = d_1 d_0'$ else $\binom{d_0+d_1}{d_0'+d_1'}$.

Comments. When $d_0 = d_1 = 1$ this is better than [4]. Here we move an element $\mathcal{O} \in \mathbb{R}^* \times \mathbb{R}^*$ along a reparameterization curve untill it intersects with one of the lines $r = 1$ or $s = 1$. Then we join it with 1 by line (see [7] for details).

Cylinder

C: $x_1^2 + x_2^2 = x_0^2$; one singular point $[0,0,0,1]$. If we put $x_0 = 1$, then C will be an affine cylinder. It will be an affine cone for an appropriate choice of the plane of infinite points in $\mathbb{R}\,\mathrm{P}^3$.

$$u_C: \mathbb{R} \times \mathbb{C} \times \mathbb{R} \to \mathbb{R} \times \mathbb{C} \times \mathbb{R} \cong \mathbb{R}^4, \quad (y_0, z, y_1) \mapsto (y_0|z|^2, y_0 z^2, y_1)$$

The group $G_C = \mathbb{R}^* \times \mathbb{R}^*$ acts on $U_C = \mathbb{R}^2 \times \mathbb{C} \setminus (\{y_0 = 0, y_1 = 0\} \cup \{z = 0, y_1 = 0\})$: $(r, s) * (y_0, z, y_1) = (r \cdot y_0, s \cdot z, rs^2 \cdot y_1)$.

(y_0, z, y_1) is C-irreducible iff $\gcd(y_0, y_1) = \gcd(\mathrm{Re}\,z, \mathrm{Im}\,z) = 1$. When $y_0 = \mathrm{const} \neq 0$, the projection Π_C can be factored:

$$\{(y_0, z, y_1) \in U_C \mid y_0 = \mathrm{const}\} \xrightarrow{\Pi_w} \mathbb{R}\,\mathrm{P}(1,1,2) \to C,$$

where $\Pi_w(y_0, z, y_1) = [\mathrm{Im}\,z, \mathrm{Re}\,z, y_1]$. This construction will be useful in the spindle torus case.

Results

Input: $\deg \xi = \langle 2, 2, 2 \rangle$. Output: triangular patch $\deg s = 2$.

Input: $\deg \xi = \langle 2, 2, 2, 2 \rangle$. Output: tensor product patch $\deg s = \binom{2}{2}$.

Comments. This is the result from [8]. Here we can use just reparameterization because for nontrivial conics $y_0 = \mathrm{const}$, and we have 1-dimensional fibers. The higher degree case will be considered in detail elsewhere.

Ring Torus

T_+: $(x_1^2 + x_2^2 + x_3^2 + (a^2 - b^2)x_0^2)^2 = 4ax_0^2(x_1^2 + x_2^2)$, $a > b > 0$.

$u_{T_+}: \mathbb{C}^4 \to \mathbb{R} \times \mathbb{C} \times \mathbb{R} \cong \mathbb{R}^4$

$$\begin{pmatrix} z_0 \\ z_1 \\ z_2 \\ z_3 \end{pmatrix} \mapsto \begin{pmatrix} |z_0|^2 \left(\frac{1}{a+b} \mathrm{Re}^2(z_1 \bar{z}_2 z_3) + \frac{1}{a-b} \mathrm{Im}^2(z_1 \bar{z}_2 z_3) \right) \\ |z_3|^2 (z_0 z_1 z_2)^2 \\ \frac{2b}{\sqrt{a^2 - b^2}} |z_0|^2 \mathrm{Re}(z_1 \bar{z}_2 z_3) \mathrm{Im}(z_1 \bar{z}_2 z_3). \end{pmatrix}$$

The group $G_{T_+} = \mathbb{R}^* \times (\mathbb{C}^*)^2 \times \mathbb{R}^*$ acts on $U_{T_+} = (\mathbb{C}^*)^4$:

$$(r, \lambda, \mu, s) * (z_0, z_1, z_2, z_3) = (r\lambda\bar{\mu} \cdot z_0, \bar{\lambda} \cdot z_1, \mu \cdot z_2, s\lambda\mu \cdot z_3).$$

A c/s $\gamma = (z_0, z_1, z_2, z_3)$ is T_+-irreducible iff $\gcd(\mathrm{Re}z_i, \mathrm{Im}z_i) = 1$ (for all i) and the following pairs of complex polynomials are also mutually prime $(z_0, \bar{z}_1)$, $(z_0, \bar{z}_2)$, $(z_3, \bar{z}_1)$, (z_3, z_2), $(z_0\bar{z}_0, z_3\bar{z}_3)$, $(z_1\bar{z}_1, z_2\bar{z}_2)$. Similarly to the hyperbolic paraboloid case, U_{T_+} can be factored $(\mathbb{C}^*)^4 \to (\mathbb{R}\,\mathrm{P}^1)^4 \to T_+$, where $z_i \mapsto [\mathrm{Re}z_i, \mathrm{Im}z_i]$. Hence for every c/s of degree d on T_+ the quadruple degree $(d_0, \ldots, d_3)$, $2(d_0 + \cdots + d_3) = d$, is defined. For example, there are 6 types of proper quartics on T_+: $(1,1,0,0), \ldots, (0,0,1,1)$.

Results

Input: ξ 3-chain of quartics of one type (resp. mixed type).

Output: triangular patch deg $s = 6$ (resp. deg $s = 8$).

Input: ξ 4-chain of quartics of one type (resp. mixed type).

Output: tensor product patch deg $s = \binom{4}{6}$ (resp. deg $s = \binom{8}{8}$).

Comments.

If we fix a type of quartics, we can directly apply the results achieved in the hyperbolic paraboloid case when $d_0 = d_1 = 1$. The mixed case is trickier [7].

Spindle Torus

T_-: $(x_1^2 + x_2^2 + x_3^2 + (a^2 - b^2)x_0^2)^2 = 4ax_0^2(x_1^2 + x_2^2)$, $0 < a < b$.

Two singular points $[0, 0, \pm\sqrt{b^2 - a^2}, 1]$.

$u_{T_-}: \mathbb{C}^3 \times \mathbb{R}^2 \to \mathbb{R} \times \mathbb{C} \times \mathbb{R} \cong \mathbb{R}^4$

$$
\begin{pmatrix} z_0 \\ z_1 \\ z_2 \\ y_0 \\ y_1 \end{pmatrix} \mapsto
\begin{pmatrix} |z_2|^2\left(\dfrac{1}{b+a}(y_0|z_0|^2 + y_1|z_1|^2)^2 + \dfrac{1}{b-a}(y_0|z_0|^2 - y_1|z_1|^2)^2\right) \\ 4y_0y_1(z_0\bar{z}_1z_2)^2 \\ \dfrac{2b}{\sqrt{b^2 - a^2}}|z_2|^2(y_0^2|z_0|^4 - y_1^2|z_1|^4) \end{pmatrix}.
$$

The group $G_{T_-} = (\mathbb{C}^*)^2 \times (\mathbb{R}^*)^2$ acts on $U_{T_-} = \mathbb{C}^3 \times \mathbb{R}^2 \setminus u_{T_-}^{-1}(0)$:

$$(l, \mu, r, s) * (z_0, z_1, z_2, y_0, y_1) = (\lambda \cdot z_0, \mu \cdot z_1, r\bar{\lambda}\mu \cdot z_2, s|\mu|^2 \cdot y_0, s|\lambda|^2 \cdot y_1).$$

$(z_0, z_1, z_2, y_0, y_1)$ is T_--irreducible iff $\gcd(\mathrm{Re}z_i, \mathrm{Im}z_i) = 1$ (for all i) and the following pairs of polynomials are also mutually prime $(z_0, \bar{z}_2)$, (z_1, z_2), $(|z_0|^2, y_1)$, $(|z_1|^2, y_0)$, (y_0, y_1), $(|z_0|^2, |z_1|^2)$. Now we can decompose the projection Π_{T_-} in two different ways

$$U_{T_-} \xrightarrow{\Pi_w^2 \times \Pi} \mathbb{R}\,\mathrm{P}(1,1,2)^2 \times \mathbb{R}\,\mathrm{P}^1 \setminus W'' \to T_-,$$

$$U_{T_-} \xrightarrow{\Pi' \times \Pi^2} \mathbb{C}\mathrm{P}^1 \times (\mathbb{R}\,\mathrm{P}^1)^2 \setminus W' \longrightarrow T_-,$$

where W' and W'' are some two couples of projective lines, and projections are defined by formulas: $\Pi_w^2 \times \Pi$: $(z_0, z_1, z_2; y_0, y_1) \mapsto ([z_0, y_1], [z_1, y_0], [z_2])$, $\Pi' \times \Pi^2$: $(z_0, z_1, z_2; y_0, y_1) \mapsto ([z_0, z_1], [y_0, y_1], [z_2])$.

Similarly to the previous case, we can detect 4 types of quartics (not mentioned degrees are 0):

  (a) $\deg z_0 = \deg z_1 = 1$, (b) $\deg z_2 = \max\{\deg y_0, \deg y_1\} = 1$

  (c) $\deg z_0 = 1$, $\deg y_1 = 1, 2$,(d) $\deg z_1 = 1$, $\deg y_0 = 1, 2$.

Results

Input: ξ 3-chain of quartics of type (a), (b), (c), (d) (resp. mixed type).

Output: triangular patch $\deg s = 8, 6, 4, 4$ (resp. $\deg s = 8$).

Input: ξ 4-chain of quartics of type (a), (b), (c), (d) (resp. mixed type).

Output: tensor product patch $\deg s = \binom{4}{8}, \binom{4}{6}, \binom{4}{4}, \binom{4}{4}$ (resp. $\deg s = \binom{8}{8}$).

Comments. If we fix a type of quartics, then we can directly apply the results related with different models: $\mathbb{C}P^1$, $(\mathbb{R}P^1)^2$, $\mathbb{R}P(1,1,2)$. The mixed case is more difficult.

References

1. Berger, M., *Géométrie, 1, Action de groupes, espaces affines et projectifs*, Cedic, F. Natan, Paris, 1977.

2. DeRose, T. D., Rational Bézier curves and surfaces on projective domains, in *NURBS for Curve and Surface Design*, G. Farin (ed.), SIAM, Philadelphia, PA, 1991, 35–45.

3. Dietz, R., J. Hoschek, and B. Jüttler, An algebraic approach to curves and surfaces on the sphere and other quadrics, Comput. Aided Geom. Design **10** (1993), 211–229.

4. Dietz, R., J. Hoschek, and B. Jüttler, Rational patches on quadric surfaces, Computer-Aided Design **27** (1995), 27–40.

5. Fiorot, J.-C. and P. Jeannin, Nouvelle description et calcul des courbes rationnelles à l'aide de points et de vecteurs de contrôle, C.R. Acad. Sci. Paris, Série I, **305** (1987), 435–440.

6. Harris, J., *Algebraic Geometry. A First Course*, Springer-Verlag, Berlin, 1992.

7. Krasauskas, R., Rational Bézier surface patches on quadrics and the torus, Preprint 95-25, Vilnius University, 1995.

8. Lü, W., Rational parameterization of quadrics and their offsets, Technical Report Nr. 24, Technische Universität Wien, 1995.

9. Mäurer, C., Rational curves and surface patches on Dupin ring cyclides, Preprint, Nr. 1806, Technical University Darmstadt, 1996.

10. Patterson, R. R., Projective transformations of the parameter of a Bernstein–Bézier curve, ACM Transactions on Graphics **4** (1985), 276–290.

Rimvydas Krasauskas
Vilnius University, Department of Mathematics
Naugarduko 24, 2600 Vilnius, LITHUANIA
rimvydas.krasauskas@maf.vu.lt

Curves on Surfaces for Computer Graphics: Theoretical Results

Yevgeniy Kuzmin and Marc Daniel

Abstract. A new method based on geometrical preprocessing of surfaces is proposed for solving some problems of Computer Graphics and Geometrical Design. Performing the preprocessing once, we obtain a set of the right number of starting points to march once along each necessary curve or to obtain specific points for any given initial condition, without any additional surface subdivisions.

§1. Introduction

Manipulation of surfaces in graphics or modelling applications involves finding a set of particular lines or points on the surfaces. These sets include contour lines, section lines and extreme points.

Initial techniques for obtaining these sets are subdivision techniques. They unfortunately require a lot of computation and a proliferation of data. Marching techniques exhibit a lack of reliability without specific assumptions. The goal of hybrid methods is to obtain patches satisfying these assumptions under which the right number of starting points can be found. Different hybrid methods are proposed in [1,2,3,6,7].

We propose a new approach based on one preprocessing stage which computes once a set of points giving the right number of starting points to march once along each curve or to obtain specific points for any given initial condition, without any additional surface subdivisions. This pure marching approach is efficient in the case of multiple applications. The purpose of this paper is to present several theoretical results, and discuss their importance.

§2. Smooth Surfaces

In this section we consider some theoretical results for *smooth parametric surfaces in general position*, which will be applied to the case of surfaces considered in CAD software in a forthcoming study.

Curves and Surfaces with Applications in CAGD
A. Le Méhauté, C. Rabut, and L. L. Schumaker (eds.), pp. 239–246.
Copyright © 1997 by Vanderbilt University Press, Nashville, TN.
ISBN 0-8265-1293-3.
All rights of reproduction in any form reserved.

239

2.1. Definitions and Notations

Let us consider a two-dimensional surface $S(u,v)$ embedded in $\mathbb{R}^3$. Let S be a C^2-smooth, compact, possibly with boundary, oriented and connected surface. These properties describe the class of surfaces which are used in surface modelling. Consider the following sets on S:

$B(S)$ - the set of boundary curves of S;
$K_V(S)$ - the set of contour lines of S for parallel projection in direction V.
$C_O(S)$ - the set of contour lines of S for stereographic projection from point O.
$O(S)$ - the set of lines on which Gaussian curvature of S is equal to 0.
$E_W(S)$ - the set of extreme points on S in direction W.
$I_P(S)$ - the set of intersection lines between S and plane P.

In this section we will also assume that *all objects involved are in general position, and that all of the above sets consist of finite number of connected sectionally-smooth curves*. Also, let us suppose that any pair of such curves have finite intersections. The main purpose of this work is the analysis of the interposition and relationships of these sets.

2.2. Normal Surface

We consider a new surface which will be useful for our analysis. The set of non-unit normal vectors $N(u,v) = S_u \times S_v$ on surface S defines a surface, which we call the *normal surface* of S and denote by $N(S)$. We will describe the relationships between some sets on S and $N(S)$. Consider the plane P which passes through the point $O = (0,0,0)$ with normal vector V and the stereographic projection with its central point in O.

Statement 1. *The sets $K_V(S)$ and $I_P(N)$ coincide in the parametric plane.*

Proof: At any point M of $K_V(S)$, $(N(M) \cdot V) = 0$. This means that the corresponding images of all normal vectors N of such points in the space of $N(S)$ lie on the plane which is orthogonal to vector V and passes through the origin, which is plane P. Therefore both sets coincide. ∎

Statement 2. *The sets $O(S)$ and $C_O(N)$ coincide in the parametric plane.*

Proof: At any point M of $O(S)$, $(N(M), N_u(M), N_v(M)) = 0$ (see appendix). This means that, in the general case, $N(M)$ lies in the plane defined by vectors $N_u(M)$ and $N_v(M)$ (see Figure 1). This plane is tangent to surface $N(S)$ at point $N(M)$, and therefore the normal vector to $N(S)$ at this point is orthogonal to vector $N(M)$. This means that point $N(M)$ on $N(S)$ is a contour point for the stereographic projection from point O. ∎

Remarks:
1) The previous determinant can vanish if one of the vectors $N(M)$, $N_u(M)$, $N_v(M)$ vanishes or if the two last vectors are collinear. This corresponds to singular points of surface S or surface $N(S)$.
2) The sets $B(S)$ and $B(N)$ coincide in the parametric plane.

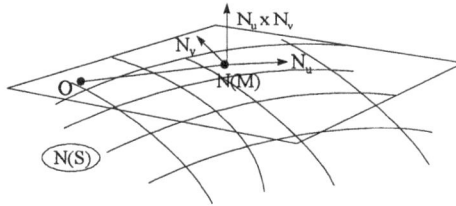

Fig. 1. Point of zero Gaussian curvature in the general case.

2.3. Contour Lines

In this section we will describe several useful properties of contour lines for parallel surface projection. Let us construct the following set on S:

$$Sp_V(S) = B(S) \cup O(S) \cup K_V(S)$$

We will call this set the *spherical net* associated with direction V. It is made up of the union of the border, all the contour lines in direction V and the zero Gaussian curvature lines.

Let us consider any connected component K of $K_U(S)$, where U is any non-zero vector.

Statement 3. $Sp_V(S) \cap K \neq \emptyset$. *In other words any contour line of $K_U(S)$ intersects the spherical net $Sp_V(S)$.*

Proof: The image of K on $N(S)$ is a connected component L which belongs to a planar section of $N(S)$ by plane P_U whose normal vector is U (statement 1). Component L may be an open or closed curve on $N(S)$. If this component is an open curve, then it must have an intersection with the border of $N(S)$ and there exists an intersection of K with $B(S)$. Therefore K has intersection with the spherical net. If L is a closed loop, we have two cases, depending on the location of the origin of the universe, inside or outside our loop (in plane P_U). Figure 2 illustrates the variation of normal vectors in both cases.

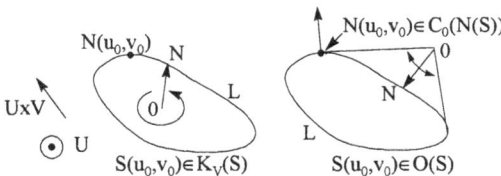

Fig. 2. Planar section of the normal surface: variation of normal vectors.

In case of internal position, loop L must intersect the image of $K_V(S)$ on $N(S)$, because there exists a point $N(u_0, v_0)$ on loop L which is collinear to vector $V \times U$. This point belongs to both lines and these lines intersect each other. Therefore there exists an intersection of K with $K_V(S)$.

In case of external position of the origin, a point $N(u_0, v_0)$ which is a contour point for the central projection from origin must lie on loop L.

Therefore the intersection of loop L and $C_O(N)$ is not an empty set (statement 2) and on surface S we also have a non-empty intersection of K and $O(S)$. In both cases, K intersects $Sp_V(S)$. ∎

Another proof of this statement can be given by using the Gaussian spherical map Φ, which maps the surface onto the unit sphere (at each point corresponds its unit normal vector).

Proof: For Gaussian mapping we have the following correspondences between sets on S and the unit sphere: the image of contour line $K_U(S)$ lies on equator E_U of sphere, and $O(S)$ corresponds to singularities of Φ mapping. We have three possibilities for the image of component K:

1) $\Phi(K)$ is an open segment of E_U. In this case K has intersection with $B(S)$;
2) $\Phi(K)$ covers E_U. In this case $\Phi(K)$ also has intersection with E_V and therefore K has intersection with $K_V(S)$;
3) $\Phi(K)$ is a closed line which doesn't cover E_U. In this case the end points of $\Phi(K)$ are singular points for Gaussian mapping, therefore K has intersection with $O(S)$;

Thus, in any case we have intersection of K and the spherical net. ∎

Corollary 1. *If the image of the normal vectors of a surface S by Φ mapping is included in one hemisphere, which is the case when any normal bounding volume is constructed for the surface, then any contour line reaches the border of S or intersects $O(S)$.*

Proof: Let us consider component K of $K_V(S)$. The image of K by Φ is included in E_V but cannot cover E_V as the image of the surface is included in one hemisphere. The problem therefore reduces to cases 1 and 3 of the previous proof. ∎

Corollary 2. *For a given vector U we can reconstruct $K_U(S)$ by marching along the lines of $Sp_V(S)$.*

The first possibility is to compute the spherical net once, then use the set of lines to find the starting points of $K_U(S)$, and finally reconstruct the whole $K_U(S)$ by marching through them. The disadvantage of this technique is that we must store a representation of the spherical net. This leads to large problems in the accuracy of the representation and a high amount of memory is required.

The second possibility is to store only one starting point per line of $Sp_V(S)$. Then we find starting points for each line of $K_U(S)$ by marching along all the lines of the spherical net.

The third possibility is to consider virtual subdivision of S onto patches S_i by lines of the spherical net, then we can have only one point per every patch S_i to reconstruct the contour lines of $K_U(S)$. Indeed, we can start from each point and reconstruct any contour line passing through this point until we reach the border of S_i. This will necessarily occur, as stated in statement 3. So we obtain a set of points of the spherical net as in the second possibility.

In any case we explore all the lines of the spherical net and obtain all intersection with $K_U(S)$. This process ensures that all the required starting points for $K_U(S)$ are obtained.

2.3.1 Extreme Points

The interposition of extreme points and contour lines is described by the following statement:

Statement 4. *Let U and W be any non-zero orthogonal vectors and $K_U(S)$ and $E_W(S)$ be the corresponding contour and set of extreme points. Then, we have $E_W(S) \subset K_U(S) \cup B(S)$. In other words any extreme point lies either on the border or on the contour for an orthogonal direction.*

Proof: This result is obvious since for each point Q not belonging to the border, we have $N(Q) = \alpha W$ ($\alpha \neq 0$) and by hypothesis on U and W, $N(Q) \cdot U = 0$. ■

Therefore if we have the contour lines we can obtain all extreme points in any orthogonal direction during the marching step along these contour lines.

On the other hand, the statement that we can reconstruct all contour lines starting in all extreme points for a given orthogonal direction isn't valid, because there exist closed contour lines which don't entirely cover the equator during Gaussian mapping, and therefore may not contain the image of some extreme points for orthogonal directions. In other words, if the image of the extreme points in a given direction lies outside the image of one contour line in an orthogonal direction we can't reach it by marching, starting from this extreme point.

2.3.2 Plane Section Lines

The interposition of plane section lines and contour lines is described by the following statement:

Statement 5. *Let U and W be any non-zero orthogonal vectors and $K_U(S)$ and $I_{P_W}(S)$ be the corresponding contour and section of surface S by plane P_W with normal vector W, then intersection $(K_U(S) \cup B(S)) \cap L \neq \emptyset$, where L is any connected component of $I_{P_W}(S)$. In other words any component of section lines intersects with the border or the contour lines in an orthogonal direction.*

Proof: If L is an open curve, L reaches the border $(B(S))$. Otherwise, L is a closed curve and there necessarily exists a tangent vector T at point M on curve L which satisfies $T = \alpha U$. In M, the dot product $N(M) \cdot U$ vanishes since vector T is included in the tangent plane of the surface in M (see Figure 3). L intersects $K_U(S)$. ■

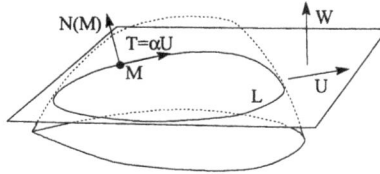

Fig. 3. Point belonging to a section line and a contour line.

This proof is different than those of statement 3 because in this one, we are working with sections of normal surface $N(S)$.

So we will reconstruct the contour lines for any U, which is perpendicular to a given vector W and simultaneously find points of intersection with the plane and these contour lines. The border of the surface must also be studied. Therefore, we can entirely reconstruct any plane section by marching.

§3. Singularities on Surface

All our previous reasoning holds if we have no degenerate regions on our surface in which Gaussian curvature is equal to zero (developable or flat regions). Images of such regions during Gaussian mapping is 3D curve (or point) on the unit sphere, and therefore the image of the border of such regions coincides with the image of the entire degenerate region.

Let us consider the problem of contour and section lines. For a degenerate region, we have the following result:

Statement 6. *For a degenerate region, three cases can be encountered: no curve exists, one curve crosses the region and the study of the border is enough to detect it, a complete flat region is a solution of the problem and the sets defined at the beginning of section 2 are not uniquely composed of lines.*

Proof: First, it is obvious that one complete region can be found as solution of the problem only for a flat region and not for a curved developable region. We have only to prove that no closed loop lies inside a flat or developable region. Let us consider such a closed curve C. There exists an infinite number of generators D which intersect C twice at two points P_1 and P_2 (see Figure 4). The normal vector of the surface is the same along the entire generator. This implies that generator D is also a solution of the line determination problem and finally that a complete region is part of the solution. This one is then necessarily a flat region. No such curve C can exist. This reasoning is valid both for contour and section lines. Thus, we can only analyze the border of such a region for detection of open curves crossing the region. ∎

Therefore we can consider the borders of the degenerate regions as zero curvature Gaussian lines and add them to our spherical net. The previous statements are thus valid for such a spherical net:

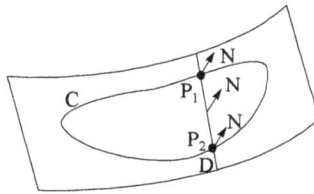

Fig. 4. A closed loop in a developable region.

Statement 7. *Statements 1-5 are valid for surfaces with degenerate parts. Interior of flat regions must be treated separately.*

Thus, we can also process surfaces with degenerate parts by marching algorithms. As the border of flat regions is given, they can easily be studied.

§4. Conclusion

In this paper a new approach for reconstruction of some line and point sets on surfaces is proposed. The main advantage of the technique considered is that after *a surface preprocessing stage*, we obtain a special point set, which will be the starting set for marching. Then we can reconstruct, without further subdivision of the surface, any contour, section lines or extreme points set by simple marching.

As with any method involving a preprocessing stage, ours is efficient in case of multiple applications. An interesting particular case is the computation of the set of parallel sections. In that case all the starting points can be obtained during one single march along one contour in an orthogonal direction and the border.

We should note that our approach is useful for surface analysis because a set of section lines can be computed ([5]) and also because we reach the lines of zero Gaussian curvature.

§5. Appendix: Expression of Gaussian Curvature

Background on differential geometry can be found in [4]. To avoid any confusion between scalars and vectors, notation with arrows will be used for vectors. Let us consider:

* surface S and its two first derivatives $\vec{S_u}$ and $\vec{S_v}$.
* normal vector $\vec{N}$ and its two first derivatives $\vec{N_u}$ and $\vec{N_v}$.
* unit normal vector $\vec{n}$ and its two first derivatives $\vec{n_u}$ and $\vec{n_v}$.
* E, F, G, L, N, M the coefficients of the two fundamental forms.
* D the norm of $\vec{N}$ ($D^2 = E \cdot G - F^2$), and its two first derivatives D_u and D_v.

The Weingarten equations expressed with $\vec{S_u}$ and $\vec{S_v}$ are
$$\vec{n_u} = 1/D^2 \left((M \cdot F - G \cdot L) \vec{S_u} + (L \cdot F - M \cdot E) \vec{S_v} \right)$$
$$\vec{n_v} = 1/D^2 \left((N \cdot F - M \cdot G) \vec{S_u} + (M \cdot F - N \cdot E) \vec{S_v} \right)$$

From these two equations, a direct computation gives $\vec{n_u} \times \vec{n_v} = K \cdot D\, \vec{n}$, where K is the Gaussian curvature. This immediately leads to $K = 1/D\,(\vec{n}, \vec{n_u}, \vec{n_v})$. Our interest is in an expression using the vectors $\vec{N}$, $\vec{N_u}$ and $\vec{N_v}$.

$$\vec{N_u} = D_u\,\vec{n} + D\,\vec{n_u}$$
$$\vec{N_v} = D_v\,\vec{n} + D\,\vec{n_v}$$

Combining the previous formulas of K and those of vectors $\vec{N_u}$ and $\vec{N_v}$, we finally obtain $K = 1/D^4\,(\vec{N}, \vec{N_u}, \vec{N_v})$. This formula proves that determinant $(\vec{N}, \vec{N_u}, \vec{N_v})$ supplies all the information about the sign of the Gaussian curvature and also allows us to characterize the zeros of the Gaussian curvature.

References

1. Barnhill R. E., and S. N. Kersey, A marching method for parametric surface/surface intersection, Comput. Aided Geom. Design **7** (1990), 257–280.

2. Cheng, K. P., Using plane vector fields to obtain all the intersection curves of two general surfaces, in *Theory and Practice of Geometric Modeling*, W. Strasser and H. P. Seidel (eds.), Springer Verlag, 1989,187–280.

3. Daniel, M., and A. Nicolas, A Surface-Surface Intersection Algorithm with a Fast Clipping Technique, in *Curves and Surfaces in Geometric Design*, P.-J. Laurent, A. Le Méhauté, and L. L. Schumaker (eds.), A. K. Peters, Wellesley MA, 1994, 105–112.

4. Do Carmo M. P., *Differential Geometry for Curves and Surfaces*, Prentice Hall, 1976.

5. Hagen H., T. Schreiber, and E. Gschwind, Methods for surface interrogation, in *Visualization'90 Proceedings*, 105–112.

6. Koparkar P., Surface intersection by switching from recursive subdivision to iterative refinement, The Visual Computer **8**, (1991) 47–63.

7. Kriezis G. A., N. M. Patrikalakis, and F. E. Wolter, Topological and differential equation methods for surface intersections, Computer-Aided Design**24** (1), (1992), 41–55.

Yevgeniy Kuzmin
Lab. for Computation Methods, Dept. of Mathematics
Moscow State University, Moscow, 119899, RUSSIA
cls@online.ru

Marc Daniel
IRIN, Ecole Centrale de Nantes
B.P. 92101, 44321 Nantes cedex 3, FRANCE
Marc.Daniel@ec-nantes.fr

Generalized Tension B-splines

B. I. Kvasov and P. Sattayatham

Abstract. Explicit formulae and recurrence relations for calculation of generalized tension B-splines of arbitrary degree are given. We derive the main properties of GB-splines and their series, i.e. partition of unity, shape preserving properties, invariance with respect to linear transformations, etc. It is shown that such splines, providing the variation diminishing property, are Chebyshev splines.

§1. Introduction

Fitting curves and surfaces to functions and data requires the availability of methods which preserve the shape of the data. In practical calculations we usually deal with data given with prescribed accuracy. Therefore we should develop methods for constructing fair-shape-preserving approximations that satisfy given tolerances and inherit geometric properties of the data such as positivity, monotonicity, convexity, presence of linear sections, etc.

Such approximation can be based on generalized B-splines. Until recently, local support bases for computations with generalized splines have been available for only some special types of splines [1,8,11]. This limits the choice of methods when using generalized splines in tension. In [4,5,9] local support basis functions for exponential splines were introduced and their application to interpolation problems was considered. A recurrence relation for rational B-splines with prescribed poles was recently obtained in [2]. In this paper we expand the main results of [6,7] on generalized tension B-splines of arbitrary degree allowing the tension parameters to vary from interval to interval.

§2. Generalized B-splines of Arbitrary Degree

Let a partition $\Delta : a = x_0 < x_1 < \cdots < x_N = b$ be given on the segment $[a, b]$ to which we associate a space of splines S_n^G whose restriction to a subinterval $[x_i, x_{i+1}]$, $i = 0, \ldots, N - 1$ is spanned by the system of linearly independent functions $\{1, x, \ldots, x^{n-2}, \Phi_{i,n}(x), \Psi_{i,n}(x)\}$, $n \geq 1$, and where any function in S_n^G has $n - 1$ continuous derivatives.

Curves and Surfaces with Applications in CAGD
A. Le Méhauté, C. Rabut, and L. L. Schumaker (eds.), pp. 247–254.
Copyright © 1997 by Vanderbilt University Press, Nashville, TN.
ISBN 0-8265-1293-3.
All rights of reproduction in any form reserved.

Definition 2.1. *The generalized spline of degree n is a function $S(x) \in S_n^G$ such that*

(1) for any $x \in [x_i, x_{i+1}]$, $i = 0, \ldots, N-1$,

$$S(x) = P_{i,n-2}(x) + S^{(n-1)}(x_i)\Phi_{i,n}(x) + S^{(n-1)}(x_{i+1})\Psi_{i,n}(x),$$

where $P_{i,n-2}(x)$ is a polynomial of degree $n-2$, and

$$\Phi_{i,n}^{(r)}(x_{i+1}) = \Psi_{i,n}^{(r)}(x_i) = 0, \quad r = 0, \ldots, n-1$$
$$\Phi_{i,n}^{(n-1)}(x_i) = \Psi_{i,n}^{(n-1)}(x_{i+1}) = 1; \tag{2.1}$$

(2) $S(x) \in C^{n-1}[a, b]$.

Consider the problem of constructing a basis in the space S_n^G consisting of functions with local support of minimal length. For this, it is convenient to extend the mesh Δ by adding points $x_{-n} < \cdots < x_{-1} < a$, $b < x_{N+1} < \cdots < x_{N+n}$. As $dim(S_n^G) = (n+1)N - n(N-1) = N + n$, it is sufficient to construct a system of linearly independent splines $B_{j,n}(x)$, $j = -n, \ldots, N-1$ in S_n^G such that $B_{j,n}(x) > 0$ if $x \in (x_j, x_{j+n+1})$ and $B_{j,n}(x) \equiv 0$ outside (x_j, x_{j+n+1}).

For $n > 1$ we require the fulfillment of the normalization condition

$$\sum_{j=-n}^{N-1} B_{j,n}(x) \equiv 1 \quad \text{for} \quad x \in [a, b]. \tag{2.2}$$

By Definition 2.1, we will seek basis splines in the form

$$B_{j,n}(x) = \begin{cases} B_{j,n}^{(n-1)}(x_{j+1})\Psi_{j,n}(x), & x_j \le x \le x_{j+1} \\ P_{j,l,n-2}(x) + B_{j,n}^{(n-1)}(x_{j+l})\Phi_{j+l,n}(x) \\ \qquad + B_{j,n}^{(n-1)}(x_{j+l+1})\Psi_{j+l,n}(x) \\ \qquad x_{j+l} \le x \le x_{j+l+1}, \quad l = 1, \ldots, n-1 \\ B_{j,n}^{(n-1)}(x_{j+n})\Phi_{j+n,n}(x), & x_{j+n} \le x \le x_{j+n+1} \\ 0, & x \notin (x_j, x_{j+n+1}). \end{cases} \tag{2.3}$$

The form of $B_{j,n}(x)$ in (2.3) for $x \in [x_{j+k}, x_{j+1+k}]$, $k = 0, n$ has been simplified in virtue of the conditions $B_{j,n}^{(r)}(x_j) = B_{j,n}^{(r)}(x_{j+n+1}) = 0$, $r = 0, \ldots, n-1$, and the properties (2.1) of functions $\Phi_{j,n}(x)$, $\Psi_{j,n}(x)$.

Taking into account the continuity conditions for polynomials $P_{j,l,n-2}(x)$, $l = 1, \ldots, n-1$, in (2.3) we have the relations

$$P_{j,l,n-2}(x) = P_{j,l-1,n-2}(x) + B_{j,n}^{(n-1)}(x_{j+l}) \sum_{r=0}^{n-2} z_{j+l,n}^{(r)}(x - x_{j+l})^r / r!$$
$$l = 1, \ldots, n, \tag{2.4}$$

with $z_{j+l,n}^{(r)} = \Psi_{j+l-1,n}^{(r)}(x_{j+l}) - \Phi_{j+l,n}^{(r)}(x_{j+l})$, $r = 0, \ldots, n - 2$.

By (2.3), polynomials $P_{j,l,n-2}(x) \equiv 0$ when $l = 0$ and $l = n$. Then by repeated application of the formula (2.4) we have

$$P_{j,l,n-2}(x) = \sum_{l'=1}^{l} B_{j,n}^{(n-1)}(x_{j+l'}) \sum_{r=0}^{n-2} z_{j+l',n}^{(r)}(x - x_{j+l'})^r/r!$$

$$= - \sum_{l'=l+1}^{n} B_{j,n}^{(n-1)}(x_{j+l'}) \sum_{r=0}^{n-2} z_{j+l',n}^{(r)}(x - x_{j+l'})^r/r!, \quad l = 1, \ldots, n - 1.$$

In particular, the following identity is valid

$$\sum_{l=1}^{n} B_{j,n}^{(n-1)}(x_{j+l}) \sum_{r=0}^{n-2} z_{j+l,n}^{(r)}(x - x_{j+l})^r/r! \equiv 0. \tag{2.5}$$

Using the expansion of polynomials (2.5) by powers of x we arrive at a system of $n-1$ linear algebraic equations which defines the unknown quantities $B_{j,n}^{(n-1)}(x_{j+l})$, $l = 1, \ldots, n$. To obtain the unique solution of this system we can use the normalization condition (2.2). We can eliminate the unknowns analogously as in [6,7].

§3. Recurrence Algorithm for Calculation of GB-splines

Let us define the function

$$B_{j,1}(x) = \begin{cases} \Psi_{j,n}^{(n-1)}(x), & x_j \leq x \leq x_{j+1} \\ \Phi_{j+1,n}^{(n-1)}(x), & x_{j+1} \leq x \leq x_{j+2} \\ 0, & x \notin (x_j, x_{j+2}), \end{cases} \tag{3.1}$$

where the functions $\Psi_{j,n}^{(n-1)}(x)$, $\Phi_{j+1,n}^{(n-1)}(x)$ are assumed to be positive and monotone on (x_j, x_{j+1}) and (x_{j+1}, x_{j+2}) respectively.

We will consider the sequence of B-splines defined by the recurrence formula

$$B_{j,k}(x) = \int_{x_j}^{x} \frac{B_{j,k-1}(\tau)}{c_{j,k-1}} d\tau - \int_{x_{j+1}}^{x} \frac{B_{j+1,k-1}(\tau)}{c_{j+1,k-1}} d\tau, \quad k = 2, \ldots, n, \tag{3.2}$$

where

$$c_{j,k-1} = \int_{x_j}^{x_{j+k}} B_{j,k-1}(\tau) d\tau.$$

Differentiating formula (3.2) we obtain

$$B_{j,k}'(x) = B_{j,k-1}(x)/c_{j,k-1} - B_{j+1,k-1}(x)/c_{j+1,k-1}, \quad k = 2, \ldots, n. \tag{3.3}$$

Theorem 3.1. *The recurrence formulae (3.1) and (3.2) define the sequence of B-splines of the form*

$$
B_{j,k}(x) = \begin{cases}
B_{j,k}^{(k-1)}(x_{j+1})\Psi_{j,n}^{(n-k)}(x), & x_j \le x \le x_{j+1} \\[2mm]
P_{j,l,k-2}(x) + B_{j,k}^{(k-1)}(x_{j+l})\Phi_{j+l,n}^{(n-k)}(x) \\[1mm]
\qquad + B_{j,k}^{(k-1)}(x_{j+l+1})\Psi_{j+l,n}^{(n-k)}(x) \\[1mm]
\qquad x_{j+l} \le x \le x_{j+l+1}, \quad l = 1, \ldots, k-1 \\[2mm]
B_{j,k}^{(k-1)}(x_{j+k})\Phi_{j+k,n}^{(n-k)}(x), & x_{j+k} \le x \le x_{j+k+1} \\[2mm]
0, & x \notin (x_j, x_{j+k+1}),
\end{cases} \tag{3.4}
$$

$k = 1, \ldots, n$, where

$$
\begin{aligned}
P_{j,l,k-2}(x) &= \sum_{l'=1}^{l} B_{j,k}^{(k-1)}(x_{j+l'}) \sum_{r=n-k}^{n-2} z_{j+l',n}^{(r)}(x - x_{j+l'})^{r-n+k}/(r-n+k)! \\
&= -\sum_{l'=l+1}^{k} B_{j,k}^{(k-1)}(x_{j+l'}) \sum_{r=n-k}^{n-2} z_{j+l',n}^{(r)}(x - x_{j+l'})^{r-n+k}/(r-n+k)!
\end{aligned} \tag{3.5}
$$

and

$$
\sum_{l=1}^{k} B_{j,k}^{(k-1)}(x_{j+l}) \sum_{r=n-k}^{n-2} z_{j+l,n}^{(r)}(x - x_{j+l})^{r-n+k}/(r-n+k)! \equiv 0,
$$
$$
k = 2, \ldots, n.
$$

This can be shown by induction using the differentiation formula (3.3).

To use the formulae (3.4) and (3.5) for calculations we first need to find the quantities $B_{j,k}^{(k-1)}(x_{j+l})$, $l = 1, \ldots, k$; $k = 2, \ldots, n$. According to (3.3),

$$
B_{j,k}^{(k-1)}(x_{j+l}) = B_{j,k-1}^{(k-2)}(x_{j+l})/c_{j,k-1} - B_{j+1,k-1}^{(k-2)}(x_{j+l})/c_{j+1,k-1} \tag{3.6}
$$
$$
l = 1, \ldots, k; \quad k = 2, \ldots, n.
$$

In particular, it follows from here with $B_{j,1}(x_{j+1}) = 1$ that

$$
B_{j,2}'(x_{j+1}) = \frac{1}{c_{j,1}}, \quad B_{j,3}''(x_{j+1}) = \frac{1}{c_{j,1}c_{j,2}},
$$
$$
B_{j,2}'(x_{j+2}) = -\frac{1}{c_{j+1,1}}, \quad B_{j,3}''(x_{j+2}) = -\frac{1}{c_{j+1,1}}\left(\frac{1}{c_{j,2}} + \frac{1}{c_{j+1,2}}\right),
$$
$$
B_{j,3}''(x_{j+3}) = \frac{1}{c_{j+2,1}c_{j+1,2}},
$$

etc. Therefore to find the necessary values of the derivatives of the basis splines in interior nodes of their interval supports, it is necessary to know the quantities $c_{j,k}$, i. e. the integrals of the B-splines $B_{j,k}(x)$, $k = 1, \ldots, n-1$.

Theorem 3.2. *The integrals* $c_{j,k} = \int_{x_j}^{x_{j+k+1}} B_{j,k}(\tau)d\tau$ *of the generalized basis splines are given by*

$$c_{j,k} = \sum_{l=1}^{k} B_{j,k}^{(k-1)}(x_{j+l}) \sum_{r=n-k-1}^{n-2} z_{j+l,n}^{(r)} \frac{(x_{j+\alpha} - x_{j+l})^{r-n+k+1}}{(r-n+k+1)!},$$

$$\alpha = 1, \ldots, k; \quad k = 1, \ldots, n-1. \tag{3.7}$$

This can be proven by induction using formulae for B-splines (3.4) and (3.5).

To construct the basis spline $B_{j,k}(x)$, $k = 2, \ldots, n$, we apply formulae (3.6) and (3.7), and consecutively calculate the quantities $B_{j,k}^{(k-1)}(x_{j+\alpha})$, $\alpha = 1, \ldots, k$, $k = 1, \ldots, n$, and $c_{j+\beta}$, $\beta = 0, \ldots, n-k$, $k = 1, \ldots, n-1$.

§4. Properties of Generalized B-splines and Their Series

Let us formulate some properties of GB-splines which are mainly analogous to the properties of polynomial B-splines [10].

Theorem 4.1. *The functions* $B_{j,k}(x)$, $k = 1, \ldots, n$ *have the following properties:*

1) $B_{j,k}(x) > 0$ *if* $x \in (x_j, x_{j+k+1})$ *and* $B_{j,k}(x) \equiv 0$ *if* $x \notin (x_j, x_{j+k+1})$;
2) *The splines* $B_{j,k}(x)$ *have* $k-1$ *continuous derivatives;*
3) *for* $k \geq 2$ *and* $x \in [a, b]$, $\sum_{j=-k}^{N-1} B_{j,k}(x) = 1$;
4) *For* $x \in [x_j, x_{j+1}]$,

$$\Psi_{j,n}^{(r)}(x) = (\prod_{k=1}^{n-r-1} c_{j,k})B_{j,n-r}(x), \quad \Phi_{j,n}^{(r)}(x) = \prod_{k=1}^{n-r-1}(-c_{j-k,k})B_{j-n+r,n-r}(x)$$

$j = 0, \ldots, N-1$, $r = 0, \ldots, n-1$, *where* $c_{j,k} = \int_{x_j}^{x_{j+k+1}} B_{j,k}(\tau)d\tau$.

We denote by S_k^G the set of splines $S(x) \in C^{k-1}[a, b]$ which are spanned by linear combinations of the functions $\{1, \ldots, x^{k-2}, \Phi_i^{(n-k)}(x), \Psi_i^{(n-k)}(x)\}$, $k = 1, \ldots, n$, in any subinterval $[x_i, x_{i+1}]$, $i = 0, \ldots, N-1$. Using the methods in [12], it is easy to show that the splines $B_{j,k}(x)$, $j = -k, \ldots, N-1$, $k = 1, \ldots, n$, have minimum-length supports, are linearly independent and form a basis in S_k^G, i.e., any generalized spline $S(x) \in S_k^G$, $k = 1, \ldots, n$ can be uniquely represented in the form

$$S(x) = \sum_{j=-k}^{N-1} b_{j,k} B_{j,k}(x) \quad \text{for} \quad x \in [a, b] \tag{4.1}$$

for some constant coefficients $b_{j,k}$.

Applying the differentiation formula (3.3), we obtain for $r \leq k - 1$

$$S^{(r)}(x) = \sum_{j=-k+r}^{N-1} b_{j,k}^{(r)} B_{j,k-r}(x),$$

where

$$b_{j,k}^{(l)} = \begin{cases} b_{j,k}, & l = 0 \\ \dfrac{b_{j,k}^{(l-1)} - b_{j-1,k}^{(l-1)}}{c_{j,k-l}}, & l = 1, 2, \dots, r. \end{cases}$$

If now $b_j^{(k)} > 0$, $k = 0, 1, 2$, $j = -3 + k, \dots, N - 1$, then the spline $S(x)$ will be a positive monotonically increasing and convex function.

Let $Z_{[a,b]}(f(x))$ be the number of isolated zeros of a function $f(x)$ on the segment $[a, b]$.

Lemma 4.1. *If the spline $S(x) = \sum_{j=-k}^{N-1} b_{j,k} B_{j,k}(x)$, $k = 1, \dots, n$ does not vanish on any subsegment of $[a, b]$, then $Z_{[a,b]}(S(x)) \le N + k - 1$.*

Denote by $supp\, B_{j,k}(x) = \{x | B_{j,k}(x) \neq 0\}$, $k = 1, \dots, n$, the support of the spline $B_{j,k}(x)$, i.e. the interval (x_j, x_{j+k+1}).

Theorem 4.2. *Assume that $\tau_{-k} < \tau_{-k+1} < \dots < \tau_{N-1}$, $k = 1, \dots, n$. Then*

$$D = det\,(B_{j,k}(\tau_i)) \neq 0, \quad i, j = -k, \dots, N - 1$$

if and only if

$$\tau_j \in supp\, B_{j,k}(x), \quad j = -k, \dots, N + 1. \tag{4.2}$$

If condition (4.2) is satisfied, then $D > 0$.

The following three statements follow immediately from the theorem 4.2.

Corollary 4.1. *The system of generalized B-splines $\{B_{j,k}(x)\}$, $j = -k, \dots, N - 1$, $k = 1, \dots, n$, is a weak Chebyshev system in the sense of [3], i.e. for any $\tau_{-k} < \tau_{-k+1} < \dots < \tau_{N-1}$ we have $D \ge 0$ and $D > 0$ if and only if condition (4.2) is satisfied. If the latter is satisfied, then the generalized spline $S(x) = \sum_{j=-k}^{N-1} b_{j,k} B_{j,k}(x)$, $k = 1, \dots, n$ has no more than $N + k - 1$ isolated zeros.*

Corollary 4.2. *If the conditions of Theorem 4.2 are satisfied, the solution of the interpolation problem $S(\tau_i) = f_i$, $i = -k, \dots, N - 1$, $f_i \in \mathbb{R}$ exists and is unique.*

Let $A = \{a_{ij}\}$, $i = 1, \dots, m$, $j = 1, \dots, n$, be a rectangular $(m \times n)$ matrix with $m \le n$. The matrix A is said to be totally nonnegative (totally positive) [3] if the minors of all orders of the matrix are all nonnegative (positive), i.e. for all $1 \le l \le m$ we have $det(a_{i_p j_q}) \ge 0 \ (> 0)$ for all $1 \le i_1 < \dots < i_l \le m$, $1 \le j_1 < \dots < j_l \le n$.

Corollary 4.3. *For arbitrary integers $-k \le \nu_{-k} < \dots < \nu_{l-k-1} \le N - 1$ and $\tau_{-k} < \tau_{-k+1} < \dots < \tau_{l-k-1}$, $k = 1, \dots, n$, we have*

$$D_l = det\{B_{\nu_j,k}(\tau_i)\} \ge 0, \quad i, j = -k, \dots, l - k - 1,$$

and $D_l > 0$ if and only if $\tau_j \in supp\, B_{\nu_j,k}(x)$, $j = -k, \dots, l - k - 1$, i.e. the matrix $\{B_{j,k}(\tau_i)\}$, $i, j = -k, \dots, N - 1$ is totally nonnegative.

Denote by $S^-(\mathbf{v})$ the number of sign changes (variations) in the sequence of components of the vector $\mathbf{v} = (v_1, \ldots, v_n)$, with zero being neglected. For a bounded real function $f(x)$, let $S^-(f) \equiv S^-(f(x))$ be the number of sign changes of the function $f(x)$ on the real axis $\mathbb{R}$ without taking into account the zeros

$$S^-(f(x)) = \sup_p S^-[f(\tau_1), \ldots, f(\tau_p)], \quad \tau_1 < \tau_2 < \ldots < \tau_p.$$

Theorem 4.3. *The spline* $S(x) = \sum_{j=-k}^{N-1} b_{j,k} B_{j,k}(x)$, $k = 1, \ldots, n$ *is a variation diminishing function, i.e. the number of sign changes* $S(x)$ *does not exceed the one in the sequence of its coefficients*

$$S_{\mathbb{R}}^- \left(\sum_{j=-k}^{N-1} b_{j,k} B_{j,k}(x) \right) \leq S^-(\mathbf{b}), \quad \mathbf{b} = (b_{-k,k}, \ldots, b_{N-1,k}).$$

Let $\hat{S}_n^G$ be a set of generalized splines on the mesh $\hat{\Delta} = \{\hat{x}_i \mid \hat{x}_i = px_i + q, i = 0, \ldots, N\}$ which is obtained from the linear space S_n^G by linear transformation of the variable $\hat{x} = px + q$ where $p \neq 0$ and q are constant.

Theorem 4.4. *An approximating generalized spline* $S(x) \in S_n^G$ *is invariant with respect to a linear transformation of the real axis* $\mathbb{R} = (-\infty, \infty)$.

The proofs of the statements above are based on the methods of [10] for polynomial B-splines.

§5. Local Approximation by Generalized Splines

Using the locality of B-splines one can reduce the representation of a spline $S(x)$ as a linear combination of B-splines (4.1) for $k = n$ to the form

$$S(x) = \sum_{j=i-n}^{i} b_{j,n} B_{j,n}(x), \quad x \in [x_i, x_{i+1}], \quad i = 0, 1, \ldots, N-1. \tag{5.1}$$

Theorem 5.1. *The restriction (5.1) of the spline* $S(x)$ *to the interval* $[x_i, x_{i+1}]$ *can be written in the form*

$$S(x) = P_{i,n-2}(x) + b_{i-1,n}^{(n-1)} \Phi_{i,n}(x) + b_{i,n}^{(n-1)} \Psi_{i,n}(x),$$

where

$$P_{i,n-2}(x) = \sum_{k=0}^{n-2} b_{i-n+1+k,n}^{(k)} Q_{i,n-2}^{(n-2-k)}(x)$$

$$Q_{i,n-2}^{(k)}(x) = \begin{cases} Q_{i,n-2}(x)/c_{i-1,1}, & k = 0 \\ \dfrac{Q_{i-1,n-2}^{(k-1)}(x) - Q_{i,n-2}^{(k-1)}(x)}{c_{i-k-1,k+1}}, & k = 1, 2, \ldots, n-2 \end{cases}$$

$$Q_{j,n-2}(x) = \sum_{l=0}^{n-2} z_{j,n}^{(r)} \frac{(x-x_j)^r}{r!}, \quad j = i-n+2,\dots,i, \quad Q_{i,n-2}^{(n-2)}(x) \equiv 1$$

$$b_{j,n}^{(k)} = \frac{b_{j,n}^{(k-1)} - b_{j-1,n}^{(k-1)}}{c_{j,n-k}}, \quad k = 1,\dots,n-1; \quad b_{j,n}^{(0)} = b_{j,n}, \quad j = i-n,\dots,i.$$

This assertion is new even for polynomial splines, and can be proven by induction.

References

1. Dyn, N. and A. Ron, Recurrence relations for Tchebycheffian B-splines, J. Analyse Math. **51** (1988), 118–138.

2. Gresbrand, A., A recurrence relation for rational B-splines with prescribed poles, preprint, 1995.

3. Karlin, S., *Total Positivity*, Vol. 1, Stanford University Press, 1968.

4. Koch, P. E. and T. Lyche, Construction of exponential tension B-splines of arbitrary order, in *Curves and Surfaces*, P.-J. Laurent, A. Le Méhauté, and L. L. Schumaker (eds.), Academic Press, New York, 1991, 255–258.

5. Koch, P. E. and T. Lyche, Interpolation with exponential B-splines in tension, in *Geometric Modelling*. Computing/Supplement 8, G. Farin et al. (eds,), Springer–Verlag, Wien, 1993, 173–190.

6. Kvasov, B. I., Local bases for generalized cubic splines, Russ. J. Numer. Anal. Math. Modelling **10** (1995), No. 1, 49–80.

7. Kvasov, B. I., GB-splines and their properties, Annals of Numerical Mathematics **3** (1996), 139–149.

8. Lyche, T., A recurrence relation for Chebyshevian B-splines, Constr. Approx. **1** (1985), 155–173.

9. McCartin, B. J., Theory of exponential splines, J. Approx. Theory **66** (1991), 1–23.

10. Schumaker, L. L., *Spline Functions: Basic Theory*, Wiley, New York, 1981.

11. Schumaker, L. L., On recurrence for generalized B-splines, J. Approx. Theory **36** (1982), 16–31.

12. Zav'yalov, Yu. S., B. I. Kvasov, and V. L. Miroshnichenko, *Methods of Spline Functions*, Nauka, Moscow, 1980 (in Russian).

Boris I. Kvasov
Institute of Computational
Technologies
Russian Academy of Sciences
6, Lavrentyev Avenue
630090 Novosibirsk, RUSSIA
boris@math.sut.ac.th

Pairote Sattayatham
School of Mathematics
Suranaree University of
Technology
Nakhon Ratchasima 30000
THAILAND
pairote@ccs.sut.ac.th

Shape Effects with Polynomial Chebyshev Splines

Pierre-Jean Laurent, Marie-Laurence Mazure
and Géraldine Morin

Abstract. We define a parametric spline curve whose restriction to each sub-interval belongs to a 4-dimensional Chebyshev space depending on a (shape) parameter and which is included in the space of polynomials of degree 4.

§1. Introduction

Shape parameters play a crucial rôle in geometric design. Considering a piecewise smooth parametric curve, there are essentially two kinds of shape parameters. The first kind is related to the junction between two sections of the curve: for example the well-known parameters $\beta_1, \beta_2, \ldots$ expressing geometric continuity at the knot t_i, or, more generally, the coefficients of the connection matrix at t_i. Another type of shape parameter is related to the nature of the representation in each section: for instance, we can replace polynomials by a space of hyperbolic or trigonometric functions depending on a parameter. In order to keep all the shape preserving properties of polynomials, it is essential to use Chebyshev subspaces. But the inconvenience is that we lose the computational simplicity of the polynomial case. In order to avoid this fault, we consider here a family of 4-dimensional Chebyshev subspaces (hyperplanes) included in the space of polynomials of degree less than or equal to 4 and depending on a parameter. This parameter plays the rôle of a shape parameter. A polynomial Chebyshev spline is then a function whose restriction to each subinterval belongs to such a Chebyshev space for different values of the parameter with convenient connections at the knots. The corresponding B-splines have a 4-interval support and have essentially all the properties of the classical B-splines of order 4. Several examples show that the parameters can induce strong shape effects on the curve.

Curves and Surfaces with Applications in CAGD
A. Le Méhauté, C. Rabut, and L. L. Schumaker (eds.), pp. 255–262.
Copyright © 1997 by Vanderbilt University Press, Nashville, TN.
ISBN 0-8265-1293-3.
All rights of reproduction in any form reserved.

§2. Polynomial Chebyshev Bézier Curves

Let us denote by $\mathcal{P}_n$ the space of polynomials of degree less than or equal to n. Suppose that p is a given element of $\mathcal{P}_1$ nonvanishing on an interval I. Let us consider the following subspace of $C^\infty(I)$:

$$\mathcal{E} := \{\, F \in C^\infty(I)\,|\,(F'/p)'''(x) \equiv 0\}. \tag{1}$$

For $F \in \mathcal{E}$, $q := F'/p$ belongs to $\mathcal{P}_2$, and hence F belongs to $\mathcal{P}_4$. Thus, E is a linear subspace of $\mathcal{P}_4$. It is easy to verify that

$$\mathcal{E} = \begin{cases} \text{span}(\mathbb{1}, p^2, p^3, p^4) & \text{if } p \notin \mathcal{P}_0, \\ \mathcal{P}_3 & \text{if } p \in \mathcal{P}_0. \end{cases} \tag{2}$$

A nonvanishing element $F \in \mathcal{E}$ has at most 3 zeros in I (with multiplicities): if $p \in \mathcal{P}_0$, this is obvious; if $p \notin \mathcal{P}_0$, then $F' = pq$ (with $q \neq 0$ belonging to $\mathcal{P}_2$) has at most 2 zeros in I, and thus F has at most 3 zeros in I (with multiplicities). In fact, $\mathcal{E}$ is the 4-dimensional extended Chebyshev space on I defined by the following weight functions:

$$\omega_0 = \mathbb{1},\ \omega_1 = p,\ \omega_2 = \mathbb{1},\ \omega_3 = \mathbb{1}. \tag{3}$$

It is classical to define the associate differential operators

$$L_0 F = F,\quad L_i F = (1/\omega_i)(L_{i-1}F)',\quad i = 1,2,3, \tag{4}$$

i.e., explicitly here

$$L_1 F = F'/p,\quad L_2 F = (F'/p)',\quad L_3 F = (F'/p)'', \tag{5}$$

and we have

$$\mathcal{E} = \{F \in C^\infty(I)\,|\,(L_3 F)' = 0\}. \tag{6}$$

For given $a, b \in I$, $(a \neq b)$, the linear functional

$$\theta\ :\ F \in \mathcal{P}_4 \to (F'/p)''(a) - (F'/p)''(b)$$

is equal to 0 on $\mathcal{E}$, but is not identically equal to zero on $\mathcal{P}_4$. As a consequence, its nullspace is a 4-dimensional subspace of $\mathcal{P}_4$ which is necessarily equal to $\mathcal{E}$. Thus, we also have

$$\mathcal{E} = \{F \in \mathcal{P}_4\,|\,(F'/p)''(a) = (F'/p)''(b)\}. \tag{7}$$

In the rest of this section we suppose for simplicity that $a = 0$ and $b = 1$. We denote by $\mathcal{E}^d$ the space of functions $I \to \mathbb{R}^d$ each component of which belongs to $\mathcal{E}$. Let $F \in \mathcal{E}^d$. Considering F as an element of $\mathcal{P}_4^d$, let us denote

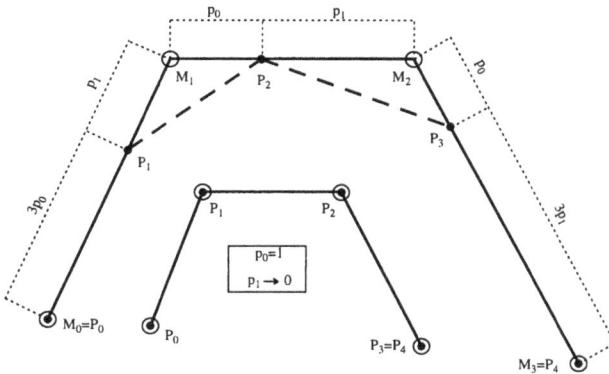

Fig. 1. Chebyshev-Bézier points M_i and Bézier points P_i.

by $P_0, \ldots, P_4$ its Bézier points with respect to $[0, 1]$. Developing the relation involved in (7) yields

$$\frac{1}{p_0} F'''(0) - \frac{2d}{p_0^2} F''(0) + \frac{2d^2}{p_0^3} F'(0) = \frac{1}{p_1} F'''(1) - \frac{2d}{p_1^2} F''(1) + \frac{2d^2}{p_1^3} F'(1),$$

with $p_0 = p(0)$, $p_1 = p(1)$, $d = p'(t) = p_1 - p_0$. Replacing the derivatives by their expressions in terms of the Bézier points $(F'0) = 4(P_1 - P_0)$, $F''(0) = 12(P_2 - 2P_1 + P_0)$, $F'''(0) = 24(P_3 - 3P_2 + 3P_1 - P_0)$, and similarly in 1), we obtain the relation

$$P_2 = \frac{p_1}{p_0 + p_1}\left(P_1 + \frac{p_1}{3p_0}(P_1 - P_0)\right) + \frac{p_0}{p_0 + p_1}\left(P_3 + \frac{p_0}{3p_1}(P_3 - P_4)\right). \qquad (8)$$

Let us set

$$M_1 := P_1 + \frac{p_1}{3p_0}(P_1 - P_0),$$

$$M_2 := P_3 + \frac{p_0}{3p_1}(P_3 - P_4),$$

and $M_0 := P_0$, $M_3 := P_4$. Conversely, we can arbitrarily choose the points M_i and then deduce the Bézier points P_i by

$$P_1 = \frac{p_1}{3p_0 + p_1} M_0 + \frac{3p_0}{3p_0 + p_1} M_1, \qquad (9)$$

$$P_2 = \frac{p_1}{p_0 + p_1} M_1 + \frac{p_0}{p_0 + p_1} M_2, \qquad (10)$$

$$P_3 = \frac{3p_1}{p_0 + 3p_1} M_2 + \frac{p_0}{p_0 + 3p_1} M_3, \qquad (11)$$

and of course $P_0 = M_0$, $P_4 = M_3$. This is illustrated in Fig. 1. The points $M_i, i = 0, \ldots, 3$ will be called the *Chebyshev-Bézier points*.

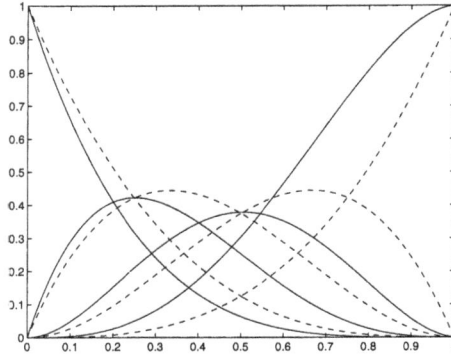

Fig. 2. (—)Chebyshev-Bernstein functions $(p_0 = 1, p_1 = 0.01)$ compared with (- - -) Bernstein polynomials of degree 3 $(p_0 = p_1 = 1)$.

An element $F \in \mathcal{E}^d$ can be written in the Bernstein basis as $F(t) = \sum_{i=0}^{4} P_i B_i(t)$, where the B_i's are the classical Bernstein polynomials in $\mathcal{P}_4$. Replacing the P_i's by their expression (9-11), we obtain

$$F(t) = \sum_{i=0}^{3} M_i \, CB_i(t), \qquad (12)$$

where

$$CB_0(t) = B_0(t) + p_1/(3p_0 + p_1) \, B_1(t),$$
$$CB_1(t) = 3p_0/(3p_0 + p_1) \, B_1(t) + p_1/(p_0 + p_1) \, B_2(t),$$
$$CB_2(t) = p_0/(p_0 + p_1) \, B_2(t) + 3p_1/(p_0 + 3p_1) \, B_3(t),$$
$$CB3(t) = p_0/(p_0 + 3p_1) \, B_3(t) + B_4(t).$$

The functions $CB_i, i = 0, \ldots, 3$, are called the *Chebyshev-Bernstein functions*. They are given in Fig. 2 for $p_0 = 1$ and $p_1 = 0.01$. Observe that for the limit case where $p_1 \to 0$, we have $CB_i = B_i, i = 0, 1, 2$ and $CB_3 = B_3 + B_4$.

§3. Polynomial Chebyshev Spline Curves

Suppose now that we have a sequence of knots t_i and a positive continuous function p whose restriction $p_{|[t_i, t_{i+1}]}$ is in $\mathcal{P}_1$. This function is well defined by the values $p_i = p(t_i)$. We want to construct a spline S whose restriction to each interval $[t_i, t_{i+1}]$ belongs to the space $\mathcal{E}_i^d$ constructed as in §2 with the weight function $p_{|[t_i, t_{i+1}]}$. Following P. J. Barry [1], we define the junction at t_i by a totally positive connection matrix applied to the differential operators L_j. More precisely, we suppose that

$$\begin{vmatrix} L_1 S(t_i^+) \\ L_2 S(t_i^+) \end{vmatrix} = \begin{pmatrix} \beta_1^i & 0 \\ \beta_2^i & (\beta_1^i)^2 \end{pmatrix} \begin{vmatrix} L_1 S(t_i^-) \\ L_2 S(t_i^-) \end{vmatrix}, \qquad (13)$$

where $\beta_1^i > 0$ and $\beta_2^i \geq 0$. Explicitly, we have

$$\left|\begin{array}{c} S'(t_i^+) \\ (\frac{S'}{p})'(t_i^+) \end{array}\right| = \left(\begin{array}{cc} \beta_1^i & 0 \\ \beta_2^i & (\beta_1^i)^2 \end{array}\right) \left|\begin{array}{c} S'(t_i^-) \\ (\frac{S'}{p})'(t_i^-) \end{array}\right|, \tag{14}$$

i.e.,

$$\left|\begin{array}{c} S'(t_i^+) \\ S''(t_i^+) \end{array}\right| = \left(\begin{array}{cc} \beta_1^i & 0 \\ \tilde{\beta}_2^i & (\beta_1^i)^2 \end{array}\right) \left|\begin{array}{c} S'(t_i^-) \\ S''(t_i^-) \end{array}\right|, \tag{15}$$

with

$$\tilde{\beta}_2^i = \beta_2^i + \frac{p'(t_i^+)\beta_1^i - p'(t_i^-)(\beta_1^i)^2}{p(t_i)}. \tag{16}$$

Observe that we have G^2-continuity at t_i but that $\tilde{\beta}_2^i$ is not necessarily greater than or equal to zero. In particular, it is not always possible to choose $\beta_1^i = 1$ and $\beta_2^i \geq 0$ so as to ensure parametric continuity of order 2, *i.e.,* $\tilde{\beta}_2^i = 0$.

The shape effects produced by the geometric continuity parameters β_1^i and β_2^i and by the weights p_i can of course be combined. But as we are mainly interested in exploring the effects of the parameters which define the Chebyshev subspaces in each section (*i.e.,* the p_i's), from now on we will choose $\beta_1^i = 1$ and $\beta_2^i = 0$. We also suppose that $t_{i+1} - t_i = 1$. In that case, we simply have (15) with

$$\beta_1^i = 1 \quad \text{and} \quad \tilde{\beta}_2^i = \frac{p_{i+1} - 2p_i + p_{i-1}}{p_i}. \tag{17}$$

Let us denote by $P_{i,j}, j = 0, \ldots, 4$, the Bézier points relative to the interval $[t_i, t_{i+1}]$ and by $M_{i,j}, j = 0, \ldots, 3$, the corresponding Chebyshev-Bézier points. Expressing conditions (15) in terms of the Bézier points, and then in terms of the Chebyshev-Bézier points, we obtain

$$M_{i-1,2} + \lambda_i(1 - \theta_{i-1})(M_{i-1,2} - M_{i-1,1}) = M_{i,1} + \lambda_i\theta_i(M_{i,1} - M_{i,2}), \tag{18}$$

with

$$\lambda_i = \frac{p_{i-1} + 6p_i + p_{i+1}}{4p_i}, \quad \theta_i = \frac{p_i}{p_i + p_{i+1}}. \tag{19}$$

The points P_i defined by the right or left side of relation (18) will be called the *poles* of the polynomial Chebyshev spline. Conversely, given these poles, we can compute the Chebyshev-Bézier points $M_{i,j}$ and deduce the Bézier points $P_{i,j}$. First, we compute all the $M_{i,1}$'s and the $M_{i,2}$'s by

$$M_{i,1} = (\, (1 + \lambda_{i+1}(1 - \theta_i))P_i + \lambda_i\theta_i P_{i+1} \,)/d_i, \tag{20}$$

$$M_{i,2} = (\, \lambda_{i+1}(1 - \theta_i)P_i + (1 + \lambda_i\theta_i)P_{i+1} \,)/d_i, \tag{21}$$

where $d_i = \lambda_i\theta_i + 1 + \lambda_{i+1}(1 - \theta_i)$. Then, we can compute $M_{i-1,3} = M_{i,0}$ by

$$M_{i-1,3} = M_{i,0} = \frac{(1 + u_i)M_{i-1,2} + (1 + v_{i-1})M_{i,1}}{2 + u_i + v_{i-1}}, \tag{22}$$

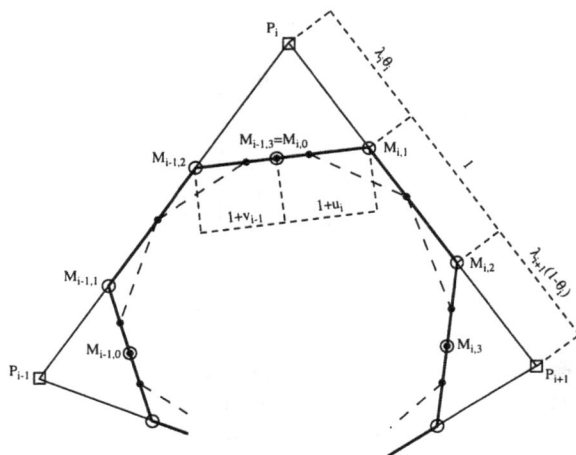

Fig. 3. Computation of the Chebyshev-Bézier points $M_{i,j}$ from the poles P_i.

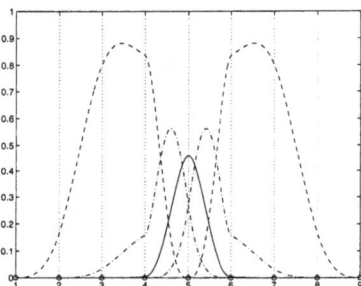

Fig. 4. Functions $CN_j, j = 3, \ldots, 7$ for all $p_j = 1$ except $p_5 = 100$.

with $u_i = p_{i+1}/3p_i$ and $v_i = p_i/3p_{i+1}$. Using in each section the formulae obtained in §2, from the $M_{i,j}$ we can deduce the Bézier points $P_{i,j}$. This algorithm is summarized in Fig. 3.

For a given weight function p, we denote by CN_i the Chebyshev B-splines, i.e., CN_i is the influence function of P_i. For computing $CN_i(t)$, put all the P_j equal to 0 except $P_i = 1$ and use the algorithm described above. The functions CN_i are nonnegative, have $[t_{i-2}, t_{i+2}]$ for support (except at the extremities) and their sum is equal to 1. For parameters p_i all equal to 1 except one which is taken equal to 100 (at the knot 5), Fig. 4 shows the graphs of the 5 Chebyshev B-splines CN_i which are not identical with the classical B-splines of order 4 (i.e., the 5 whose support intersects the interval $]4, 6[$ where the weight function p is not equal to 1). Observe that the two "exterior" C-B-splines CN_3 and CN_7 (with supports $[1, 5]$ and $[5, 9]$) are becoming dominant when p_5 is growing. On the other hand, the three "central" C-B-splines CN_4,

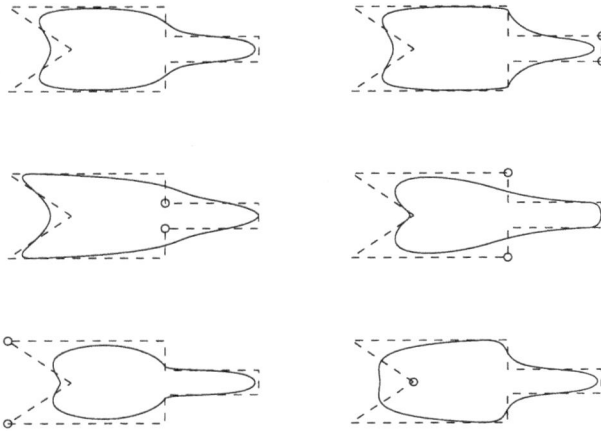

Fig. 5. A Chebyshev spline curve with $p_j = 100$ at the poles marked with o, the other p_i's being equal to 1.

CN_5 and CN_6 are becoming smaller. For $p_5 = 100$, the support of CN_5 is "almost" reduced to $[4, 6]$. At the limit ($p_5 \to \infty$), the supports of CN_4, CN_5 and CN_6 are reduced to $[4, 6]$. The functions CN_i seem to have a discontinuity of the first derivative at 4 and 6. Zooming in at that place would show that this is not the case (the CN_i are C^1, but not C^2). Nevertheless, the limits of these functions when $p_5 \to \infty$ have such a discontinuity.

Figure 5 shows the effects of the weights on a closed curve defined by 9 poles. The weights p_i are all taken equal to 1 except those indicated by "o" which are equal to 100. We see that the curve becomes flatter around these places and gets closer to the poles that are two sections away. This shape effect is consistent with the shape of the C-B-splines.

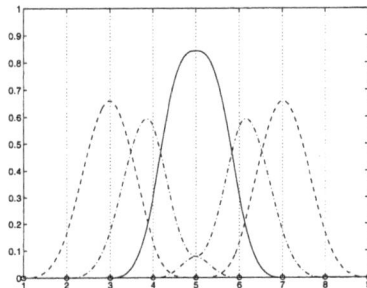

Fig. 6. Functions $CN_j, j = 3, \ldots, 7$ for all $p_j = 1$ except $p_5 = 0.01$.

Let us consider another interesting example for which all the weights p_j are equal to 1 except one equal to 0.01. The 5 Chebyshev B-splines which are different from the classical B-splines are shown in Fig. 6. We see that the "central" function CN_5 becomes dominant and that the other 4 functions become smaller. The consequence is that the curve comes closer to the poles for which the weights are small.

References

1. Barry P. J., de Boor-Fix dual functionals and algorithms for Tchebycheffian B-spline curves, Constr. Approx. **12**, No. 3 (1996), 385–408.

2. Costantini P., Variable degree polynomial splines, in *Curves and Surfaces with Applications in CAGD*, A. Le Méhauté, C. Rabut, and L. L. Schumaker (eds.), Vanderbilt Univ. Press, Nashville, 1997, 85–94.

3. Dyn N. and A. Ron, Recurrence relations for tchebycheffian B-splines, Journal d'Analyse Mathématique **51** (1988), 118–138.

4. Karlin S. and Z. Ziegler, Chebyshevian spline functions, SIAM J. Numer. Anal. **3** (1966), 514–543.

5. Kulkarni R. and P. J. Laurent, Q-splines, Numer. Algorithms **1** (1991), 45–74.

6. Kulkarni R., P. J. Laurent and M. L. Mazure, Non affine blossoms and subdivision for Q-splines, *Mathematical Methods in Computer Aided Geometric Design II*, T. Lyche and L. Schumaker (eds.), Academic Press, New York, 1992, 367–380.

7. Lyche T., A recurrence relation for chebyshevian B-splines, Constr. Approx. **1** (1985), 155–173.

8. Mazure M. L., Blossoming of Chebyshev splines, *Mathematical Methods for Curves and Surfaces*, Morten Dæhlen, Tom Lyche, Larry L. Schumaker (eds.), Vanderbilt University Press, Nashville & London, 1995, 355–364.

9. Mazure M. L., Chebyshev spaces, RR 952M, IMAG, Université Joseph Fourier, Grenoble, 1996.

10. Mazure M. L. and P. J. Laurent, Affine and nonaffine blossoms, in *Computational Geometry*, World Scientific, (1993), 201–230.

11. Mazure, M. L. and H. Pottmann, Tchebycheff curves, *Total Positivity and its Applications*, Kluwer Acad. Pub., (1996), 187–218.

12. Pottmann H., The geometry of tchebycheffian splines, Comput. Aided Geom. Design **10** (1993), 181–210.

13. Schumaker L. L., *Spline Functions: Basic Theory*, Wiley, New York, (1981).

Pierre-Jean Laurent, Marie-Laurence Mazure and Géraldine Morin
LMC-IMAG, Université Joseph Fourier
BP 53X, 38041 Grenoble, FRANCE
pjl@imag.fr, mazure@imag.fr

Interpolation with Triangulated Surfaces and Curvature Minimization

Michel Léger

Abstract. In oil exploration, geological surfaces are usually known from scattered data and therefore interpolation or approximation is needed. Triangulation is a simple and versatile method for modeling surfaces. The local paraboloid method gives the curvature matrix accurately. A subset of the vertices of the triangulation being fixed, the integral of the squared norm of the second fundamental form can be computed and minimized by a conjugate gradient method. As a numerical test, we interpolate four points building a regular tetrahedron with a surface of the same genus as a sphere. The initial model is a 36-vertex triangulation of a cube, with triangles of various sizes and shapes. After convergence, the solution is very close to the theoretically expected sphere. The effectiveness of the method is confirmed with larger triangulations obtained by subdivision. The local paraboloid method makes it possible to smooth irregular triangulations with a "smoothing effect" which does not depend on their idiosyncrasies.

§1. Introduction

In oil exploration, geological layers are usually known from scattered data; therefore, interpolation or approximation procedures are needed. Triangulation is a popular method for modeling surfaces because it is simple and versatile. However, triangulations are only continuous surfaces and, if smoothing purposes involve curvature minimization, this is a severe drawback since two main problems arise: curvature definition and optimization.

In the second section we state the interpolation problem as a curvature minimization procedure.

For curvature to make sense on a triangulation, an attractive idea is to use the facets to approximate the plane tangent to the triangulated surface, and to obtain curvature by differentiating the orientation of these facets with respect to suitable coordinates. This approach is investigated in Section 3, and turns out to be unsuccessful.

In Section 4 we use the local paraboloid method introduced by Sander & Zucker ([3]). A local quasi-tangent coordinate system at each vertex being

Curves and Surfaces with Applications in CAGD
A. Le Méhauté, C. Rabut, and L. L. Schumaker (eds.), pp. 263–270.
Copyright ℗ 1997 by Vanderbilt University Press, Nashville, TN.
ISBN 0-8265-1293-3.
All rights of reproduction in any form reserved.

chosen, the method consists in constructing the paraboloid which interpolates five suitable neighbours of that vertex. This gives a second order numerical scheme for the curvature matrix, at least in the explicit case. Then, the known vertices of the triangulation being fixed, the optimization of the curvature criterion gives the interpolating triangulation.

J.-M. Morvan has shown that spheres minimize this criterion among surfaces of the same genus as a sphere. From this theorem stated in the fifth section, both nontrivial and easy-to-check numerical tests can be designed.

In Section 6 we present numerical results.

§2. The Interpolation Problem

The interpolation problem consists in finding surface S in E such that

$$Q(S) = \frac{1}{2} \int_S q^2 dS$$

is minimized, where $q = \sqrt{k_1^2 + k_2^2}$ is the quadratic curvature and k_1 and k_2 are the principal curvatures. In the context of the present paper, the space E is the space of C^2 graphs of the unit sphere S^2. A surface Σ is a C^2 graph of S^2 iff there exists a map ϕ in $C^2(S^2,]-1, \infty[)$ such that the map

$$\Phi : s \in S^2 \mapsto p = s + \phi(s)\mathbf{v}(O, s) \in \mathbb{R}^3_*$$

is a C^2 diffeomorphism from S^2 onto Σ, $\mathbf{v}(O, s)$ being the vector linking origin O to point s.

The space E is discretized into a space of triangulations E_d generated by a particular N-vertex triangulation T_S of S^2 and an element $\mathbf{r}$ in $]-1, \infty[^N$, that is, each vertex v_i of some T in E_d is such that $\mathbf{v}(O, v_i) = r_i \mathbf{v}(O, v_{Si})$, with v_{Si} being the ith vertex of T_S. The integral $Q(S)$ becomes

$$Q(T) = \frac{1}{2} \sum_{i=1}^{N} q_i^2 a_i,$$

with q_i being the quadratic curvature at vertex i, and a_i equal to one third of the area of the neighbouring triangles. Optimization will be achieved by using a conjugate gradient procedure ([1]).

Let us now evaluate the curvature at each vertex of a triangulation.

§3. The Curvature via the Normals

The basic idea underlying the approach "the curvature via the normals" refers to the very simple case of 1D second order numerical schemes on regular grids. Assuming unit step, $u_?^{(1)} = u_{i+1} - u_i$ is a second order scheme of first derivative $u^{(1)}$ of function u only if $u_?^{(1)} = u_{i+1/2}^{(1)}$. Then, a numerical scheme for second derivative is easily derived: $u_i^{(2)} = u_{i+1/2}^{(1)} - u_{i-1/2}^{(1)} = u_{i-1} - 2u_i + u_{i+1}$.

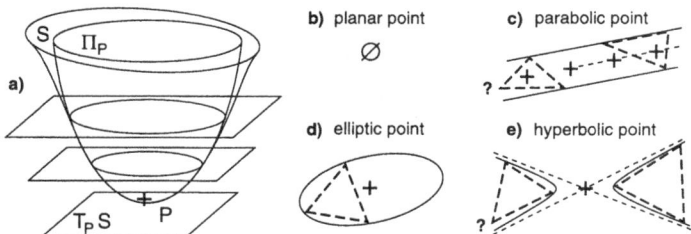

Fig. 1. Circumscribing a triangle with Dupin indicatrices.

Analogously, for triangulations, the unit normal to a triangle would stand for a numerical scheme of the orientation of the tangent plane. To become second order and allow for curvature computation, the normal needs to be properly located in the plane of its associated triangle. If the triangulated surface is a sphere, this correct location is clearly the center of the circumcircle of the considered triangle. Indeed, the projection of this point on the sphere gives a point at which the plane tangent to the sphere is parallel to the plane of the triangle. For more general surfaces, the analogue of the circumcircle, if any, is the Dupin indicatrix that circumscribes the triangle (Figure 1), and the center of symmetry of that curve is the proper location for the normal to the triangle.

The Dupin indicatrix at point P on surface S is a conic which results from the intersection of osculating paraboloid Π_P to S at P with a plane parallel to tangent plane T_PS ([2,4]). Its equation is $k_1 x^2 + k_2 y^2 = \pm 1$. Gaussian curvature is $K = k_1.k_2$ and mean curvature is $H = (k_1 + k_2)/2$. If $K > 0$, the point is elliptic and the Dupin indicatrix is an ellipse. If $K < 0$, the point is hyperbolic and the Dupin indicatrix is a hyperbola. Note that a hyperbola cannot circumscribe a triangle if one of its sides is parallel to one of the asymptotes. If $K = 0$ and $H \neq 0$, the point is parabolic and the Dupin indicatrix is a pair of parallel straight lines. Note that two parallel lines of a given direction cannot circumscribe a triangle if none of its sides lies on one of the lines. This suggests that developable surfaces are better triangulated if all triangles have a side parallel to the local generator. If $K = 0 = H$, the point is planar and there is no Dupin indicatrix.

We are facing a vicious circle, since to be second order, the normal to a triangle of a triangulated surface must be located at the center of its circumscribing Dupin indicatrix, and since the Dupin indicatrix requires the knowledge of the curvature. Therefore, we now investigate another approach for curvature evaluation.

§4. The Curvature via the Local Paraboloid

The local paraboloid method ([3]) consists in the following three steps, for each vertex v_i of the triangulation:

1) Constructing a local frame. The local origin is vertex v_i and the local basis is orthonormal. One of the local basis vectors is $\mathbf{v}(O, v_i)/ \parallel \mathbf{v}(O, v_i) \parallel$.

2) Choosing five suitable neighbours. They are chosen in the first or second order neighbourhood of the considered vertex and they are such that the condition number is acceptable at the next step.

3) Computing local paraboloid $w(u, v) = a_{10}u + a_{01}v + a_{20}u^2 + a_{11}uv + a_{02}v^2$. It is defined by $w(u_k, v_k) = w_k$, with (u_k, v_k, w_k) being the local coordinates of the kth suitable neighbour. Solving the linear system for the a_{ij} leads to the second derivatives and then to the curvature.

Next, the quadratic curvature criterion Q is computed and optimized. The gradient is computed by local finite differences, that is,

$$g_i = \frac{\partial Q}{\partial r_i} \simeq \frac{Q(\mathbf{r} + \varepsilon\mathbf{p}_i) - Q(\mathbf{r} - \varepsilon\mathbf{p}_i)}{2\varepsilon},$$

where g_i is the partial derivative of Q with respect to the ith radial distance $r_i = \parallel \mathbf{v}(O, v_i) \parallel$, and $\mathbf{p}_i$ is the null vector except its ith component $(\mathbf{p}_i)_i = 1$. Since the sums $Q(\mathbf{r} + \varepsilon\mathbf{p}_i)$, $Q(\mathbf{r} - \varepsilon\mathbf{p}_i)$ and $Q(\mathbf{r})$ have many terms in common, only the few terms modified by $\varepsilon\mathbf{p}_i$ need be computed again for each component of the gradient.

§5. The Test of the Sphere

To check the effectiveness of the local paraboloid method, we need a test having the following properties. The solution should be theoretically known, simple, easy-to-check and nontrivial, that is, it is not a plane. Moreover, the test should discriminate between geometrically correct interpolation procedures and others. In this purpose, J.-M. Morvan has proven the following theorem:

Theorem 1. *Suppose we are given four points in the Euclidean space* $\mathbb{R}^3$ *which do not belong to the same plane. Let S be a closed surface passing through these points. Then $\int_S H^2 dS \geq 4\pi$, and equality holds iff S is the sphere circumscribing the four points.*

Then, we have $Q(S) \geq 4\pi$ for any S in E since $4H^2 = q^2 + 2K$ and since $\int_S K dS = 4\pi$ by virtue of the Gauss-Bonnet theorem. As a consequence, starting from triangulation T_0 in E_d, minimizing $Q(T)$ should give a triangulation such that all the vertices belong to the sphere circumscribing the four fixed points.

Figure 2a shows the initial triangulation T_0. The number of vertices (36) is rather small. Therefore, this triangulation can easily be subdivided several times (scheme in Figure 2b). The triangles have various sizes and shapes, hence interpolation methods sensitive to this would probably fail to recover the expected sphere. Of course, the theoretical solution of the interpolation problem depends on the chosen smoothing criterion. However, in any case,

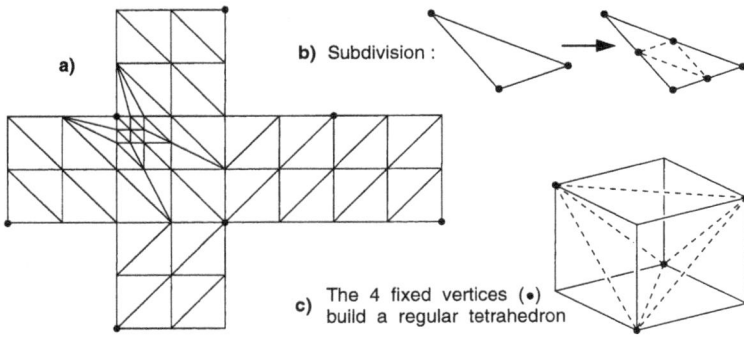

Fig. 2. A 36 vertex triangulation of a cube.

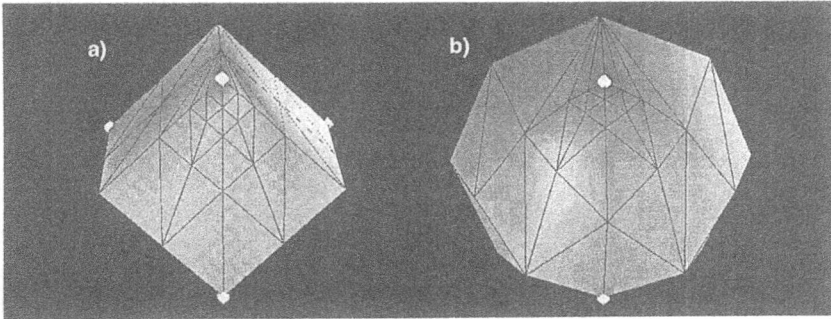

Fig. 3. a) Initial 36 vertex triangulation. **b)** Result of the interpolation.

possible symmetries in the fixed point subset should be preserved after interpolation for any geometrically correct criterion. In particular, the interpolated surface should match the symmetries of the regular tetrahedron built by the four fixed points in Figures 2a and 2c.

§6. Numerical Results

Figure 3 shows the result of the interpolation, and Table 1 gives numerical information about it. Computation time refers to a Sun Sparc 20 workstation. The theoretical values of objective function $Q(S)$ and invariant $\int K\,dS$ are both $4\pi \simeq 12.57$. The orientation error is the mean angle between the true normal and the normal computed from the local paraboloid at each vertex. The theoretical value of the quadratic curvature is $\sqrt{2} \simeq 1.41$.

After two standard subdivisions, 36 vertex triangulation in Figure 3a becomes the 546 vertex triangulation displayed in Figure 4a. The result of the interpolation is displayed in Figure 4e. Intermediate results show that

	Fig. 3	Fig. 4	Fig. 6a & 6b	Fig. 6c
number of points	36	546	2178	2178
number of iterations	161	892	0	3 x 30
computation time	1 mn 17 s	1 h 44 min	1 min 25 s	15 min
objective function	14.80	12.81	20.42	12.83
invariant	14.06	12.64	12.66	12.51
distance to origin	1.031 ± 0.026	1.005 ± 0.036	1.016 ± 0.018	1.014 ± 0.016
orientation	2.8°	2.4°	3.9°	2.4°
quadratic curvature	1.48 ± 0.25	1.39 ± 0.20	1.72 ± 1.64	1.38 ± 0.22

Table 1. Characteristics of the results displayed in Figures 3, 4 and 6.

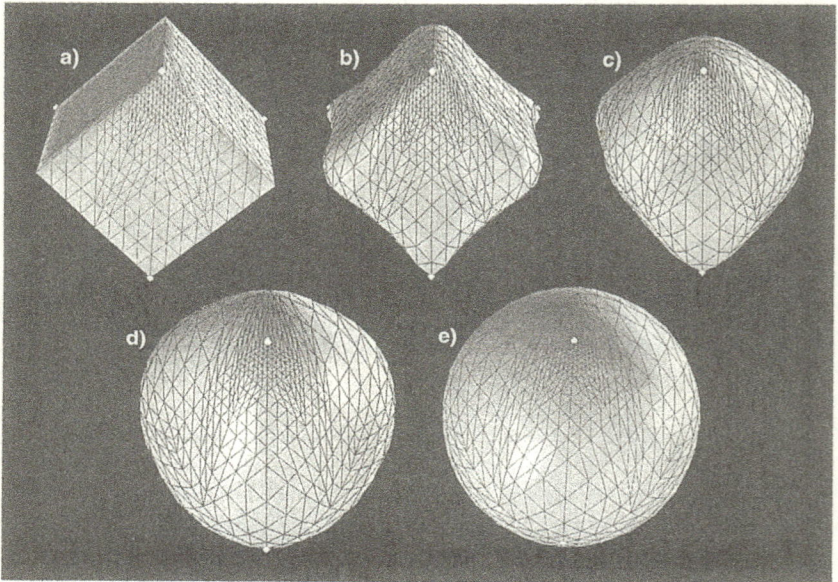

Fig. 4. a) Initial 546 vertex triangulation. Intermediate results after 30 (**b**), 100 (**c**), 300 (**d**) iterations. **e)** Final result after 892 iterations.

smoothing seems to *propagate* from the vertices and edges. For this reason, the number of iterations needed increases with the number of vertices, and starting the interpolation from the three times subdivided, 2178 vertex triangulation of the cube is unmanageable.

To solve this problem, a "parabolic" subdivision is used (Figure 5). The first step consists in computing the barycentric coordinates of each new vertex V. Then, V is projected on the local paraboloids related to the three vertices

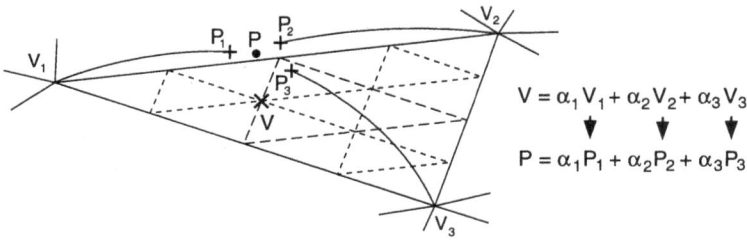

$$V = \alpha_1 V_1 + \alpha_2 V_2 + \alpha_3 V_3$$
$$\Downarrow \quad \Downarrow \quad \Downarrow$$
$$P = \alpha_1 P_1 + \alpha_2 P_2 + \alpha_3 P_3$$

Fig. 5. Parabolic subdivision.

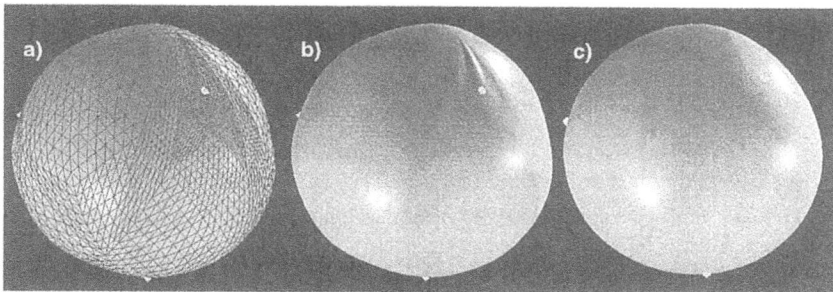

Fig. 6. a) The same as Figure 3b after threefold parabolic subdivision.
b) The same, without the lines. **c)** The same as Figure 3b after applying
three times a "parabolic subdivision plus 30 iteration" procedure.

V_1, V_2 and V_3 of the considered triangle, which gives points P_1, P_2 and P_3,
respectively. Using the barycentric coordinates of T, these three points are
finally combined into point P.

The threefold parabolic subdivision of the triangulation in Figure 3b gives
Figure 6a and 6b. Figure 6b shows folds where triangles are elongated. Iterating the interpolation from this initial model works, but the computation
time is excessive (2 h 45 min for 300 iterations). Starting with Figure 3b,
after three applications of a two step procedure we get the result displayed
in Figure 6c. The first step is a parabolic subdivision, and the second is an
interpolation with 30 conjugate gradient iterations. The quality of this result
is satisfactory and the computation time is reasonable (Table 1).

§7. Conclusions

The approach "curvature via the normals", which seems *a priori* attractive,
leads to a vicious circle since the proper location of these normals in the
plane of the triangle requires the curvature to be known. In contrast, the local paraboloid method of Sander & Zucker gives accurate information about
curvature. The test of the 36 vertex triangulation of a cube is designed to dis-

tinguish geometrically correct interpolations. The optimization of a quadratic curvature criterion makes it possible to smooth irregular triangulations with a "smoothing effect" which do not depend on these irregularities. Starting from a coarse triangulation and alternating smoothing and subdivision improves considerably the ratio between final quality and computation time. These preliminary results should be extended to more practical problems in geosciences involving complicated multi-surface models.

Acknowledgments. I thank J.-M. Morvan (`morvan@jonas.univ-lyon1.fr`) for his proof of Theorem 1 while he was supported by the Institut Français du Pétrole. I also thank Chakib Bennis and Lang Nguyên for their help in graphics matters. The figures were obtained with the Gocad and IslandDraw software.

References

1. Fletcher, R., *Practical Methods of Optimization.*, 2 volumes, John Wiley & Sons, Chichester, 1980.
2. Hicks, N. J., *Notes on Differential Geometry*, Van Nostrand Reinhold, New York, 1965.
3. Sander, P. T., and S. W. Zucker, Inferring surface trace and differential structure from 3-D images, IEEE Trans. Pattern Anal. and Machine Intelligence PAMI-12 (1990), 833–854.
4. Spivak, M., *A Comprehensive Introduction to Differential Geometry*, 5 volumes, Publish or Perish, Houston, 1979.

Michel Léger
Institut Français du Pétrole
1 & 4, avenue de Bois-Préau,
92506 Rueil-Malmaison, FRANCE
`Michel.LEGER@ifp.fr`

On Convexity and Subharmonicity
of Some Functions on Triangles

Jerónimo Lorente-Pardo, Paul Sablonnière,
and M. Carmen Serrano-Pérez

Abstract. This paper is a survey on some new results on convexity, axial convexity and polyhedral convexity for Bézier-nets and Bernstein polynomials. We also study subharmonicity property in a similar way and present the relationship between all these properties for continuous functions on triangles.

§1. Introduction

This work is basically a comparative study of different properties of bivariate functions defined on a triangle such as convexity, axial convexity, polyhedral convexity and subharmonicity. Some of these concepts have been already studied in various papers (e.g. [2,4,7]. Our purpose is to unify the notations of these papers with the aim of giving a general idea of the known results and of solving some open problems in order to complete this study.

Thus, in Section 2 we recall some known results of convexity for univariate functions defined on a interval, and for bivariate functions defined on a triangle. Section 3 is devoted to stating two problems and to solving the first one by using Bézier nets. In Section 4, we set the notation and we present the properties of bivariate functions that are useful in order to solve the second problem. Finally, in Section 5, we present the relationship between the above properties.

§2. Convexity of Univariate and Bivariate Functions

Let $f : [0,1] \mapsto \mathbb{R}$ be a continuous function and $n \geq 1$. Let

$$\mathbf{B_n f}(x) = \sum_{0 \leq i \leq n} f\left(\frac{i}{n}\right) b_i^n(x), \quad x \in [0,1],$$

be the n-th Bernstein polynomial of f.

Curves and Surfaces with Applications in CAGD
A. Le Méhauté, C. Rabut, and L. L. Schumaker (eds.), pp. 271–278.
Copyright © 1997 by Vanderbilt University Press, Nashville, TN.
ISBN 0-8265-1293-3.
All rights of reproduction in any form reserved.

Let $\mathbf{L_n f}$ denotes the *n-th Bézier net of* f, i.e., the unique continuous and affine linear function interpolating the points $(\frac{i}{n}, f(\frac{i}{n})) \in \mathbb{R}^2, i = 0, \ldots, n$. It is known (cf. [4]) that

$$(1) \quad f \text{ convex} \iff \forall n \in \mathbb{N}^*, \quad \underbrace{L_n f \text{ convex}}$$
$$\Downarrow$$
$$f \text{ convex} \iff \forall n \in \mathbb{N}^*, \quad \overbrace{B_n f \text{ convex}}$$

$$(2) \quad f \text{ convex} \iff \underbrace{\forall n \in \mathbb{N}^*, B_n f \geq B_{n+1} f}$$
$$\Updownarrow$$
$$f \text{ convex} \iff \overbrace{\forall n \in \mathbb{N}^*, B_n f \geq f}$$

Let T be a nondegenerate triangle, $u : T \mapsto \mathbb{R}$ a continuous function and $n \geq 1$. Let

$$\mathbf{B_n u}(\lambda) = \sum_{|\alpha|=n} u\left(\frac{\alpha}{n}\right) b_\alpha^n(\lambda) = \sum_{|\alpha|=n} c(\alpha) b_\alpha^n(\lambda), \ \lambda \in T,$$

be *the n-th Bernstein polynomial of* u, where

$\lambda \equiv (\lambda_1, \lambda_2, \lambda_3)$ (barycentric coordinates with respect to T);
$\alpha \equiv (\alpha_1, \alpha_2, \alpha_3)$, $|\alpha| = \alpha_1 + \alpha_2 + \alpha_3$, $\alpha_i \in \mathbb{N}$, $i = 1, 2, 3$;
$b_\alpha^n(\lambda) = \frac{n!}{\alpha_1! \alpha_2! \alpha_3!} \lambda_1^{\alpha_1} \lambda_2^{\alpha_2} \lambda_3^{\alpha_3}$ (Bernstein basis).

The elements of $\{c(\alpha) : |\alpha| = n\}$ are called *B-coefficients of* $B_n u$.

$\mathbf{L_n u}$ denotes the *n-th Bézier net of* u, namely, the unique continuous function which interpolates the points $\varphi_n = \{(\frac{\alpha}{n}, c(\alpha)) \in \mathbb{R}^3 : |\alpha| = n\}$ and is affine on every $T_i \in \tau_n(T)$, where $\tau_n(T)$ is the triangulation of T whose edges are parallel to the sides of T, and whose vertices are obtained by projecting the set φ_n over the triangle T (*the triangulation on* T *induced by* φ_n). For every *n*-th Bézier net we can design its planar representation by plotting at each vertex of $\tau_n(T)$ the height of $L_n u$ at this vertex.

It is known (cf. [2,3]) that

$$(1) \quad u \text{ convex} \xleftarrow{\quad}\xrightarrow{+\!\!+} \text{ for all } n \in \mathbb{N}^*, \quad \underbrace{L_n u \text{ convex}}$$
$$\Downarrow$$
$$u \text{ convex} \xleftarrow{\quad}\xrightarrow{+\!\!+} \text{ for all } n \in \mathbb{N}^*, \quad \overbrace{B_n u \text{ convex}}$$

$$(2) \quad u \text{ convex} \xleftarrow{+\!\!\!} \xrightarrow{\quad} \text{ for all } n \in \mathbb{N}^*, \underbrace{B_n u \geq B_{n+1} u}$$
$$\Downarrow$$
$$u \text{ convex} \xleftarrow{+\!\!\!} \xrightarrow{\quad} \text{ for all } n \in \mathbb{N}^*, \overbrace{B_n u \geq u}$$

and, hence, the situation is not the same as for univariate functions.

§3. Two Problems for Bivariate Functions

In view of the previous section, two questions arise:

1) *Are there any bivariate functions with the same behaviour as univariate functions with respect to convexity property?.*

2) *Are there weaker or stronger properties than convexity which improve the scheme of convexity properties for bivariate functions?.*

In this context, we analyse the following relations for various properties (P):

$$u \text{ satisfies (P)} \quad \xleftrightarrow{?} \quad \forall\, n \in \mathbb{N}^*, \quad \underbrace{L_n u \text{ satisfies (P)}}$$

$$\updownarrow ?$$

$$u \text{ satisfies (P)} \quad \xleftrightarrow{?} \quad \forall\, n \in \mathbb{N}^*, \quad \overbrace{B_n u \text{ satisfies (P)}}$$

$$u \text{ satisfies (P)} \quad \xleftrightarrow{?} \quad \underbrace{\forall\, n \in \mathbb{N}^*,\ B_n u \geq B_{n+1} u}$$

$$\updownarrow ?$$

$$u \text{ satisfies (P)} \quad \xleftrightarrow{?} \quad \overbrace{\forall\, n \in \mathbb{N}^*,\ B_n u \geq u}$$

We start with the first question: in effect there exist bivariate functions defined on triangles reproducing the same scheme as univariate functions with respect to the property of convexity. These functions are the Bézier nets $L_n u$ for which the following properties hold:

$$(1) \quad L_n u \text{ convex} \quad \Longleftrightarrow \quad \forall\, m \in \mathbb{N}^*, \quad \underbrace{L_m(L_n u) \text{ convex}}$$

$$\Downarrow$$

$$L_n u \text{ convex} \quad \Longleftrightarrow \quad \forall\, m \in \mathbb{N}^*, \quad \overbrace{B_m(L_n u) \text{ convex}}$$

$$(2) \quad L_n u \text{ convex} \quad \Longleftrightarrow \quad \underbrace{\forall\, m \in \mathbb{N}^*,\ B_m(L_n u) \geq B_{m+1}(L_n u)}$$

$$\Updownarrow$$

$$L_n u \text{ convex} \quad \Longleftrightarrow \quad \overbrace{\forall\, m \in \mathbb{N}^*,\ B_m(L_n u) \geq L_n u}$$

where the first equivalence of (1) is a consequence of the following properties:

(*i*) If $L_n u$ is convex then $L_q(L_n u)$ is convex, whenever q is a divisor of n.

(*ii*) If p is a multiple of n then $L_p(L_n u) \equiv L_n u$.

The other implications of (1) follow from the preceding results together with the scheme of convexity for bivariate functions. Finally, the equivalences of (2) are known (cf. [3] and the scheme for the convexity property).

The next section is devoted to answering the second question.

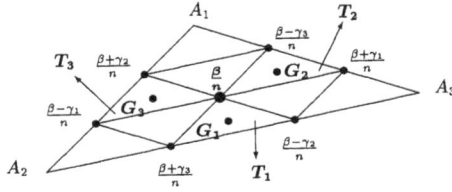

Fig. 1. Geometric interpretation of weak axial convexity.

§4. Convexity and Subharmonicity of Bivariate Functions

Let T be a nondegenerate triangle in $\mathbb{R}^2$ with vertices A_1, A_2 and A_3. We denote $\gamma_1 = e_3 - e_2$, $\gamma_2 = e_1 - e_3$, $\gamma_3 = e_2 - e_1$ the directions of sides of T, where e_i are the barycentric coordinates of A_i with respect to T. For $i = 1, 2, 3$, l_i denotes the length of the vector $\overrightarrow{A_j A_k}$, $i \neq j \neq k \neq i$.

Let $u : T \mapsto \mathbb{R}$ be a continuous function. We say that

- u is *axially convex* (cf. [7]) if u is convex with respect to the directions parallel to the sides of T.
- u is *polyhedrally convex* if for each $n \in \mathbb{N}^*$, $L_n u$ is convex (cf. [7] for $m = 2$).
- u is *subharmonic* [6] if $u \in C(T)$ and satisfies

$$u(x_0, y_0) \leq \frac{1}{2\pi} \int_0^{2\pi} u(x_0 + r \cos\theta, y_0 + r \sin\theta) \, d\theta$$

for each $(x_0, y_0) \in T$, $r > 0$ such that $\overline{B((x_0, y_0), r)} \subset T$.

- u is *weakly axially convex* if $\forall\, n \in \mathbb{N}^*$ and $\forall\, \beta : |\beta| = n$,

$$0 \leq \beta_1 \beta_2 \, \delta^2_{\gamma_3} \, u\left(\frac{\beta}{n}\right) + \beta_1 \beta_3 \, \delta^2_{\gamma_2} \, u\left(\frac{\beta}{n}\right) + \beta_2 \beta_3 \, \delta^2_{\gamma_1} \, u\left(\frac{\beta}{n}\right)$$

where $\delta^2_{\gamma_i} u(\frac{\beta}{n}) = u(\frac{\beta - \gamma_i}{n}) - 2u(\frac{\beta}{n}) + u(\frac{\beta + \gamma_i}{n})$, $i = 1, 2, 3$.

This notion has the following geometric interpretation: the point $(\frac{\beta}{n}, u(\frac{\beta}{n}))$ is under the plane that contains the points $(G_i, L_n u(G_i))$, $i = 1, 2, 3$, where G_i is the point of T whose barycentric coordinates with respect to T_i are $(\beta_1, \beta_2, \beta_3)$ for $i = 1, 2, 3$ (see Figure 1 for $n = 3$). In fact, we think that this concept is the same as the concept given with this name in [1] at least for functions of class C^2 on T (cf. [5]).

We note that axial convexity and subharmonicity are weaker properties than convexity, that polyhedral convexity is a stronger notion than convexity, and that weak axial convexity is a weaker property than axial convexity.

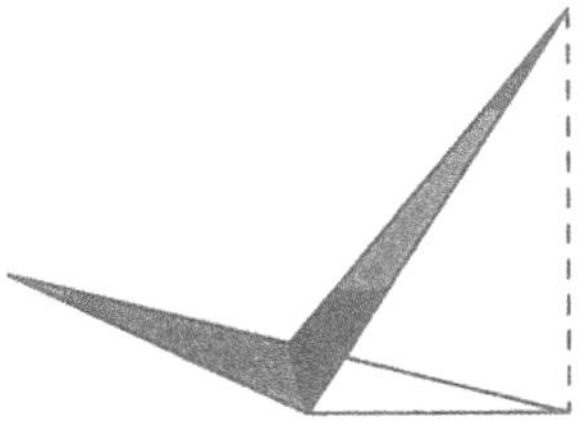

Fig. 2a. The function u 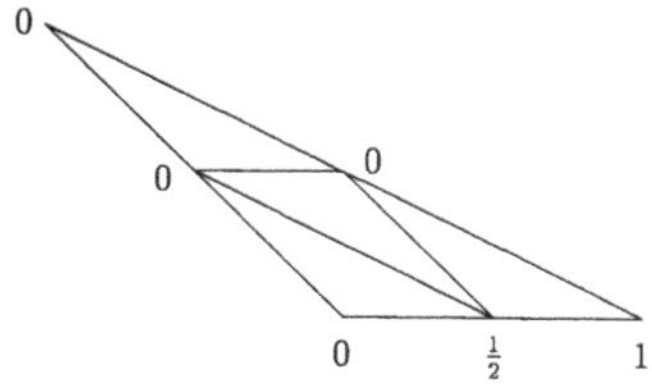   **Fig. 2b.** Planar representation of $L_2 u$.

Moreover, if $u \in C^2(T)$, then

- u is *axially convex* (cf. [7]) if and only if for each $P \in T$, $D^2_{\gamma_i} u(P) \geq 0$, $i = 1, 2, 3$,

- u is *polyhedrally convex* if and only if for each $P \in T$, $D^2_{\gamma_j \gamma_k} u(P) \leq 0$ with $1 \leq j < k \leq 3$, (cf. [7])

- u is *subharmonic* if and only if $\Delta u = \frac{\partial^2 u}{\partial x^2} + \frac{\partial^2 u}{\partial y^2} \geq 0$, or equivalently, for each $P \in T$, $l_1^2 \, D^2_{\gamma_2 \gamma_3} u(P) + l_2^2 \, D^2_{\gamma_1 \gamma_3} u(P) + l_3^2 \, D^2_{\gamma_1 \gamma_2} u(P) \leq 0$,

- u *weakly axially convex* implies that for each $P = (\lambda_1, \lambda_2, \lambda_3)$ in T , $0 \leq \lambda_1 \lambda_2 \, D^2_{\gamma_3} u(P) + \lambda_1 \lambda_3 \, D^2_{\gamma_2} u(P) + \lambda_2 \lambda_3 \, D^2_{\gamma_1} u(P)$,

with the notations $D^2_\omega u = D_\omega D_\omega u$, $D^2_{\gamma \, \omega} u = D_\gamma D_\omega u$ for any directions γ and ω, where $D_\omega u(P) = \omega_1 \, D_1 u(P) + \omega_2 \, D_2 u(P) + \omega_3 \, D_3 u(P), \forall P \in T$, is the derivative of u in the direction $\omega = (\omega_i)_{i=1}^3$, and $D_i(u) = \frac{\partial u}{\partial \lambda_i}$.

Axial Convexity

It is known (cf. [7]) that the axial convexity of a continuous function guarantees the axial convexity of all its Bernstein polynomials and the decreasing of the sequence $\{B_n u\}_{n \in \mathbb{N}^*}$. We have completed this study and we have obtained the following scheme:

(1) u axially convex $\xoverleftrightarrow{}$ for all $n \in \mathbb{N}^*$, $\underbrace{L_n u \text{ axially convex}}$

$$\Downarrow$$

u axially convex $\Longleftrightarrow$ for all $n \in \mathbb{N}^*$, $\overbrace{B_n u \text{ axially convex}}$

(2) u axially convex $\xrightarrow{\ \ \ }$ for all $n \in \mathbb{N}^*$, $\underbrace{B_n u \geq B_{n+1} u}$

$$\Downarrow$$

u axially convex $\xrightarrow{\ \ \ }$ $\overbrace{\text{for all } n \in \mathbb{N}^*, \ B_n u \geq u}$

We can observe that there is an improvement with respect to convexity, namely, the axial convexity of u is equivalent to the axial convexity of all its Bernstein polynomials. The rest of the relationships remain the same as for convexity:

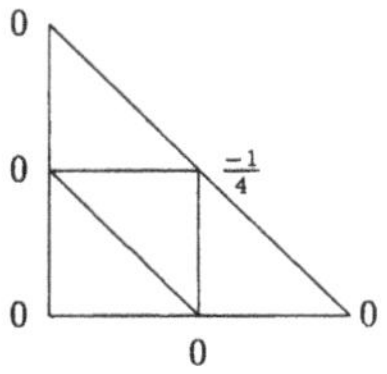

Fig. 3. Planar represent. of $L_2 u$ **Fig. 4.** Planar represent. of $L_4 u$.

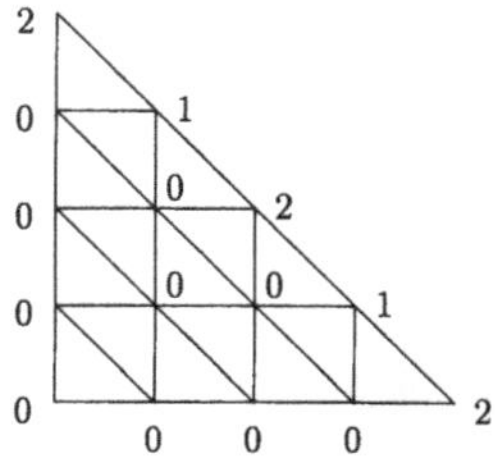

a) There are axially convex functions whose Bézier nets are not axially convex (see Figures 2a, 2b).

b) There are functions not axially convex but with $B_n u \geq B_{n+1} u$ and $B_n u \geq u$, for all $n \in \mathbb{N}^*$. For instance, $u : T \mapsto \mathbb{R}$ given by $u(x, y) = -xy$, where T is the triangle with vertices $A_1 = (-1, 0), A_2 = (1, 0)$ and $A_3 = (0, 1)$.

c) It is possible to have $B_n u$ axially convex and $L_n u$ not axially convex: we can consider $n = 2$, $c(1, 1, 0) = -\frac{1}{2}$ and $c(i, j, k) = 0$ for the rest.

Polyhedral Convexity

Polyhedral convexity improves the above scheme. The only known result in the literature is that the polyhedral convexity of a continuous function guarantees the polyhedral convexity of all its Bernstein polynomials (cf. [7]). We have completed the rest of the scheme as follows:

(1) u polyh. convex $\iff$ for all $n \in \mathbb{N}^*$, $\underbrace{L_n u \text{ polyh. convex}}$

$\Downarrow$

u polyh. convex $\iff$ for all $n \in \mathbb{N}^*$, $\overbrace{B_n u \text{ polyh. convex}}$

(2) u polyh. convex $\xrightarrow{\;+\!\!+\!\!\!\!-}$ for all $n \in \mathbb{N}^*$, $\underbrace{B_n u \geq B_{n+1} u}$

$\Downarrow$

u polyh. convex $\xrightarrow{\;+\!\!+\!\!\!\!-}$ $\overbrace{\text{for all } n \in \mathbb{N}^*, B_n u \geq u}$

Again part (2) is the same as for convexity and axial convexity:

a) There are functions u not polyhedrally convex such that $B_n u \geq B_{n+1} u$ and $B_n u \geq u$, for all $n \in \mathbb{N}^*$. It is enough to consider $u : T \mapsto \mathbb{R}$ given by $u(x, y) = -xy$, where T is the triangle with vertices $(0, 0), (1, 0)$ and $(0, 1)$ (in Figure 3 we can observe that $L_2 u$ is not convex).

b) It is possible to have $B_n u$ polyhedrally convex and $L_n u$ not polyhedrally convex (see Figure 4 for $n = 4$).

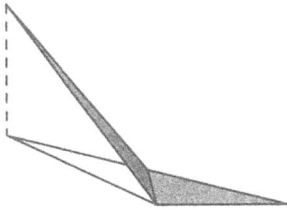

Fig. 5a. The function u

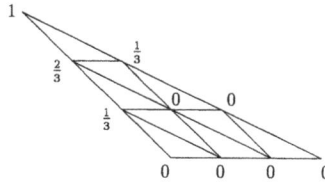

Fig. 5b. Planar Representation of $L_3 u$.

Subharmonicity

Finally, we present the scheme for subharmonicity. In this case, almost nothing is given in the literature, and the following seems to be new:

(1) u subharmonic $\overset{\longleftarrow}{-\!\!\!/\!\!\!\!\longrightarrow}$ for all $n \in \mathbb{N}^*$, $\underbrace{L_n u \text{ subharmonic}}$

$$\Downarrow$$

u subharmonic $\overset{\longleftarrow}{-\!\!\!/\!\!\!\!\longrightarrow}$ for all $n \in \mathbb{N}^*$, $\overbrace{B_n u \text{ subharmonic}}$

(2) u subharmonic $\overset{\longleftarrow/\!\!\!\!-}{-\!\!\!/\!\!\!\!\longrightarrow}$ for all $n \in \mathbb{N}^*$, $\underbrace{B_n u \geq B_{n+1} u}$

$$\Downarrow$$

u subharmonic $\overset{\longleftarrow/\!\!\!\!-}{-\!\!\!/\!\!\!\!\longrightarrow}$ $\overbrace{\text{for all } n \in \mathbb{N}^*, B_n u \geq u}$

Hence, we can conclude that subharmonicity behaves worse than even convexity. In effect:

a) We have a subharmonic function $u : T \mapsto \mathbb{R}$ given by $u(\lambda_1, \lambda_2, \lambda_3) = 2\lambda_1^2 - \lambda_2^2$, where T is the triangle with vertices $(1,0),(0,1)$ and $(0,0)$, such that all its Bernstein polynomials are subharmonic but none of its Bézier nets is subharmonic, and also $B_2 u(\lambda) > B_1 u(\lambda) < u(\lambda)$ for $\lambda = (\frac{1}{16}, \frac{3}{4}, \frac{3}{16})$.

b) There is a subharmonic function u with $B_3 u$ not subharmonic (see Figures 5a, 5b).

c) There is a non-subharmonic function $u : T \mapsto \mathbb{R}$ given by $u(\lambda_1, \lambda_2, \lambda_3) = -\lambda_1 \lambda_2 - \lambda_1^2 \lambda_2^2$, where T is the triangle with vertices $(1,0)$, $(0,1)$ and $(0,0)$, such that $B_n u \geq B_{n+1} u$ and $B_n u \geq u$, for all $n \in \mathbb{N}^*$.

§5. Relationship Between the Above Properties

Finally, we give the relationship between the properties studied above:

• For *univariate functions,*

convexity $\equiv$ axial convexity $\equiv$ polyhedral convexity $\equiv$ subharmonicity

• For *bivariate functions on triangles* the relationship between the above properties depend on the geometry of domain T (cf. [Serrano-Pérez'96]). Thus,

(a) If T is not obtuse-angled, then

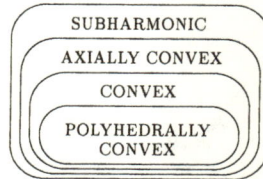

(b) If T is obtuse-angled, then

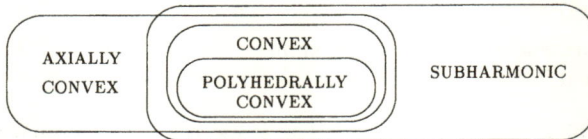

In particular, for Bézier nets,

$L_n u$ convex $\iff$ $L_n u$ pol. convex $\iff$ $L_n u$ ax. convex $\iff$ $L_n u$ subharmonic.

Acknowledgments. The authors would like to thank Professor Jesús Carnicer for his suggestions about the subharmonicity of the n-th Bernstein polynomial.

References

1. Beśka, M., Geometric properties on Bernstein polynomials over simplices, in *Mathematical Methods in CAGD and Image Processing*, T. Lyche and L. L. Schumaker (eds.), Academic Press, Boston, 1992, 1–4.

2. Chang, G. Z., and Davis, P. J., The convexity of Bernstein polynomials over triangles, J. Approx. Theory **40** (1984), 11–28.

3. Chang, G. Z., and Zhang, J. Z., Converse theorems of convexity for Bernstein polynomials over triangles, J. Approx. Theory **61** (1990), 265–278.

4. Dahmen, W., Convexity and Bernstein-Bézier polynomials, in *Curves and Surfaces*, P.-J. Laurent, A. Le Méhauté, and L. L. Schumaker (eds.), Academic Press, New York, 1991, 107–134.

5. Lorente-Pardo, J., Sablonnière, P., and Serrano-Pérez, M. C., Subharmonicity and convexity properties of triangular Bernstein polynomials and Bézier nets, submitted to Comput. Aided Geom. Design.

6. Montel, P., Sur les fonctions convexes et les fonctions sousharmoniques, J. Math. pures appl., series 9, **7** (1928), 29–60.

7. Sauer, T., Multivariate Bernstein polynomials and convexity, Comput. Aided Geom. Design **8** (1991), 465–478.

8. Serrano-Pérez, M. C., Sobre la conservación de ciertas propiedades geométricas por interpolantes polinomiales a trozos, doctoral thesis in preparation, University of Granada.

Marching Methods in
Surface-Surface Intersection

Emmanuel Malgras

Abstract. In previous research we have implemented a surface-surface intersection method. Fine analysis showed us that most of the computing time is spent in the marching stage. In this paper, we focus on this stage and investigate various improvements. We present suitable methods for marching along the border of a degenerate case (identical regions or tangency).

§1. Introduction

Most geometric modeling systems use Bézier or B-spline parametric curves or surfaces (sometimes rational: NURBS). A frequent problem is to determine the intersection between two such surfaces. Usually, the intersection consists of curves (open or closed); one region can be found in a degenerate case. Isolated points can be found. Singularities on intersection curves can also be encountered (cups, branches, etc.).

Many methods exist but do not fulfill the three criteria of robustness, accuracy and rapidity which are essential for CAD software.

The recursive subdivisions method appeared first [7]. Flatness of surfaces, obtained after subdivisions, is necessary to compute the segments which constitute the intersection curves. These methods are slow and have numerical problems when too many subdivisions are made.

The marching method allows us to compute an entire curve if one of its points is known ([2] for example). This method is faster than the previous one and does not require a final connection process of the elementary segments. The fundamental problem is to find one and only one point per intersection curve.

Hybrid methods overcome this problem [1,6,11]. The use of a detection stage allows us to split the intersection curves into elements computable by

Curves and Surfaces with Applications in CAGD
A. Le Méhauté, C. Rabut, and L. L. Schumaker (eds.), pp. 279–286.
Copyright © 1997 by Vanderbilt University Press, Nashville, TN.
ISBN 0-8265-1293-3.
All rights of reproduction in any form reserved.

marching. Recursive subdivisions, marching and possibly other methods (vector fields, bounding volumes, etc.) are implemented for this purpose, leading generally to algorithms fulfilling our three previous criteria.

We implement such a method. After different improvements, we show that most of the computing time is spent on the marching process [9]. After briefly describing our general method in Section 2, we present results on marching in Section 3. Section 4 deals with marching along the border of a region where both surfaces are identical.

§2. Preparing Surfaces for Marching

We succinctly present the first stages of our proposed method. More details can be obtained in [6] and [11]. This method is based on B-spline surfaces but can be extended to NURBS surfaces.

The first step is called *localization* and its goal is to reduce the parametric area of both surfaces in order to improve overall performance. It does not require recursive subdivision.

The second step is called *detection* and splits all the closed curves into open ones. After this step, all the required first points have to be found on the border of the surfaces [11].

Singular cases are also detected during the process when two colinear normals are obtained for an intersection point. They have to be separately computed.

§3. Marching Along Intersection Curves

Here we assume that one first point has been found for each intersection curve.

A marching process requires two stages. The first one is called *prediction*. It provides an approximation of the next intersection point. This prediction must be quickly computed to preserve the efficiency of the method, but accuracy is also required so that the next stage (the *correction* stage) can run in optimal conditions. Several solutions exist for estimating a next point, in Euclidean space or in the parametric planes. Tangency to the intersection curve [2] or its curvature [4] can be evaluated. In the first approach, we compute simple extensions [5] on each surface, i.e. on each parametric plane providing two prediction points. These extensions are rectilinear, or functions of a rough estimation of the curvature if the correction stage fails.

To obtain a real intersection point, we have to correct the two prediction points previously found. The correction stage is a minimization problem between these two points. To be sure of staying on the surfaces, the correction must give us new points on each parametric area. With this technique we avoid projections which spread errors. Computing time in marching is principally spent on this second stage. We thus decided to improve it first.

3.1. Correction Stage

The initial algorithm is based on a heuristic minimization. It consists in building two sequences of crosses (one in each parametric plane) whose centers converge towards an intersection point [8]. We selected this type of algorithm because such a cross can always be defined on a surface. Moreover, we can always find a minimal distance between two points (one on each cross). With this method, C^0 surfaces with broken lines or peaks can be computed.

We also implemented the Hookes and Jeeves algorithm. The difference consists of another choice for the center of the new crosses which generally improve the convergence. Unfortunately, the prediction points are too close to the real intersection point. This method is inefficient.

We also considered descent methods. Two attitudes are possible. First we can decide to study the approximate quadratic forms of the surfaces rather than the surfaces themselves. In this case, an exact result (1) for each descent step λ_i exists:

$$\lambda_i = \frac{\langle \nabla F(\omega_i), d_i \rangle}{\langle H_F(\omega_0) d_i, d_i \rangle}, \tag{1}$$

where F is the distance between the two surfaces to be minimized, d_i the descent direction and ω_i the four parametric coordinates.

Studying the surfaces themselves leads to the Fletcher-Reeves method and another minimization algorithm (2) to find parameters λ_i:

$$\lambda_i = \min_{\lambda \in \mathbb{R}} F(\omega_i + \lambda d_i) \tag{2}$$

This minimization can be computed with the Golden Section algorithm which reduces the size of an interval by moving an extremity at each step. Parameter λ_i can also be determined with "economical" algorithms such as Goldstein's or Wolfe-Powell's. In fact, there is always the same idea of reducing the size of the interval.

Newton methods are well known for their quadratic convergence and can be implemented in a marching method. Two kinds of algorithms are possible. First, the Newton method can be applied on a rectangular system defined by

$$\begin{cases} S_{1,x}(u,v) - S_{2,x}(s,t) = 0 \\ S_{1,y}(u,v) - S_{2,y}(s,t) = 0 \\ S_{1,z}(u,v) - S_{2,z}(s,t) = 0, \end{cases} \tag{3}$$

S_1 and S_2 being respectively the first and the second surface. Here (u,v) and (s,t), respectively, are the parametric coordinates of a point of the first and the second surface. Pseudo-inverse matrices, also called *Moore-Penrose matrices*, must be computed.

An equation can be added to the system to obtain a 4×4 system. Real inverse matrices are now considered. The constraint can bring on a constant step (at least locally) or it can bring on orthogonal convergence with respect to the direction of prediction (to avoid turning back).

With or without constraints, we never want to compute inverse matrices (or pseudo-inverse matrices). For that reason, we carried out a QR decomposition with Householder's algorithm which preserves the numerical stability of the correction stage.

The Improved Marquardt method [10], suitable for minimizing a sum of squares, was also tested. It requires that the square of the distance between the two surfaces be studied. In fact, this method can be considered as a compromise between Descent and Newton methods. The goal of this algorithm is to reduce the gradient function to zero. Applying the Taylor formula and several heuristic contributions (to ensure numerical stability) yields

$$\left({}^t\!J J + \alpha D^2\right) \delta = -\,{}^t\!J F, \tag{4}$$

where F is the square of the distance, J is its Jacobian matrix, δ is the correction term to apply to points, and D is the matrix determined by the system

$$\begin{cases} D^2_{i,i} = \left({}^t\!J J\right)_{i,i} + \Phi \\ D^2_{i,j} = 0 \quad i \neq j, \end{cases} \tag{5}$$

where $\Phi \in I\!R$ (in practice $\Phi = 1$ seems to be a good value for many cases).

In this method, the matrices are by construction symmetric positive definite. Cholesky decomposition can be used. This decomposition is faster than Householder's, but when we are near a singular point it may be better to use QR decomposition for more stability. In practice we implement both techniques. The QR algorithm is computed when Cholesky decomposition fails.

3.2. Prediction Stage

The first prediction we have used consists of simple extensions (straight line or with a rough estimate of the curvature). The advantage of this algorithm is that it can always be applied as soon as two or three intersection points have been found. Moreover, the algorithm is very fast.

Other algorithms are also possible. Differential equations can be used as proposed by Brackhage [3]. In this case, the tangency on each parametric plane of the intersection curve must be computed, which requires C^1 continuity.

An osculating circle as proposed by Chen and Ozsoy [4] can also be implemented. In this case, C^2 continuty is necessary to obtain a new prediction point.

These two last methods are slower than extensions, and are not suitable for all surfaces. That is the reason the correction algorithms were tested with prediction points computed by extensions.

3.3. Some Results

We applied our surface-surface intersection to different surfaces. In particular, those proposed in Figure 1 were tested.

The first pair of surfaces (left) intersects in only three curves which are nearly straight lines. The second pair leads to eleven curves with large curvatures and curvature variations.

Fig. 1. Examples of intersection between 2 surfaces.

For all methods we evaluated the computing time to obtain all the curves and then deduced the average for computing one intersection point. The reference time taken is equated to unit time for the crosses method in the case of intersection curves which are nearly straight lines.

The first overview of Figure 2 shows us that heuristic methods are not promising even if the crosses method always found all the intersection curves. We notice that the Newton method with a constraint equation can fail if the intersection curves have large variations of curvature. In this particular case the Newton method only finds 10 curves out of 11 in the second example (right) in Figure 2. The Improved Marquardt and Newton methods (without constraint) seem to be the best. In practice both algorithms are implemented, which allows us to switch from one algorithm to the other if necessary. At worst, if both algorithms fail, the crosses method can be activated. It is very slow but it works without continuity conditions on surfaces and succeeds even if singularities on the surfaces exist.

§4. Marching Along the Border of Degenerate Cases

A degenerate case occurs when the two surfaces are locally identical. Then the intersection corresponds to a region. The limit case is the case of tangency between two surfaces [6]. Our problem is to march along the border of such a region.

The first implemented technique scans along isoparametric curves. It provides good results, but is very slow. So we decided to test marching methods to improve performance.

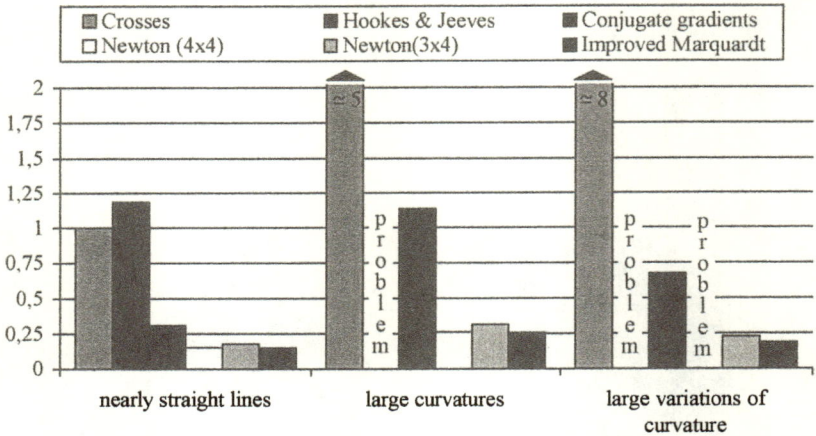

Fig. 2. Results obtained by testing all the proposed methods of correction.

In a degenerate region all the normal vectors of both surfaces are colinear and the distance between the two surfaces vanishes. As soon as this distance (or the angle between two normal vectors) increases, we know that we are leaving the region. So, the border of such a region can be determined as the set of the points for which the distance between the two surfaces is very small.

If we decide to implement the same algorithms as in Section 3.2, the problem consists in finding an equation which characterizes the required border. For example, we can impose the condition that the border is a curve where the normal vectors of both surfaces form a very small angle θ_0. Added to system (3) we then obtain a square system and all the previous algorithms (Section 3.2) can be implemented.

Other algorithms based on minimization with constraints can be applied. For example, by considering angles, we obtain the following minimization problem:

$$\min_{\substack{(u,v)\times(s,t)\in D_{S_1}\times D_{S_2} \\ |\theta(u,v,s,t)|\geq\theta_0}} S_1(u,v) - S_2(s,t) \tag{6}$$

Minimizing a distance between two surfaces without obtaining too small a value (in order to stay on the border of the degenerate region) is not easy to implement. It leads to numerical problems in several examples.

The systems corresponding to the methods disucussed above are complex. That is the reason we investigated a third kind of algorithm based on specific algorithms. One idea is to choose an orientation of the border. Then we know that the degenerate region always lies on the same side of this curve.

We can make a prediction (with extensions). If this point is outside the region, we have to search for an internal point on the perpendicular line, in the general case, through the prediction point (respectivelly if this point is inside the region, we have to search for an external point). These two points define a line segment which intersects the border. Then the correction can be achieved within this segment (Figure 3). The process is easier because it takes

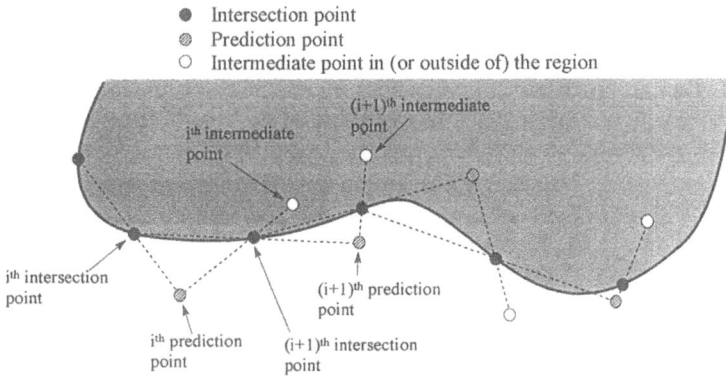

Fig. 3. Prediction-correction for a degenerate case.

place on a manifold of dimension 1. In practice, a bisection is implemented. This algorithm provides the best results of all the proposed methods when the borders are regular curves.

§5. Conclusion and Perspectives

Very significant improvements have been made in marching methods along an intersection curve. Two methods attract our attention: the Improved Marquardt method and the Newton method (without constraint). These methods work well, but we have to study whether it is worthwhile to implement more expensive prediction algorithms: tests are currently in progress.

We have implemented marching methods on degenerate regions. It is a new approach to quickly surround a region where two surfaces are identical. One problem consists in finding a "hole" in such a region but it is not a marching problem.

Our most interesting perspective is the implementation of an algorithm which allows for automatic choice and adaptation of all the parameters required in marching methods.

References

1. R.E. Barnhill and S.N. Kersey, A marching method for parametric Surface/Surface Intersection, Computer Aided Geometric Design **7** (1990), 257–280.

2. R. E. Barnhill, G. Farin, M. Jordan, and B. R. Piper, Surface/Surface intersection, Computer Aided Geometric Design **4** (1987), 3–16.

3. K.-H. Brakhage, Numerical Treatment of Surface-Surface Intersection and Contouring, in *Mathematical Methods for Curves and Surfaces*, Morten Dæhlen, Tom Lyche, Larry L. Schumaker (eds.), Vanderbilt University Press, Nashville & London, 1995, 19–28.

4. J. J. Chen and T.M. Ozsoy, Predictor-corrector type of intersection algorithm for C^2 parametric surfaces, Computer-Aided Design **20** (1988), 347–352.

5. M. Daniel, Modélisation de courbes et surfaces par des B-splines. Application à la conception et à la visualisation de formes, Thèse de doctorat, Université de Nantes, 1989.

6. M. Daniel, A. Nicolas, A Surface-Surface Intersection Algorithm with a fast Clipping Technique, in *Curves and Surfaces in Geometric Design*, P.-J. Laurent, A. Le Méhauté, and L. L. Schumaker (eds.), A. K. Peters, Wellesley MA, 1994, 105–112.

7. J. M. Lane and R. F. Riesenfeld, A Theoretical development for the computer generation and display of piecewise polynomial surfaces, IEEE Transaction on Pattern Analysis and Machine Intelligence, volume PAMI-2, No 1, 1980, 35–46.

8. E. Malgras, Une étude comparative de différentes méthodes de correction dans un suivi de contours, Rapport de recherche IRIN-110, Université de Nantes, 1996.

9. E. Malgras and M. Daniel, Performance Improvement of a Surface-Surface Intersection Method, First International Conference IDMME'96, Nantes, France, vol. 1, 15-17 April 1996, 475–484.

10. J.C. Nash, Compact Numerical Methods for Computers. Linear Algebra and Function Minimisation, Adam Hilger, 1990.

11. A. Nicolas, Contribution à l'étude de l'intersection de surfaces gauches, Thèse de doctorat, Université de Nantes, 1995.

Emmanuel Malgras
Institut de Recherche en Informatiques de Nantes
Faculté des Sciences et des Techniques
2, rue de la Houssinière
44072 Nantes cedex 03, FRANCE
Emmanuel.Malgras@irin.univ-nantes.fr

Algorithms from Blossoms

Stephen Mann

Abstract. Blossoming is a theoretical technique that has been used to develop new CAGD theory. However, once we have developed this theory, we need to devise algorithms to implement it. In this paper, I will discuss converting blossoming equations into code, noting techniques that can be used to develop efficient algorithms. These techniques are illustrated by considering the operation of polynomial composition.

§1. Introduction

Blossoming has been used successfully to analyze and develop new CAGD theory. Once we have developed a new theory, we must convert our blossom equations into algorithms. Although there is usually an obvious transformation, the resulting code will be extremely inefficient. In this paper, I will show methods for creating efficient code from blossom equations.

Blossoming analysis is based on the *blossom*, which is defined as a symmetric and multi-affine (affine in each argument) map of n arguments. The following theorem [9] states that polynomials and blossoms are essentially the same.

The Blossoming Principle. *There is a one-to-one correspondence between degree n polynomials, $F : X \to Y$, and n–affine blossoms, $f : X^n \to Y$, such that*

$$F(u) = f(\underbrace{u, \ldots, u}_{n}),$$

where X and Y are spaces of arbitrary dimension.

Ramshaw and others have successfully used blossoming to analyze existing CAGD algorithms and to develop new theory. These theoretical results are expressed in equations involving sums of the blossom at various sets of arguments. The heart of blossom equations are evaluations of the blossom. To convert these equations into code, we must perform two tasks. First, we

Curves and Surfaces with Applications in CAGD
A. Le Méhauté, C. Rabut, and L. L. Schumaker (eds.), pp. 287–294.
Copyright ⓒ 1997 by Vanderbilt University Press, Nashville, TN.
ISBN 0-8265-1293-3.
All rights of reproduction in any form reserved.

f(u,u,u)   f(t1,t2,t3)

f(u,u,0)   f(u,u,1)   f(t1,t2, u2)   f(t1,t2, u3)

f(u,0,0)   f(u,0,1)   f(u,1,1)   f(t1, u1,u2)   f(t1, u2,u3)   f(t1, u3,u4)

f(0,0,0)   f(0,0,1)   f(0,1,1)   f(1,1,1)   f(u0,u1,u2)   f(u1,u2,u3)   f(u2,u3,u4)   f(u3,u4,u5)

(a)   (b)

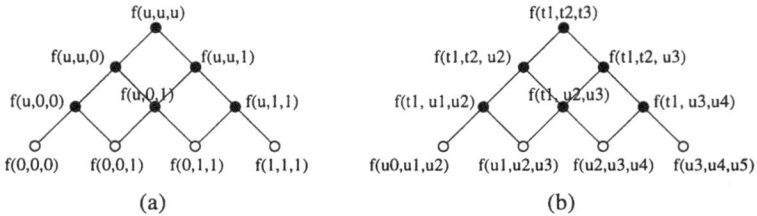

Fig. 1. de Casteljau evaluation of (a) a Bézier curve and (b) a blossom..

must iterate over the required argument sets, and second, we must evaluate the blossom at these argument sets.

Normally, we evaluate the blossom using a variation of de Casteljau's algorithm. The de Casteljau's algorithm uses repeated linear interpolation to evaluate Bézier simplices. Figure 1(a) shows the data flow diagram for a de Casteljau evaluation of a cubic Bézier curve parameterized over $[0, 1]$.

We can extend the de Casteljau algorithm to blossoms by starting with the blossom values $f(u_0, u_1, u_2)$, $f(u_1, u_2, u_3)$, $f(u_2, u_3, u_4)$, and $f(u_3, u_4, u_5)$, and by computing $f(t_1, t_2, t_3)$ as shown in Figure 1(b), where $u_1 \leq u_2 \leq u_3 < u_4 \leq u_5 \leq u_6$. Note that these knots and control points are the knots and control points for a single segment B-spline. (Although the examples given in this figure are for cubic curves, these ideas generalize to polynomials of any degree with domain simplices of any dimension.)

Although we often want to perform a complete evaluation of the blossom, sometimes we only want a *partial evaluation*. For example, in Figure 1(b), we might want to stop the de Casteljau evaluation after having evaluated f at t_1, yielding the light gray control points. Note the close relationship between partial evaluation and knot insertion: In Figure 1(b), the light gray points together with $f(u_0, u_1, u_2)$ and $f(u_3, u_4, u_5)$ are the B-spline control points resulting from inserting the knot t_1 into the B-spline specified by the u_i and corresponding blossom values.

The way to make efficient algorithms from blossom equations is by making effective use of these intermediate blossom values. In particular, the most expensive step of the de Casteljau algorithm is the first one. It is these values that we need to best reuse in order to devise efficient algorithms.

In this paper, I will illustrate the following three techniques for devising efficient algorithms from blossom equations:

(1) Evaluate the blossom only once for each permutation of its arguments.
(2) Reuse of partial evaluations.
(3) Converting to a better basis.

The first technique is well known; the other two have been used in an ad hoc fashion for a variety of algorithms. I will present these techniques by illustrating their use for polynomial composition. The point of this paper is to illustrate the techniques. Thus, I will not give many details about these composition algorithms, and I will restrict my discussion to curves of degrees

2 and 3 (although all the algorithms discussed in this paper generalize both to higher degrees and higher dimensional domains). A second example analysing these techniques for the basis conversion algorithms of Barry and Goldman [1] can be found in [7].

§2. Polynomial Composition

The univariate polynomial composition problem is the following:

Given. *Affine spaces $\mathcal{X}$, $\mathcal{Y}$, and $\mathcal{Z}$ (of dimensions 1, 1, and K_Z respectively), control points $\{G_i\}$, $i = 0 \ldots \ell$, defining a Bézier curve $G : \mathcal{X} \to \mathcal{Y}$ of degree ℓ relative to a domain simplex $\Delta_\mathcal{X} \subset \mathcal{X}$, and control points $\{F_j\}$, $j = 0 \ldots m$ defining a degree m Bézier curve $F : \mathcal{Y} \to \mathcal{Z}$ relative to a domain simplex $\Delta_\mathcal{Y} \subset \mathcal{Y}$.*

Find. *The control points $\{H_k\}$, $k = 0 \ldots m\ell$ of the degree $m\ell$ Bézier curve $H = F \circ G$ relative to $\Delta_\mathcal{X}$.*

Solution. *If f denotes the blossom of F, then*

$$H_k = \sum_{\substack{I \in \mathbb{Z}_\ell^m, \\ |I| = k}} \mathcal{C}(I) f(G_I), \tag{1}$$

where G_I with $I = (i_1, ..., i_m)$ is an abbreviation for $(G_{i_1}, \ldots, G_{i_m})$, and $\mathcal{C}(I)$ is a combinatorial function. Here, the I are known as hyper-indices.

Note that when computing all H_k, we will need to evaluate f at all combinations of G's control points.

A proof of this result can be found in the paper by DeRose et al. [2], along with a reasonably efficient algorithm, and a discussion of applications of polynomial composition. Mann and Liu later developed two improved algorithms [8]. In this paper, I will discuss all three algorithms, and analyze the blossoming techniques used.

As a first comment, note that for each blossom value $f(a_1, \ldots, a_n)$, equation (1) sums the blossom evaluated at all permutations of these arguments. Since all permutations of any particular hyper-index appear in the equation for a single H_k, we can evaluate f at just one permutation of a hyper-index, and weight it by the number of permutations of its arguments. Figure 2 shows the 60 affine combinations used when composing a cubic curve with a quadratic curve when we evaluate at only one permutation of each set of blossom arguments.

§3. 1993 Algorithm

DeRose et al. observed the following: When computing the required blossom values using de Casteljau's algorithm, many of the intermediate values are the same. To reuse these partial evaluations, they imposed a lexicographical

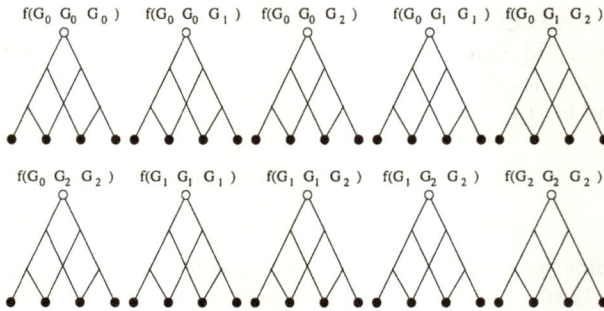

Fig. 2. Polynomial Composition Algorithm.

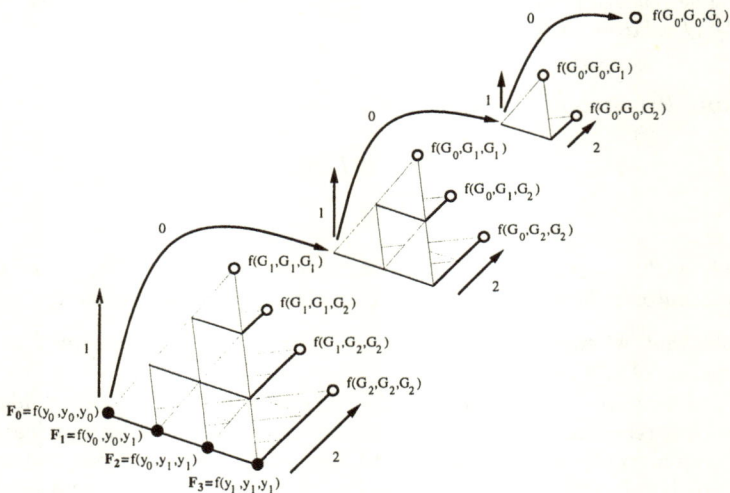

Fig. 3. 1993 Polynomial Composition Algorithm.

ordering on the hyper-indices. They then evaluated the blossom in the order imposed on the hyper-indices. Figure 3 shows the 31 affine combinations used when composing a cubic curve with a quadratic curve. Note the high level of reuse of the intermediate blossom values. Details on this algorithm can be found in several papers [2,5,8].

§4. Recursive Algorithm

The 1993 algorithm orders the indices within a hyper-index in increasing order, and uses a lexicographical ordering of the hyper-indices. We can improve the speed of the 1993 algorithm if we make optimal reuse of the intermediate

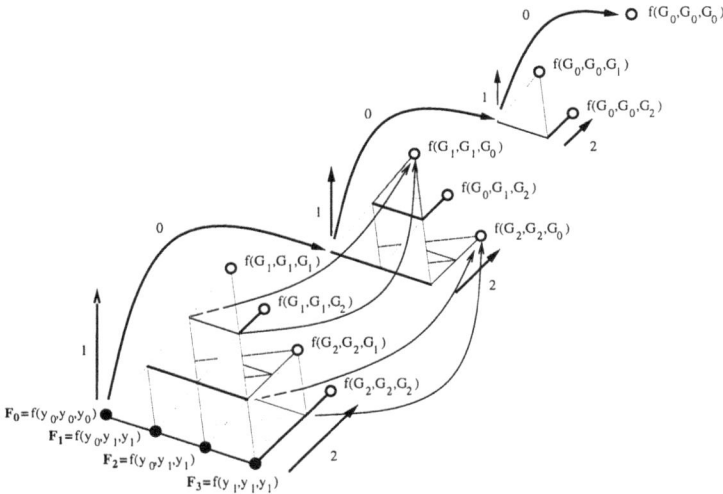

Fig. 4. Recursive Algorithm.

blossom values by using a different ordering on both the indices within a hyper-index and of the hyper-indices.

The observation to exploit is the following: In the steps of the de Casteljau algorithm, the early levels of evaluation are more expensive than later ones. Thus, to make the best reuse of partial evaluations, we want to minimize the number of early evaluations. We can achieve this minimization by ordering the indices within a hyper-index from most to least repetitions, i.e., our hyper-indices will be

$$I = (i_1^{r_1}, \ldots, i_k^{r_k})$$

where $r_j \leq r_{j+1}$. To make our hyper-indices unique up to permutations, we add the additional condition that $i_j < i_{j+1}$ when $r_j = r_{j+1}$.

Figure 4 shows the 27 affine combinations used when composing a cubic curve with a quadratic curve; the dotted lines show the evaluations that would have been used by the 1993 Algorithm. Details on this recursive algorithm can be found in the Mann-Liu technical report [8].

§5. Optimal Algorithm

If before running the 1993 algorithm we perform a change of basis to a basis consisting of G's control points, then we get a remarkable result: *every* intermediate value computed (after the basis conversion) is a blossom value used in equation (1). Further, each of these values is computed exactly once.

Figure 5 shows the 18 affine combinations used when composing a cubic curve with a quadratic curve. Details on this algorithm can be found in several papers [8,5].

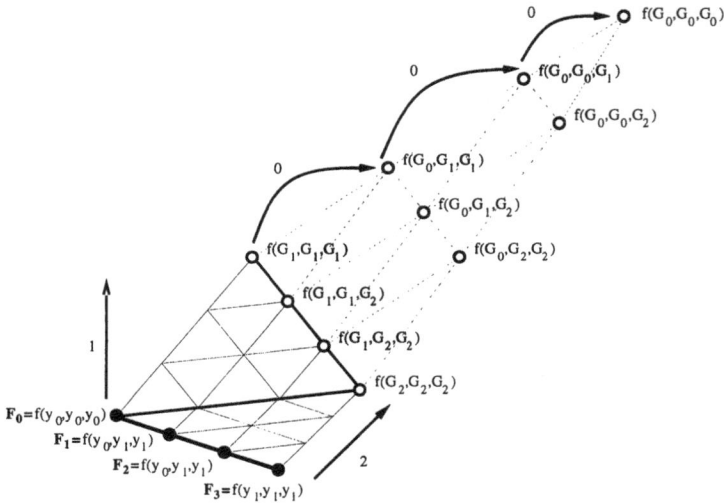

Fig. 5. Optimal Composition Algorithm. Solid lines show the basis conversion.

§6. Knot Insertion, Partial Evaluation, and Knot Swapping

In the previous sections we have seen illustrations of some techniques to make efficient algorithms from blossom equations. These algorithms involve evaluating the blossom and reusing these partial evaluations. The operation of partial evaluation is similar to that of knot insertion. A third, similar technique is *knot swapping*, where essentially we take the light gray points of Figure 1(b) together with one of the $f(u_0, u_1, u_2)$ and $f(u_3, u_4, u_5)$. In this section I will discuss these three techniques.

All three operations have the same computational cost. For curves, knot insertion is somewhat more natural, as it gives us exactly the set of knots we commonly want. However, knot insertion does not cleanly generalize to domains of higher dimension, while partial evaluation and knot swapping both do. When we generalize to dimension k, a degree n patch has $k + 1$ sets of n knots [10]. Consider the diagram in Figure 6. On the left are the control points of a B-patch with knots $\{a_0, a_1, a_2; b_0, b_1, b_2; c_0, c_1, c_2\}$. If we perform one level of the de Casteljau algorithm, we get the shaded panels on the right. In this right-hand figure, I have drawn the original control points underneath, but have only labeled the new points.

Looking at this diagram, it is clear that the shaded panels represent the partial evaluation of the original B-patch. If we take the shaded panel together with any edge of the original control net, we get the knot-swapped representation of the patch, where t has replaced one of a_2, b_2, or c_2 (in the figure, the shaded points are the control points for the patch with knot net $\{t, a_0, a_1; b_0, b_1, b_2; c_0, c_1, c_2\}$). However, it is unclear what the idea of knot

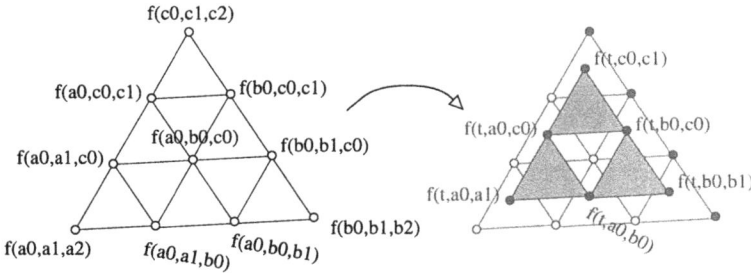

Fig. 6. Evaluation of a B-patch.

insertion even means in this context. Thus, except when restricted to curves, one should use either partial evaluation or knot swapping. See Liu's thesis [4] for a more complete discussion of all three techniques.

A C++ library implementing these ideas is discussed in [4,6].

§7. Conclusions

Blossoming is a useful technique for developing CAGD theory. Once we have developed the theory, however, we need to convert blossom formulas into program code. This primarily involves evaluating the blossom at argument sets specified by our blossom equations. In this paper, we have seen three techniques to write efficient code: evaluating the blossom only once for each permutation of its arguments; reuse of partial evaluations; and conversion to a basis where the computation requires fewer blossom evaluations.

A few notes are in order. Based on efficiency considerations, there is no point in using the 1993 algorithm for polynomial composition since more efficient algorithms are available. However, the 1993 algorithm has a couple of advantages over the more efficient algorithms: it is easier to code and (in comparison to the Optimal composition algorithm) it is numerically more stable.

Thus, when converting from blossom formulas to code, it is almost always advantageous to consider a straight-forward reuse of intermediate blossom values. This is readily found by looking for a natural ordering on the blossom indices we are evaluating at. Once this first improvement is made, we should check to see if we can get a further improvement by either using a better ordering of the blossom arguments, or by first converting to a better basis.

Finally, note that these ideas are just starting points for optimizing blossom equations. Further manipulation of the blossom may be necessary.

Acknowledgments. This work was supported by the Natural Sciences and Engineering Research Council of Canada.

References

1. Barry, P. J. and R. N. Goldman, Knot insertion algorithms, in *Knot Insertion and Deletion Algorithms for B-spline Modeling*, R. N. Goldman and T. Lyche (eds), SIAM, Philadelphia, 1993, 89–133.

2. DeRose, T, R. Goldman, H. Hagen, and S. Mann, Functional composition algorithms via blossoming, ACM Trans. on Graphics **12** (1993), 113–135.

3. Farin, G., *Curves and Surfaces for Computer Aided Geometric Design*, Third Edition, Academic Press, NY, 1992.

4. Liu, W, Programming support for blossoming: The Blossom Classes, dissertation, Univ. Waterloo, Waterloo, 1996.

5. Liu, W., Mann, S., An optimal algorithm for expanding the composition of polynomials, submitted for publication.

6. Liu, W., Mann, S., Programming support for blossoming, Proceedings of Graphics Interface '96, (Toronto), 1996, 95–106.

7. Mann, S., Algorithms from Blossoms, University of Waterloo, Computer Science Dept. Report CS-96-29, 1996.

8. Mann, S. and W. Liu, An analysis of polynomial composition algorithms, University of Waterloo, Computer Science Dept. Report CS-95-24, 1995.

9. Ramshaw, L., Blossoming: a connect-the-dots approach to splines, Techn. Rep., Digital Systems Research Center, Palo Alto, 1987.

10. Seidel, H.-P., Symmetric recursive algorithms for surfaces: B-patches and the de Boor algorithm for polynomials, Constructive Approximation **7** (1991), 259–279.

Computer Science Department
University of Waterloo
200 University Ave W
Waterloo, Ontario N2L 3G1
CANADA
smann@cgl.uwaterloo.ca

Generalized Parameter Representations of Tori, Dupin Cyclides and Supercyclides

Abstract. An explicit representation for any rational Bézier curve and Bézier surface on a Dupin ring cyclide and on special supercyclides is given. The construction is based on an algebraic result concerning diophantine equations in polynomial rings and can be interpreted as a rational mapping from a seven-dimensional projective space onto the cyclide. Due to a line geometric interpretation of the mapping, simple algorithms can be used to construct rational curves and patches on cyclides.

§1. Introduction

In the last years Dupin cyclides have been mentioned as surfaces with important virtues for surface modeling. They have been analyzed by many authors (see [2,10] and references given there) and they are recognized as blending surfaces with nice geometric properties [2,10]. Cyclides are also used in applications as motion planing [13] and geometric modeling [13]. It seems that also a more general class of surfaces, called supercyclides, is useful for CAGD-applications [1]. So far, only special biquadratic representations of cyclides have been studied. But it may be necessary to deal with other shapes of surface patches. In this paper we analyze *arbitrary* rational curves and patches on cyclides. In Section 3 we introduce a rational mapping called a *generalized parameter representation*, in order to reach this goal. The geometric interpretation allows simple constructions of curves and surface patches on cyclides.

§2. Tori, Dupin Cyclides and Supercyclides

A normal-form (in homogeneous coordinates) of an algebraic equation of a torus $\mathcal{T}$ is given by

$$(pqx_0^2 + x_1^2 + x_2^2 + x_3^2)^2 = (p+q)^2 x_0^2 (x_1^2 + x_2^2). \tag{1}$$

Curves and Surfaces with Applications in CAGD
A. Le Méhauté, C. Rabut, and L. L. Schumaker (eds.), pp. 295–302.
Copyright © 1997 by Vanderbilt University Press, Nashville, TN.
ISBN 0-8265-1293-3.

All rights of reproduction in any form reserved.

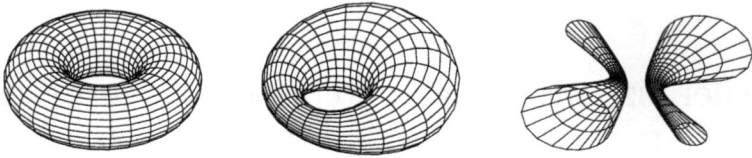

Fig. 1a-c. Torus, Dupin cyclide and supercyclide.

The constants p, q describe the sum and the difference of the mid-circle radius and the meridian-circle radius. The meridian-circles and the parallel-circles are curvature lines of the surface. It is well known from classical geometry (see e.g. [3,6]), that a reflection with respect to a sphere (called *inversion* ι) maps the torus to an algebraic surface of degree 4 (if the center of ι is not lying on the torus) which is called a *Dupin cyclide*. With

$$(x_0, x_1, x_2, x_3)^T \longrightarrow (x_1^2 + x_2^2 + x_3^2, r^2 x_0 x_1, r^2 x_0 x_2, r^2 x_0 x_3)^T$$

an inversion with center in the origin is described. The inversion ι preserves circles and lines of curvature. So a Dupin cyclide $\mathcal{C}$ is a surface whose lines of curvature are circles as well. A normal form of the algebraic equation reads as follows:

$$(x_1^2 + x_2^2 + x_3^2 + x_0^2(b^2 - \mu^2))^2 = 4x_0^2(ax_1 - c\mu x_0)^2 + 4b^2 x_0^2 x_2^2. \qquad (2)$$

a, b, c, μ are constants, only three of which are independent since $b^2 = a^2 - c^2$. Tori as well as Dupin cyclides are intersecting the plane at infinity $x_0 = 0$ in the absolute conic $x_1^2 + x_2^2 + x_3^2 = 0$, which is a set of basepoints of ι. Since any rational curve of degree n on $\mathcal{T}$ and $\mathcal{C}$ is intersecting the absolute conic in n points we have:

Lemma 1. *If the center of the inversion ι is not lying on the torus $\mathcal{T}$ then ι defines a bijection between the rational curves of degree n on $\mathcal{T}$ and the rational curves of degree n on $\mathcal{C}$.*

So there is the same structure of real rational curves on $\mathcal{T}$ and on $\mathcal{C}$. But the surfaces which can be obtained by a nonsingular projective transformation π from a cyclide have also this structure. These surfaces belong to a surface class called *supercyclides* $\mathcal{SC}$. In the beginning they were investigated by Kummer [8]. He gave an algebraic equation of the surface:

$$(\hat{s}\hat{t} - \hat{q}\hat{r} - \hat{p}^2)^2 - 4\hat{p}^2\hat{q}\hat{r} = 0. \qquad (3)$$

$\hat{p}, \hat{q}, \hat{r}, \hat{s}, \hat{t}$ are linear forms in the coordinates. The surface has a singular conic and two pairs of isolated singularities. Each plane of the pencil defined by $\hat{q} = \hat{r} = 0$ (resp. $\hat{s} = \hat{t} = 0$) intersects $\mathcal{SC}$ in two conics. Not every supercyclide is a real projective image of a Dupin cyclide. For more details on supercyclides and their relation to Dupin cyclides see [4,11].

Remark 1. *Note, that in this paper we consider only ring tori, Dupin ring cyclides and those supercyclides which are real projective images of ring cyclides. The other types of tori, cyclides and supercyclides can be analyzed with the present method as well. But there is a different structure of real rational curves on these surfaces of course!*

§3. Generalized Parameter Representation (GPR)

If a rational curve $x = (x_0, x_1, x_2, x_3)^T$ is lying on a torus $\mathcal{T}$ the polynomials $x_i \in \mathbb{R}[t]$ have to fulfill the algebraic equation (1). To discover all rational curves and patches on $\mathcal{T}$, (1) has to be solved in a polynomial ring $\mathbb{R}[t]$ (resp. $\mathbb{R}[u, v]$). This Diophantine equation was solved independently in [7] and [9]:

Theorem 1. *If $\mathbb{R}[-]$ is one of the polynomial rings $\mathbb{R}[t]$ or $\mathbb{R}[u, v]$ and the polynomials x_0, x_1, x_2, $x_3 \in \mathbb{R}[-]$ fulfill the Diophantine equation $(p, q > 0)$*

$$(pqx_0^2 + x_1^2 + x_2^2 + x_3^2)^2 = (p + q)^2 x_0^2 (x_1^2 + x_2^2),$$

then x_0, x_1, x_2, x_3 have the form

$$\begin{aligned}
x_0 &= \pm(y_0^2 + y_3^2)y_5 \\
x_1 &= (y_1^2 - y_2^2)y_4 \\
x_2 &= 2y_1 y_2 y_4 \\
x_3 &= (q - p)y_0 y_3 y_5
\end{aligned} \tag{4}$$

with

$$\begin{aligned}
y_0 &= \frac{1}{\sqrt{q}}(p_0 p_2 p_6 - p_0 p_3 p_7 - p_1 p_2 p_7 - p_1 p_3 p_6) \\
y_1 &= (p_0 p_2 p_4 + p_0 p_3 p_5 - p_1 p_2 p_5 + p_1 p_3 p_4) \\
y_2 &= (p_0 p_3 p_4 - p_1 p_3 p_5 - p_1 p_2 p_4 - p_0 p_2 p_5) \\
y_3 &= \frac{1}{\sqrt{p}}(p_1 p_3 p_7 - p_1 p_2 p_6 - p_0 p_3 p_6 - p_0 p_2 p_7) \\
y_4 &= p_6^2 + p_7^2 \\
y_5 &= p_4^2 + p_5^2
\end{aligned} \tag{5}$$

and with $p_i \in \mathbb{R}[-]$.

The equations (4) and (5) define a rational mapping τ of degree 8 from a seven-dimensional projective space on the ring torus $\mathcal{T}$. Any rational curve (surface patch) on $\mathcal{T}$ can be obtained under τ as image of a suitable rational curve (surface patch) in $\mathbb{P}^7$. Due to this fact we call τ a *generalized parameter representation* (GPR) of $\mathcal{T}$. With the mapping from Section 2

$$\mathbb{P}^7 \xrightarrow{\tau} \mathcal{T} \xrightarrow{\iota} \mathcal{C} \xrightarrow{\pi} \mathcal{SC},$$

we can generalize the GPR to cyclides and supercyclides: The mapping $\gamma := \iota \circ \tau$ $(\sigma := \pi \circ \iota \circ \tau)$ is a GPR of $\mathcal{C}$ $(\mathcal{SC})$.

§4. Properties of GPR

The points on the surface define fibres in the parameter space $\mathbb{P}^7$. A point on $\mathcal{T}$ which is described by the parameters ϕ and ψ (from the trigonometric standard parameter representation) corresponds to a five-dimensional manifold in $\mathbb{P}^7$ given by the cubic equations ($k_1 := \tan \frac{\phi}{2}, k_2 := \tan \frac{\psi}{2}$):

$$
\begin{aligned}
(p_0 p_3 - p_1 p_2)(p_4 - k_1 p_5) - (p_0 p_2 + p_1 p_3)(k_1 p_4 + p_5) = 0 \\
(p_1 p_3 - p_0 p_2)(k_2 \sqrt{p} p_6 + \sqrt{q} p_7) - (p_1 p_2 + p_0 p_3)(\sqrt{q} p_6 - k_2 \sqrt{p} p_7) = 0.
\end{aligned}
\tag{6}
$$

The generalized parameter representation σ has the polynomial degree 8. So a curve of degree n in the parameter space is mapped under σ to a curve of formal degree $8n$. To obtain also lower degree curves, the basepoints of σ have to be analyzed as follows:

Theorem 2. *Any conic on the supercyclide $\mathcal{SC}$ can be obtained under σ as an image of a line. Any irreducible rational curve of degree $2n$ ($n > 1$) on $\mathcal{SC}$ can be obtained under σ as an image of a rational curve of degree $n-1$.*

Proof: The basepoints of σ (resp. γ, τ) are given by the following 4 projective subspaces: $\mathcal{B}_0 : p_0 = p_1 = 0$, $\mathcal{B}_2 : p_2 = p_3 = 0$, $\mathcal{B}_4 : p_4 = p_5 = 0$, $\mathcal{B}_6 : p_6 = p_7 = 0$. Any basepoint has the multiplicity 2. So the relation between the degree d of the preimage curve, the number b of basepoints of the preimage curve and the degree n of the image curve under σ is given by $n = 8d - 2b$ ($\star$). The image curve is irreducible if and only if each pair of the polynomials $\{p_i, p_{i+1}\}$ $i \in \{0, 2, 4, 6\}$ has the trivial greatest common divisor. Assume $d_i := \deg(p_i)$. To apply ($\star$) $\{p_i, p_{i+1}\}$ can be interpreted as a polynomial curve of degree $g = \max\{d_0, d_2, d_4, d_6\}$ which contains $g - d_i$ basepoints. Thus,

$$
n = 2(4g - b) = 2\left(4g - \sum(g - d_i)\right) = 2(d_0 + d_2 + d_4 + d_6) \geq 2g \tag{7}
$$

For $2g = n$, three d_i are vanishing. But then the preimage curve lies in a one-dimensional projective subspace (projective line). Because of (7), in this case the degree of the image curve is 2. ∎

So any conic and any quartic curve on $\mathcal{SC}$ can be obtained as image of a line under σ. Formula (7) defines a classification of rational curves on $\mathcal{SC}$:

Type of curve	II_1	II_2	II_3	II_4	IV_1	IV_2	IV_3	IV_4	IV_5	IV_6
d_0	0	0	0	1	1	0	0	0	1	1
d_2	0	0	1	0	1	0	1	1	0	0
d_4	0	1	0	0	0	1	0	1	0	1
d_6	1	0	0	0	0	1	1	0	1	0
degree n	2	2	2	2	4	4	4	4	4	4

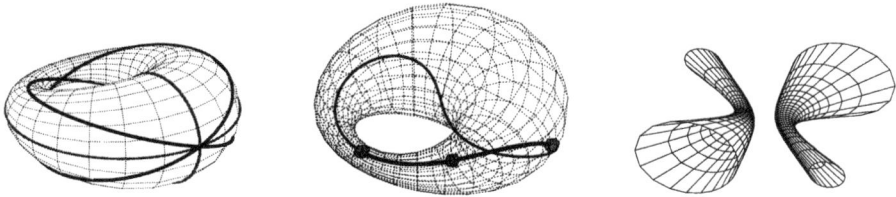

Fig. 2a-c. Rational curves of type II_1–II_4, IV_1 and IV_2 on ring cyclides.

For cyclides, the first four types of curves are well known: II_1 and II_2 are *circles of longitude* and *circles of latitude* ; II_3 and II_4 are the two types of *Villarceau circles*.

Without any loss of generality in the following part the results are presented for the cyclide surface: The previous table shows that the six types IV_1–IV_6 of rational curves on Dupin ring cyclides are dependent on just 4 parameters $\{p_i, p_{i+1}, p_j, p_{j+1}\}$ with $i \neq j \in \{0, 2, 4, 6\}$. To construct these quartic curves it is sufficient to work in 3-dimensional projective subspaces of $\mathbb{P}^7$. For each type, this defines a 4-parameter homogeneous rational mapping

$$\gamma_i : \mathbb{P}^3 \longrightarrow \mathcal{C}, \quad i \in \{1, .., 6\},$$

which is of degree 4. Any rational quartic curve of type IV_i can be obtained as an image of a straight line under γ_i, $i \in \{1, .., 6\}$.

§4. Geometric Interpretation

The advantage of working with the mappings γ_i instead of γ is motivated by the following simple geometric interpretation of the fibres in the 3-dimensional parameter spaces:

Theorem 3. *The inverse images of cyclide points under γ_i (supercyclide points under σ_i) are lines in $\mathbb{P}^3$. These lines form a linear hyperbolic congruence of lines, i.e., any preimage line intersects two real focus lines.*

Proof: Straightforward but technical, see [9]. ∎

The line geometric interpretation of the fibres is very similar to analogous constructions on quadrics [4] and yields to simple geometric algorithms to construct quartic curves and patches on cyclides: To construct a quartic curve of type IV_i through 3 points on $\mathcal{C}$, a line by intersecting the 3 preimage lines of the cyclide points has to be mapped with γ_i on $\mathcal{C}$. There exist ∞^1 lines which fulfill this condition (see Figure 3b), but they are all mapped to the same image curve.

Theorem 4. *There is a unique rational quartic curve of type IV_i $i \in \{2, .., 6\}$ through three arbitrary points on the ring cyclide $\mathcal{C}$. There are exactly 4 rational quartic curves of type IV_1 through three arbitrary points on $\mathcal{C}$.*

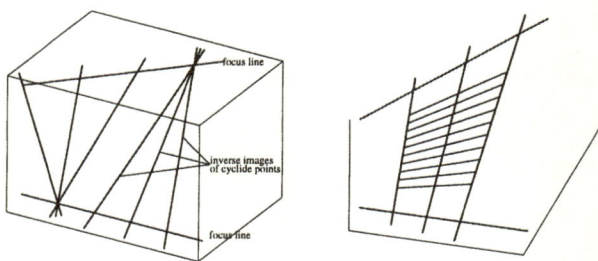

Fig. 3a,b. Linear hyperbolic congruence and preimages of a quartic curve.

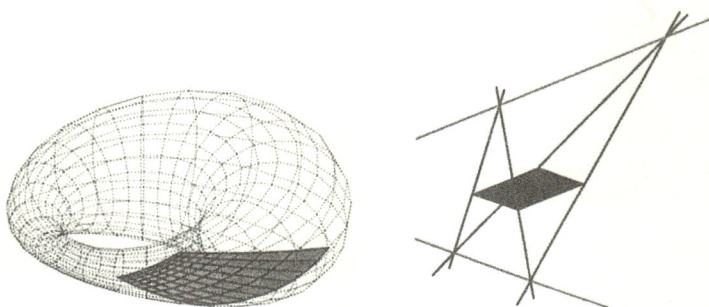

Fig. 4a,b. Biquadratic TP-patch and corresponding preimage patch.

Proof: This follows directly from properties of linear congruences. Note that the curve of type IV_1 is not unique since a point on $\mathcal{C}$ corresponds to *two* lines (not to *one* as in the other cases) of the congruence in the parameter space.
∎

The focus lines of the linear congruence are the set of basepoints of γ_i. Due to this fact a line which intersects exactly one focus line is mapped to a circle on $\mathcal{C}$. In any mapping γ_i two types of circles are included as special cases. This can be used to construct biquadratic TP- (tensor-product) patches on $\mathcal{C}$: Two lines of the congruence are inverse images of cyclide points lying on the same circle, if and only if the congruence lines intersect a focus line in the same point. Figure 4b shows a bilinear patch in the parameterspace. Any parameter line intersects one focus line. So under γ_i the patch is mapped to a biquadratic patch on $\mathcal{C}$. Figure 4a shows the image patch under γ_1. The parameter lines are all Villarceau circles.

The construction yields the following classification of surface patches on Dupin ring cyclides:

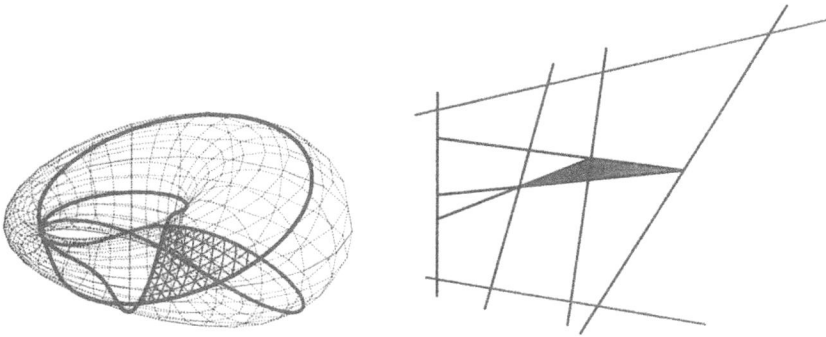

Fig. 5a,b. Triangular quartic patch and corresponding preimage patch.

Corollary 1. *There are 6 kinds of rational biquadratic TP-patches on $\mathcal{C}$. There are 6 kinds of rational triangular patches of degree 4 with parameter lines of the same type IV_i on $\mathcal{C}$. There are 6 kinds of rational TP-patches of degree (4,4) with parameterlines of the same type IV_i on $\mathcal{C}$.*

Figure 5b shows a triangular linear patch in the parameter space. It defines a plane which contains exactly one line of the congruence. So any parameter curve of the quartic image patch is intersecting one fixed point on the ring cyclide. Using this fact we can proof [9]:

Theorem 5. *There does exist a rational triangular quartic Bézier patch with parameterlines of the same type IV_i $i \in \{2, .., 6\}$ on $\mathcal{C}$ with irreducible (quartic) boundary curves if and only if the boundary curves are intersecting each other on a fixed point. For $i = 1$ the existing condition is just necessary.*

§5. Conclusion

In this paper we have presented an explicit representation formula for rational curves and surface patches on ring tori, Dupin ring cyclides and real projective images of these surfaces. The line geometric interpretation of the structure in the parameter space yields to a powerful method for constructing rational curves and patches on cyclides and supercyclides. In particular, in the quadratic and the quartic case, existence and uniqueness can be done. The method can also be used to solve interpolation problems (e.g. to compute a rational curve of degree $4n$ through $2n+1$ points on a cyclide) or to construct spline curves. For example, to construct a quartic G^1–continuous spline curve on $\mathcal{C}$ with the mapping γ_i yields a *linear* condition in the preimage space which can be easily solved. The behaviour of rational curves (resp. patches) on a cyclide $\mathcal{C}$ under *offsetting* of $\mathcal{C}$ can be analyzed with the generalized parameter representation as well. For the torus case one can show that the IV_2 curves are the only quartics which are mapped to quartics under offsetting.

References

1. Allen, S., Dutta, D., Supercyclides and blending, Report No. UM-MEAM-96-02, University of Michigan, 1996.

2. Boehm, W., On cyclides in geometric modelling, Comput. Aided Geom. Design **7** (1990), 243–255.

3. Bühler, K.: Rationale algebraische Kurven auf Dupinschen Zykliden, Diplomarbeit, Institut für graphische Datenverarbeitung, Universität Karlsruhe, 1995.

4. Degen, W. L. F., Die zweifachen Blutelschen Kegelschnittsflächen, Manuscripta Mathematica **55** (1986), 9–38.

5. Dietz, R., Hoschek, J., and B. Jüttler, An algebraic aproach to curves and surfaces on the sphere and on other quadrics, Comput. Aided Geom. Design **10** (1993), 211–229.

6. Fladt, K., and A. Baur, *Analytische Geometrie spezieller Flächen und Raumkurven*, Vieweg, Braunschweig, 1975.

7. Krasauskas, R., Rational Bézier surface patches on quadrics and the torus, preprint 95-25, Vilniaus Universitetas, 1996.

8. Kummer, E., Über die die Flächen vierten Grades, auf welchen Scharen von Kegelschnitten liegen, Journal für die reine und angewandte Mathematik **64** (1863), 66–76.

9. Mäurer, C., Rational curves and surfaces on Dupin ring cyclides, preprint. Nr. 1806, Technical University Darmstadt, 1996.

10. Pratt, M. J., Cyclides in computer aided design II, Comput. Aided Geom. Design **12** (1995), 221–242.

11. Pratt, M. J., Quartic supercyclides I: Basic theory, in: *The Mathematic of Surfaces VI* (Proc. of the VI IMA Conference on the Mathematics of Surfaces, Brunel University, 1994).

12. Srinivas, Y. L., and D. Dutta, Motion planning in three dimensions using cyclides, Proc. Computer Graphics Int. Tokyo, Japan (1992), 781–791.

13. Srinivas, Y. L., and D. Dutta, Intuitive procedure for constructing geometrically complex objects using cyclides, Computer-Aided Design**26** (1994), 327–335.

Christoph Mäurer
Fachbereich Mathematik
Technische Hochschule Darmstadt
Schloßgartenstraße 7
D–64289 Darmstadt, GERMANY
cmaeurer@mathematik.th-darmstadt.de

Korovkin Type Results for
Shape Preserving Operators

F.-J. Muñoz-Delgado and D. Cárdenas-Morales

Abstract. In this paper we present qualitative and quantitative results on shape preserving approximation in the space of all real-valued and k-times continuously differentiable functions of one real variable.

§1. Introduction

In approximating a function by a linear positive operator, the well known result of Bohman-Korovkin [1,5] provides a very useful method to guarantee the convergence. However, if we want the approximation to preserve other properties different from positivity, it is more difficult to find such a nice result. The aim of this paper is to discuss operators which preserve more general shape properties related to the sign of one or more of the derivatives of functions.

As a starting point, we recall two generalizations of the classical Bohman-Korovkin theorem (Theorems 1.1 and 1.2 below) which were given in [6]. First we need some definitions and notations.

Let $X = [0,1]$, and let X' be a compact subinterval of X. Let $C^k(X)$, $k \geq 0$ be the space of all real-valued and k-times continuously differentiable functions on X. D^i is the i-th differential operator and $\| \cdot \|$ denotes the sup-norm in $C(X) = C^0(X)$. Moreover, Π_k is the subspace of $C(X)$ spanned by $\{e_0, \ldots, e_k\}$, where $e_i(x) = x^i$, i.e. $\Pi_k = \langle e_0, \ldots, e_k \rangle$.

Let $\sigma = \{\sigma_i\}_{i \geq 0}$ be a sequence with $\sigma_i \in \{-1, 0, 1\}$, and let h, k be two integers with $0 \leq h < k$ and $\sigma_h \sigma_k \neq 0$. We denote

$$C_X^{h,k}(\sigma) = \{f \in C(X) : \sigma_i f[x_0, \ldots, x_i] \geq 0, h \leq i \leq k\},$$

where $f[x_0, \ldots, x_i]$ is the i-th order divided difference of f. Let $\Gamma = \{i : h \leq i < k, \sigma_i \neq 0, \sigma_{i+1} = 0 \text{ and } \sigma_i \sigma_{i+2} \neq -1\}$. When $\Gamma = \emptyset$ then we call $C_X^{h,k}(\sigma)$ a *cone of type I*. When $\Gamma \neq \emptyset$ then we call $C_X^{h,k}(\sigma)$ a *cone of type II*.

Curves and Surfaces with Applications in CAGD
A. Le Méhauté, C. Rabut, and L. L. Schumaker (eds.), pp. 303–310.

Copyright ⓒ 1997 by Vanderbilt University Press, Nashville, TN.
ISBN 0-8265-1293-3.
All rights of reproduction in any form reserved.

Theorem 1.1. ([6, Theorem 2]) *Let $C_X^{h,k}(\sigma)$ be a cone of type I or II, and let $L_n : C^k(X) \to C^k(X)$ be a sequence of linear operators such that*

$$L_n\left(C_X^{h,k}(\sigma) \cap C^k(X)\right) \subset C_X^{k,k}(\sigma).$$

If $\|D^k(L_n e_j) - D^k e_j\| \to 0$ as $n \to \infty$ for every $j = h, ..., k+2$, then $\|D^k(L_n f) - D^k f\| \to 0$ as $n \to \infty$ for all $f \in C^k(X)$.

Theorem 1.2. ([6, Theorem 3]) *Let $C_X^{h,k}(\sigma)$ be a cone of type II with $r \in \Gamma$, and let $L_n : C^k(X) \to C^r(X)$ be a sequence of linear operators such that*

$$L_n\left(C_X^{h,k}(\sigma) \cap C^k(X)\right) \subset C_X^{r,r}(\sigma).$$

If $\|D^r(L_n e_j) - D^r e_j\| \to 0$ as $n \to \infty$ for every $j = h, ..., k$, then $\|D^r(L_n f) - D^r f\| \to 0$ as $n \to \infty$ for all $f \in C^k(X)$.

On the other hand, Knoop and Pottinger [4], extending the well known result of Shisha and Mond [8], prove the following quantitative version of Theorem 1.1 under more restricted hypotheses:

Theorem 1.3. ([4, Satz 2.1]) *Let $L_n : C^k(X) \to C^k(X')$ be a sequence of linear operators. Assume that $L_n(\Pi_{k-1}) \subset \Pi_{k-1}$, and assume that there exists a cone $C_X^{h,k}(\sigma)$ of type I or II with $\sigma_i \in \{0, 1\}$ such that*

$$L_n\left(C_X^{h,k}(\sigma) \cap C^k(X)\right) \subset C_{X'}^{k,k}(\sigma).$$

Then, for $f \in C^k(X), x \in X', \delta_n > 0$,

$$\left|D^k f(x) - D^k L_n f(x)\right| \leq \frac{1}{k!}\left|D^k f(x)\right|\left|D^k e_k(x) - D^k L_n e_k(x)\right|$$

$$+ \left(\frac{1}{k!}D^k L_n e_k(x) + \delta_n^{-2}\beta_n^2(x)\right)\omega(D^k f, \delta_n), \tag{1}$$

where $\beta_n^2(x) = D^k L_n\left(\frac{2}{(k+2)!}e_{k+2} - \frac{2}{(k+1)!}x e_{k+1} + \frac{1}{k!}x^2 e_k\right)(x)$.

Gonska [3], refining a Freud's inequality [2] and involving mainly the second order modulus of smoothness, proves a similar result to the one of Shisha and Mond which he uses to improve the estimate of Theorem 1.3.

The aim of this paper is twofold. First, in Section 2 we derive a quantitative version of Theorem 1.1 with weaker hypotheses than the ones of Knoop and Pottinger. We also show that, although from a practical point of view, while both results are perhaps useful, they work under much too strong assumptions. Then in Section 3, using a different method of proof, we present a quantitative version of Theorem 1.2 and give an example.

§2. A Quantitative Version of Theorem 1.1

We now present a generalization of Theorem 1.3 by removing the assumption on σ_i.

Theorem 2.1. *Let $L_n : C^k(X) \to C^k(X')$ be a sequence of linear operators with $L_n(\Pi_{k-1}) \subset \Pi_{k-1}$, and assume there exists a cone $C_X'^{h,k}(\sigma)$ of type I or II such that*

$$L_n \left(C_X^{h,k}(\sigma) \cap C^k(X) \right) \subset C_{X'}^{k,k}(\sigma).$$

Then (1) holds.

Proof: The statement is a direct consequence of the following proposition which shows that the hypotheses of this theorem and consequently those of Theorem 1.3 imply

$$L_n \left(C_X^{k,k}(\sigma) \cap C^k(X) \right) \subset C_{X'}^{k,k}(\sigma). \tag{2}$$

Thus, the result is in essence that of Shisha and Mond, because it suffices to rewrite carefully their proof to obtain (1) using only (2). ∎

Remark. Note that as we pointed out in the introduction, Theorems 1.3 and 2.1 can be useful in practice because in some cases it is easier to check their assumptions than to check (2).

Proposition 2.2. *The hypotheses of Theorem 2.1 imply (2).*

Proof: Let $f \in C_X^{k,k}(\sigma) \cap C^k(X)$. If we consider a function $g \in \Pi_{k-1}$ such that $D^{k-1}g(x) = \sigma_{k-1}\|D^{k-1}f\|$ $\forall x \in X$ and successively $D^i g(x) = \int_0^x D^{i+1}g(z)dz + \sigma_i(\|D^{i+1}g\| + \|D^i f\|)$ for $i = k-2, k-3, \ldots, h$, then $f + g \in C_X^{h,k}(\sigma) \cap C^k(X)$. Indeed, $\sigma_i D^i(f+g)(x) = \sigma_i D^i f(x) + \sigma_i \int_o^x D^{i+1}g(z)dz + \sigma_i^2 \left(\|D^{i+1}g\| + \|D^i f\| \right) \geq 0$ for $i = k-2, k-3, \ldots, h$. Consequently $L_n(f + g) = L_n f + L_n g \in C_{X'}^{k,k}(\sigma) \cap C^k(X')$. Now using that $L_n(\Pi_{k-1}) \subset \Pi_{k-1}$, we have $0 \leq \sigma_k D^k L_n(f+g) = \sigma_k D^k L_n f$ and so $L_n f \in C_{X'}^{k,k}(\sigma)$. ∎

§3. A Quantitative Version of Theorem 1.2

Next we derive an estimate for the degree of convergence for sequences of operators satisfying the conditions of Theorem 1.2. Notice that we are not assuming any hypothesis on the spaces of polynomials.

Theorem 3.1. *Let $L_n : C^k(X) \to C^r(X')$, be a sequence of linear operators and let $C_X^{h,k}(\sigma)$ be a cone of type II with $r \in \Gamma$ such that*

$$L_n \left(C_X^{h,k}(\sigma) \cap C^k(X) \right) \subset C_X^{r,r}(\sigma)$$

(due to linearity we suppose without loss of generality that $\sigma_r = 1$). Then for $f \in C^k(X), x \in X', \delta_n > 0$,

$$|D^r f(x) - D^r L_n f(x)| \leq \frac{1}{r!} |D^r f(x)| \, |D^r e_r(x) - D^r L_n e_r(x)|$$

$$+ \left(\frac{1}{r!} D^r L_n e_r(x) + N \frac{\beta_n^2(x)}{\delta_n^2} \right) \omega(D^r f, \delta_n).$$

Here $\beta_n^2(x) = D^r L_n g_x(x)$, where g_x is the unique function in $\langle e_h, \ldots, e_k \rangle$ satisfying

$$D^k g_x(t) = 2\sigma_k \; \forall t \in X,$$
$$D^i g_x(0) = \sigma_i (2 + \|D^{i+1} g_x\|) \text{ for } i = k-1, \ldots, r+3 \text{ (if } k > r+3),$$
$$D^{r+2} g_x(0) = 2 + \|D^{r+3} g_x\|,$$
$$D^{r+1} g_x(x) = D^r g_x(x) = 0,$$
$$D^i g_x(0) = \sigma_i (2 + \|D^{i+1} g_x\|) \text{ for } i = r-1, r-2, \ldots, h \text{ (if } r > h).$$

Moreover, $N = \max\{1, N_1, N_2\}$, where

$$N_1 = \max_{i \in \Omega, i > r+1} \left\{ \frac{\delta_n^2 \|D^i f\|}{2\omega(D^r f, \delta_n)} \right\},$$

$$N_2 = \max_{i \in \Omega, i < r} \left\{ \frac{\delta_n^2}{2\omega(D^r f, \delta_n)} \left(\|D^i f\| + \frac{|D^r f(x)| + \omega(D^r f, \delta_n)}{(r-i)!} \right) \right\},$$

and $\Omega = \{ i \in \{h, \ldots, r-1, r+2, \ldots, k\} : \sigma_i \neq 0 \}$. Note that $D^r L_n g_x(x)$ is nonnegative since $g_x \in C_X^{h,k}(\sigma) \cap C^k(X)$ by definition, and that N depends on both n and f.

Proof: Let $f \in C^k(X), x \in X'$ and $\delta_n > 0$. Following the proof of the result of Shisha and Mond [8], we have

$$|D^r f(t) - D^r f(x)| \leq \left(1 + \frac{(t-x)^2}{\delta_n^2} \right) \omega(D^r f, \delta_n) \quad \forall t \in X. \tag{3}$$

On the other hand, as $D^{r+2} g_x(t) = \int_0^t D^{r+3} g_x(z) dz + D^{r+2} g_x(0) = 2 + \|D^{r+3} g_x\| + \int_0^t D^{r+3} g_x(z) dz \geq 2 \; \forall t \in X$ and $D^r g_x(x) = D^{r+1} g_x(x) = 0$, then $D^r g_x(t) \geq (t-x)^2 \; \forall t \in X$ and so since $N \geq 1$

$$D^r \left(f - D^r f(x) \frac{1}{r!} e_r + \omega(D^r f, \delta_n) \frac{1}{r!} e_r + N \frac{\omega(D^r f, \delta_n)}{\delta_n^2} g_x \right) \geq 0$$

and

$$D^r \left(-f + D^r f(x) \frac{1}{r!} e_r + \omega(D^r f, \delta_n) \frac{1}{r!} e_r + N \frac{\omega(D^r f, \delta_n)}{\delta_n^2} g_x \right) \geq 0.$$

Besides, for $i \in \Omega$, using that $\sigma_i D^i g_x(t) = \sigma_i \int_0^t D^{i+1} g_x(z) dz + \sigma_i D^i g_x(0) = \sigma_i \int_0^t D^{i+1} g_x(z) dz + 2 + \|D^{i+1} g_x\| \geq 2$, we have that for $i > r + 1$

$$\sigma_i D^i \left(f - D^r f(x) \frac{1}{r!} e_r + \omega(D^r f, \delta_n) \frac{1}{r!} e_r + N \frac{\omega(D^r f, \delta_n)}{\delta_n^2} g_x \right)$$

$$= \sigma_i D^i f + N \frac{\omega(D^r f, \delta_n)}{\delta_n^2} \sigma_i D^i g_x \geq 0$$

and

$$\sigma_i D^i \left(-f + D^r f(x) \frac{1}{r!} e_r + \omega(D^r f, \delta_n) \frac{1}{r!} e_r + N \frac{\omega(D^r f, \delta_n)}{\delta_n^2} g_x \right)$$

$$= -\sigma_i D^i f + N \frac{\omega(D^r f, \delta_n)}{\delta_n^2} \sigma_i D^i g_x \geq 0,$$

and for $i < r$

$$\sigma_i D^i \left(f - D^r f(x) \frac{1}{r!} e_r + \omega(D^r f, \delta_n) \frac{1}{r!} e_r + N \frac{\omega(D^r f, \delta_n)}{\delta_n^2} g_x \right)$$

$$= \sigma_i D^i f - \sigma_i \frac{D^r f(x) e_{r-i}}{(r-i)!} + \sigma_i \omega(D^r f, \delta_n) \frac{e_{r-i}}{(r-i)!} + N \frac{\omega(D^r f, \delta_n)}{\delta_n^2} \sigma_i D^i g_x \geq 0$$

and

$$\sigma_i D^i \left(-f + D^r f(x) \frac{1}{r!} e_r + \omega(D^r f, \delta_n) \frac{1}{r!} e_r + N \frac{\omega(D^r f, \delta_n)}{\delta_n^2} g_x \right)$$

$$= -\sigma_i D^i f + \sigma_i \frac{D^r f(x) e_{r-i}}{(r-i)!} + \sigma_i \omega(D^r f, \delta_n) \frac{e_{r-i}}{(r-i)!} + N \frac{\omega(D^r f, \delta_n)}{\delta_n^2} \sigma_i D^i g_x \geq 0.$$

In other words,

$$f - D^r f(x) \frac{1}{r!} e_r + \omega(D^r f, \delta_n) \frac{1}{r!} e_r + N \frac{\omega(D^r f, \delta_n)}{\delta_n^2} g_x \in C_X^{h,k}(\sigma) \cap C^k(X)$$

and

$$-f + D^r f(x) \frac{1}{r!} e_r + \omega(D^r f, \delta_n) \frac{1}{r!} e_r + N \frac{\omega(D^r f, \delta_n)}{\delta_n^2} g_x \in C_X^{h,k}(\sigma) \cap C^k(X).$$

Then using the hypothesis on L_n and the definition of $\beta_n^2(x)$ we have that evaluating at the point x

$$\left| D^r L_n f(x) - D^r f(x) \frac{1}{r!} D^r L_n e_r(x) \right| \leq \omega(D^r f, \delta_n) \left(\frac{D^r L_n e_r(x)}{r!} + N \frac{\beta_n^2(x)}{\delta_n^2} \right).$$

Adding the previous equation to

$$\left| D^r f(x) \frac{1}{r!} D^r L_n e_r(x) - D^r f(x) \right| = |D^r f(x)| \left| \frac{1}{r!} D^r L_n e_r(x) - 1 \right|,$$

we have the result. ∎

Remark. If we let $C_X^{h,k}(\sigma)$ be a cone of type I and we take any $r \in \{h, h + 1, \ldots, k\}$ with $\sigma_r \neq 0$, then we cannot estimate the rate of convergence to zero of $|D^r L_n f(x) - D^r f(x)|$ for any $f \in C^k(X)$ in terms of those of $|D^r L_n e_i(x) - D^r e_i(x)|$ for $i = h, \ldots, k$. Indeed, a linear polynomial operator that preserves the cone $C_X^{h,k}(\sigma)$ and holds the space Π_{k+1} fixed can be constructed (see [7]).

Example 3.2. ([6, Example 3]) Let $K_n : C^2[0,1] \to C[0,1]$, be a sequence of linear operators where $K_n f(x)$ is defined by

$$\begin{cases} f(0) + f\left[0, \frac{1}{n}\right] x + f\left[0, \frac{1}{n}, \frac{2}{n}\right] x^2, & x \in [0, \frac{1}{n}] \\ K_n f(\frac{i}{n}) + DK_n f(\frac{i}{n})(x - \frac{i}{n}) + f\left[\frac{i}{n}, \frac{i+1}{n}, \frac{i+2}{n}\right](x - \frac{i}{n})^2, & x \in (\frac{i}{n}, \frac{i+1}{n}] \\ & i = 1, \ldots, n-3 \\ K_n f(\frac{n-2}{n}) + DK_n f(\frac{n-2}{n})(x - \frac{n-2}{n}) \\ \quad + f\left[\frac{n-2}{n}, \frac{n-1}{n}, 1\right](x - \frac{n-2}{n})^2, & x \in (\frac{n-2}{n}, 1]. \end{cases}$$

K_n are not positive operators (if $f(t) = t(1-t)$ then $K_n f(x) = x(1-x) - \frac{x}{n}$) but they are such that if $f \geq 0$ and $f[x_0, x_1, x_2] \geq 0$ for all $x_0, x_1, x_2 \in [0,1]$, then $K_n f \geq 0$. On the other hand, it can be easily proved by recurrence that

$$K_n e_0 = e_0, \quad K_n e_1 = e_1 \text{ and } K_n e_2 = \frac{1}{n} e_1 + e_2. \tag{4}$$

Hence, we can apply Theorem 1.2 with $h = r = 0, k = 2$ and $\sigma = \{1, 0, 1\}$ to obtain that $K_n f$ converges uniformly to f as $n \to \infty$ for all $f \in C^2(X)$.

Moreover, in order to get an estimate for the degree of convergence we can apply Theorem 3.1. Actually, taking $\delta_n = \frac{1}{\sqrt{n}}$ and using (4) we have

$$|K_n f(x) - f(x)| \leq \omega(f, \frac{1}{\sqrt{n}}) + \max\left\{\frac{x\|D^2 f\|}{2n}, x\omega(f, \frac{1}{\sqrt{n}})\right\}.$$

for $n = 1, 2, \ldots$ The operator K_n approximates functions well at points with large second derivatives. In Fig. 1 we compare $B_n f$ (line with dots), $K_n f$ (dashed line), and the original function f (solid line), where B_n is the Bernstein operator.

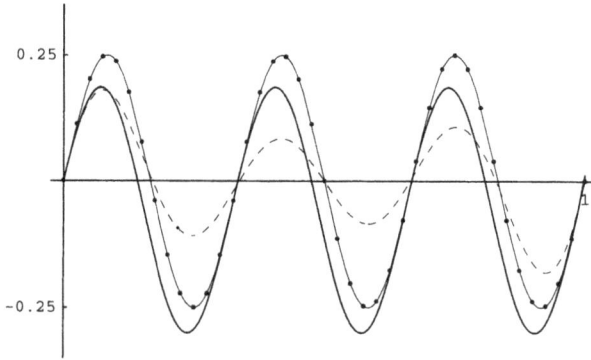

Fig. 1. $f(x) = \frac{1}{4}\sin(6\pi x)$, $n = 40$.

Acknowledgments. This work is partially supported by DGICYT. Proyecto de Investigación Número: PB94-0460.

References

1. Bohman, H., On approximation of continuous and of analytic functions, Ark. Mat., **2** (1952), 43–56.

2. Freud, G., On approximation by positive linear methods, II, Studia Sci. Math. Hungar. **3** (1968), 365–370.

3. Gonska, H. H., Quantitative Korovkin type theorems on simultaneous approximation, Math. Z. **186** (1984), 419–433.

4. Knoop, H.-B., and P. Pottinger, Ein satz vom Korovin-typ für C^k räume, Math Z., **148** (1976), 23–32.

5. Korovkin, P. P., On convergence of linear positive operators in the space of continuous functions (Russian), Doklady Akad. Nauk SSSR (N.S.) **90** (1953), 961–964.

6. Muñoz-Delgado, F.-J., V. Ramírez-González, and D. Cárdenas-Morales, An extension of Korovkin's theorem. Convergence for sequences of operators preserving shape properties, C. R. Acad. Sci. Paris, t. **323**, serie I, 1996, 421–426.

7. Muñoz-Delgado, F.-J., V. Ramírez-González, and P. Sablonnière, On conservative approximation by linear polynomial operators. An extension of the Bernstein operator, Approx. Theory Appl. **No. 1** (1995), 62–71.

8. Shisha, O., and B. Mond, The degree of convergence of linear positive operators, Proc. Nat. Acad. Sci. U.S.A. **60** (1968), 1196–1200.

Fransciso-Javier Muñoz-Delgado
Departamento de Matematicas
Universidad de Jaén
Escuela Politécnica Superior
23071, Jaén, SPAIN
fdelgado@piturda.ujaen.es

Daniel Cárdenas-Morales
Departamento de Matematicas
Universidad de Jaén
Escuela Politécnica Superior
23071, Jaén, SPAIN
cardenas@piturda.ujaen.es

Stable Progressive Smoothing

A. Nigro

Abstract. A new algorithm is proposed for smoothing data progressively by a piecewise quadratic parametric curve with G^1 continuity. The parameter β_1 of geometric continuity and the amount of data taken ahead are used jointly to ensure the stability as well as correct reactions to discontinuities.

§1. Introduction

In [1,2], we studied several interpolating and smoothing methods for data which are known as "progressive". These algorithms are governed by a recurrence relation, and one of our goals was to study their stability. A recurrence relation is stable if the spectral radius of the associated matrix is less than one. The iteration matrices depend on parameters, for instance the coefficients defining the connection at the knots (the classical parameters β_1, β_2, ... of geometric continuity). We obtained various stability domains for these coefficients.

For the progressive smoothing of data, F. Yoshimoto [4] proposed another method not only using the data in the current interval, but also a number of data ahead.

In the present paper we suggest combining the two approaches, i.e. essentially to consider two parameters: the parameter β_1 of G^1 continuity and the number n_2 of data ahead. This way we obtain stability but also a correct reaction to discontinuities in the data.

§2. Discrete Least Squares Fitting

Let us consider a sequence of equidistant knots $\{\,t_i\,\}$ such that $t_{i+1} - t_i = h$. Suppose that on each interval $]t_i, t_{i+1}]$ we have a set of n data denoted by $\{z_i^k \in \mathbb{R}^d , \ k = 1, \dots, n\}$ corresponding to the uniform subdivision defined by

$$t_i^k \;=\; t_i + k\,\frac{h}{n}\,, \qquad k = 1, \dots, n \;.$$

Curves and Surfaces with Applications in CAGD
A. Le Méhauté, C. Rabut, and L. L. Schumaker (eds.), pp. 311–318.
311
Copyright © 1997 by Vanderbilt University Press, Nashville, TN.
ISBN 0-8265-1293-3.
All rights of reproduction in any form reserved.

For convenience, we will use indices k outside of $\{1,\ldots,n\}$ with the meaning $t_i^k = t_i + k\frac{h}{n}$, $k \in \mathbb{Z}$, and we define the corresponding data by $z_i^k = z_{i+j}^{k'}$ if $k = k' + jn$ with $k' \in \{1,\ldots,n\}$, $j \in \mathbb{Z}$.

We suppose that the data are of the form

$$z_i^k = f(t_i^k) + \varepsilon_i^k ,$$

where f is an unknown function and ε_i^k are independent errors with mean value zero and variance σ^2.

We wish to fit the data (step by step from the beginning to the end) by a piecewise quadratic polynomial function $P : \mathbb{R} \longrightarrow \mathbb{R}^d$ of continuity G^1. Let us denote by P_i the restriction of P to the interval $[t_i, t_{i+1}]$. P_i is determined as a solution of a least squares problem. For the sum of the squares of the residuals we not only use data in the interval $[t_i, t_{i+1}]$, but also n_1 data before t_i and n_2 data after t_{i+1}.

The section P_{i-1} being already computed, we obtain P_i by the following conditions:

$$\begin{cases} P_i(t_i) = P_{i-1}(t_i), \\ P_i'(t_i) = \beta_1 P_{i-1}'(t_i) \quad , \quad \beta_1 > 0, \\ P_i \text{ minimizes } \displaystyle\sum_{k=-n_1}^{n+n_2} \left\| P_i(t_i^k) - z_i^k \right\|^2 , \end{cases} \qquad (1)$$

where $\|\,.\,\|$ means the Euclidian norm in $\mathbb{R}^d$.

A key point is to choose the parameters β_1, n_1, n_2 in order to ensure the stability of the algorithm. Let us write the ith P_i as follows:

$$P_i(t) = y_i + (t - t_i)\, m_i + (t - t_i)^2\, \gamma_i ,$$

where y_i, m_i and γ_i are determined by relations (1). By setting $Y_i = (y_i, m_i)^{\mathrm{T}}$, we obtain the following recurrence relation:

$$Y_i = \begin{pmatrix} 1+\alpha_1 & h(1+\alpha_2) \\ 2h\beta_1\alpha_1 & \beta_1(1+2\alpha_2) \end{pmatrix} Y_{i-1} + \begin{pmatrix} 1 \\ 2\beta_1/h \end{pmatrix} \Gamma_{i-1} ,$$

where

$$\alpha_1 = -n^2 \frac{\displaystyle\sum_{k=-n_1}^{n+n_2} k^2}{\displaystyle\sum_{k=-n_1}^{n+n_2} k^4} \quad , \quad \alpha_2 = -n \frac{\displaystyle\sum_{k=-n_1}^{n+n_2} k^3}{\displaystyle\sum_{k=-n_1}^{n+n_2} k^4} \quad , \quad \Gamma_i = n^2 \frac{\displaystyle\sum_{k=-n_1}^{n+n_2} k^2 z_i^k}{\displaystyle\sum_{k=-n_1}^{n+n_2} k^4} .$$

The algorithm is stable if and only if the spectral radius of the iteration matrix is less than one, i.e., the roots of the following equation are in the interior of the unit circle:

$$\lambda^2 - \left(\alpha_1 + 1 + \beta_1(1 + 2\alpha_2)\right)\lambda + \beta_1\left(2\alpha_2 + 1 - \alpha_1\right) = 0 , \qquad (2)$$

As the coefficients of equation (2) depend on four parameters (i.e., n, n_1, n_2 and β_1), the general study of the stability is very difficult.

So, we will restrict the analysis to the parameters n_2 and β_1 (for given values of n and n_1). These two parameters are interesting for the stability as well as for the shape of the solution.

2.1. Effect of β_1

It is easy to prove that the algorithm is stable if and only if

$$\beta_1 < \varepsilon_0 \frac{1}{2\alpha_2 - \alpha_1 + 1} + (1-\varepsilon_0) \min\left\{\frac{\alpha_1 + 2}{\alpha_1 - 4\alpha_2 - 2}, \frac{1}{2\alpha_2 - \alpha_1 + 1}\right\}, \quad (3)$$

where

$$\varepsilon_0 = \begin{cases} 0 & \text{if } \alpha_1 - 4\alpha_2 - 2 > 0, \\ 1 & \text{elsewhere .} \end{cases}$$

Let us remark that for $n_1 = n_2 = 0$, condition (3) becomes

$$\beta_1 < \frac{(1 + 2n)(n^2 + 6n - 2)}{8n^3 + 7n^2 - 2n + 2}. \quad (4)$$

It gives a stability interval depending on the number n of data points in each interval $]t_i, t_{i+1}]$.

$n =$	1	2	3	4	5	10	20	50	100	500
$\beta_1 <$	1.0	0.77	0.63	0.55	0.49	0.38	0.31	0.27	0.26	0.25

We observe that the length of the stability interval decreases when the number n of data per interval increases. We can then choose the parameter β_1 in this interval to modify the shape of the curve. The numerical results are in general satisfactory (see [1]) except for some examples of data, for instance, when there is a jump. In this case, the oscillations after the jump are difficult to avoid; see Fig. 1, where $n_1 = n_2 = 0$ and $n = 4$.

In order to improve this result, we extend the stability interval by increasing the value of n_2 whilst maintaining $n_1 = 0$ for simplicity. This means that n_2 data in the next intervals are used for determining the current polynomial. We obtain a stability domain larger than the one defined in (4). In that case, the stability condition (3) becomes

$$\beta_1 < \frac{2n^3 + 13n^2 + 26n^2 n_2 + 36nn_2 + 36nn_2^2 + 2n + 12n_2^3 + 2n_2 - 2 + 18n_2^2}{8n^3 + 14n^2 n_2 + 7n^2 - 2n - 6nn_2 - 6nn_2^2 - 12n_2^3 - 2n_2 + 2 - 18n_2^2}.$$
$$(5)$$

In Fig. 2, we show the effect on the solution of the tension parameter β_1 for $n = 4$, $n_1 = 0$, $n_2 = 2$. In that case, the algorithm is stable if $\beta_1 < 2.94$.

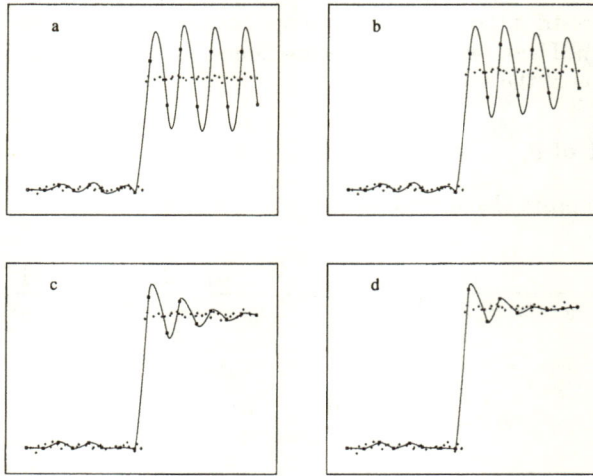

Fig. 1. a) $\beta_1 = 0.55$, b) $\beta_1 = 0.5$, c) $\beta_1 = 0.25$, d) $\beta_1 = 0.125$.

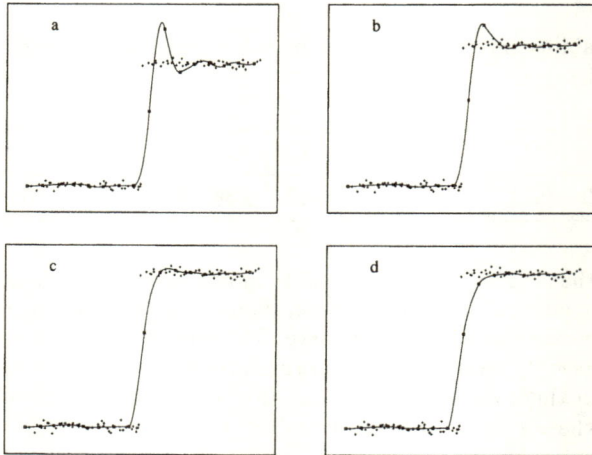

Fig. 2. a) $\beta_1 = 1.5$, b) $\beta_1 = 1.0$, c) $\beta_1 = 0.5$, d) $\beta_1 = 0.25$.

2.2. Effect of n_2

Now, supposing that n_1 and β_1 are fixed, we are interested in the following question: How many data in the next intervals (i.e., on the right-hand side of t_{i+1}) are necessary to ensure the stability of the algorithm ? It is difficult to compute the value of n_2 because it is in general a root of a fourth degree equation. In fact, n_2 belongs to the domain defined by the following system

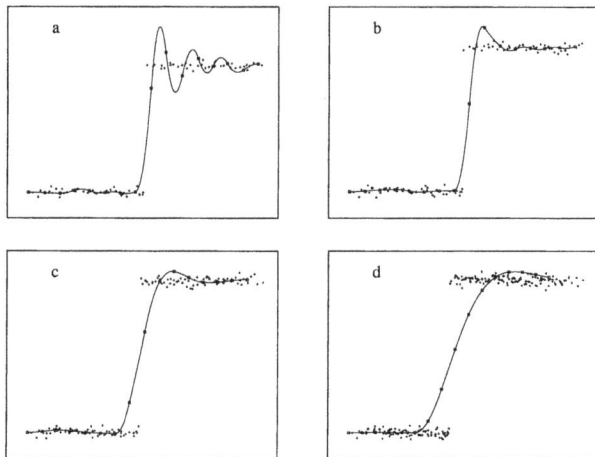

Fig. 3. a) $n_2 = 1$, b) $n_2 = 2$, c) $n_2 = 4$, d) $n_2 = 10$.

of inequalities:

$$\begin{cases} a_4\, n_2^4 + a_3\, n_2^3 + a_2\, n_2^2 + a_1\, n_2 + a_0 \;<\; 0\,, \\ b_4\, n_2^4 + b_3\, n_2^3 + b_2\, n_2^2 + b_1\, n_2 + b_0 \;<\; 0\,, \end{cases} \qquad (6)$$

where a_i and b_i are given by rather complicated expressions depending on n, n_1 and β_1. When n increases, it is the second condition in (6) that defines the stability domain. In that case, the stability domain becomes

$$n_2 > \frac{n}{4}\,. \qquad (7)$$

In order to illustrate this result, let us give the following example: assume that $n_1 = 0$ and $\beta_1 = 1$, then the stability condition is given by the second inequality of (6) and in that case it is simply given by

$$-12n_2^3 - (18 + 21n)n_2^2 - (6n^2 + 21n + 2)n_2 + 3n^3 - 3n^2 + 2 - 2n \;<\; 0\,. \quad (8)$$

For different values of n, we give the number n_2 of data ahead necessary to ensure stability:

$n =$	4	5	6	7	8	10	20	50	100	500
$n_2 >$	0.56	0.8	1.04	1.2	1.5	2.02	4.5	12	24.5	124.5

In Fig. 3, we give an example of the effect of n_2 for $n_1 = 0$, $\beta_1 = 1$ and $n = 4$. In that case, the algorithm is stable as soon as $n_2 \geq 1$. We note that when n_2 is too large, the solution is far from the data points and too smooth.

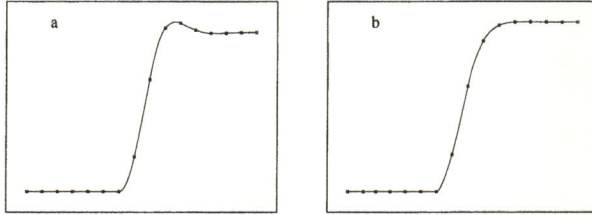

Fig. 4. a) $\beta_1 = 1$, b) $\beta_1 = 0.5$.

§3. Continuous Least Squares Fitting

Now, we consider the case where the data points are dense enough so that the summation in (1) can be replaced by the integral

$$\int_{t_i - \theta_1 h}^{t_{i+1} + \theta_2 h} \left\| P_i(t) - f(t) \right\|^2 dt \;, \tag{9}$$

where θ_1 and θ_2 are non negative real numbers. Values of α_1 and α_2 in (2) become

$$\alpha_1 \;=\; -\frac{5}{3} \frac{(1+\theta_2)^3 + \theta_1^3}{(1+\theta_2)^5 + \theta_1^5} \;,\qquad \alpha_2 \;=\; -\frac{5}{4} \frac{(1+\theta_2)^4 - \theta_1^4}{(1+\theta_2)^5 + \theta_1^5} \;. \tag{10}$$

3.1. Stability condition for β_1

Suppose that θ_1 and θ_2 are fixed. Then the algorithm is stable if and only if β_1 satisfies

$$\beta_1 \;<\; \varepsilon_0 \frac{x_2}{x_1} \;+\; (1-\varepsilon_0) \min \left\{ \frac{x_2}{x_1} \,,\, \frac{y_2}{y_1} \right\} \;, \tag{11}$$

where

$$x_1 = \frac{1}{6}\theta_1^2 - \frac{1}{6}\theta_1 + \frac{3}{2}\theta_2^3 + \frac{3}{2}\theta_1^3 - \frac{1}{6}\theta_2 + \frac{1}{6}\theta_2^2 + \theta_2^4 + \theta_1^4 + \frac{1}{6} - \theta_2^3\theta_1 + \theta_2^2\theta_1^2$$
$$+ \frac{1}{3}\theta_2\theta_1 - \frac{1}{2}\theta_2\theta_1^2 - \theta_2\theta_1^3 - \frac{1}{2}\theta_2^2\theta_1$$

$$x_2 = \theta_2^4 + 4\theta_2^3 - \theta_2^3\theta_1 + 6\theta_2^2 - 3\theta_2^2\theta_1 + \theta_2^2\theta_1^2 + 4\theta_2 - 3\theta_2\theta_1 + 2\theta_2\theta_1^2 - \theta_2\theta_1^3$$
$$+ 1 - \theta_1 + \theta_1^2 - \theta_1^3 + \theta_1^4$$

$$y_1 = \frac{4}{3}\theta_1^2 - \frac{4}{3}\theta_1 - 3\theta_2^3 + \frac{11}{3}\theta_2 + \frac{4}{3}\theta_2^2 - 2\theta_2^4 - 3\theta_1^3 + \frac{4}{3} + 2\theta_2^3\theta_1 - 2\theta_2^2\theta_1^2 - 2\theta_1^4$$
$$+ \theta_2\theta_1^2 + 2\theta_2\theta_1^3 + \theta_2^2\theta_1 - \frac{7}{3}\theta_2\theta_1$$

$$y_2 = 2\theta_2^4 + 8\theta_2^3 - 2\theta_2^3\theta_1 + 2\theta_2^2\theta_1^2 - 2\theta_2\theta_1^3 + 4\theta_2\theta_1^2 + \frac{1}{3}\theta_1^2 - 2\theta_1^3 + \frac{31}{3}\theta_2^2 + \frac{14}{3}\theta_2$$
$$- \frac{1}{3}\theta_1 - \frac{13}{3}\theta_2\theta_1 - 6\theta_2^2\theta_1 + 2\theta_1^4 + \frac{1}{3} \;.$$

In Figure 4, we take $\theta_1 = 0$, $\theta_2 = 1.0$. Then condition (11) leads to $\beta_1 < 5.99$. We observe that reducing the value of β_1 can lead to a monotone curve.

3.2. Stability condition for θ_2

Finally, if θ_1 and β_1 are fixed, we can obtain a stability domain in terms of θ_2. This corresponds to the limit case (when n goes to infinity) of conditions (6), i.e., θ_2 must satisfy

$$\begin{cases} a_4\,\theta_2^4 + a_3\,\theta_2^3 + a_2\,\theta_2^2 + a_1\,\theta_2 + a_0 \;<\; 0, \\ b_4\,\theta_2^4 + b_3\,\theta_2^3 + b_2\,\theta_2^2 + b_1\,\theta_2 + b_0 \;<\; 0, \end{cases} \tag{12}$$

where coefficients a_i and b_i are given by :

$$a_0 = -\theta_1^4 + \theta_1^3 - \frac{1}{6}\theta_1\beta_1 + \beta_1\theta_1^4 + \frac{3}{2}\theta_1^3\beta_1 + \frac{1}{6}\theta_1^2\beta_1 + \frac{1}{6}\beta_1 - \theta_1^2 + \theta_1 - 1$$

$$a_1 = -\theta_1^3\beta_1 + \theta_1^3 - \frac{1}{2}\theta_1^2\beta_1 - 4 + \frac{1}{3}\theta_1\beta_1 - \frac{1}{6}\beta_1 - 2\theta_1^2 + 3\theta_1$$

$$a_2 = \theta_1^2\beta_1 + 3\theta_1 + \frac{1}{6}\beta_1 - \frac{1}{2}\theta_1\beta_1 - \theta_1^2 - 6$$

$$a_3 = -\theta_1\beta_1 + \frac{3}{2}\beta_1 - 4 + \theta_1$$

$$a_4 = \beta_1 - 1$$

$$b_0 = -2\theta_1^4 + 2\theta_1^3 - \frac{4}{3}\theta_1\beta_1 - 2\beta_1\theta_1^4 - 3\theta_1^3\beta_1 + \frac{4}{3}\theta_1^2\beta_1 + \frac{4}{3}\beta_1 - \frac{1}{3}\theta_1^2 + \frac{1}{3}\theta_1 - \frac{1}{3}$$

$$b_1 = 2\theta_1^3\beta_1 + 2\theta_1^3 + \theta_1^2\beta_1 - \frac{14}{3} - \frac{7}{3}\theta_1\beta_1 + 1\frac{1}{3}\beta_1 - 4\theta_1^2 + \frac{13}{3}\theta_1$$

$$b_2 = -2\theta_1^2\beta_1 + 6\theta_1 + \frac{4}{3}\beta_1 + \theta_1\beta_1 - 2\theta_1^2 - \frac{31}{3}$$

$$b_3 = 2\theta_1\beta_1 - 3\beta_1 - 8 + 2\theta_1$$

$$b_4 = -2\beta_1 - 2 \ .$$

Let us consider the case where $\theta_1 = 0$. Then the algorithm is stable in the domain

$$\begin{cases} \theta_2 > r_1 & \text{if } \beta_1 \in \,]0,1] \\ \theta_2 \in \,]r_1,r_2[& \text{if } \beta_1 \in \,]1,7] \ , \end{cases} \tag{13}$$

where

$$r_1 = \frac{-12 + 3\beta_1 + \sqrt{120 + 105\beta_1^2}}{12(\beta_1 + 1)} \quad , \quad r_2 = \frac{12 + 3\beta_1 + \sqrt{240\beta_1 - 15\beta_1^2}}{12(\beta_1 - 1)} \ .$$

This condition extends the result given in [4] for $\beta_1 = 1$ which is simply $\theta_2 > \frac{1}{4}$. In Fig. 5, we illustrate the effect of θ_2 for $\theta_1 = 0$ and $\beta_1 = 0.5$ for which condition (13) is $\theta_2 > 0.089$.

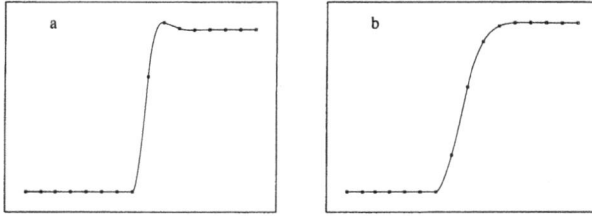

Fig. 5. a) $\theta_2 = 0.5$, b) $\theta_2 = 1.0$.

§4. Conclusion

In the proposed progressive smoothing algorithms, we have two parameters n_2 and β_1. In case of discontinuities, taking n_2 too large leads to a curve which is smooth, but too far from the data. On the contrary, for $n_2 = 0$, it is difficult to avoid oscillations by the only choice of β_1. Our conclusion is that the best results are obtained for a small value of n_2 relative to n (say n_2 between $\dfrac{n}{4}$ and $\dfrac{n}{2}$) by choosing an appropriate value of β_1 (say β_1 between 0.25 and 0.5) to ensure the stability as well as to reduce oscillations, see Fig. 2.

Acknowledgments. I wish to thank the "Association Française d'approximation" for its financial support.

References

1. Nigro, A., Algorithmes progressifs stables pour l'approximation de courbes et surfaces, thèse de l'Université J. Fourier, Grenoble, France, 1995.

2. Nigro, A., and P. J. Laurent, Stable progressive interpolation, to appear in Numer. Algorithms.

3. Seredinski, A., Principe de la méthode des *A*-splines récurrentes pour l'interpolation et la compression des signaux, Traitement de signal, Vol 9, No 2, (1992), 175–191.

4. Yoshimoto, F., Least squares approximation by one-pass methods with piecewise polynomials, in *Algorithms for approximation*, J. C. Mason and M. G. Cox (eds.), Clarendon Press, Oxford, 1987.

A. Nigro
LMC-IMAG, Université Joseph Fourier,
BP53, F38041 Grenoble (cedex 9), FRANCE
Abdelmalek.Nigro@imag.fr

Riemannian Quadratics

Lyle Noakes

Abstract. A well-known corner-cutting algorithm for quadratic curves is generalized to a Riemannian setting and shown to produce a differentiable curve whose derivative is Lipschitz. The differentiability is surprising in view of [2,3,4,5,6], and [1], but consistent with [7] which applies to the cubic case.

§1. Background

Let N be a C^∞ manifold of finite dimension n, equipped with a Riemannian metric $\langle \ , \ \rangle$. Then for each $x \in N$ an inner product $\langle \ , \ \rangle_x$ is defined on the tangent space TN_x of N at x. Let $\| \ \|_x$ be the associated norm. Given $a < b \in \mathbb{R}$ and a pair $(x_a, x_b) \in N \times N$ let Ω be the set of piecewise C^∞ curves $\omega : [a, b] \to N$ satisfying $\omega(a) = x_a, \omega(b) = x_b$. The *energy functional* $E : \Omega \to \mathbb{R}$ is defined by

$$E(\omega) = \int_{[a,b]} \|\dot\omega(t)\|^2_{\omega(t)} dt.$$

When any $x_a, x_b \in N$ are joined by a curve ω of minimum energy, the Riemannian manifold $(N, \langle \ , \ \rangle)$ is said to be *complete*. We suppose from now on that this is the case.

A necessary condition for $\omega \in \Omega$ to minimise $E(\omega)$ is that ω should satisfy a second order (usually nonlinear) autonomous system of ordinary differential equations, called the *geodesic equations*. In local coordinates the geodesic equations take the form

$$\ddot\omega(t) = -\Gamma_{\omega(t)}(\dot\omega(t), \dot\omega(t)),$$

where $\Gamma_x : \mathbb{R}^n \times \mathbb{R}^n \to \mathbb{R}^n$ is a symmetric bilinear transformation called the *Christoffel transformation*, whose dependence on $x \in N$ is C^∞. Any curve satisfying the geodesic equations is called a *geodesic*. So curves that minimise $E(\omega)$ are necessarily geodesics. However, not all geodesics minimise E: those

Curves and Surfaces with Applications in CAGD
A. Le Méhauté, C. Rabut, and L. L. Schumaker (eds.), pp. 319–328.
Copyright © 1997 by Vanderbilt University Press, Nashville, TN.
ISBN 0-8265-1293-3.
All rights of reproduction in any form reserved.

that do are called *minimal geodesics*. When looking for curves that minimise E we often consider geodesics first, just as when optimising a real-valued function critical points are considered as candidates for extreme points.

When x_a, x_b are far apart it is usually nontrivial to find a geodesic joining them [8]. However here we are mainly interested in joining *nearby* points by geodesics. In practice this is a simple task which can be efficiently implemented, and piecewise-geodesic paths in N can be considered natural generalizations of polygons in $\mathbb{R}^n$ both conceptually and from a computational point of view. To motivate what we do with piecewise-geodesic paths, recall a well-known construction with polygons in $\mathbb{R}^n$.

Define a *scaled triple* in $\mathbb{R}^n$ to be a quadruple Y be of the form

$$(y_0, y_1, y_2, h) \in (\mathbb{R}^n)^3 \times \mathbb{R}_+ .$$

The *scale* of Y is h. The *fundamental polygon* $p(Y)$ is the piecewise-linear curve $p : [0, 2h] \to \mathbb{R}^n$ given by

$$p(t) = (1 - t/h)y_0 + (t/h)y_1 \text{ or } (2 - t/h)y_1 + (t/h - 1)y_2$$

according as $t \leq h$ or not. There is also a *splitting construction* which produces a *left triple* Y^L and a *right triple* Y^R associated with Y. To define splitting, let y_3 and y_4 be the midpoints of the line segments $y_0 y_1$ and $y_1 y_2$ respectively. Let y_5 be the midpoint of $y_3 y_4$. Then Y^L is the scaled triple $(y_0, y_3, y_5, h/2)$ and $Y^R = (y_5, y_4, y_2, h/2)$.

Splitting can also be applied to Y^L and then Y^R, producing scaled triples $Y^{LL}, Y^{LR}, Y^{RL}, Y^{RR}$ of scales $h/4$. Continuing, after m iterations we obtain, for every word w of length m in the symbols L, R, a scaled triple Y^w of scale $h/2^m$ called *descendants* of Y in generation m. Writing the Y^w in dictionary order, the track-sum p_m of their fundamental polygons has a smoother appearance than the fundamental polygon p, and as $m \to \infty$ the p_m converge uniformly to the quadratic curve $p_\infty : [0, 2h] \to \mathbb{R}^n$ from from y_0 to y_2 with initial velocity $(y_1 - y_0)/h$. This construction is the *Euclidean Quadratic Algorithm*. The *Riemannian Quadratic Algorithm* is an extension of the Euclidean case, with interesting mathematical properties. The main result of the present paper is Theorem 2.1 below.

§2. The Riemannian Quadratic Algorithm

Let d be the metric on N defined by the Riemannian distance. A subset C of the Riemannian manifold $(N, \langle \ , \ \rangle)$ is said to be *convex* when any $x_a, x_b \in C$ are joined by a unique minimal geodesic $\omega : [a, b] \to C$ depending on x_a, x_b in a C^∞ fashion. Let U be a convex open subset of N whose closure is compact and contained in another convex open subset V of N. Finally, suppose that there is a diffeomorphism $\phi : V \to \mathbb{R}^n$ namely V is a coordinate chart of N. In §3 calculations are made in local coordinates defined by ϕ. Then we will be able to assume that $d(x_a, x_b)$ is bounded for all $x_a, x_b \in U$ and that the Christoffel transformations

$$\Gamma_x : \mathbb{R}^n \times \mathbb{R}^n \to \mathbb{R}^n$$

are also bounded for $x \in U$. For the moment it is enough to bear in mind that U is convex.

For $a < b$ let $C[a, b]$ be the complete metric space of continuous curves $\omega : [a, b] \to U$ with respect to the uniform metric d_U, where

$$d_U(\omega, \omega') = Max_{t \in [a,b]} d(\omega(t), \omega'(t)).$$

Extending the Euclidean definition, define a *Riemannian scaled triple* to be a quadruple $Y = (y_0, y_1, y_2, h) \in U^3 \times \mathbb{R}_+$. The fundamental polygon $p : [0, 2h] \to U$ of Y is the track sum of the geodesic segments joining y_0, y_1 and y_1, y_2, parameterized by $[0, h]$ and $[h, 2h]$ respectively.

To define *splitting*, replace line segments by minimal geodesics in the construction for the Euclidean case. Because U is convex, splitting is well-defined, and so are the Y^w, where w is any word of finite length m in the symbols L, R. The track-sum of the $p(Y^w)$ is a piecewise-geodesic $p_m : [0, 2h] \to U$, and because the uniform limit $p_\infty = lim_{m \to \infty} p_m$ is extremely regular in the Euclidean case, it is natural to enquire after p_∞ in general. However in the Riemannian case, even showing that the uniform limit p_∞ exists requires expenditure of effort. First define the *mesh* $\mu(Y)$ as the larger of $d(y_i, y_{i+1})$ where $i = 0, 1$.

Lemma 2.1. $\mu(Y^L), \mu(Y^R) \leq \mu(Y)/2$.

Proof: Geodesics are parameterized proportionally to arc-length, and so $d(y_0, y_3) = d(y_0, y_1)/2 \leq \mu(Y)/2$. Similarly

$$2d(y_3, y_5) = d(y_3, y_4) \leq d(y_3, y_1) + d(y_1, y_4)$$
$$= d(y_0, y_1)/2 + d(y_1, y_2)/2 \leq \mu(Y).$$

Therefore $\mu(Y^L) \leq \mu(Y)/2$. The argument for $\mu(Y^R)$ is similar in all respects.
∎

The *mesh* $\mu(p_m)$ is the largest $\mu(Y^w)$ where w is a word of length m. Applying Lemma 2.1 repeatedly gives

Lemma 2.2. $\mu(p_m) \leq \mu(Y)/2^m$.

Proposition 2.1. $\{p_m : m \geq 1\}$ *converges in* $C[0, 2h]$.

Proof: Consider the scaled triples Y, Y^L, Y^R as well as their respective fundamental polygons p, p^L, p^R. For $s \in [h/2, h]$,

$$d(p(s), p^L(h/2)) \leq d(y_1, y_3) = d(y_0, y_1)/2 \leq \mu(Y)/2.$$

From the definition of the mesh of a scaled triple,

$$d(p^L(h/2), p^L(s)) \leq \mu(Y^L) \leq \mu(Y)/2$$

by Lemma 2.1. Therefore $d(p(s), p^L(s)) \leq \mu(Y)$, and since p^L agrees with p on $[0, h/2]$, the inequality holds for all $s \in [0, h]$. A similar argument with p^R gives

$$d_U(p, p_1) \leq \mu(Y).$$

Applying this to the scaled triples Y^w instead of Y gives

$$d_U(p_m, p_{m+1}) \leq \mu(p_m) \leq \mu(Y)/2^m$$

by Lemma 2.2. Similarly for $k \geq 1$,

$$d_U(p_{m+k}, p_{m+k+1}) \leq \mu(Y)/2^{m+k},$$

and therefore

$$d_U(p_m, p_{m+r}) \leq \mu(Y)(1 + 2^{-1} + \cdots + 2^{-r})/2^m \leq 2\mu(Y)/2^m.$$

So $\{p_m : m \geq 1\}$ is Cauchy. ∎

The analytic properties of p_∞ are much more problematic. There is almost no chance that p_∞ has a nice algebraic description, because the Riemannian quadratic construction is geometrical in nature and there is no well-accepted generalization of a quadratic curve in a general Riemannian manifold (indeed the present paper is one of a series designed to study possible generalizations). So we focus on relatively coarse properties such as continuity and differentiability. The main result of the present paper is

Theorem 2.1. *p_∞ is right-differentiable at 0, left-differentiable at $2h$, and differentiable on $(0, 2h)$. Furthermore both left and right derivatives of p_∞ are Lipschitz.*

The proof is not immediate and in §3, §4 quite a lot of effort is spent preparing the ground. It turns out that scaled triples have *accelerations* related to the accelerations along fundamental polygons. A key result is Lemma 4.4 which says that the acceleration tends to be inherited by descendants of a scaled triple. By the time Lemma 4.4 is proved, Theorem 2.1 is starting to look plausible.

The proof is completed in two stages. First in §5 a candidate for the right-derivative of p_∞ is found by proving that the sequence of right-derivatives $\{p_m : m \geq 1\}$ converges. The candidate is the limit, and in §6 is proved to be the right-derivative. These proofs also require some effort, but once they are done, Theorem 2.1 follows very quickly.

§3. Calculating Midpoints

We need to make some calculations in the local coordinate system for U described at the beginning of §2 (now is a good time to review the description). Note that U has finite diameter with respect to the Riemannian distance d, and that all derivatives to any finite order of the Christoffel transformations

$$\Gamma_x : \mathbb{R}^n \times \mathbb{R}^n \to \mathbb{R}^n$$

are bounded for $x \in U$. Such quantities are said to be $O(1)$.

For $h > 0, a \in \mathbb{R}$ and $b = a + h$ let $\omega : [a, b] \to U$ be a minimal geodesic and let δ be the diameter of its image. Because geodesics are parameterized proportionally to arc-length, $h\dot{\omega}(t) = O(1)\delta$ for all $t \in (a, b)$. By the geodesic equations $h^2\ddot{\omega}(t) = O(1)\delta^2$. Differentiating the geodesic equations $h^3 d^3\omega/dt^3 = O(1)\delta^3$, and continuing in this way we soon obtain

Lemma 3.1. *For any* $m \in \mathbb{N}$ *and* $t \in (a, b)$

$$h^m d^m\omega/dt^m = O(1)\delta^m.$$

Lemma 3.1 with Taylor's theorem gives

$$\omega(b) = \omega(a) + h\dot{\omega}(a) - h^2\Gamma_{\omega(a)}(\dot{\omega}(a), \dot{\omega}(a))/2 + O(1)\delta^3.$$

Resubstituting for $\dot{\omega}(a)$ in the right hand side of this expression gives the velocity estimate

Lemma 3.2. $\dot{\omega}(a) = (\omega(b) - \omega(a))/h + \Gamma_{\omega(a)}(\omega(b) - \omega(a), \omega(b) - \omega(a))/(2h) + O(1)\delta^3.$

Taylor's theorem also says

$$\omega((a + b)/2) = \omega(a) + h\dot{\omega}(a)/2 - h^2\Gamma_{\omega(a)}(\dot{\omega}(a), \dot{\omega}(a))/8 + O(1)\delta^3$$

which with Lemma 3.2 gives the following estimate for the midpoint $M(x_a, x_b)$ of the geodesic ω joining $x_a, x_b \in U$

Lemma 3.3. $M(x_a, x_b) = (x_a + x_b)/2 - \Gamma_{x_a}(x_b - x_a, x_b - x_a)/8 + O(1)\delta^3.$

Let Y be a scaled triple and take ω to be a minimal geodesic joining either of the pairs (y_0, y_1) or (y_1, y_2). Define a quadratic function $g : \mathbb{R}^n \to \mathbb{R}^n$ by

$$g(v) = -\Gamma_{y_0}(v, v) = -\Gamma_{\omega(t)}(v, v) + O(1)\mu(Y)\|v\|$$

for any t in the domain of ω. Then comparison of Lemma 3.3 with the splitting construction gives

Lemma 3.4. *For* $i = 0, 1$

$$y_{i+3} = (y_i + y_{i+1})/2 + g(y_{i+1} - y_i)/8 + O(1)\mu(Y)^3.$$

Calculating y_5 with Lemma 3.3, and substituting for y_3, y_4 with Lemma 3.4, gives

Lemma 3.5.

$$y_5 = (y_0 + 2y_1 + y_2)/4 + g(y_1 - y_0)/16 + g(y_2 - y_1)/16 + g(y_2 - y_0)/32 + O(1)\mu(Y)^3.$$

These estimates are used to study *accelerations* of scaled triples, especially the phenomenon of *inheritance*.

§4. Accelerations, Inheritance, and Velocity Mesh

Let $Y = (y_0, y_1, y_2, h)$ be a scaled triple as in §3. For $i = 0, 1$ let $\omega_{i,i+1} :$ $[ih, (i + 1)h] \to U$ be the minimal geodesic from y_i to y_{i+1}. The *acceleration* of Y is the vector

$$\alpha(Y) = (\dot{\omega}_{1,2}(h) - \dot{\omega}_{0,1}(h))/(2h)$$

based at $y_1 \in U$. From Lemma 3.2 we obtain

Lemma 4.1. $\alpha(Y)$ *is equal to*

$$(y_2 - 2y_1 + y_0)/(2h^2) + g(y_2 - y_1)/(4h^2) + g(y_1 - y_0)/(4h^2) + O(1)\mu(Y)^3.$$

Here $\dot{\omega}_{0,1}(h)$ is estimated by applying Lemma 3.2 to the *reversal* of $\omega_{0,1}$, which accounts for the $+$ sign in front of $g(y_1 - y_0)$.

Applying Lemma 4.1 to Y^L, Y^R gives estimates for $\alpha(Y^L)$ and $\alpha(Y^L)$ in terms of y_i and h where $0 \le i \le 5$. Lemmas 3.4, 3.5 permit a rewrite in terms of y_0, y_1, y_2 and h: using MAPLE to carry out this routine task gives an *inheritance* property of scalar triples:

Lemma 4.2. $\alpha(Y^L) = \alpha(Y) + O(1)\mu(Y) = \alpha(Y^R).$

This property can be greatly strengthened. In order to strengthen it somehat define the *velocity-mesh* of Y to be $\nu(Y) = \|2h\alpha(Y)\|_{y_1}$. For $m \ge 1$ the *velocity-mesh* $\nu(p_m)$ is the largest $\nu(Y^w)$ where w is a word of length m in the symbols L, R. Applying Lemma 4.2 to descendants of Y in generation $m - 1$ gives

Lemma 4.3. $\nu(p_m) \le \nu(p_{m-1})/2 + O(1)h\mu(Y)/2^{2m-2}.$

Here Lemma 2.1 is used to bound the mesh in generation $m-1$. Replacing m in Lemma 4.3 by $m - 1, m - 2, \ldots 1$ in turn eventually gives

Lemma 4.4. $\nu(p_m) \le (\nu(Y) + O(1)h\mu(Y))/2^m.$

This is a sufficiently strong version of inheritance for the present paper. As $m \to \infty$ differences in left and right velocities at junctions of the track-sum p_m decrease as $1/2^m$. The parameter difference between successive junctions decreases at the same rate, which makes it plausible that the limiting curve p_∞ is differentiable. It remains to give details.

§5. Convergence of Right-Derivatives

Let V be the Banach space of functions $v : [0, 2h] \to \mathbb{R}^n$ with the uniform norm e_U defined from the Euclidean norm $\| \quad \|$ on $\mathbb{R}^n$. Because of the assumptions made about U at the beginning of §2, $\| \quad \|$ defines the same topology on V as the Riemannian norm $\| \quad \|_x$ at any $x \in U$. So the Riemannian norms implicit in the estimates of §4 can be replaced by the Euclidean norm: consider this done.

Definition D. For $s \in [0, 2h)$, let $\dot{p}_{m+}(s) \in \mathbb{R}^n$ be the right-derivative of p_m at s. Let $\dot{p}_{m+}(2h)$ be the left-derivative $\dot{p}_{m-}(2h)$. This defines a function

$$\dot{p}_{m+} : [0, 2h] \to \mathbb{R}^n .$$

The left-derivative $\dot{p}_{m-} : [0, 2h] \to \mathbb{R}^n$ is defined similarly as a function over the whole of $[0, 2h]$.

The next lemma suggests a candidate for the right derivative of p_∞. The proof uses Lemmas 2.1, 4.4 to make statements about the behaviour of μ and ν on descendants of Y.

Lemma 5.1. $\{\dot{p}_{m+} : m \geq 1\} \subset V$ converges.

Proof: By Lemma 3.1 the acceleration at any point along the geodesic segment $p|(0, h)$ is $O(1)\mu(Y)^2/h^2$. So for $s \in [h/2, h]$

$$\|\dot{p}_+(s) - \dot{p}_+(0)\| \leq O(1)\nu(Y) + O(1)\mu(Y)^2/h.$$

Similarly,

$$\|\dot{p}_+^L(s) - \dot{p}_+^L(0)\| \leq O(1)\nu(Y^L) + 2O(1)\mu(Y^L)^2/h$$
$$\leq O(1)\nu(Y)/2 + O(1)\mu(Y)h/2 + O(1)\mu(Y)^2/(2h)$$

by Lemmas 4.4, 2.1 respectively. Because $\dot{p}_+^L(0) = \dot{p}_+(0)$,

$$\|\dot{p}_+(s) - \dot{p}_+^L(s)\| \leq \|\dot{p}_+(s) - \dot{p}_+(0)\| + \|\dot{p}_+^L(0) - \dot{p}_+^L(s)\|$$
$$\leq (3/2)O(1)\nu(Y) + O(1)\mu(Y)h/2 + (3/2)O(1)\mu(Y)^2/h.$$

A similar estimate applies to $\dot{p}_+^R$ so that

$$e_U(\dot{p}_+, \dot{p}_{1+}) \leq (3/2)O(1)\nu(Y) + O(1)\mu(Y)h/2 + (3/2)O(1)\mu(Y)^2/h.$$

Applying this to the fundamental polygons in generation $r - 1$ in place of p,

$$e_U(\dot{p}_{r-1+}, \dot{p}_{r+})$$
$$\leq (3/2)O(1)\nu(p_{r-1}) + O(1)\mu(p_{r-1})h/2^r + (3/2)O(1)\mu(p_{r-1})^2 2^{r-1}/h.$$

Again by Lemmas 4.4, 2.1, the quantity on the right is bounded above by

$$(3/2)O(1)\nu(Y)/2^{r-1} + (3/2)O(1)\mu(Y)h/2^{r-1}$$
$$+ O(1)\mu(Y)h/2^{2r-1} + (3/2)O(1)\mu(Y)^2/(2^{r-1}h).$$

So we can certainly say

$$e_U(\dot{p}_{r-1+}, \dot{p}_{r+})$$
$$\leq 2O(1)\nu(Y)/2^{r-1} + 3O(1)\mu(Y)h/2^{r-1} + 2O(1)\mu(Y)^2/(2^{r-1}h).$$

Replacing r by $r-1, r-2, \ldots 1$ in turn and then summing, we have

$$e_U(\dot{p}_+, \dot{p}_{r+})$$
$$\leq (2O(1)\nu(Y) + 3O(1)\mu(Y)h + 2O(1)\mu(Y)^2/h)(1 + 1/2 + \ldots 1/2^{r-1})$$
$$\leq 2(2O(1)\nu(Y) + 3O(1)\mu(Y)h + 2O(1)\mu(Y)^2/h).$$

Applying this to the fundamental polygons in generation m in place of p,

$$e_U(\dot{p}_{m+}, \dot{p}_{m+r+})$$
$$\leq 2(2O(1)\nu(p_m) + 3O(1)\mu(p_m)h/2^m + 2O(1)\mu(p_m)^2 2^m/h)$$
$$\leq 2(2O(1)(\nu(Y) + O(1)h\mu(Y)) + 3O(1)\mu(Y)h/2^m + 2O(1)\mu(Y)^2/h)/2^m$$

again by Lemmas 4.4, 2.1. So $\{\dot{p}_{m+} : m \geq 1\}$ is Cauchy. ∎

§6. Right-Differentiability

Let $\dot{p}_{\infty+} : [0, 2h] \to \mathbb{R}^n$ be the uniform limit of the sequence $\{\dot{p}_{m+} : m \geq 1\}$ considered in Lemma 5.1. The notation is justified by

Lemma 6.1. p_∞ *is right-differentiable at any* $s \in [0, 2h)$ *with right-derivative* $\dot{p}_{\infty+}(s)$.

Proof: Choose $t > s$ in whichever interval $[0, h)$ or $[h, 2h)$ contains s. For $m \geq 1$ let i be the largest integer such that $(i-1)h/2^m \leq s$. Let j be the smallest integer for which $t \leq jh/2^m$. Then $i \leq j$, and because s and t are separated by at least $j - i - 1$ subintervals of width $h/2^m$

$$(j - i - 1)h \leq 2^m(t - s). \tag{1}$$

Write $Q \equiv p_m(t) - p_m(s)$ in the form

$$(p_m(t) - p_m((j-1)h/2^m))$$
$$+ \Sigma_{i<k<j}(p_m(kh/2^m) - p_m((k-1)h/2^m)) + (p_m(ih/2^m) - p_m(s)).$$

Use Taylor's theorem to write a typical term $(p_m(kh/2^m) - p_m((k-1)h/2^m))$ in the form

$$(h/2^m)\dot{p}_{m+}((k-1)h/2^m) + R_k.$$

By Lemma 3.1 the remainder R_k is bounded above in norm by $O(1)\mu(Y^w)^2$ for some word w of length m in the symbols L, R. By Lemma 2.1 this quantity is bounded above by

$$O(1)\mu(Y)^2/2^{2m}.$$

Applying such estimates to each of its terms Q becomes

$$(t - (j-1)h/2^m)\dot{p}_{m+}((j-1)h/2^m)$$
$$+ h/2^m \Sigma_{i<k<j}\dot{p}_{m+}((k-1)h/2^m) + (ih/2^m - s)\dot{p}_{m+}(s) + R,$$

where R is bounded in norm by

$$(j - i + 1)O(1)\mu(Y)^2/2^{2m}.$$

Now $\dot{p}_{m+}$ is C^∞ except at $ih/2^m$ for $i \equiv 1, 2 \quad mod(3)$. The norms of the jumps at these parameter values are at most

$$O(1)\nu(p_m) \le O(1)(\nu(Y) + O(1)h\mu(Y))/2^m$$

by Lemma 4.4.

For $u \in [s, t]$ at most $j - i - 1$ jumps separate s and u. So

$$\|\dot{p}_{m+}(u) - \dot{p}_{m+}(s)\|$$
$$\le (j - i - 1)\nu(p_m) + (j - i + 1)O(1)\mu(Y)^2/(2^m h).$$

(The last term bounds velocity changes within geodesic track-summands, resembling the bound for $\|R\|$.) By Lemma 4.4 this quantity is bounded above by

$$(j - i - 1)(\nu(Y) + O(1)h\mu(Y))/2^m + (j - i + 1)O(1)\mu(Y)^2/(2^m h) \le$$

$$(t - s)(\nu(Y)/h + O(1)\mu(Y) + O(1)\mu(Y)^2/h^2) + 2O(1)\mu(Y)^2/(2^m h)$$

by (1). This bound has the form $(t - s)K + L/2^m$ where K, L are independent of m.

Now

$$\|p_m(t) - p_m(s) - (t-s)\dot{p}_{m+}(s)\| = \|Q - (t-s)\dot{p}_{m+}(s)\|$$
$$\le (j - i)h/2^m((t-s)K + L/2^m) + (j - i + 1)O(1)\mu(Y)^2/2^{2m}.$$

Applying (1) twice, this bound is replaced by

$$(t - s)((t-s)K + L/2^m + O(1)\mu(Y)^2/(2^m h))$$
$$+ h/2^m((t-s)K + L/2^m + 2O(1)\mu(Y)^2/(2^m h)).$$

In the limit as $m \to \infty$,

$$\|p_\infty(t) - p_\infty(s) - (t-s)\dot{p}_{\infty+}(s)\| \le (t-s)^2 K. \quad \blacksquare$$

Lemma 6.1 can also be applied to the scaled triple $Y^{-1} \equiv (y_2, y_1, y_0, h)$ in place of Y. Then p_∞ is seen to be left-differentiable on $(0, 2h]$ with left-derivative

$$\dot{p}_{\infty-} = lim_{m \to \infty} \dot{p}_{m-}.$$

Now

$$e_U(\dot{p}_{m-}, \dot{p}_{m+}) = O(1)\nu(p_m) \to 0$$

as $m \to \infty$, by Lemma 4.4. So $\dot{p}_{\infty-} = \dot{p}_{\infty+}$ and p_∞ is differentiable on $(0, 2h)$. In the proof of Lemma 6.1

$$\|\dot{p}_{m+}(t) - \dot{p}_{m+}(s)\| \leq (t - s)K + L/2^m,$$

where $s < t$. So

$$\|\dot{p}_\infty(t) - \dot{p}_\infty(s)\| \leq (t - s)K,$$

and $\dot{p}_\infty$ is Lipschitz. Theorem 2.1 is proved.

Acknowledgments. It is a pleasure to record my appreciation to Charles Micchelli for enlightenment and encouragement over a number of years.

References

1. Cavaretta, A. S., W. Dahmen, and C. A. Micchelli, *Stationary Subdivision*, Mem. Amer. Math. Soc. No. 453, Providence, 1991.

2. de Rham, G., Sur quelques fonctions différentiables dont toutes les valeurs sont des valeurs critiques, Celebrazioni archimedee de secolo XX, Siracusa, II:61–65, 11-16 aprile 1961.

3. de Rham, G., Un peu de mathématiques à propos d'une courbe plane, Revue de Mathématiques Elémentaires, II, 1947.

4. de Rham, G., Sur certaines équations fonctionnelles, Ouvrage publié à l'occasion de son centenaire par l'Ecole Polytechnique de l'Université de Lausanne, 1953, 95–97.

5. de Rham, G., Sur une courbe plane, J. de Math. Pures et Appliquées **35** (1956), 25–42.

6. de Rham, G., Sur les courbes limites de polygones obtenus par trisection, L'enseignement Mathématique **5** (1959), 29–43.

7. Noakes, L., Asymptotically smooth splines, Advances in Computational Math. **4** (1994), 131–137.

8. Noakes, L., A global algorithm for geodesics, submitted to IMA J. Math. Control & Information.

Department of Mathematics
The University of Western Australia
Nedlands, WA 6907, AUSTRALIA.
lyle@maths.uwa.edu.au

Approximation of Offset Curves and Surfaces by Discrete Smoothing D^m-splines with Tangent Conditions

M. Pasadas, J. J. Torrens and M. C. López de Silanes

Abstract. In this paper we present a method for the approximation of an offset Υ_0 of a given curve or surface Υ from finite sets of data points and tangent spaces (or, equivalently, normal vectors) to Υ. First, we construct an approximant $\tilde{\Upsilon}$ of Υ by solving a quadratic minimization problem in a parametric finite element space, whose solution is called *discrete smoothing D^m-spline with tangent conditions*. An approximant $\tilde{\Upsilon}_0$ of Υ_0 is then obtained by offsetting $\tilde{\Upsilon}$. We study the convergence of this approximation method and, finally, we give some numerical and graphical examples.

§1. Introduction

It is well known that offset curves and surfaces (sometimes referred as parallel curves and surfaces, see M. P. Do Carmo [5]) frequently appear in many fields, such as CAGD, mechanical geometric modelling, geophysics and structural geology. In this last field, for example, the two layers that delimit a geological stratum can be reconstructed in the following way: one by using a fitting method from a set of points and normal vectors to the layer, and the other as an offset surface of the first layer, by taking the mean thickness of the stratum as the offset magnitude.

In many practical situations, it is not possible to obtain exact representations of offsets, due to the complexity of the computations involved or to the lack of data. In such cases, offsets need to be approximated (see, for example, R. Farouki [6] and J. Hoschek [7]). In this paper we cope with this problem. We propose a method, based on the theory of D^m-splines over a bounded domain (cf. M. Attéia [2], R. Arcangéli [1]), for finding offset approximants from sets of data points and tangent spaces (or normal vectors) to the original curves or surfaces.

Curves and Surfaces with Applications in CAGD   329
A. Le Méhauté, C. Rabut, and L. L. Schumaker (eds.), pp. 329–336.
Copyright © 1997 by Vanderbilt University Press, Nashville, TN.
ISBN 0-8265-1293-3.
All rights of reproduction in any form reserved.

The remainder of this paper is organized as follows. In Section 2, we briefly recall some preliminary notation and results. Sections 3 and 4 are devoted, respectively, to state the approximation problem we consider and to present a method for solving it. The convergence of this method is analyzed in Section 5. Finally, in Section 6 we provide some numerical and graphical examples.

§2. Preliminaries

Let $n \in \mathbb{N}$. We denote by $\langle \cdot \rangle$ and $\langle \cdot , \cdot \rangle$, respectively, the Euclidean norm and inner product in $\mathbb{R}^n$.

For any $m, n, p \in \mathbb{N}$ and for any open subset Ω in $\mathbb{R}^p$, we denote by $H^m(\Omega; \mathbb{R}^n)$ the usual Sobolev space of functions with values in $\mathbb{R}^n$, endowed with the norm

$$\|v\|_{m,\Omega,\mathbf{R}^n} = \left(\sum_{|\alpha| \leq m} \int_\Omega \langle \partial^\alpha v(x) \rangle^2 \, dx \right)^{1/2} .,$$

where $\alpha = (\alpha_1, \ldots, \alpha_p) \in \mathbb{N}^p$, $|\alpha| = \alpha_1 + \cdots + \alpha_p$, and $\partial^\alpha v$ is the α-th derivative of v in the distribution sense. We shall also use the inner semi-products

$$(u, v)_{l,\Omega,\mathbf{R}^n} = \sum_{|\alpha|=l} \int_\Omega \langle \partial^\alpha u(x), \partial^\alpha v(x) \rangle \, dx , \quad l = 0, \ldots, m,$$

and its associated semi-norms

$$|v|_{l,\Omega,\mathbf{R}^n} = (v, v)^{1/2}_{l,\Omega,\mathbf{R}^n} , \quad l = 0, \ldots, m.$$

When Ω is a bounded, open subset of $\mathbb{R}^p$ with Lipschitz–continuous boundary (in the J. Nečas [8] sense), it follows from the Sobolev's imbedding theorem that, if $m > k + p/2$,

$$H^m(\Omega; \mathbb{R}^n) \text{ is a subset of } C^k(\overline{\Omega}; \mathbb{R}^n) \text{ with continuous injection}, \quad (1)$$

where $C^k(\overline{\Omega}; \mathbb{R}^n)$, with $k \in \mathbb{N}$, stands for the space of functions with values in $\mathbb{R}^n$ which are bounded and uniformly continuous on Ω, together with all their partial derivatives of order $\leq k$.

Finally, for any $k \in \mathbb{N}$, we denote by P_k the space of all polynomials of degree $\leq k$.

§3. Modelling the Problem

Let $p = 1$ or 2 and $n = p + 1$. Let $\Upsilon \subset \mathbb{R}^n$ be a curve (if $p = 1$) or a surface (if $p = 2$) parameterized by a function f that belongs to the Sobolev space $H^m(\Omega; \mathbb{R}^n)$, where m is an integer such that

$$m > \frac{p}{2} + 1, \tag{2}$$

and $\Omega \subset \mathbb{R}^p$ is a bounded, open interval if $p = 1$, or a polygonal, bounded, open set if $p = 2$. By (1) and (2), f is also a continuously differentiable function on $\overline{\Omega}$. Hence, we can suppose that

for all $x \in \overline{\Omega}$, the total derivative of f at x is injective.

This hypothesis implies the existence, for all $x \in \Omega$, of a unit vector $\nu(x)$ normal to Υ at the point $f(x)$, which can be computed, if $p = 1$, as

$$\nu(x) = \left\langle \frac{df}{dx}(x) \right\rangle^{-1} \left((p_2 \circ \frac{df}{dx})(x), -(p_1 \circ \frac{df}{dx})(x) \right), \tag{3}$$

where p_1 and p_2 are the two canonical projections of $\mathbb{R}^2$ on $\mathbb{R}$, or, if $p = 2$, as

$$\nu(x) = \left\langle \frac{\partial f}{\partial x_1}(x) \times \frac{\partial f}{\partial x_2}(x) \right\rangle^{-1} \left(\frac{\partial f}{\partial x_1}(x) \times \frac{\partial f}{\partial x_2}(x) \right), \tag{4}$$

where the symbol $\times$ stands for the vector product in $\mathbb{R}^3$.

Given a real number r, the offset of Υ, with offset magnitude r, is a curve or surface Υ_0 parameterized by the function $f_0 : x \in \overline{\Omega} \mapsto f(x) + r\nu(x) \in \mathbb{R}^n$. Geometrically Υ_0 can be considered as the curve or surface generated by the center of a circumference ($p = 1$) or a sphere ($p = 2$) which wheel on all points of Υ.

Let $\mathcal{D}$ be a set of real, positive numbers for which 0 is an accumulation point. Suppose that, for any $d \in \mathcal{D}$, we are given:

(i) two ordered subsets A_1^d and A_2^d of $\overline{\Omega}$, not necessarily distinct;
(ii) the set $\{f(a) \mid a \in A_1^d\} \subset \Upsilon$;
(iii) for all $a \in A_2^d$, a basis of the tangent linear space $\mathrm{span}\{\partial^\alpha f(a) \mid |\alpha| = 1\}$, or, equivalently, the unit normal vector $\nu(a)$, given by (3) or (4).

In this situation, we consider the following problem:

$$\left\{ \begin{array}{l} \text{For a given offset magnitude } r \text{ and for any } d \in \mathcal{D}, \text{ find} \\ \text{an approximating curve or surface of the offset } \Upsilon_0 \text{ from} \\ \text{the data specified in (i), (ii) and (iii).} \end{array} \right. \tag{5}$$

§4. The Approximation Method

To solve problem (5), we proceed in two steps. First, we construct an approximant $\tilde{\Upsilon}$ of the original curve or surface Υ from the given data. $\tilde{\Upsilon}$ is parameterized by a discrete smoothing D^m–spline with tangent conditions, whose definition we recall below. In the second step, we compute the offset $\tilde{\Upsilon}_0$ of $\tilde{\Upsilon}$ for the offset magnitude r, which is the searched approximant of the offset Υ_0.

Let us begin with the first step. Let $\mathcal{H}$ be a set of real, positive numbers of which 0 is an accumulation point. For any $h \in \mathcal{H}$, we consider a tesselation $\mathcal{T}_h$ of $\overline{\Omega}$ by elements K of diameter $h_K \leq h$ which are intervals, if $p = 1$, or triangles or rectangles, if $p = 2$. We construct on $\mathcal{T}_h$ a finite element space X_h from a generic finite element (K, P_K, Σ_K) of class C^k, with $m \leq k + 1$, and such that $P_K \subset H^m(K)$ (see P. G. Ciarlet [3] for notation). It is easily shown that the parametric finite element space $V_h = (X_h)^n$ satisfies

$$V_h \subset H^m(\Omega; \mathbb{R}^n) \cap C^k(\overline{\Omega}; \mathbb{R}^n). \tag{6}$$

For any $d \in \mathcal{D}$, for any $\varepsilon > 0$ and for any $\tau \geq 0$, let $J_{\varepsilon\tau}^d$ be the functional defined on $H^m(\Omega; \mathbb{R}^n)$ by

$$J_{\varepsilon\tau}^d(v) = \sum_{a \in A_1^d} \langle v(a) - f(a) \rangle^2 + \tau \sum_{a \in A_2^d} \sum_{|\alpha|=1} \langle \partial^\alpha v(a), \nu(a) \rangle^2 + \varepsilon |v|_{m,\Omega,\mathbb{R}^n}^2 .$$

Then we consider the following minimization problem: find $\sigma_{\varepsilon\tau}^{dh}$ such that

$$\begin{cases} \sigma_{\varepsilon\tau}^{dh} \in V_h, \\ \forall v \in V_h, \ J_{\varepsilon\tau}^d(\sigma_{\varepsilon\tau}^{dh}) \leq J_{\varepsilon\tau}^d(v). \end{cases} \tag{7}$$

Assuming that A_1^d contains a P_{m-1}-unisolvent subset, we have the following result.

Theorem 1. *Problem (7) has a unique solution, called a discrete smoothing D^m–spline with tangent conditions, which is also the unique solution of the following variational problem: find $\sigma_{\varepsilon\tau}^{dh}$ such that*

$$\begin{cases} \sigma_{\varepsilon\tau}^{dh} \in V_h, \\ \forall v \in V_h, \ \displaystyle\sum_{a \in A_1^d} \langle \sigma_{\varepsilon\tau}^{dh}(a), v(a) \rangle + \tau \sum_{a \in A_2^d} \sum_{|\alpha|=1} \langle \partial^\alpha \sigma_{\varepsilon\tau}^{dh}(a), \nu(a) \rangle \langle \partial^\alpha v(a), \nu(a) \rangle \\ \quad + \varepsilon (\sigma_{\varepsilon\tau}^{dh}, v)_{m,\Omega,\mathbb{R}^n} = \displaystyle\sum_{a \in A_1^d} \langle f(a), v(a) \rangle. \end{cases}$$

The proof is sketched in [10], where an interpretation of the functional $J_{\varepsilon\tau}^d$ is also found. The reader is referred to [9] for an indepth treatment of these topics and related questions. In particular, it is proved there that, under suitable conditions, $\sigma_{\varepsilon\tau}^{dh} \to f$ in $H^m(\Omega; \mathbb{R}^n)$ as card $A_1^d \to +\infty$. Hence, the curve or surface $\tilde{\Upsilon}$ parameterized by $\sigma_{\varepsilon\tau}^{dh}$ is, in some sense, a good approximant of Υ.

Theorem 1 makes it possible to reduce the computation of $\sigma_{\varepsilon\tau}^{dh}$ to the solution of a linear system whose order depends on the dimension of the space V_h, but not on the number of data points or tangent conditions. For details, see again [9] or [10].

For the second step of the method, let $\nu_{\varepsilon\tau}^{dh} : \overline{\Omega} \to \mathbb{R}^n$ be the field of unit normal vectors to $\tilde{\Upsilon}$, given by (3) or (4) with $\sigma_{\varepsilon\tau}^{dh}$ instead of f. The function $\tilde{\sigma}_{\varepsilon\tau}^{dh} : \overline{\Omega} \to \mathbb{R}^n$, defined by

$$\tilde{\sigma}_{\varepsilon\tau}^{dh}(x) = \sigma_{\varepsilon\tau}^{dh}(x) + r\,\nu_{\varepsilon\tau}^{dh}(x),$$

is then a parameterization of the offset $\tilde{\Upsilon}_0$ of $\tilde{\Upsilon}$. As we shall state in Section 5, $\tilde{\sigma}_{\varepsilon\tau}^{dh}$ converges to f_0, parameterization of Υ_0, in a suitable functional space. In this sense, $\tilde{\Upsilon}_0$ is an approximant of Υ_0, and hence, it provides a solution to the initial approximation problem (5).

§5. Convergence of the Method

In addition to the hypotheses introduced in Sections 3 and 4, we suppose that

$$\forall d \in \mathcal{D}, \ \sup_{x \in \Omega} \delta(x, A_1^d) = d\,,$$

where δ denotes the Euclidean distance in $\mathbb{R}^p$, and that

$$\exists C > 0, \ \forall d \in \mathcal{D}, \ \max(N_1(d), N_2(d)) \le C d^{-p}\,,$$

$N_1(d)$ and $N_2(d)$ being, respectively, the number of points in A_1^d and A_2^d. We also assume that the following result, due to P. Clément [4]: there exist an integer $m' > m$, a constant $C > 0$ and, for all $h \in \mathcal{H}$, a linear operator $\Pi_h : L^2(\Omega; \mathbb{R}^n) \to V_h$ such that

$$\forall h \in \mathcal{H}, \ \forall l = 0, \ldots, m'-1, \ \forall v \in H^{m'}(\Omega; \mathbb{R}^n),$$

$$\left(\sum_{K \in \mathcal{T}_h} |v - \Pi_h v|_{l,K,\mathbb{R}^n}^2 \right)^{1/2} \le C\, h^{m'-l} |v|_{m',\Omega,\mathbb{R}^n}\,. \tag{8}$$

Remark 2. *The result (8) assumes that the family of tesselations $(\mathcal{T}_h)_{h \in \mathcal{H}}$ is regular (cf. P. G. Ciarlet [3]), and that the generic finite element (K, P_K, Σ_K) of the family $(V_h)_{h \in \mathcal{H}}$ satisfies the condition $P_{m'-1}(K) \subset P_K \subset H^{m'}(K)$ and an "uniformity"property of the basis functions (cf. P. Clément [4], G. Strang [11]) fulfilled in the usual cases (for example, for the elements of Bogner–Fox–Schmit, Argyris or Bell).*

Finally, we suppose given three functions $\varepsilon : \mathcal{D} \to (0, +\infty)$, $h : \mathcal{D} \to \mathcal{H}$ and $\tau : \mathcal{D} \to [0, +\infty)$ (i.e., ε, h and τ are functions of d) such that

$$\varepsilon = o(d^{-p}), \ d \to 0,$$

and that

$$\frac{h^{2(m'-m)}(1+\tau)}{\varepsilon d^p} = o(1), \ d \to 0.$$

Under these conditions, one has the following result, proved in [9].

Theorem 3. *If the parameterization* f *belongs to the space* $H^{m'}(\Omega; \mathbb{R}^n)$, *then*

$$\lim_{d \to 0} \| f_0 - \tilde{\sigma}_{\varepsilon\tau}^{dh} \|_{C^{m-p-1}(\overline{\Omega}; \mathbb{R}^n)} = 0.$$

Remark 4. *The functions* f_0 *and* $\tilde{\sigma}_{\varepsilon\tau}^{dh}$ *really belong to* $C^{m-p-1}(\overline{\Omega}; \mathbb{R}^n)$, *since, by (1), (6) and the hypotheses on* p, m *and* m', *the functions* f *and* $\sigma_{\varepsilon\tau}^{dh}$ *belong to* $C^{m-p}(\overline{\Omega}; \mathbb{R}^n)$ *and* $m - p - 1 \geq 0$. *As a consequence of Theorem 3, there always exists uniform convergence of* $\tilde{\sigma}_{\varepsilon\tau}^{dh}$ *to* f_0.

§6. Numerical and Graphical Examples

We have tested our offset approximation method with different curves and surfaces. In all cases, the data sets are composed of N_1 data points and N_2 normal vectors randomly distributed on the curve or surface Υ. We have set $N_1 = 30$ and $N_2 = 100$ if $p = 1$, and $N_1 = 900$ and $N_2 = 2500$ if $p = 2$.

Likewise, the parametric finite element spaces V_h have been constructed on uniform tesselations $\mathcal{T}_h$ of $\overline{\Omega}$, which have 16 elements, if $p = 1$, or 36 elements, if $p = 2$. The generic finite element (K, P_K, Σ_K) of the space V_h, if $p = 2$, is the rectangle of Bogner–Fox–Schmit of class C^2 (see, for example, [9] or [10]), and, if $p = 1$, it is an Hermite finite element of class C^1, where $K = [a, b]$ is any interval in $\mathcal{T}_h$, P_K is the space of restrictions to K of polynomials of degree ≤ 3 and $\Sigma_K = \{v \mapsto v(a),\ v \mapsto v'(a),\ v \mapsto v(b),\ v \mapsto v'(b)\}$.

The approximants $\tilde{\Upsilon}$ of Υ have been parameterized, in any case, by discrete smoothing D^{p+1}–splines with tangent conditions for $\varepsilon = 10^{-6}$ and $\tau = 1$.

The quality of the offset approximation is measured by an estimate E_{tol} of the tolerance error, computed by the formula

$$E_{tol} = \Big(\frac{1}{10000} \sum_{i=1}^{10000} \frac{1}{r^2} \langle \tilde{\sigma}_{\varepsilon\tau}^{dh}(x_i) - f_0(x_i) \rangle^2 \Big)^{\frac{1}{2}},$$

where $x_1, \ldots, x_{10000}$ are random points in $\overline{\Omega}$.

As a first example, we consider the curve defined by the parameterization $f : \overline{\Omega} \to \mathbb{R}^2$, with $\Omega = (-1, 1)$ and

$$f(x) = \big(\tfrac{1}{3}(5 + 4x + 6x^2 - x^3 - 8x^4)e^{\cos x}, 4 - 3x + 3x^2 - 4x^4 \big).$$

In Figure 1, the 1st, 2nd, 4th and 5th curves, counting from top to bottom, are, repectively, the approximants $\tilde{\Upsilon}_0$ of the offsets Υ_0 for the offset magnitudes $r = -0.6, -0.3, 0.3$ and 0.6. The corresponding values of the estimate E_{tol} of the tolerance error are, respectively, 1.1642×10^{-3}, 2.3846×10^{-3}, 2.1492×10^{-3} and 1.0806×10^{-3}. In this figure, the 3rd curve is the approximant $\tilde{\Upsilon}$.

A new example is provided by the surface defined by the parameterization $f : \overline{\Omega} \to \mathbb{R}^3$, with $\Omega = (0, 1) \times (0, 1)$ and

$$f(u, v) = \big(\cos(\pi u - \tfrac{3\pi}{2})v, \sin(\pi u - \tfrac{3\pi}{2})v, u \big).$$

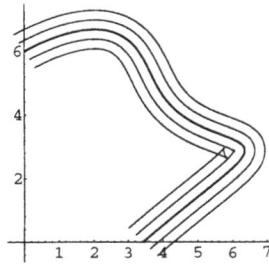

Fig. 1. Approximants of offset curves.

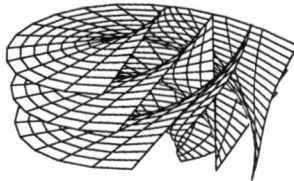

Fig. 2. Approximants of offset surfaces.

Figure 2 shows, from top to bottom, the approximant of the offset surface with offset magnitude $r = -0.2$, the approximant $\tilde{\Upsilon}$ of the original surface and the approximant of the offset surface with offset magnitude $r = 0.2$. The values of E_{tol} corresponding to $r = -0.2$ and $r = 0.2$ are, respectively, 3.0201×10^{-2} and 2.9817×10^{-2}.

Acknowledgments. The second and third authors have been supported in part by the Gobierno de Navarra (Research Project BON n. 33–18/3/94) and by the European Union (Research Project INTAS 94–4070) respectively.

References

1. Arcangéli, R., D^m-splines sur un domaine borné de $\mathbb{R}^n$, Publication UA 1204 CNRS n° 1986/2.

2. Attéia, M., Fonctions splines définies sur un ensemble convexe, Numer. Math. **12** (1968), 192–210.

3. Ciarlet, P. G., *The Finite Element Method for Elliptic Problems*, North–Holland, 1978.

4. Clément, P., Approximation by finite element functions using local regularization, RAIRO R–2 (1975), 77–84.

5. Do Carmo, M. P., *Differential Geometry of Curves and Surfaces*, Prentice–Hall, Englewood Cliffs, 1976.

6. Farouki, R. T., The approximation of non–degenerate offset surfaces, Comput. Aided Geom. Design **3** (1986), 15–44.

7. Hoscheck, J., Spline approximation of offset curves, Comput. Aided Geom. Design **5** (1988), 33–40.

8. Nečas, J., *Les Méthodes Directes en Théorie des Équations Elliptiques*, Masson, 1967.

9. Pasadas, M., *Aproximación de Curvas y Superficies Paramétricas con Condiciones de Tangencia*, Thesis, Universidad de Granada, 1995.

10. Pasadas, M., M. C. López de Silanes, and J. J. Torrens, Approximation of parametric surfaces by discrete smoothing D^m–splines with tangent conditions, in *Mathematical Methods for Curves and Surfaces*, Morten Dæhlen, Tom Lyche, Larry L. Schumaker (eds.), Vanderbilt University Press, Nashville & London, 1995, 403–412.

11. Strang, G., Approximation in the finite element method, Numer. Math. **19** (1972), 81–98.

M. Pasadas
Departamento de Matemática Aplicada
E.U.A.T.
Universidad de Granada
Severo Ochoa s/n
18001 Granada, SPAIN
mpasadas@goliat.ugr.es

J. J. Torrens
Departamento de Matemática e Informática
Universidad Pública de Navarra
Campus de Arrosadía s/n
31006 Pamplona, SPAIN
jjtorrens@upna.es

M. C. López de Silanes
Departamento de Matemática Aplicada
C.P.S.
Universidad de Zaragoza
María de Luna 3
50015 Zaragoza, SPAIN
mcsilanes@mcps.unizar.es

Algorithm for Computing the Product
of two B-splines

Les Piegl and Wayne Tiller

Abstract. An algorithm for computing the product of two B-spline functions is presented. The idea is to decompose the B-spline functions into their Bézier components, using a variation of knot insertion, compute the product of the Bézier functions, and to recompose the results, using a variation of knot removal. Full details of the algorithm along with a precise analysis are presented.

§1. Introduction

The purpose of this paper is to present a complete algorithm for the computation of the product of two B-spline functions. This operation is useful in geometric computing to create B-spline based entities symbolically. Some applications are given in [5], and an entire array of numeric operators, based on symbolic B-spline algebra, is presented in [1,2].

Very few papers have been published in this area, and no complete algorithm is available. Three major approaches can be identified. The first computes the product by solving an interpolation problem [1,2,5]. The second converts the B-spline into polynomial form, computes the product by convolution of the coefficients, and obtains the B-spline coefficients by the de Boor-Fix formula [4,6,8]. The third approach is that of Morken [7]. It is based on discrete B-splines and uses a recurrence relation to compute B-spline coefficients. Some remarks pertinent to the above approaches can be found in [6] and [11].

The approach presented in this paper is similar to our method of degree elevation [9]: (1) decompose the B-splines into their Bézier components using knot insertion; (2) compute the product of the Bézier functions; and (3) recompose the Bézier product functions into B-spline form using knot removal.

Full knot insertions and removals are not required; the decomposition and recomposition are done on the fly using a variation of knot insertion and knot removal, respectively.

Curves and Surfaces with Applications in CAGD
A. Le Méhauté, C. Rabut, and L. L. Schumaker (eds.), pp. 337–344.
Copyright © 1997 by Vanderbilt University Press, Nashville, TN.
ISBN 0-8265-1293-3.
All rights of reproduction in any form reserved.

The organization of the paper is as follows. Section 2 gives some notation and states the precise problem. In Section 3 the computation of the product of two Bézier functions is discussed and analyzed. Section 4 presents the complete algorithm, followed by a complexity analysis in Section 5.

§2. Notation and Problem Statement

The problem is as follows: given two B-spline functions

$$F(u) = \sum_{i=0}^{n_f} N_{i,p}(u) f_i \qquad R = \{\underbrace{a, ..., a}_{p+1}, r_{p+1}, ..., r_{n_f}, \underbrace{b, ..., b}_{p+1}\}$$

$$G(u) = \sum_{j=0}^{n_g} N_{j,q}(u) g_j \qquad S = \{\underbrace{c, ..., c}_{q+1}, s_{q+1}, ..., s_{n_g}, \underbrace{d, ..., d}_{q+1}\}$$

of degree p and q, respectively, compute the degree $p + q$ product function

$$F(u)G(u) = H(u) = \sum_{k=0}^{n_h} N_{k,p+q}(u) h_k$$

defined over the knot vector

$$T = \{\underbrace{e, ..., e}_{p+q+1}, t_{p+q+1}, ..., t_{n_h}, \underbrace{f, ..., f}_{p+q+1}\}.$$

To compute $H(u)$, we need to compute the coefficients $h_0, ..., h_{n_h}$, and the knot vector T. Based on the continuity of the product of two functions, the computation of T is fairly simple: (1) scale R and S to a common interval, say $[0,1]$; (2) insert all internal r knots, not present in S, into S; (3) insert all internal s knots, not present in R, into R; and (4) copy all new internal r or s knots into T with the following multiplicities: $m_\ell = q + m_\ell^r$ if $m_\ell^s = 0$, $m_\ell = p + m_\ell^s$ if $m_\ell^r = 0$, and $m_\ell = max(q + m_\ell^r, p + m_\ell^s)$ otherwise, where m_ℓ^r and m_ℓ^s are the *original* multiplicities of $r_\ell = s_\ell$, in R and S, respectively.

To compute $h_0, ..., h_{n_h}$, we need some Bézier tools discussed in the following section.

§3. Product of Bézier Functions

Given two Bézier functions

$$F^B(u) = \sum_{i=0}^{p} B_{i,p}(u) f_i^b$$

$$G^B(u) = \sum_{j=0}^{q} B_{j,q}(u) g_j^b,$$

where f_i^b and g_i^b are the Bézier coefficients, and $B_{i,p}(u)$ and $B_{j,q}(u)$ are the Bernstein polynomials of degree p and q, respectively. Their product is computed as follows [1,3]:

$$H^B(u) = \sum_{k=0}^{p+q} B_{k,p+q}(u) h_k^b$$

where the coefficients are

$$h_k^b = \sum_{\ell=max(0,k-q)}^{min(p,k)} \frac{\binom{p}{\ell}\binom{q}{k-\ell} f_\ell^b g_{k-\ell}^b}{\binom{p+q}{k}} \quad k = 0, ..., p+q.$$

A few remarks are in order. The coefficients are the degree elevation coefficients [9]. This should not be surprising because setting $g_i = 1$ would produce degree elevation. In each row the coefficients sum to 1, i.e. the product algorithm is a convex combination scheme. The matrix is symmetric (diagonally) with respect to the row $\frac{p+q}{2}$, i.e. only $\frac{1}{4}$ of the full matrix needs to be computed. The computational complexity is $\frac{min(p,q)(p+q)}{2}$ (assuming a precomputed matrix).

In terms of B-splines, the above formula has the following advantages. The product matrix can be precomputed and stored as it is the same for each span. The computational cost is a function of the number of non-zero knot spans and is constant for each Bézier piece. And finally, the Bézier product computation is numerically stable as only convex combinations are involved.

The following section shows how the Bézier product algorithm can be integrated into a scheme, containing knot insertion and refinement, to obtain an efficient algorithm for computing the product of B-splines.

§4. The Algorithm

In this section we give a detailed pseudocode of the product algorithm. It can be readily implemented in any language. For convenience, we assume a function

$hb \leftarrow$ **BezierProduct**$(fb, p, gb, q, M, first, last)$

that takes, as arguments, the Bézier coefficients fb and gb, and the product matrix M, computed by

$M \leftarrow$ **ProductMatrix**(p, q)

and computes the coefficients

$$h_i^b \quad i = first, ..., last \quad first \geq 0 \quad last \leq p+q$$

of the product function. There are four major components of the algorithm. The first is decomposing the functions into their Bézier pieces. This is done by inserting knots on the fly. There are several simplifications that make this very efficient. The second is computing the product of two Bézier functions.

Because of the decomposition/recomposition paradigm, not all Bézier coefficients need to be computed. The third component is the recomposition of two Bézier functions into B-splines. This is done by removing knots without error checking. Again this is a special case of general knot removal. The final step is to prepare for the next pass through by initializing for the next decomposition.

We give full details of the algorithm, however, for detailed explanations on the first, second and fourth steps, the reader is referred to [9,10]. The following local arrays are required: $fb[0...p]$, $nextfb[0...p]$, $gb[0...q]$, $nextgb[0...q]$, $hb[0...p+q]$, $\alpha[0...p]$, $\beta[0...q]$. The usage of these arrays should be clear from the pseudocode.

INPUT:
  $f[...]$: B-spline coefficients of $F(u)$
  $R[...], p$: Knots of $F(u)$ and the degree
  $g[...]$: B-spline coefficients of $G(u)$
  $S[...], q$: Knots of $G(u)$ and the degree

OUTPUT:
  $h[0...nh]$: B-spline coefficients of $H(u)$
  $T[0...mt]$: Knots of $H(u)$

ALGORITHM:

```
(* refine functions *)
R[0...mr] ← insert s-knots not in R;
S[0...ms] ← insert r-knots not in S;
f[0...nf] ← refine F(u);
g[0...ng] ← refine G(u);
(* initialize some variables *)
ar = p;   br = p + 1;   as = q;   bs = q + 1;   rr = -1;
rs = -1;   rem = -1;   nh = 1;   mt = p + q + 1;
h[0] = f[0] * g[0];
(* compute left end of output knot vector *)
for (i = 0 to p + q by 1) T[i] = S[0];
(* initialize first Bézier segments *)
for (i = 0 to p by 1) fb[i] = f[i];
for (j = 0 to q by 1) gb[j] = g[j];
(* precompute Bézier product matrix *)
M ← ProductMatrix(p, q);
(* loop through the knot vectors *)
while (br < mr and bs < ms )
.   (* compute knot multiplicities *)
.   i = br;
.   while (br < mr and R[br] = R[br + 1]) br = br + 1;
.   mlr = mir = br - i + 1;
```

```
.   j = bs;
.   while (bs < ms and S[bs] = S[bs + 1]) bs = bs + 1;
.   mls = mis = bs − j + 1;
.   (* adjust multiplicities *)
.   if (R[br] is inserted knot) mir = 0;
.   if (S[bs] is inserted knot) mis = 0;
.   (* compute multiplicity of output knot *)
.   if (mis = 0) then mi = q + mir; else
.   if (mir = 0) then mi = p + mis; else mi = max(q + mir, p + mis);
.   (* insert knots R[br] and S[bs] to get Bézier functions *)
.   rr = p − mlr;    rs = q − mls;    t = p + q − mi;
.   if (rem > 0) then lbz = (rem + 2)/2; else lbz = 1;
.   if (t > 0) then rbz = p + q − (t + 1)/2; else rbz = p + q;
.   if (rr > 0)
. .     num = R[br] − R[ar];
. .     for (k = p to mlr + 1 by −1)
. .        α[k − mlr − 1] = num/(R[ar + k] − R[ar]);
. .     for (j = 1 to rr by 1)
. . .       save = rr − j;    s = mlr + j;
. . .       for (k = p to s by −1)
. . .          fb[k] = α[k − s] ∗ fb[k] + (1 − α[k − s]) ∗ fb[k − 1];
. . .       nextfb[save] = fb[p];
. .     endfor
.   endif
.   if (rs > 0)
. .     num = S[bs] − S[as];
. .     for (k = q to mls + 1 by −1)
. .        β[k − mls − 1] = num/(S[as + k] − S[as]);
. .     for (j = 1 to rs by 1)
. . .       save = rs − j;    s = mls + j;
. . .       for (k = q to s by −1)
. . .          gb[k] = β[k − s] ∗ gb[k] + (1 − β[k − s]) ∗ gb[k − 1];
. . .       nextgb[save] = gb[q];
. .     endfor
.   endif
.   (* compute product of Bézier segments *)
.   hb ← BezierProduct(fb, p, gb, q, M, lbz, p + q);
.   (* remove knot R[ar] = S[as] *)
.   if (rem > 1)
. .     first = mt − 2;    last = mt;
. .     den = S[bs] − S[as];    β = (S[bs] − T[kind − 1])/den;
. .     for (k = 1 to rem − 1 by 1)
. . .       i = first;    j = last;    ℓ = j − mt + 1;
. . .       while (j − i > k)
. . . .         if (i < nh)
. . . . .          δ = (S[bs] − T[i])/(S[as] − T[i]);
```

```
. . . . .  h[i] = δ * h[i] + (1 − δ) * h[i − 1];
. . . .  endif
. . . .  if (j ≥ lbz)
. . . . .  if (j − k ≤ kind − p − q + rem)
. . . . . .  γ = (S[bs] − T[j − k])/den;
. . . . . .  hb[ℓ] = γ * hb[ℓ] + (1 − γ) * hb[ℓ + 1];
. . . . .  else
. . . . . .  hb[ℓ] = β * hb[ℓ] + (1 − β) * hb[ℓ + 1];
. . . . .  endif
. . . .  endif
. . . .  i = i + 1;    j = j − 1;    ℓ = ℓ − 1;
. . .  endwhile
. . .  first = first − 1;    last = last + 1;
. .  endfor
.  endif
.  (* load knots and coefficients *)
.  if (ar ≠ p and as ≠ q)
. .  for (i = 0 to p + q − rem − 1 by 1)
. . .  T[mt] = S[as];    mt = mt + 1;
. .  endfor
.  endif
.  for (i = lbz to rbz by 1)
. .  h[nh] = hb[i];    nh = nh + 1;
.  endfor
.  (* prepare for next pass through *)
.  if (br < mr and bs < ms)
. .  for (i = 0 to rr − 1 by 1) fb[i] = nextfb[i];
. .  for (i = rr to p by 1) fb[i] = f[br − p + i];
. .  for (j = 0 to rs − 1 by 1) gb[j] = nextgb[j];
. .  for (j = rs to q by 1) gb[j] = g[bs − q + j];
. .  ar = br;    br = br + 1;
. .  as = bs;    bs = bs + 1;
. .  rem = p + q − mi;
.  else
. .  for (i = 0 to p + q by 1) T[mt + i] = S[bs];
.  endif
endwhile
```

§5. Computational Complexity

The algorithm consists of three major components: (1) decomposition, (2) product, and (3) recomposition. Our aim is to perform a worst case analysis. The worst case is obtained by summing the worst case of each component. Since the Bézier product is independent of knot multiplicities, the worst case occurs when all knots in the R and S knot vectors appear with multiplicity one (this would require the maximum number of knot insertions and removals).

To make things even worse, assume that all r knots have to be inserted into S, and that all s knots must be inserted into R. Since the algorithm performs decomposition, product and recomposition number-of-knot-span times, the maximum number of times the main *while* loop runs is

$$n_f + n_g - p - q + 1$$

In the following analysis, we count the computational cost of performing arithmetic operations during the decomposition-product-recomposition cycle. We assume that the functions have been refined and that assigning values to memory cells is insignificant compared to arithmetic operations.

To decompose the functions, at most $p - 1$ r knots and $q - 1$ s knots are inserted. The cost is $(p - 2)$ α computations, $(q - 2)$ β computations, and $\frac{(p-1)(p-2)}{2}$ and $\frac{(q-1)(q-2)}{2}$ convex combinations. Putting it all together, the total cost of decomposition is

$$\frac{(p + 1)(p - 2) + (q + 1)(q - 2)}{2}.$$

The Bézier product, as indicated above, requires

$$\frac{min(p, q)(p + q)}{2}$$

steps to compute the new coefficients.

The recomposition is a bit more involved. The worst multiplicity is $min(p, q) + 1$. The maximum number of knots to be removed is therefore $p + q - min(p, q) - 1$, which is $max(p, q) - 1$. The outer loop runs from 1 to $max(p, q) - 2$. The gap between i and j is initially 2, and it is increased to 4, 6, 8, etc. However, because $(j - i) > k$ must hold, only half of the triangular array is computed. Because in this half two combinations are evaluated, the total cost comes to

$$\frac{(max(p, q) - 2)(max(p, q) - 1)}{2}.$$

Using the notations $N = n_f + n_g - p - q + 1$, $r = min(p, q)$ and $s = max(p, q)$, the three complexities are combined as follows:

$$N \left[\underbrace{\frac{(p + 1)(p - 2) + (q + 1)(q - 2)}{2}}_{DECOMPOSITION} + \underbrace{\frac{r(p + q)}{2}}_{PRODUCT} + \underbrace{\frac{(s - 2)(s - 1)}{2}}_{RECOMPOSITION} \right]$$

In terms of orders of growth, it simplifies to

$$N \left[\underbrace{O(p^2) + O(q^2)}_{DECOMPOSITION} + \underbrace{O(pq)}_{PRODUCT} + \underbrace{O(s^2)}_{RECOMPOSITION} \right] = NO(s^2).$$

If the total number of coefficients are much larger than the maximum degree, the algorithm becomes linear in the number of coefficients. If not, it is quadratic in the maximum degree.

Acknowledgments. This research was supported by the National Science Foundation under grant No: DMI 9526119.

References

1. Elber, G., *Free form surface analysis using a hybrid of symbolic and numeric computation*, PhD Thesis, University of Utah, 1992.

2. Elber, G. and Cohen E., Second-order surface analysis using hybrid symbolic and numeric operators, ACM Transactions on Graphics **12** (1993), 160–178.

3. Farouki, R. T. and Rajan, V. T., Algorithms for polynomials in Bernstein form, Computer Aided Geometric Design **5** (1988), 1–26.

4. Fuhr, R. D., Representations of NURBS curves and surfaces using de Boor-Fix functionals, Presented at the *SIAM Conference on Geometric Design*, Tempe, AZ, 1989.

5. Fuhr, R. D., Hsieh, L. and Kallay, M., Object-oriented paradigm for NURBS curve and surface design, Computer-Aided Design **27**, 2 (1995), 95–100.

6. Lee, E. T. Y., Computing a chain of blossoms, with application to products of splines, Computer Aided Geometric Design **11** (1994), 597–620.

7. Morken, K., Some identities for products and degree raising of splines, Constructive Approximation **7** (1991), 195–208.

8. Ueda, K., Multiplication as a general operation for splines, in *Curves and Surfaces in Geometric Design*, Laurent, P. J., Le Méhauté, A. and Schumaker, L. L. (eds.), A. K. Peters, Wellesley, MA, 1994, 475–482.

9. Piegl, L. and Tiller, W., Software engineering approach to degree elevation of B-spline curves, Computer-Aided Design **26** 1 (1994), 17–28.

10. Piegl, L. and Tiller, W., *The NURBS Book*, Springer-Verlag, New York, NY, 1995.

11. Vermuelen, A. H., Bartels, R. H. and Heppler, G. R., Integrating products of B-splines, SIAM J. Sci. Stat. Comput. **13** (1992), 1025–1038.

Les Piegl
Department of Computer Science & Engineering
University of South Florida
4202 Fowler Avenue, ENG 118
Tampa, FL 33620
piegl@babbage.csee.usf.edu

Wayne Tiller
GeomWare, Inc.
3036 Ridgetop Road
Tyler, TX 75703
76504.3045@compuserve.com

Smoothing Spatial Cubic B-splines under Shape Constraints

Konstantinos G. Pigounakis and Panagiotis D. Kaklis

Abstract. A method for improving the shape of a spatial C^2 cubic $B-$spline under tolerance constraints is proposed. This is achieved by specifying the sign of torsion and controlling the direction of the binormal and normal vectors of the curve, which leads to a Non-Linear Programming problem with a quadratic objective function, multilinear shape constraints, and quadratic tolerance ones. The numerical performance of the scheme is tested on an industrial data set.

§1. Introduction

According to differential geometry, a spatial curve is, up to a rigid motion, fully determined by its curvature and torsion. Also, it is well known that the sign of curvature in plane determines whether a curve preserves its shape or not. Unlike the planar case, the curvature of a spatial curve is sign indifferent, so it is not suitable for shape interrogation. One must then mainly rely on torsion in order to extract some basic shape information and to tell whether a spatial curve 'turns' right (positive values of torsion) or left (negative values). For a more detailed analysis a good background in differential geometry is needed in order to properly interpret the behaviour of the curve and to propose methods for shape improvement.

On the other hand, differential geometry uses also the *Frenet frame (Ff)* for studying spatial curves (see, e.g., [8]), an approach that is becoming popular in the CAGD community; see [2,3]. The authors of this paper are of the opinion that the easiest way to realise the behaviour of a 3D curve is through the vectors of Ff. For instance, since Ff is well defined also for planar curves, the notion of *inflection* can be introduced in exactly the same way for 2D and 3D curves. More specifically, when the curvature of a planar curve changes sign, two vectors of Ff (normal, binormal) change their direction. So, a unified notion of inflection can be introduced by replacing the *sign of curvature* by the *direction of the binormal vector* (see Def. 1).

Curves and Surfaces with Applications in CAGD
A. Le Méhauté, C. Rabut, and L. L. Schumaker (eds.), pp. 345–354.
Copyright ⓒ 1997 by Vanderbilt University Press, Nashville, TN.
ISBN 0-8265-1293-3.

All rights of reproduction in any form reserved.

345

In CAD/CAM practice the importance of the quality of 3D curves has been recognised along with surface modelling (see §1 of [5]). Following the trend, this work aims to combine the use of Ff and the sign of torsion, so that the curve exhibits the proper shape and its behaviour becomes predictable. More precisely, we propose a method for improving, within a tolerance, the shape of a C^2 cubic B−spline, by specifying the sign of torsion and controlling the direction of the binormal and normal vectors; see §5. This control is achieved through a stronger notion of inflection, that of the *latent inflection* (see Def. 2 in §2). Aiming to avoid latent inflections, sufficient conditions are set for the control polygon; see Propositions 1 and 2 in §4. The performance of the method is tested for an industrial data set.

A final remark is due: the choice of cubic splines leads to discontinuous torsion at the nodal points; nonetheless this defficiency does not affect the shape features which are adopted as principal (torsion sign, Ff).

§2. Frenet Frame and Shape of a Curve

In three dimensional space, a curve $\mathbf{Q}(u)$ is, up to a rigid motion fully defined through its *curvature* $\kappa(u)$ and *torsion* $\tau(u)$:

$$\kappa(u) = \frac{\|\mathbf{L}(u)\|}{\|\dot{\mathbf{Q}}(u)\|^3}, \qquad \tau(u) = \frac{|\dot{\mathbf{Q}}(u)\ddot{\mathbf{Q}}(u)\dddot{\mathbf{Q}}(u)|}{\|\mathbf{L}(u)\|^2}, \qquad (1)$$

with $\mathbf{L}(u) = \dot{\mathbf{Q}}(u) \times \ddot{\mathbf{Q}}(u)$. The dot denotes derivation with respect to the parameter u, '$\|\cdot\|$' is the euclidean norm in $\mathbb{R}^3$, and '$|\mathbf{a}\,\mathbf{b}\,\mathbf{c}|$' is the triple scalar product of the vectors $\mathbf{a}$, $\mathbf{b}$ and $\mathbf{c}$.

At any position of the curve $\mathbf{Q}(u)$, Ff consists of three unit vectors: the *tangent* $\mathbf{t}(u)$, the *binormal* $\mathbf{b}(u)$, which is perpendicular to the osculating plane of the curve, and the *normal* $\mathbf{n}(u)$:

$$\mathbf{t}(u) = \frac{\dot{\mathbf{Q}}(u)}{\|\dot{\mathbf{Q}}(u)\|}, \qquad \mathbf{b}(u) = \frac{\mathbf{L}^{(k)}(u)}{\|\mathbf{L}^{(k)}(u)\|}, \qquad \mathbf{n}(u) = \mathbf{b}(u) \times \mathbf{t}(u), \qquad (2)$$

where $\mathbf{L}^{(k)}(u) = \dot{\mathbf{Q}}(u) \times \mathbf{Q}^{(k)}(u)$ and $k \geq 2$ is the order of the first non-zero derivative of the curve (see [9], p.9).

When $\mathbf{L}(u^*) = \mathbf{0}$ and $\dot{\mathbf{Q}}(u) \neq \mathbf{0}$ (regular curve), curvature vanishes at u^* (see eq. (1)) and the curve should exhibit a straight−line behaviour there. In such cases, if the tangent straight-line at u^* intersects the curve, then it must be $\mathbf{b}(u^*_-) \cdot \mathbf{b}(u^*_+) = -1$. Since planar curves do not differ from spatial ones in this regard, the notion of inflection can be introduced in a unified way for both 2D and 3D curves, as follows:

Definition 1. *The curve $\mathbf{Q}(u)$ exhibits an inflection at u^* iff there is a discontinuity of the binormal vector there so that $\mathbf{b}(u^*_-) \cdot \mathbf{b}(u^*_+) = -1$ or, equivalently, $\angle(\mathbf{b}(u^*_-), \mathbf{b}(u^*_+)) = \pi$, where $\angle(\mathbf{a}, \mathbf{b})$ denotes the angle between the vectors $\mathbf{a}$ and $\mathbf{b}$, which is less than or equal to π.*

The requirement $\angle(\mathbf{b}(u_-^*), \mathbf{b}(u_+^*)) = \pi$ holds true for inflections in both the plane and the space, because the curve must change the orientation of its osculating plane there, but not the osculating plane itself. At u^* the curvature vanishes, while the value of torsion can be calculated through higher derivatives of the curve, which is not always efficient for a CAD system.

In 3D space, the binormal vector of a curve can change its direction pretty much, but not so abruptly as at an inflection. That causes a substantial change of the osculating plane, though the sign of torsion can remain constant. For classifying this undesirable case, which is similar to an inflection, we introduce a new notion:

Definition 2. *The curve* $\mathbf{Q}(u)$ *exhibits a* latent inflection, *iff there exist an interval* (u_a, u_b) *so that* $\mathbf{b}(u_a) \cdot \mathbf{b}(u_b) \leq 0$ *or, equivalently,* $\angle(\mathbf{b}(u_a), \mathbf{b}(u_b)) \geq \pi/2$.

Definition 2 differs from Definition 1 in two ways. First, the positions $\mathbf{Q}(u_a)$ and $\mathbf{Q}(u_b)$ can be close, but not necessarily in the same neighbourhood of the curve. Second, the angle $\angle(\mathbf{b}(u_a), \mathbf{b}(u_b))$, though greater than $\pi/2$, does not necessarily equal π. As a result, Definition 2 is more general than Definition 1, and one can easily establish

Lemma 1. *If a curve exhibits no latent inflections in an interval, then there exist no inflections therein.*

§3. Local Properties of C^2 Cubic B-splines

Let $\mathcal{D} = \{\mathbf{d}_i, \ i = 0(1)M \ ^{(*)}\}$ be the control polygon of $\mathbf{Q}(u)$ with $\mathbf{d}_i \neq \mathbf{d}_{i+1}, \ i = 0(1)M - 1$, and $\mathcal{U} = \{u_0, u_1, \ldots, u_N; u_i < u_{i+1}, i = 3(1)N - 4\}$, $N = M + 4$, a knot vector (parametrization). If $\mathbf{Q}(u)$ is an open B-spline, then the knots u_3 and u_{M+1} can be of multiplicity four, i.e., $u_0 = u_1 = u_2 = u_3$ and $u_{M+1} = u_{M+2} = u_{M+3} = u_{M+4}$. The restriction of $\mathbf{Q}(u)$ on $[u_i, u_{i+1}]$, $i = 3(1)M$, can be written as

$$\mathbf{Q}(u) = \sum_{j=i-3}^{i} \mathbf{d}_j N_j^3(u), \tag{3}$$

where $N_j^3(u)$ are the normalized B-spline basis functions of degree 3, calculated by the recursive formula:

$$N_i^n(u) = \frac{u - u_i}{u_{i+n} - u_i} N_i^{n-1}(u) + \frac{u_{i+n+1} - u}{u_{i+n+1} - u_{i+1}} N_{i+1}^{n-1}(u), \qquad u \in \mathbb{R}, \tag{4}$$

$$\text{with} \qquad N_i^0(u) = \begin{cases} 1 & \text{if } u \in [u_i, u_{i+1}); \\ 0 & \text{elsewhere.} \end{cases}$$

$^{(*)}$ The abbreviation $i = n_1(p)n_2$ means that the index i takes all values from n_1 to n_2 with step p; $n_1, n_2, p \in \mathbb{N}$, $n_1 < n_2$, $n_2 = n_1(mod\,p)$.

The first derivative of the curve $\mathbf{Q}(u)$ in $[u_i, u_{i+1}]$ can be written

$$\dot{\mathbf{Q}}(u) = 3 \sum_{j=i-2}^{i} \frac{\Delta \mathbf{d}_{j-1}}{u_{j+3} - u_j} N_j^2(u) = \alpha_1(u)\Delta \mathbf{d}_{i-3} + \alpha_2(u)\Delta \mathbf{d}_{i-2} + \alpha_3(u)\Delta \mathbf{d}_{i-1},$$

(5)

where $\Delta \mathbf{d}_{j-1} = \mathbf{d}_j - \mathbf{d}_{j-1}$ and $a_j(u)$, $j = 1, 2, 3$, are non-negative in $[u_i, u_{i+1}]$. The second derivative of $\mathbf{Q}(u)$ is given by

$$\ddot{\mathbf{Q}}(u) = 6 \sum_{j=i-1}^{i} \left[\frac{\Delta \mathbf{d}_{j-1}}{u_{j+3} - u_j} - \frac{\Delta \mathbf{d}_{j-2}}{u_{j+2} - u_{j-1}} \right] \frac{N_j^1(u)}{u_{j+2} - u_j},$$

(6)

Taking into consideration (5) and (6), $\mathbf{L}(u)$ can be written as

$$18\{\lambda_1(u)(\Delta \mathbf{d}_{i-3} \times \Delta \mathbf{d}_{i-2}) + \lambda_2(u)(\Delta \mathbf{d}_{i-3} \times \Delta \mathbf{d}_{i-1}) + \lambda_3(u)(\Delta \mathbf{d}_{i-2} \times \Delta \mathbf{d}_{i-1})\},$$

(7)

where

$$\lambda_1(u) = \frac{1}{(u_{i+1} - u_{i-2})(u_{i+2} - u_{i-1})} \cdot$$
$$\left[\frac{(u_{i+2} - u_{i+1})N_i^1(u)N_{i-1}^1(u)}{(u_{i+2} - u_i)(u_{i+1} - u_{i-1})} + \frac{(N_{i-1}^1(u))^2}{(u_{i+1} - u_{i-1})} \right],$$

$$\lambda_2(u) = \frac{1}{(u_{i+1} - u_{i-2})(u_{i+3} - u_i)} \cdot \frac{(u_{i+1} - u_i)N_i^1(u)N_{i-1}^1(u)}{(u_{i+2} - u_i)(u_{i+1} - u_{i-1})},$$

$$\lambda_3(u) = \frac{1}{(u_{i+2} - u_{i-1})(u_{i+3} - u_i)} \cdot$$
$$\left[\frac{(u_i - u_{i-1})N_i^1(u)N_{i-1}^1(u)}{(u_{i+2} - u_i)(u_{i+1} - u_{i-1})} + \frac{(N_i^1(u))^2}{(u_{i+2} - u_i)} \right].$$

It can be easily shown that the coefficients $\lambda_j(u)$, $j = 1, 2, 3$, are also non-negative in $[u_i, u_{i+1}]$, and, for uniform parametrization, $\lambda_1(u)$ and $\lambda_3(u)$ are always greater than $\lambda_2(u)$ (see also [4]).

From (2), we observe that the normal vector lies in the direction of the quantity $\mathbf{L}(u) \times \dot{\mathbf{Q}}(u)$, which, with the aid of (5) and (7), is written as a sum of nine factors, namely:

$$\sum_{j=i-3}^{i-2} \sum_{k=j+1}^{i-1} \sum_{r=i-3}^{i-1} \lambda_{6+j+k-2i}(u) \cdot \alpha_{4+r-i}(u)[(\Delta \mathbf{d}_j \times \Delta \mathbf{d}_k) \times \Delta \mathbf{d}_r].$$

(8)

For $u \in [u_i, u_{i+1}]$, the above sum is a linear combination of the double cross products of the vectors $\Delta \mathbf{d}_j$, $j = i - 3(1)i - 1$, and the coefficients are non-negative.

Another quantity of high interest is the numerator of torsion (see (1)), since it controls the sign of torsion. The numerator $|\dot{\mathbf{Q}}(u)\ddot{\mathbf{Q}}(u)\dddot{\mathbf{Q}}(u)|$, is expressed, after some algebra, as:

$$C_i \cdot |\Delta\mathbf{d}_{j-3}\Delta\mathbf{d}_{j-2}\Delta\mathbf{d}_{j-1}|, \tag{9}$$

where C_i is a constant positive number depending on the knots u_j, $j = i - 2(1)i + 3$.

Remark: It is well known that each cross product of the type $\Delta\mathbf{d}_{k-1} \times \Delta\mathbf{d}_k$, divided by the lengths of the two vectors involved, expresses a vector, whose magnitude equals to the sine of the angle between $\Delta\mathbf{d}_{k-1}$ and $\Delta\mathbf{d}_k$. This sine is also the numerator of the so-called *discrete curvature* of the polygon arc $\mathbf{d}_{k-1}\mathbf{d}_k\mathbf{d}_{k+1}$; see [6]. Additionally, the triple scalar product in (9), divided by the lengths of the vectors involved, expresses the sine of the bihedral angle between the planes $(\mathbf{d}_{j-3}\mathbf{d}_{j-2}\mathbf{d}_{j-1})$ and $(\mathbf{d}_{j-2}\mathbf{d}_{j-1}\mathbf{d}_j)$. That sine is the numerator of the *discrete torsion* of the polygon $\mathbf{d}_{j-3}\mathbf{d}_{j-2}\mathbf{d}_{j-1}\mathbf{d}_j$. So, one can observe how the discrete properties of the control polygon affect the continuous ones of the corresponding curve.

§4. Acuteness and Local Shape

Taking advantage of the simple formulae, derived in the previous section, we shall try to predict the local behaviour of Ff of a C^2 cubic $B-$spline when its control polygon possesses some attributes. But first, some new definitions are necessary.

Definition 3. *A planar polygon* $\mathbf{p}_0\mathbf{p}_1\mathbf{p}_2$ *is called acute, iff it is convex and* $0 \leq \omega_1, \omega_2 < \pi/2$, *where* $\omega_1 = \angle(\mathbf{p}_0\mathbf{p}_1)$, *and* $\omega_2 = \angle(\mathbf{p}_1\mathbf{p}_2)$; *see Fig.1.*

A planar acute polygon is also *regular* according to [1].

Definition 4. *A spatial polygon* $\mathbf{p}_0\mathbf{p}_1\mathbf{p}_2$ *is called acute, iff the projection of* $\mathbf{p}_2$ *on the plane of* $(\mathbf{p}_0, \mathbf{p}_1)$ *forms an acute polygon with three legs or, equivalently, iff* $0 \leq \omega_1, \omega_2, \phi < \pi/2$, *where* ϕ *is the bihedral angle of the planes* $(\mathbf{p}_0, \mathbf{p}_1)$ *and* $(\mathbf{p}_1, \mathbf{p}_2)$; *see also Fig. 1.*

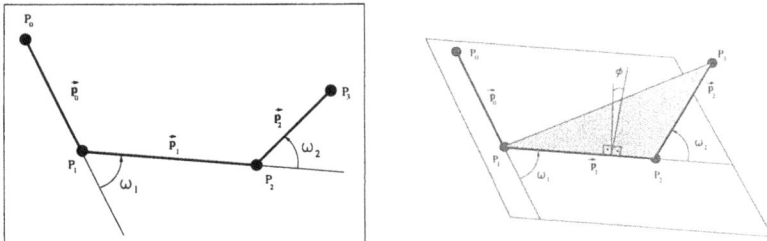

Fig. 1. A planar (left) and a spatial (right) acute polygon.

Then, we can prove the following propositions:

Proposition 1. *If* $\Delta d_{i-3}\Delta d_{i-2}\Delta d_{i-1}$ *is an acute 3D polygon corresponding to a segment of* $\mathbf{Q}(u)$, $u \in [u_i, u_{i+1}]$, *then the change of direction of the binormal vector is always less than* $\pi/2$ *therein, i.e., there is no latent inflection.*

Proposition 2. *If* $\Delta d_{i-3}\Delta d_{i-2}\Delta d_{i-1}$ *is an acute 3D polygon corresponding to a segment of* $\mathbf{Q}(u)$, $u \in [u_i, u_{i+1}]$, *then the change of direction of the normal vector is always less than* π *therein.*

Based on the results above, we can say that, if the control polygon of a B−spline curve is locally acute within the segment it influences, neither an inflection nor even a latent inflection occurs. Furthermore, Ff changes predictably therein.

§5. Formulation of the Problem

In the earlier sections we investigated conditions on a control polygon so that the manipulation of the shape of the curve is guaranteed in any segment. It is also well known how one can create planar segments (zero torsion) or straight-line segments (zero curvature) by enforcing a special configuration on the control polygon. Having all these in mind, we set the following problem:

Problem $(\mathcal{P})$: Let $\mathbf{Q}^{(0)}(u)$, $u \in [u_3, u_{N-3}]$, be a C^2 cubic B−spline with a knot vector $\mathcal{U} = \{u_i,\ i = 0(1)N\}$ and a control polygon $\mathcal{D}^{(0)} = \{d_i^{(0)},\ i = 0(1)M\}$, $N = M + 4$. Construct a cubic B−spline $\mathbf{Q}(u)$ with the same knot vector $\mathcal{U}$ and a new control polygon $\mathcal{D}$, which satisfies the *fidelity criterion*:

$$\sum_{i=0}^{M} \|d_i - d_i^{(0)}\|^2 = \min_{\hat{d}_i \in \mathbf{R}^3} \sum_{i=0}^{M} \|\hat{d}_i - d_i^{(0)}\|^2, \qquad (10)$$

and fulfills the following shape requirements (i), (ii) and the following tolerance constraints (iii):

i. Torsion constraints: Let $\mathcal{I}_\tau = \{r, 3 \leq r \leq N - 4\}$ be a set of indices specifying the segments $[u_r, u_{r+1}]$, where the sign of torsion can be predetermined. For those segments, torsion can be *positive, negative* or even *zero (planar case)*, i.e.,

$$|\Delta d_{r-3}\Delta d_{r-2}\Delta d_{r-1}| >< = 0. \qquad (11)$$

ii. Ff constraints: Let $\mathcal{I}_{Ff} = \{l, 3 \leq l \leq N - 4\}$ be a set of indices specifying the segments $[u_l, u_{l+1}]$ where the user desires a controlled change of the Ff. For those segments, the corresponding control vertices can either form a *spatial acute polygon*:

$$\Delta d_{l-3} \cdot \Delta d_{l-2} > 0,\ \Delta d_{l-2} \cdot \Delta d_{l-1} > 0,\ (\Delta d_{l-3} \times \Delta d_{l-2}) \cdot (\Delta d_{l-2} \times \Delta d_{l-1}) > 0, \qquad (12)$$

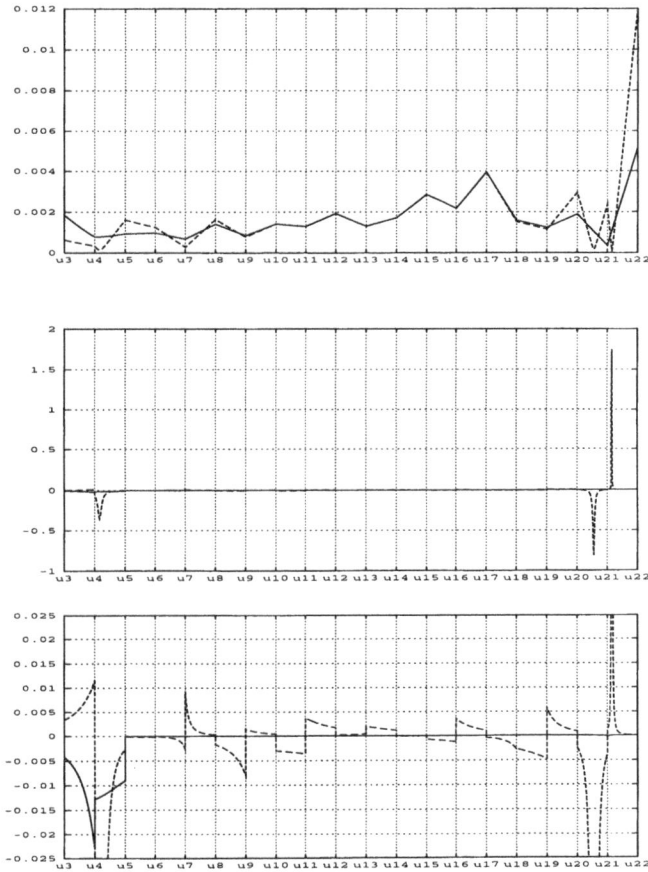

Fig. 2. Example: Curvature (upper) and torsion (middle-lower) plots.

or be *planar*(=zero torsion) and form an *acute polygon*:

$$\Delta\mathbf{d}_{l-3}\cdot\Delta\mathbf{d}_{l-2} > 0, \Delta\mathbf{d}_{l-2}\cdot\Delta\mathbf{d}_{l-1} > 0, (\Delta\mathbf{d}_{l-3}\times\Delta\mathbf{d}_{l-2})\cdot(\Delta\mathbf{d}_{l-2}\times\Delta\mathbf{d}_{l-1}) \geq 0,$$
(13)

or even be *collinear*:

$$\Delta\mathbf{d}_{l-3} \times \Delta\mathbf{d}_{l-2} = 0, \text{ and } \Delta\mathbf{d}_{l-2} \times \Delta\mathbf{d}_{l-1} = 0.$$
(14)

iii. Tolerance constraints: Each control point $\mathbf{d}_i$ should lie within a sphere with centre at $\mathbf{d}_i^{(0)}$ and user-specified radius, r_i:

$$\|\mathbf{d}_i - \mathbf{d}_i^{(0)}\|^2 \leq r_i{}^2, \quad i = 0(1)M.$$
(15)

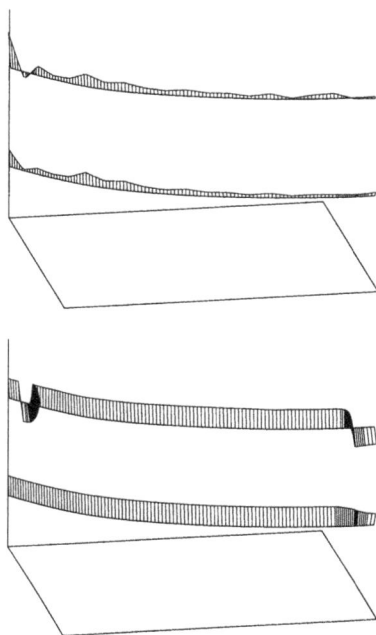

Fig. 3. Example: Distributions of the curvature and the binormal vector.

$(\mathcal{P})$ is a Non-Linear Programming (NLP) problem, consisting of a quadratic objective function, bi- tri- and quarti-linear shape constraints and quadratic tolerance constraints. This problem can be solved with the Sequential Quadratic Programming technique (SQL, see [7]).

Strictness in inequalities (12) and (13) secures the acuteness of the polygon and can be realized by replacing zero with sufficiently small numbers, ϵ. These quantities along with the tolerance radii, r_i, are parameters of the problem. Because of the tolerance constraints, $(\mathcal{P})$ is not always solvable. In such a case, the designer should permit larger deviations from the initial control points and/or decrease of ϵ's.

§6. An Industrial Example

In order to test the performance of the algorithm, we use a data set provided by General Motors Design Center, (Warren, MI). The results of the method are depicted in Figures 2 and 3. In Figure 2 the curvature and the torsion plots of the initial (dashed) and the resulted (solid) curves are given, while in Figure 3 we show the distribution of the curvature vector, $\kappa\mathbf{n}$, along the initial and the final curve (top), as well as the distribution of the binormal vector for the two cases (lower).

The initial curve is a feature line of the hood of a car. The control polygon consists of twenty-two (22) points, and the parametrization is equidistant (uniform). The quality of the curve is good. No serious problems can be detected from the curvature plot, though in the torsion plot there exist peaks near the two ends (Fig. 2). These problematic areas are quite obvious in the distributions of the curvature vector and the binormal one (Fig. 3).

For this example, the permissible deviation for the control points was set to $r_i = 1.00mm$, $i = 2(1)21$ (reference length: $800mm$), and the sign of torsion had to be negative. The resulting curve exhibits maximum deviation equal to $0.69mm$, maximum nodal deviation equal to $0.66mm$ and sum of nodal deviations equal to $4.73mm$. The quality of the resulting curve can be checked through the torsion plot of Figure 2, as well as the curvature and binormal distributions of the final curve in Figure 3.

Acknowledgments. The first author is indebted to Professor J. Hoschek for providing him the priviledge to work in the environment of the *Zentrum für Praktische Mathematik* of the *Technische Hochschule Darmstadt*. Both authors would like to thank Dr. N. Sapidis for many helpful comments. The work was partially supported by the "Human Capital and Mobility" project: FAIRSHAPE (CHRX-CT94-0522)

References

1. Goodman, T. N. T., Inflections on curves in two and three dimensions, Comput. Aided Geom. Design **8** (1991), 37-49.

2. Greiner, G., A. Kolb, R. Pfeifle, H.- P. Seidel, P. Slusallek, M. Encarnação, R. Klein, A platform for visualizing curves and surfaces, Computer-Aided Design **27** (1995), 559-566.

3. Moreton, H. P., Simplified curve and surface interrogation via mathematical packages and graphics libraries and hardware, Computer-Aided Design **27** (1995), 523-543.

4. Pigounakis, K. G. and P. D. Kaklis, Convexity-preserving fairing, to appear in Computer-Aided Design, NTUA 1994.

5. Pigounakis, K., N. Sapidis, P. Kaklis, Fairing spatial $B-$spline curves, to appear in Journal of Ship Research, NTUA 1995.

6. Sauer, R., *Differenzengeometrie*, Springer Verlag, Germany, 1970.

7. Spellucci, P., *donlp*: DO Non-Linear Programming, code and users-guide, ftp-site: *netlib/opt/donlp.shar*, THD 1995.

8. Struik, D. J., *Lectures on Classical Differential Geometry*, Dover, 1988.

9. Willmore, T. J., *An Introduction to Differential Geometry*, Oxford University Press, 1993.

Konstantinos G. Pigounakis and Panagiotis D. Kaklis
Ship-Design Laboratory
Dept. of Naval Architecture and Marine Engineering
National Technical University of Athens
Heroon Polytechneiou 9
Zografou 15773, Athens, Greece
kpig@deslab.naval.ntua.gr
kaklis@deslab.naval.ntua.gr

Some Error Estimates for Periodic Interpolation on Full and Sparse Grids

Gisela Pöplau and Frauke Sprengel

Abstract. We give a unified approach to error estimates for periodic interpolation on full and sparse grids in certain Sobolev spaces. We impose 'periodic' Strang–Fix conditions on the underlying functions in order to obtain error bounds with explicit constants.

§1. Introduction

The approximation and interpolation of bivariate periodic functions have been studied for some time. While periodic interpolation by translates on full grids is well investigated for several function spaces, especially Sobolev spaces $H_2^\alpha(\mathbb{T}^2)$ (see e.g. [2,3,8]), error estimates for periodic interpolation on sparse grids are studied for functions from Korobov spaces $E^\alpha(\mathbb{T}^2)$ (see e.g. [1,4]).

In this paper, we give a unified approach to error estimates for periodic interpolation on full and sparse grids for functions from the usual Sobolev spaces $H_2^\alpha(\mathbb{T}^2)$ and Sobolev spaces $S_{2,2}^{\alpha,\beta}H(\mathbb{T}^2)$ with dominating mixed smoothness. We consider tensor product interpolation and j-th order blending interpolation. Applying the concept of 'periodic' Strang–Fix conditions on the underlying univariate fundamental interpolant, we are able to give error estimates with explicit constants.

§2. Univariate Periodic Interpolation by Translates

Let $d \in \mathbb{N}$ be fixed. Put $d_j := 2^j d$, $j \in \mathbb{N}_0$. We consider continuous 2π-periodic functions Λ_j defined as fundamental interpolants on the equidistant interpolation grid $\mathcal{T}_j := \{t_{j,k} := 2\pi k/d_j \ : \ k = 0, \ldots, d_j - 1\}$, that means, $\Lambda_j(t_{j,k}) = \delta_{k,0}$, $k = 0, 1, \ldots, d_j - 1$. The corresponding interpolation operator L_j is given as

$$L_j f := \sum_{k=0}^{d_j - 1} f(t_{j,k})\, \Lambda_j(\cdot - t_{j,k}).$$

Curves and Surfaces with Applications in CAGD
A. Le Méhauté, C. Rabut, and L. L. Schumaker (eds.), pp. 355–362.
Copyright © 1997 by Vanderbilt University Press, Nashville, TN.
ISBN 0-8265-1293-3.
All rights of reproduction in any form reserved.

Furthermore, we assume the functions Λ_j, $j \in \mathbb{N}_0$ satisfy $\operatorname{Im} L_j \subset \operatorname{Im} L_{j+1}$. Let the remainder of an arbitrary operator P be defined as $P^c := I - P$, where I denotes the identity operator in the function space under consideration. For instance, we investigate the error for functions from Sobolev spaces of order $\alpha \geq 0$ of periodic functions

$$H_2^\alpha(\mathbb{T}) := \{ f \in L_2(\mathbb{T}) \ : \ \|f\|_{H_2^\alpha(\mathbb{T})} < \infty \}$$

with

$$\|f\|_{H_2^\alpha(\mathbb{T})}^2 := \sum_{k \in \mathbb{Z}} (1 + k^2)^\alpha \, |c_k(f)|^2,$$

where $c_k(f) := \langle f, e^{ik\cdot} \rangle_{L_2(\mathbb{T})}$ denote the Fourier coefficients. Let I_j for $j \in \mathbb{N}_0$ denote the index set $I_j := \{ k \in \mathbb{Z} \ : \ -d_j/2 < k \leq d_j/2 \}$.

To obtain error estimates in the case of multivariate periodic interpolation by translates, the 'periodic' Strang–Fix conditions were introduced in [2]. In our setting the univariate fundamental interpolants Λ_j ($j \in \mathbb{N}_0$) satisfy the 'periodic' Strang–Fix conditions of order $m > 0$, if for $k \in I_j$ and all $j \in \mathbb{N}_0$ the following inequalities hold:

$$\begin{aligned}
|1 - d_j c_k(\Lambda_j)| &\leq a_0 \left| \frac{2\pi k}{d_j} \right|^m \\
|d_j c_{k+d_j \ell}(\Lambda_j)| &\leq a_\ell \left| \frac{2\pi k}{d_j} \right|^m, \qquad \ell \in \mathbb{Z} \setminus \{0\},
\end{aligned} \tag{1}$$

where $\{a_\ell\}_{\ell \in \mathbb{Z}}$ with $\gamma_1^2 := \sum_{\ell \in \mathbb{Z}} a_\ell^2 < \infty$ is a series of non-negative constants independent of j.

Applying the estimates for multivariate periodic interpolation by translates proved in [2,8], we obtain in our special case the following theorem.

Theorem 1. *Let* $f \in H_2^\alpha(\mathbb{T})$ *with* $\alpha > 1/2$ *be given. Further, let the fundamental interpolants* Λ_j ($j \in \mathbb{N}_0$) *with* $\{c_k(\Lambda_j)\}_{k \in \mathbb{Z}} \in \ell_1(\mathbb{Z})$ *satisfy the 'periodic' Strang–Fix conditions* (1) *of order* m. *Let* $\varrho := \min\{m, \alpha\}$. *Then the following estimate holds*

$$\|f - L_j f\|_{L_2(\mathbb{T})} \leq c_\varrho \, d_j^{-\varrho} \, \|f\|_{H_2^\alpha(\mathbb{T})},$$

where

$$c_\varrho := (2\pi)^\varrho \begin{cases} \gamma_1 + \pi^{-\alpha} + \gamma_2 \gamma_3 & \text{for } \alpha \geq m, \\ (\sqrt{2}\,\pi)^{m-\alpha} \gamma_1 + \pi^{-\alpha} + \gamma_2 \, \gamma_3 & \text{for } \alpha < m. \end{cases}$$

The constant γ_1 *is given by* (1) *and* γ_2, γ_3 *are defined by*

$$\gamma_2^2 := \sup_{j \in \mathbb{N}_0} d_j^2 \max_{k \in I_j} \sum_{\ell \in \mathbb{Z}} |c_{k+d_j \ell}(\Lambda_j)|^2, \qquad \gamma_3^2 := \max_{k \in I_j} \sum_{\ell \in \mathbb{Z} \setminus \{0\}} |2\pi(k/d_j + \ell)|^{-2\alpha}.$$

All constants are independent of j.

Remark: The constant γ_3 is independent of d_j $(j \in \mathbb{N}_0)$, because for $\ell \in \mathbb{Z}\backslash\{0\}$ we obtain $\max_{k \in I_j} |2\pi(k/d_j + \ell)|^{-2\alpha} = (\pi(2|\ell| - 1))^{-2\alpha}$. In [8] the estimate $\gamma_3^2 \le (\pi^{2\alpha})^{-1}(4\alpha - 1)(2\alpha - 1)^{-1}$ was given for $\alpha > 1/2$. If $\alpha \in \mathbb{N}$, then we get the sharper estimate $\gamma_3^2 \le (2^{2\alpha} - 1)((2\alpha)!)^{-1}B_\alpha$, where B_α denotes the Bernoulli number.

Corollary 2. *Assume the fundamental interpolants Λ_j with $\{c_k(\Lambda_j)\}_{k \in \mathbb{Z}} \in \ell_1(\mathbb{Z})$ $(j \in \mathbb{N}_0)$ satisfy the 'periodic' Strang–Fix conditions (1) of order m. Let $\alpha > 1/2$, $\varrho := \min\{m, \alpha\}$. Then*

$$\|L_j^c\|_{H_2^\alpha(\mathbb{T}) \to L_2(\mathbb{T})} \le c_\varrho \, d_j^{-\varrho}$$

with c_ϱ defined in Theorem 1.

§3. Examples

Lagrange functions constructed from B–Splines or trigonometric polynomials are often used for bivariate interpolation on full and sparse grids which we consider in the next section. Therefore, we want to investigate the 'periodic' Strang–Fix conditions (1) for B–Splines and special de la Vallée Poussin means. The constant γ_1 we obtain here occurs in the error estimates of Theorem 1 for univariate interpolation as well as those of Theorem 5 and 6 for the bivariate case.

Let $M_{m,j}$ $(j \in \mathbb{N}_0)$ denote the 2π–periodic centred B–Spline of order $m \in \mathbb{N}$ given by its Fourier coefficients as $c_k(M_{m,j}) = d_j^{-1} \left(\text{sinc}\,(\pi k/d_j)\right)^m$ $(k \in \mathbb{Z}, j \in \mathbb{N}_0)$ with $\text{sinc}\,t := \sin t/t$. Then the corresponding fundamental interpolants Λ_j are known [7] to be

$$c_k(\Lambda_j) = \frac{c_k(M_{m,j})}{\sum_{\ell \in I_j} M_{m,j}(2\pi\ell/d_j)e^{-2\pi i\,k\ell/d_j}}, \qquad k \in \mathbb{Z}, \; j \in \mathbb{N}_0. \qquad (2)$$

The following 'periodic' Strang–Fix conditions for B–Splines were shown in [8]. The order of the Strang–Fix conditions coincides with the known approximation order of B–Splines. Theorem 3 also provides explicit constants in the error estimates.

Theorem 3. *Let Λ_j $(j \in \mathbb{N}_0)$ be the fundamental interpolant of the 2π–periodic B–Spline of order $m \in \mathbb{N}$ given by (2). Then Λ_j satisfies the 'periodic' Strang–Fix conditions (1) of order m with the constants*

$$a_0 = \begin{cases} 2^{-(m+1)} & \text{for } m \text{ odd,} \\ (2(2^m - 1))^{-1} & \text{for } m \text{ even,} \end{cases}$$

$$a_\ell = \frac{1}{\pi^m(2|\ell| - 1)^m} \times \begin{cases} 1 & \text{for } m = 1, \\[2mm] \dfrac{(m-1)!}{E_{(m-1)/2}} & \text{for } m > 1 \text{ odd,} \\[4mm] \dfrac{m!}{2^m(2^m - 1)B_{m/2}} & \text{for } m \text{ even} \end{cases}$$

and

$$\gamma_1^2 = \begin{cases} \dfrac{5}{16} & \text{for } m = 1, \\[2ex] \dfrac{1}{2^{2(m+1)}} + \left(\dfrac{(m-1)!}{E_{(m-1)/2}}\right)^2 \dfrac{2^{2m}-1}{(2m)!} B_m & \text{for } m > 1 \text{ odd}, \\[2ex] \dfrac{1}{4(2^m-1)^2} + \left(\dfrac{m!}{2^m(2^m-1)B_{m/2}}\right)^2 \dfrac{2^{2m}-1}{(2m)!} B_m & \text{for } m \text{ even}. \end{cases}$$

Here, B_n and E_n ($n \in \mathbb{N}$) denote the corresponding Bernoulli and Euler number, respectively.

The interpolation by trigonometric polynomials may serve as another example. For $N, M \in \mathbb{N}, N > M$, the de la Vallée Poussin means φ_N^M of the Dirichlet kernel are given as

$$\varphi_N^M(x) := \frac{1}{4MN} \sum_{\ell=N-M}^{N+M-1} \sum_{k=-\ell}^{\ell} e^{ikx}.$$

For $j \in \mathbb{N}_0$, we define the fundamental polynomial Λ_j as a special de la Vallée Poussin mean

$$\Lambda_j := \begin{cases} 1 & \text{for } j = 0, d = 1, \\ \varphi_{N_j}^{M_j} & \text{otherwise}, \end{cases} \tag{3}$$

where

$$N_j := d\, 2^{j-1} \quad \text{and} \quad M_j := \begin{cases} 2^{j-1-\lambda} & \text{for } j > \lambda, \\ 1 & \text{for } j \leq \lambda, \end{cases} \quad \lambda \in \mathbb{N}, \lambda \geq 2.$$

Theorem 4. *Let Λ_j ($j \in \mathbb{N}_0$) be the special de la Vallée Poussin mean given by (3). Then Λ_j satisfies the 'periodic' Strang–Fix conditions (1) of order m for arbitrary m with the constants*

$$a_\ell = \begin{cases} (4/3\pi)^m/2 & \text{for } \ell = -1, 0, 1, \\ 0 & \text{otherwise} \end{cases}$$

and $\gamma_1^2 = 3\,(4/3\pi)^{2m}/4$.

§4. Bivariate Periodic Interpolation

Starting from univariate interpolation, we want to investigate bivariate interpolation on full and sparse grids. Therefore, we define the corresponding interpolation projectors. The tensor product operator $L_j \otimes L_k$ realizes the interpolation on a full grid $\mathcal{T}_j \times \mathcal{T}_k$ with $d^2 2^{j+k}$ interpolation points. The j-th order Boolean sum operator is defined by

$$B_j := \bigoplus_{r=0}^{j} L_r \otimes L_{j-r},$$

where $A \oplus B := A + B - AB$. The j-th order Boolean sum B_j is an interpolation projector on the sparse grid $\bigcup_{r=0}^{j} \mathcal{T}_r \times \mathcal{T}_{j-r}$ which contains $d^2(j2^{j-1} + 2^{j+1})$ interpolation points.

We want to measure the error of interpolation for functions from certain Sobolev spaces. With the Fourier coefficients $c_{k,\ell}(f) := \langle f, e^{ik\cdot} \otimes e^{i\ell\cdot} \rangle_{L_2(\mathbb{T}^2)}$, the usual Sobolev space of order $\alpha \geq 0$ of bivariate periodic functions is defined by

$$H_2^\alpha(\mathbb{T}^2) := \{ f \in L_2(\mathbb{T}^2) \ : \ \|f\|_{H_2^\alpha(\mathbb{T}^2)} < \infty \}$$

with

$$\|f\|_{H_2^\alpha(\mathbb{T}^2)}^2 := \sum_{k,\ell \in \mathbb{Z}} (1 + k^2 + \ell^2)^\alpha \, |c_{k,\ell}(f)|^2.$$

Another important function space is the Sobolev space of order (α, β), $\alpha, \beta \geq 0$ of bivariate periodic functions with mixed smoothness. It is defined (see [9]) as

$$S_{2,2}^{\alpha,\beta} H(\mathbb{T}^2) := \{ f \in L_2(\mathbb{T}^2) \ : \ \|f\|_{S_{2,2}^{\alpha,\beta} H(\mathbb{T}^2)} < \infty \}$$

with

$$\|f\|_{S_{2,2}^{\alpha,\beta} H(\mathbb{T}^2)}^2 := \sum_{k,\ell \in \mathbb{Z}} (1 + k^2)^\alpha \, (1 + \ell^2)^\beta \, |c_{k,\ell}(f)|^2.$$

The spaces $S_{2,2}^{\alpha,\alpha} H(\mathbb{T}^2)$ are strongly related to the Korobov spaces [6]

$$E^\alpha(\mathbb{T}^2) := \left\{ f \in L_2(\mathbb{T}^2) \ : \ |c_{k,\ell}(f)| = \mathcal{O}\left(((1 + |k|)(1 + |\ell|))^{-\alpha} \right), \right.$$

$$\left. \text{for } |k|, |\ell| \longrightarrow \infty \right\},$$

which are often used in error estimates for j-th order blending interpolation (e.g. [1,4]) and numerical integration based on number theoretical methods (e.g. [6]). It holds that

$$S_{2,2}^{\alpha,\alpha} H(\mathbb{T}^2) \subset E^\alpha(\mathbb{T}^2) \subset S_{2,2}^{\alpha-1/2-\varepsilon,\alpha-1/2-\varepsilon} H(\mathbb{T}^2)$$

for arbitrary $\varepsilon > 0$. With the help of these definitions, we obtain immediately $L_2(\mathbb{T}^2) = H_2^0(\mathbb{T}^2) = S_{2,2}^{0,0} H(\mathbb{T}^2)$ and $S_{2,2}^{\alpha,\beta} H(\mathbb{T}^2) = H_2^\alpha(\mathbb{T}) \hat{\otimes} H_2^\beta(\mathbb{T}) \subset H_2^{\min\{\alpha,\beta\}}(\mathbb{T}^2) \subset L_2(\mathbb{T}^2)$.

Theorem 5. *Let $f \in S_{2,2}^{\alpha,\beta} H(\mathbb{T}^2)$ with $\alpha, \beta > 1/2$ be given. Further, let the fundamental interpolants Λ_j ($j \in \mathbb{N}_0$) with $\{c_k(\Lambda_j)\}_{k \in \mathbb{Z}} \in \ell_1(\mathbb{Z})$ satisfy the 'periodic' Strang–Fix conditions (1) of order m. Let $\varrho := \min\{m, \alpha\}$, $\sigma := \min\{m, \beta\}$, $\tau := \min\{m, \alpha, \beta\}$, $\ell := \min\{\varrho j, \sigma k\}$. Then the following estimates hold:*

$$\|f - (L_j \otimes L_k)f\|_{L_2(\mathbb{T}^2)} \leq (c_\varrho \, d_j^{-\varrho} + c_\sigma \, d_k^{-\sigma} + c_\varrho \, c_\sigma \, d_j^{-\varrho} \, d_k^{-\sigma}) \, \|f\|_{S_{2,2}^{\alpha,\beta} H(\mathbb{T}^2)},$$

$$\|f - B_j f\|_{L_2(\mathbb{T}^2)} \leq \left(c_\varrho \, d_j^{-\varrho} + c_\sigma \, d_j^{-\sigma} + c_\varrho \, c_\sigma \sum_{r=0}^{j} d_r^{-\varrho} \, d_{j-r}^{-\sigma} \right.$$

$$\left. + c_\varrho \, c_\sigma \sum_{r=0}^{j-1} d_r^{-\varrho} \, d_{j-r-1}^{-\sigma} \right) \|f\|_{S_{2,2}^{\alpha,\beta} H(\mathbb{T}^2)},$$

with c_ϱ, c_σ as defined in Theorem 1. Summarizing, we obtain

$$\|f - (L_j \otimes L_k)f\|_{L_2(\mathbb{T}^2)} \le c_{\tau,T} \, d^{-\tau} \, 2^{-\ell} \, \|f\|_{S_{2,2}^{\alpha,\beta}H(\mathbb{T}^2)},$$

$$\|f - B_j f\|_{L_2(\mathbb{T}^2)} \le c_{\tau,B} \, d_j^{-\tau} \, \|f\|_{S_{2,2}^{\alpha,\beta}H(\mathbb{T}^2)},$$

where

$$c_{\tau,T} := c_\varrho \, d^{\tau-\varrho} \, 2^{\ell-\varrho j} + c_\sigma \, d^{\tau-\sigma} \, 2^{\ell-\sigma k} + c_\varrho \, c_\sigma \, d^{\tau-\varrho-\sigma} \, 2^{\ell-\varrho j - \sigma k}$$

$$c_{\tau,B} = c_{\tau,B}(j) := c_\varrho \, d_j^{\tau-\varrho} + c_\sigma \, d_j^{\tau-\sigma} + c_\varrho \, c_\sigma \sum_{r=0}^{j} d_j^\tau \, d_r^{-\varrho} \, d_{j-r}^{-\sigma}$$

$$+ c_\varrho \, c_\sigma \sum_{r=0}^{j-1} d_j^\tau \, d_r^{-\varrho} \, d_{j-r-1}^{-\sigma}.$$

We note that $c_{\tau,B}(j)$ depends on j. For $\varrho = \sigma$, we obtain $c_{\tau,B}(j) = c_\varrho^2 \, d^{-\varrho}(1 + 2^\varrho)j + 2c_\varrho - c_\varrho^2 \, d^{-\varrho} \, 2^\varrho = \mathcal{O}(j)$. That means that in this case, we lose only a logarithmic term in approximation order if we compare the interpolation by B_j on a sparse grid with $\mathcal{O}(j2^j)$ points and the interpolation by $L_j \otimes L_j$ on a full grid with $\mathcal{O}(2^{2j})$ points.

Proof: The remainder of the tensor product interpolation operator can be written as

$$(L_j \otimes L_k)^c = L_j^c \otimes I + I \otimes L_k^c - L_j^c \otimes L_k^c.$$

Since $S_{2,2}^{\alpha,\beta}H(\mathbb{T}^2) = H_2^\alpha(\mathbb{T}) \hat{\otimes} H_2^\beta(\mathbb{T})$ we can estimate the operator norm (compare [10, Chap. 8]) of the remainder

$$\|(L_j \otimes L_k)^c\|_{S_{2,2}^{\alpha,\beta}H(\mathbb{T}^2) \to L_2(\mathbb{T}^2)} \le \|L_j^c\|_{H_2^\alpha(\mathbb{T}) \to L_2(\mathbb{T})} + \|L_k^c\|_{H_2^\beta(\mathbb{T}) \to L_2(\mathbb{T})}$$

$$+ \|L_j^c\|_{H_2^\alpha(\mathbb{T}) \to L_2(\mathbb{T})} \|L_k^c\|_{H_2^\beta(\mathbb{T}) \to L_2(\mathbb{T})}.$$

Applying Corollary 2, we obtain the assertion for tensor product interpolation. The assertion for j-th order blending interpolation follows from the representation of the remainder of the j-th order Boolean sum (see [4, Chap. 1])

$$B_j^c = L_j^c \otimes I + I \otimes L_j^c - \sum_{r=0}^{j} L_r^c \otimes L_{j-r}^c + \sum_{r=0}^{j-1} L_r^c \otimes L_{j-r-1}^c$$

and the same arguments as before. ∎

With the next theorem, we give error estimates for functions from Sobolev spaces $H_2^\alpha(\mathbb{T}^2)$. The estimate for tensor product interpolation can be derived from a result on multivariate interpolation by translates [2,8]. Using the tensor product structure of the involved operators and function spaces, we can prove it here in a shorter way and obtain a similar result for the j-th order blending interpolation.

Theorem 6. *Let* $f \in H_2^\alpha(\mathbb{T}^2)$ *with* $\alpha > 1$ *be given. Further, let the fundamental interpolants* Λ_j $(j \in \mathbb{N}_0)$ *with* $\{c_k(\Lambda_j)\}_{k \in \mathbb{Z}} \in \ell_1(\mathbb{Z})$ *satisfy the 'periodic' Strang–Fix conditions* (1) *of order* m. *Let* $\varrho := \min\{m, \alpha\}$, $\sigma := \min\{m, \alpha/2\}$, $\ell := \min\{j, k\}$. *Then the following estimates hold:*

$$\|f - (L_j \otimes L_k)f\|_{L_2(\mathbb{T}^2)} \leq \tilde{c}_{\varrho,T}\, d_\ell^{-\varrho}\, \|f\|_{H_2^\alpha(\mathbb{T}^2)},$$

$$\|f - B_j f\|_{L_2(\mathbb{T}^2)} \leq \tilde{c}_{\varrho,B}\, d_j^{-\sigma}\, \|f\|_{H_2^\alpha(\mathbb{T}^2)},$$

where

$$\tilde{c}_{\varrho,T} := c_\varrho\, 2^{\varrho(\ell-j)} + c_\varrho\, 2^{\varrho(\ell-k)} + c_\sigma\, d^{-2\sigma}\, 2^{\varrho\ell - \sigma k - \sigma \ell},$$

$$\tilde{c}_{\varrho,B} = \tilde{c}_{\varrho,B}(j) := 2\, c_\varrho\, d_j^{\sigma-\varrho} + 2\, c_\sigma\, d^{-\sigma}(j + (j-1)\, 2^\sigma)$$

and c_ϱ *is as in Theorem 1.*

Proof: We will use that $H^\alpha(\mathbb{T}^2) = S_{2,2}^{\alpha,0} H(\mathbb{T}^2) \cap S_{2,2}^{0,\alpha} H(\mathbb{T}^2)$ (compare [5]) with the following norm equivalence

$$2^{1-3\alpha} \| \cdot \|_{H^\alpha(\mathbb{T}^2)}^2 \leq \| \cdot \|_{S_{2,2}^{\alpha,0} H(\mathbb{T}^2)}^2 + \| \cdot \|_{S_{2,2}^{0,\alpha} H(\mathbb{T}^2)}^2 \leq 2 \| \cdot \|_{H^\alpha(\mathbb{T}^2)}^2$$

and the inequalities

$$\| \cdot \|_{S_{2,2}^{\alpha,0} H(\mathbb{T}^2)} \leq \| \cdot \|_{H^\alpha(\mathbb{T}^2)} \qquad \text{and} \qquad \| \cdot \|_{S_{2,2}^{0,\alpha} H(\mathbb{T}^2)} \leq \| \cdot \|_{H^\alpha(\mathbb{T}^2)}.$$

Furthermore, $H^\alpha(\mathbb{T}^2) \subset S_{2,2}^{\alpha/2,\alpha/2} H(\mathbb{T}^2)$ and

$$\| \cdot \|_{S_{2,2}^{\alpha/2,\alpha/2} H(\mathbb{T}^2)} \leq \| \cdot \|_{H^\alpha(\mathbb{T}^2)}.$$

This leads to the following estimates

$$\|f - (L_j \otimes L_k)f\|_{L_2(\mathbb{T}^2)}$$
$$= \|(L_j^c \otimes I + I \otimes L_k^c - (L_j^c \otimes I)(I \otimes L_k^c))\, f\|_{L_2(\mathbb{T}^2)}$$
$$\leq (\|L_j^c\|_{H_2^\alpha(\mathbb{T}) \to L_2(\mathbb{T})} + \|L_k^c\|_{H_2^\alpha(\mathbb{T}) \to L_2(\mathbb{T})}$$
$$+ \|L_j^c\|_{H_2^{\alpha/2}(\mathbb{T}) \to L_2(\mathbb{T})} \|L_k^c\|_{H_2^{\alpha/2}(\mathbb{T}) \to L_2(\mathbb{T})})\, \|f\|_{H_2^\alpha(\mathbb{T}^2)}.$$

Applying Corollary 2, we obtain the assertion in the case of tensor products. The Boolean sum case can be handled analogously. ■

While it might be quite useful in tensor product interpolation to use different grids $\mathcal{T}_j$ and $\mathcal{T}_k$ in x- and y-direction, respectively, for $f \in S_{2,2}^{\alpha,\beta} H(\mathbb{T}^2)$, the previous result shows that it doesn't make any sense for $f \in H_2^\alpha(\mathbb{T}^2)$. If we interpolate on the grid $\mathcal{T}_j \times \mathcal{T}_j$ we achieve the same approximation order already for functions in $H_2^\alpha(\mathbb{T}^2)$ as we have for the smoother functions in $S_{2,2}^{\alpha,\alpha} H(\mathbb{T}^2)$.

Furthermore, we can use j-th order blending interpolation with essentially less interpolation knots than for tensor product interpolation. We obtain reasonable approximation order only for functions which are smooth enough, but for the classes $S_{2,2}^{\alpha,\alpha} H(\mathbb{T}^2)$, the approximation order for interpolation on sparse and full grid only differ in a logarithmic term.

References

1. Baszenski, G., and F.-J. Delvos, A discrete Fourier transform scheme for Boolean sums of trigonometric operators, in *Multivariate Approximation Theory IV*, C. K. Chui, W. Schempp, and K. Zeller (eds.), Birkhäuser, Basel, 1989, 15–24.

2. Brumme, G., Error estimates for periodic interpolation by translates, in *Wavelets, Images, and Surface Fitting*, P.-J. Laurent, A. Le Méhauté, and L. L. Schumaker (eds.), A. K. Peters, Wellesley, 1994, 75–82.

3. Delvos, F.-J., Mean approximation via interpolation by translates, in *Constructive Theory of Functions*, B. Sendov, P. Petrushev, K. Ivanov, and R. Maleev (eds.), Publishing House of the Bulgarian Academy of Science, Sofia, 1988, 97–103.

4. Delvos, F.-J. and W. Schempp, *Boolean Methods in Interpolation and Approximation*, Pitman Research Notes in Mathematics Series, Longman Scientific & Technical, Harlow, 1989.

5. Griebel, M. and P. Oswald, Tensor product type subspace splitting and multilevel iterative methods for anisotropic problems, Advances in Comp. Math. **4** (1995), 171–206.

6. Korobov, N. M., The approximate computation of multiple integrals, Dokl. Akad. Nauk SSSR **124** (1959), 1207–1210.

7. Locher, F., Interpolation on uniform meshes by translates of one function and related attenuation factors, Math. Comp. **37** (1981), 404–416.

8. Pöplau, G., Multivariate periodic interpolation and its application, PhD dissertation, Universität Rostock, 1995.

9. Schmeißer, H.-J. and H. Triebel, *Topics in Fourier Analysis and Function Spaces*, Akadem. Verlagsgesellschaft Geest & Portig, Leipzig, 1987.

10. Weidmann, J., *Linear Operators in Hilbert Spaces*, B. G. Teubner, Stuttgart, 1976.

Gisela Pöplau
Fachbereich Mathematik, Universität Rostock
D-18051 Rostock, GERMANY
gisela.poeplau@mathematik.uni-rostock.de

Frauke Sprengel
Fachbereich Mathematik, Universität Rostock
D-18051 Rostock, GERMANY
frauke.sprengel@mathematik.uni-rostock.de

Curved Surface Reconstruction
Based on Parallels

William Puech, Jean-Marc Chassery and Ioannis Pitas

Abstract. In this paper we present a new approach for reconstructing images mapped or painted on curved surfaces. In the case of monocular vision, by using *a priori* knowledge about the support surface of the pictures and projected parallels in the image, first we derive the surface localization in the camera coordinate system. Next, we reconstruct the surface using the information of parallel curves in order to backproject the image on the surface. Finally, we simulate a camera to obtain new images from different viewpoints.

§1. Introduction

A depicted scene may be represented on curved surfaces, as shown in Figure 1. Our reconstruction method will be limited to surfaces with curvature different than zero in only one direction. The curves drawn on the scene of constant altitude in the coordinate system of the surface will be called parallels. In case of monocular vision, we need to use certain a priori knowledge to perform the 3D reconstruction of a surface. So, we use the information given by the projections of at least two parallels contained on the scene. Before reconstructing the curved surface it is necessary to locate this surface in the camera coordinate system. Next, we shall present the method of surface reconstruction using the information of the parallel curves.

An overview on the general domain of 3D reconstruction is found in the papers [4,8]. Based on the perspective projection of the curved surface and parallels on the image plane, in Section 2 we shall present briefly the method of localization by finding the projection of the revolution axis in the image [6,7]. In Section 3, we describe with more details the method of reconstruction. We use the two axes found in the localization method to backproject the parallel curves in the space with a given scale factor. In our case, the equation of the surface is derived by interpolation based on the least mean squares method. With the equation of the curved surface and its localization in the camera coordinate system, the image is backprojected on the surface in order to obtain

Curves and Surfaces with Applications in CAGD
A. Le Méhauté, C. Rabut, and L. L. Schumaker (eds.), pp. 363–370.
Copyright ⓒ 1997 by Vanderbilt University Press, Nashville, TN.
ISBN 0-8265-1293-3.
All rights of reproduction in any form reserved.

Fig. 1. View of an arch of a Byzantine church.

the 3D representation of the scene. In Section 4, simulation results will first be illustrated on synthetic images, and then on an image of a mural painting shown in Figure 1. This scene could be visualized from various viewpoints.

§2. Localization

In this section we show how to locate a curved surface in the camera coordinate system from a single perspective view, as shown in Figure 2. First we want to find the projection of the revolution axis in the image. Other work has already been done on this topic, finding local symmetries [5] or by using a method based on expectation-maximization [3].

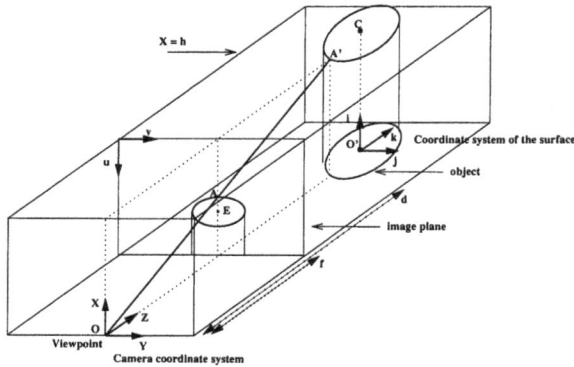

Fig. 2. A circle from a cylinder projected in the image plane.

Our approach is based on the perspective projection of the curved surface on the image as described in [7]. We identify first the common normal $P_1 P_2$ of two curves, as shown in Figure 3(a). The slope of the straight line $P_1 P_2$ gives us the direction of the revolution axis. In the coordinate system of the

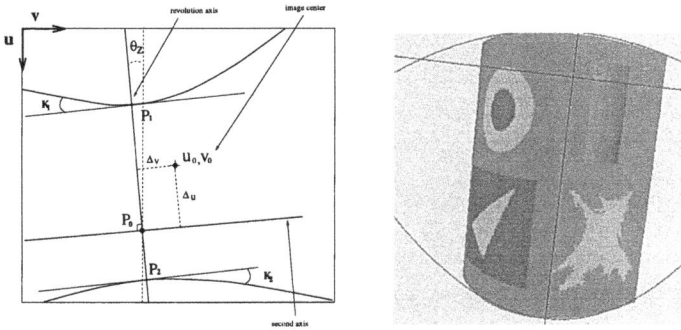

Fig. 3. (a) The revolution and the second axes detected from two curves, (b) The two axes detected on an image mapped on a cylinder.

image, the equation of this axis is

$$v = A_1.u + B_1, \tag{1}$$

where A_1 and B_1 are the coefficients of the straight line determining this axis. The second axis is defined as the intersection of the image plane with the plane perpendicular to the revolution axis containing the viewpoint [6]. We define P_0 as the intersection between the revolution axis and the second axis, as shown in Figure 3(a). From equation (1), we obtain the equation of the second axis:

$$v = -\frac{1}{A_1}.u + (v_0 + \frac{u_0}{A_1}). \tag{2}$$

In order to locate the object in the camera coordinate system we want to match the two vectors $\vec{x}$ and $\vec{i}$, shown in Figure 2. As described in [6] we obtain the three rotation angles:

$$\begin{cases} \theta_x &= \arctan(\frac{\Delta_v}{f.k}) \\ \theta_y &= \arctan(\frac{\Delta_u}{f.k}) \\ \theta_z &= \arctan A_1, \end{cases} \tag{3}$$

where f is the focal distance, k the resolution factor, and (Δ_u, Δ_v) the vector translation between the image center and P_0, as described in Figure 3(a). A result of the localization is shown in the Figure 3(b).

§3. 3D Reconstruction

The proposed 3D reconstruction method is based on the backprojection of two parallel curves. In Section 3.1 we analyse the curvature evolution of a cylinder when it is projected on the image plane. In Section 3.2 we explain how to reconstruct a curved surface based on the detection of parallel curves approximated by parabolic curves and ellipses.

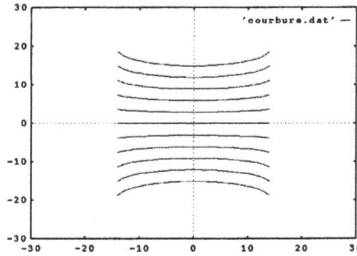

Fig. 4. Evolution of the curvature K, with $R = 100mm$, $f = 60mm$, $d = 500mm$ and $H = 250mm$ for $x = \frac{-5H}{5}, \frac{-4H}{5}, ..., \frac{4H}{5}, \frac{5H}{5}$.

3.1 Evolution of the curvature

When parallel circles from a cylinder are projected on the image plane ellipses with different parameters are obtained. We want to analyse the curvature of points belonging to the common normal as described Section 2. In the camera coordinate system shown in Figure 2, a cylinder is given by

$$\begin{cases} (z - z_c)^2 + y^2 = R^2 \\ -H \leq x \leq H, \end{cases} \tag{4}$$

where R is the radius, $2H$ the height of the cylinder and $z_c = d + R$, where d is the minimal distance between the viewpoint and the cylinder. If we take a value h for x, we obtain one circle C belonging to the cylinder. Let us denote the cone passing through C and the viewpoint O:

$$(\frac{zh}{x} - z_c)^2 + \frac{y^2 h^2}{x^2} = R^2. \tag{5}$$

The intersection of this cone described in (5) and the image plane $z = f$ gives us the equation of an ellipse:

$$\begin{cases} \frac{(x-x_e)^2}{a^2} + \frac{y^2}{b^2} = 1 \\ z = f, \end{cases} \tag{6}$$

where $a = \frac{Rfh}{d^2+2Rd}$, $b = \frac{Rf}{\sqrt{d^2+2Rd}}$ and $x_e = \frac{fh(d+R)}{d^2+2Rd}$ are the three parameters of this ellipse. The curvature K for the point $A(x_a, 0, f)$, where $x_a = x_e + a$, belonging to the ellipse and to the projection of the revolution axis is

$$K = \frac{a}{b^2}. \tag{7}$$

Equations (6) and (7) prove that the curvature depends of the altitude h of the circle in the space. We obtain two linear relationships:

$$\begin{cases} K = \frac{h}{Rf} = \frac{x_a d}{Rf^2} \\ x_a = \frac{fh}{d} = x_e + a, \end{cases} \tag{8}$$

with $-H \leq h \leq H$, and x_a the first coordinate of A. Figure 4 illustrates the curvature evolution for different h.

3.2 Curved surface reconstruction

In this section we explain how to reconstruct a curved surface based on the detection of parallel curves used in the Section 2. Two parallel curves back-projected in the space should be mutually parallel and parallel to the plane $(y, 0, z)$. We can match the intersection P_0 of the two axes with the center of a virtual image in order to obtain a vertical position of the revolution axis [6].

After an edge detection, we should approximate the interesting curves by parabolic curves and ellipses. The equation of a parabolic curve in the image plane is

$$\begin{cases} x = ay^2 + b \\ \quad z = f, \end{cases} \tag{9}$$

where z is the focal distance. The equation of the surface passing through the parabola and the viewpoint O, as shown in Figure 2, is

$$x = \frac{fa}{z}y^2 + \frac{z}{f}b. \tag{10}$$

The projection of this parabola in the space belongs to the surface (10) and to the plane $x = h$. The intersection of the equation (10) with the plane $x = h$ gives us the equation of an ellipse:

$$\begin{cases} \frac{\left(z - \frac{fh}{2b}\right)^2}{\left(\frac{fh}{2b}\right)^2} + \frac{y^2}{\frac{h^2}{4ab}} = 1 \\ \qquad x = h. \end{cases} \tag{11}$$

With $h = \frac{bd}{f}$ and $-H \le x \le H$, we obtain an object of revolution:

$$\begin{cases} \frac{\left(z - \frac{d}{2}\right)^2}{\left(\frac{d}{2}\right)^2} + \frac{y^2}{\frac{bd^2}{4af^2}} = 1 \\ \qquad -H \le x \le H. \end{cases} \tag{12}$$

Finally, to backproject our image on the curved surface we need only the half of the ellipse ($\frac{d}{2} \le z$):

$$\begin{cases} z = \frac{d}{2}\left(1 + \sqrt{1 - \frac{4afy^2}{bd^2}}\right) \\ \qquad -H \le x \le H. \end{cases} \tag{13}$$

Now we present the method of 3D reconstruction when the curves in the image are approximated by ellipses. As shown in Figure 2, the equation of the ellipse in the image is

$$\begin{cases} \frac{(x - x_e)^2}{a^2} + \frac{y^2}{b^2} = 1 \\ \qquad z = f. \end{cases} \tag{14}$$

Thus, the equation of the surface passing through the ellipse and the viewpoint O is

$$\frac{\left(x - \frac{x_e z}{f}\right)^2}{a^2} + \frac{y^2}{b^2} = \frac{z^2}{f^2}. \tag{15}$$

At the distance d we have $x_{a'} = \frac{x_a d}{f}$, where $x_{a'}$ is the backprojection of x_a in the space. The projection of this ellipse in the space belongs to this surface

Fig. 5. (a)–(c) An image mapped on parabolic surfaces with different curvature.

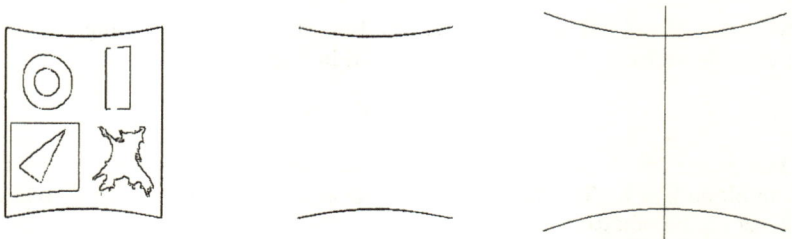

Fig. 6. (a) Edge detection, (b) Selection of two projected parallels, (c) Projection of the revolution axis.

(15) and to the plane $x = \frac{x_a d}{f}$. Their intersection gives us the equation of an ellipse:

$$\begin{cases} \frac{(z-z_c)^2}{a'^2} + \frac{y^2}{b'^2} = 1 \\ x = \frac{x_a d}{f}, \end{cases} \tag{16}$$

where $a' = \frac{da}{x_e - a}$, $b' = \frac{bd}{f}\sqrt{\frac{x_e + a}{x_e - a}}$ and $z_c = \frac{dx_e}{x_e - a}$ are the parameters of this ellipse. Its center is $(\frac{x_a d}{f}, 0, \frac{dx_e}{x_e - a})$. If this equation is a circle, then we should have $a' = b'$:

$$f^2 = \frac{b^2(x_e^2 - a^2)}{a^2}. \tag{17}$$

We obtain a new relationship to find the focal distance f. Thus for each ellipse from the image, we obtain an ellipse in the space with a given scale factor. In the case of our surface having curvature different than zero in only one direction, we derive the equation of this surface from these curves by interpolation based on the least mean squares method. Next we can backproject the image on the surface in order to obtain the representation of the scene in the space.

§4. Simulation Results

In this section we illustrate the efficiency of the reconstruction method on two examples. The first example consists of a synthetic image mapped on different curved surfaces in order to validate our method. The second example consists of a painting on a vault in a church. For this purpose, we map the original

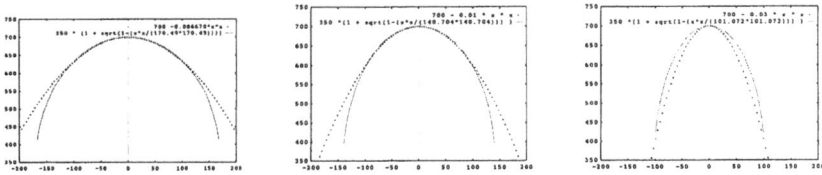

Fig. 7. (a), (b), (c) The reconstruction of the curvature for the three curves.

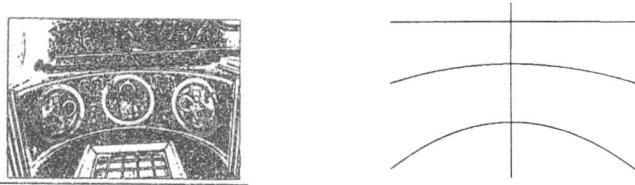

Fig. 8. (a) Edge detection of the Figure 1, (b) Curves and axes.

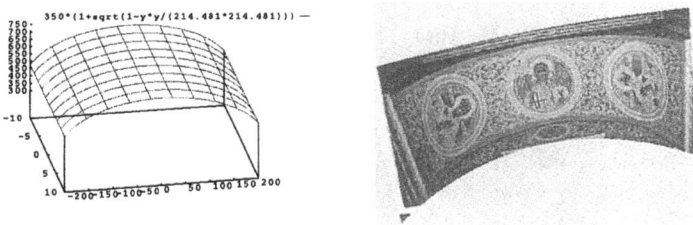

Fig. 9. (a) Reconstruction of the curved surface, (b) Synthetic visualization.

image on three parabolic surfaces, as shown in Figures 5(a), (b), (c). After an edge detection shown in Figure 6(a), we select two parallel curves as shown Figure 6(b). The common normal is shown in Figure 6(c). In the Figures 7(a), (b) and (c), we compare the curvature of the real surface (dotted line) with the surface reconstructed with our method.

Figure 8(a) illustrates the result of an edge detection on the image shown Figure 1. The two axes are shown in the Figure 8(b). The reconstructed surface is shown in Figure 9(a). We are then able to obtain different views of this painting as depicted in Figures 9(b), 10(a) and (b).

§5. Conclusion

In this paper we propose a new approach for curved surface reconstruction based on a single perspective view. In the first step, we need an a priori knowledge about the projection in the image of the surface characterized by only one revolution axis for finding the localization of the curved surface. By using two projected parallels, we are able to locate the curved surface and then to reconstruct the shape in the space. We have shown how to find the curvature of the surface based on detecting parabolas and ellipses. This work

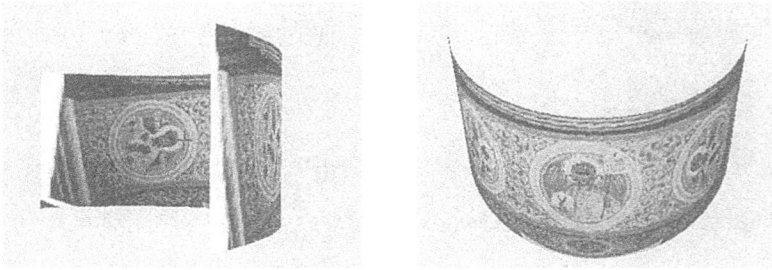

Fig. 10. (a), (b) Two other synthetic visualizations.

is applied for the visualization of painting on columns or vaults. In this case we simulate a camera to obtain new images from different viewpoints. This method will be applied for artwork restoration [2] or mosaicing [1] for a mural painting on a vault or a scene around a column.

References

1. Bors, A., Puech, W., Pitas, I. and Chassery, J.-M., Perspective Distortion Analysis for Mosaicing Images from Cylindrical Surfaces. *International Conference on Acoustics, Speech, and Signal Processing*, Munich, Germany, (1997), in press.
2. Chassery, J.-M., Image Processing and Analysis. A Challenge for Art Paintings. *In Proc. Art and Technology*, Athens, Greece, (1993), 366–377.
3. Glachet, R., Dhome, M. and Lapresté, J.-T., Finding the Perspective Projection of an Axis of Revolution. Pattern Recognition Letters, 12(11), (1991), 693–700.
4. Lavallee, S., Szeliski, R., Brunie, L., Matching 3-D Smooth Surfaces with their 2-D Projections unsing 3-D Distance Maps. *In Lecture Notes in Computer Science*, 708, (1991), 217–239.
5. Ponce, J., Chelberg, D. and Mann, W., Invariant Properties of Straight Homogeneous Generalized Cylinders and their Contours. IEEE Trans. on Pattern Analysis and Machine Intelligence, 11(9), (1989), 951–966.
6. Puech, W., Chassery, J.-M., and Pitas, I., A New Method to Localize a Curved Surface with a Single Perspective View, Pattern Recognition Letters, (1996), submitted.
7. Puech, W. and Chassery, J.-M., Curved Surface Reconstruction Using Monocular Vision. *In 8th European Signal Processing Conference*, Trieste, Italy, (1996), 9–12.
8. Richetin, M., Dhome, M., Lapresté, J.-T., Rives, G., Inverse Perspective Transform Using Zero-Curvature Contour Points: Application to the Localization of Some Generalized Cylinders from a Single view, IEEE Trans. on Pattern Analysis and Machine Intelligence **13** (1991), 185–192.

Zonal Kernels, Approximations and Positive Definiteness on Spheres and Compact Homogeneous Spaces

David L. Ragozin and Jeremy Levesley

Abstract. Several results on approximation or (quasi-)interpolation by translates of radial functions on spheres are generalized to translates of zonal functions on compact sub-manifolds of a Euclidean space which are homogeneous under some (compact) group of orthogonal matrices. Techniques which eliminate any dependence on special facts about spherical harmonics are employed, and the geometric homogeneity is exploited to simplify and clarify the previously known results for spheres.

§1. Introduction

We wish to lay the foundations for a simple approach to radial function approximation and (quasi-) interpolation on spheres, including problems related to positive definiteness. Many recent papers on these problems, e.g. [1, 5,6], seem to rely on special facts about spherical harmonics or coordinate based representations of functions. However, the geometric homogeneity of the sphere and some basic functional analysis can be used to replace the dependence on classical facts about spherical harmonics by more easily remembered facts about ordinary polynomials. This simplifies many derivations. Earlier use of these techniques can be found, *e.g.*, in [2,3]. Our goal is to illustrate the usefulness and simplicity of this approach and to generalize some recent theorems from radial kernels on spheres to their natural analogs for so-called zonal kernels on homogeneous symmetric spaces.

The unit sphere, $S^l \subset \mathbb{R}^{l+1}$ is one of the simplest examples of a *compact symmetric space* embedded in Euclidean space. All of our work will be concerned with functions and kernels defined on similar spaces.

Curves and Surfaces with Applications in CAGD
A. Le Méhauté, C. Rabut, and L. L. Schumaker (eds.), pp. 371–378.

Copyright © 1997 by Vanderbilt University Press, Nashville, TN.
ISBN 0-8265-1293-3.
All rights of reproduction in any form reserved.

Definition 1. *A compact manifold* $M \subset \mathbb{R}^l$ *is an embedded compact homogeneous space provided*

1) *There is a group* $G \subseteq O(l)$ *of orthogonal matrices with* $g \cdot M = M$ *for all* $g \in G$ *which acts transitively on* M, *i.e. for all* $\mathbf{x}, \mathbf{y} \in M$, *there is a* $g \in G$ *with* $g \cdot \mathbf{x} = \mathbf{y}$.

M is an embedded compact symmetric space provided it also satisfies

2) *For all* $\mathbf{x}, \mathbf{y} \in M$ *there is a reflection* $g \in G$ *with* $g \cdot \mathbf{x} = \mathbf{y}$ *and* $g \cdot \mathbf{y} = \mathbf{x}$.

Obvious examples of embedded compact symmetric spaces are the spheres S^{l-1} with $G = O(l)$ and the tori $(S^1)^k = T^k \subset \mathbb{R}^{2k} = \{\mathbb{R}^2\}^k$ with $G = O(2)^k$. Many other examples exist, for instance the Grassmannians, *i.e.*, the space of k-planes in $\mathbb{R}^n$ realized as those $n \times n$ matrices of rank k in $\mathbb{R}^{n \times n}$ of the form $\mathbf{v}_1 \mathbf{v}_1^t + \cdots + \mathbf{v}_k \mathbf{v}_k^t$, with $\{\mathbf{v}_i : 1 \leq i \leq k\}$ any orthonormal basis for a given k-plane. $G = O(n)$ acts by multiplying an orthonormal basis for one plane into that for another plane.

All the relevant geometric structure of a homogeneous M is encoded in the group G and its action as isometries of M. By analogy with radial kernels in $\mathbb{R}^d$ there is much interest in studying approximations on M via "translates" of a single kernel $k(\mathbf{x}, \mathbf{y})$ which reflects this geometric information. For this to be the case, k must by *G-invariant* in the sense of

Definition 2.

1) *A kernel function* k *on* $M \times M$ *is* G*-invariant if* $k(\mathbf{x}, \mathbf{y}) = k(g \cdot \mathbf{x}, g \cdot \mathbf{y})$ *for all* $g \in G$.

2) *A zonal function on* M *with pole* $\mathbf{p} \in M$ *is any function* f *with* $f(h^{-1} \cdot \mathbf{x}) = f(\mathbf{x})$ *for all* $h \in G$ *with* $h \cdot \mathbf{p} = \mathbf{p}$ *and all* $\mathbf{x} \in M$.

Each G-invariant kernel, k, determines a zonal function: $k_{\mathbf{p}}(\mathbf{x}) = k(\mathbf{x}, \mathbf{p})$ and conversely. Since we are assuming M is *symmetric*, any *real valued* G-invariant kernel k is automatically *self-adjoint*. In fact, the use of the *reflection* g from 2) of Definition 1 shows $k(\mathbf{x}, \mathbf{y}) = k(g \cdot \mathbf{y}, g \cdot \mathbf{x}) = k(\mathbf{y}, \mathbf{x})$.

Probably the simplest G-invariant kernels are those that depend only on the distance between the arguments. Specifically, we denote by $\rho(\mathbf{x}, \mathbf{y})$ the *arc length metric* on M obtained by restricting the usual (Riemannian) metric on $\mathbb{R}^l$. Then, since the metrics on $\mathbb{R}^l$ and M are invariant under the orthogonal action of G, any function F on $\mathbb{R}$ determines a G-invariant kernel $k_F(\mathbf{x}, \mathbf{y}) = F(\rho(\mathbf{x}, \mathbf{y}))$ and a zonal function with pole $\mathbf{p}$: $f_{k_F}(\mathbf{x}) = k_F(\mathbf{x}, \mathbf{p})$. In the case of S^{l-1}, all G-invariant kernels have this form, and so coincide with the *radial* kernels. In fact, any G-invariant kernel on a sphere is a radial kernel.

Each M also carries a natural *measure* μ, obtained by splitting the Lebesgue measure on $\mathbb{R}^l$ as a local product $dx = dx_M \times dx_{M^{\perp}}$ of its components tangential and normal to M. Then $d\mu = dx_M$ is just the tangential component. In order to simplify many integration arguments, we shall assume the μ has been normalized so $\int_M d\mu = 1$. Just as for the metric, since G consists of orthogonal matrices, Lebesgue measure and μ are invariant under the action of any $g \in G$.

All the natural function spaces on the homogeneous space M, such as $C(M)$ and $L^2(M;\mu)$, admit a norm preserving action of G, often referred to as translation by g, given by $g{\cdot}f(\mathbf{x}) = f(g^{-1}{\cdot}\mathbf{x})$. Any (continuous) kernel determines an operator on $L^2(M;\mu)$ defined by

$$T_k f(\mathbf{x}) = \int_M k(\mathbf{x}, \mathbf{y}) f(\mathbf{y}) d\mu(\mathbf{y}) . \tag{1}$$

Definition 3. *An operator T on functions is G-equivariant provided it commutes with the action of G so $T(g{\cdot}f) = g{\cdot}T(f)$.*

A G-invariant kernel k determines T_k which is G-equivariant. The verification of this fact follows from (1) and the G-invariance of the measure μ.

One of the most important consequences of the symmetry part of our definition of compact homogeneous spaces is the commutativity of the set of the self-adjoint operators arising from G-invariant kernels.

Proposition 4. *The set $\{T_k : k$ any G-invariant continuous kernel$\}$ is a commuting family of self-adjoint operators.*

Proof: We have already noted how the g in Definition 1 shows $k(\mathbf{x}, \mathbf{y}) = k(\mathbf{y}, \mathbf{x})$. When this is used in the inner product it shows $\langle T_k(f), h \rangle = \langle f, T_k(h) \rangle$ for all $f, h \in L^2(M;\mu)$, i.e., T_k is self-adjoint. Moreover the G-invariance of μ, k_1, and k_2 yields

$$T_{k_1} T_{k_2}(f)(\mathbf{x}) = \int_M \int_M k_1(\mathbf{x}, g^{-1}{\cdot}\mathbf{z}) k_2(g^{-1}{\cdot}\mathbf{z}, \mathbf{y}) f(\mathbf{y}) d\mu(\mathbf{y}) d\mu(\mathbf{z})$$

$$= \int_M \int_M k_1(g{\cdot}\mathbf{x}, \mathbf{z}) k_2(\mathbf{z}, g{\cdot}\mathbf{y}) f(\mathbf{y}) d\mu(\mathbf{y}) d\mu(\mathbf{z})$$

$$= \int_M \int_M k_1(\mathbf{y}, \mathbf{z}) k_2(\mathbf{z}, \mathbf{x}) f(\mathbf{y}) d\mu(\mathbf{y}) d\mu(\mathbf{z})$$

$$= T_{k_2} T_{k_1}(f)(\mathbf{x}),$$

i.e., $T_{k_1} T_{k_2} = T_{k_2} T_{k_1}$. ∎

Now that our basic notation has been defined and we understand some of the significance of the homogeneous structure on M arising from the orthogonal matrices in G, we can pass on to study some of the consequences for (polynomial) approximation and interpolation.

§2. Polynomial Decompositions of L^2 of a Compact Homogeneous Space

In any context where a compact group acts on a space of functions, the finite dimensional subspaces which are invariant under the group play a critical role in analyzing arbitrary functions. In our present setting of embedded compact symmetric spaces the polynomials $\mathcal{P}_n(M)$ provide a very rich collection of such spaces. In fact our main goal in this section is to show that **every** finite dimensional G-invariant subspace of functions (in $L^2(M;\mu)$ or $C(M)$) is contained inside some $\mathcal{P}_n$. For our purposes it is useful to define $\mathcal{P}_n$ without reference to coordinates.

Definition 5. *The polynomials of degree at most n are*

$$\mathcal{P}_n(M) = Span\left\{ \prod_{i=1}^{m} \langle \mathbf{y}_i, \cdot \rangle : \mathbf{y}_i \in \mathbb{R}^l, m \leq n \right\}.$$

In the case of S^{l-1} the orthogonal complement of $\mathcal{P}_{n-1}$ in $\mathcal{P}_n$ determines the classical *spherical harmonics* of *degree* n, so the following definition is quite natural.

Definition 6. *The Harmonics of degree n are given by* $\mathcal{H}_n(M) = \mathcal{P}_n \cap \mathcal{P}_{n-1}^\perp$.

Since a linear function $\langle \mathbf{y}_i, \cdot \rangle$ is transformed by the action of $g \in G$ into another linear function, $\langle g \cdot \mathbf{y}_i, \cdot \rangle$, the following invariance and orthogonal decomposition results are immediate, given the density of the polynomials in $C(M)$.

Proposition 7.
1) *The spaces $\mathcal{P}_n$ and $\mathcal{H}_n$ are G-invariant.*
2) *The span of the Harmonics is dense in $C(M)$.*
3) $L^2(M; \mu) = \bigoplus_{n=0}^{\infty} \mathcal{H}_n$.

Now we need to connect G-invariant subspaces and kernels. Here the critical notion is that of a *reproducing kernel* for a subspace $\mathcal{V}$:

Definition 8. *A kernel $k(\mathbf{x}, \mathbf{y})$ is a reproducing kernel for a (closed) subspace $\mathcal{V}$ provided*

$$T_k(f) = \begin{cases} f, & \text{for all } f \in \mathcal{V} \\ 0, & \text{for all } f \in \mathcal{V}^\perp. \end{cases}$$

For finite dimensional spaces the critical facts about reproducing kernels are as follows.

Theorem 9. *Let $\mathcal{V}$ be any finite dimensional subspace of $L^2(M; \mu)$. Then*

1) $\mathcal{V}$ *has a unique reproducing kernel $k_\mathcal{V}(\mathbf{x}, \mathbf{y}) = \sum_{i=1}^{\dim(\mathcal{V})} f_i(\mathbf{x}) f_i(\mathbf{y})$ where $\{f_i\}$ is any orthonormal basis for $\mathcal{V}$.*

2) $T_{k_\mathcal{V}}$ *is the orthogonal projection on $\mathcal{V}$.*

3) $k_\mathcal{V}$ *is G-invariant if and only if $\mathcal{V}$ is G-invariant.*

4) *If $\mathcal{V}$ is G-invariant, $k_\mathcal{V}(\mathbf{x}, \mathbf{x}) = \dim(\mathcal{V})$, for all $\mathbf{x} \in M$.*

Proof: The formula for $k_\mathcal{V}$ satisfies Definition 8 by the definition of an orthonormal basis. Moreover, this definition is exactly the definition of the orthogonal projection on $\mathcal{V}$. The uniqueness of $k_\mathcal{V}$ follows from the fact that the orthogonal projection is unique. Thus 1) and 2) are proved.

For 3) suppose $k_\mathcal{V}$ is G-invariant. Then for $f \in \mathcal{V}$, $T_{k_\mathcal{V}}(g \cdot f) = g \cdot T_{k_\mathcal{V}}(f) = g \cdot f$, since $T_{k_\mathcal{V}}$ is the orthogonal projection on $\mathcal{V}$. Hence $g \cdot f \in \mathcal{V}$, i.e., $\mathcal{V}$ is G-invariant. Conversely, if $\mathcal{V}$ is G-invariant, then $\{g \cdot f_i\}$ is also an orthonormal basis and the uniqueness of $k_\mathcal{V}$ shows

$$k_\mathcal{V}(g^{-1} \cdot \mathbf{x}, g^{-1} \cdot \mathbf{y}) = \sum_{i=1}^{\dim(\mathcal{V})} g \cdot f_i(\mathbf{x}) g \cdot f_i(\mathbf{y}) = k_\mathcal{V}(\mathbf{x}, \mathbf{y}).$$

Thus $k_\mathcal{V}$ is G-invariant.

For 4) just substitute $\mathbf{x}$ for $\mathbf{y}$ in the previous formula and use the homogeneity of M under G to conclude $k_\mathcal{V}(\mathbf{x}, \mathbf{x})$ is constant. Integration over M yields 4). ∎

The commutativity of G-invariant kernel operators (Proposition 4) immediately leads to

Corollary 10. *Any finite dimensional G-invariant subspace $\mathcal{V}$ is contained in some $\mathcal{P}_n$.*

Proof: Let us denote the orthogonal projection on $\mathcal{H}_n$ by H_n. This is given by a kernel operator with G-invariant kernel and so $H_n T_{k_\mathcal{V}} = T_{k_\mathcal{V}} H_n$. Thus the ranges of the projectors $T_{k_\mathcal{V}} H_n, n = 0, 1, \ldots$, are mutually orthogonal subspaces of $\mathcal{V}$, which sum to $\mathcal{V}$ by Proposition 7. Since only finitely many of these can be non-zero, $\mathcal{V} \subseteq \mathcal{P}_n$, $n = \max\{m : T_{k_\mathcal{V}} H_m \neq 0\}$. ∎

Finally we can refine 3) of Proposition 7 and obtain the most natural analog of Fourier series expansions for (zonal) functions on M.

Theorem 11.

1) Each $\mathcal{H}_n$ has a (unique) decomposition $\mathcal{H}_n = \bigoplus_{i=1}^{d_n} \mathcal{H}_{n,i}$ where each $\mathcal{H}_{n,i}$ is a G-irreducible space of polynomials.

2) Each $f \in L^2(M; \mu)$ has an L^2-expansion $f = \sum_n \sum_i f_{n,i}$ with $f_{n,i} = T_{k_{\mathcal{H}_{n,i}}}(f) \in \mathcal{H}_{n,i}$.

3) If $f = k_\mathbf{p}$ is a zonal function with pole $\mathbf{p}$ associated to a G-invariant kernel k, then
   i) There is a constant $a_{n,i}$ such that $T_k(h) = a_{n,i} h$ for all $h \in \mathcal{H}_{n,i}$.
   ii) Each $f_{n,i}$ is an eigenfunction for T_k with eigenvalue

$$a_{n,i} = \int_M k(\mathbf{x}, \mathbf{p}) k_{\mathcal{H}_{n,i}}(\mathbf{x}, \mathbf{p}) d\mu(\mathbf{x}) / \dim(\mathcal{H}_{n,i}).$$

   iii) $k(\mathbf{x}, \mathbf{p}) = \sum_{n=0}^\infty \sum_{i=0}^{d_n} a_{n,i} k_{\mathcal{H}_{n,i}}(\mathbf{x}, \mathbf{p})$.

Proof: Suppose $\mathcal{H}_n$ is not G-irreducible, *i.e.*, has some proper G-invariant subspace $\mathcal{V}$. Then $\mathcal{V}^\perp \cap \mathcal{H}_n$ is a G-invariant orthogonal complement. Since $\dim(\mathcal{V}) < \dim(\mathcal{H}_n) < \infty$, repetition of this argument must eventually reach the irreducible decomposition claimed in 1).

Now 2) refines 4) of Proposition 7, and follows immediately from the representation of the reproducing kernel for $\mathcal{H}_{n,i}$ given in Theorem 9.

As for 3(i), just note that the self-adjoint G-equivariant operator T_k maps $\mathcal{H}_{n,i}$ into itself since $T_k T_{k_{\mathcal{H}_{n,i}}} = T_{k_{\mathcal{H}_{n,i}}} T_k$ and $T_{k_{\mathcal{H}_{n,i}}}$ is the orthogonal projection on $\mathcal{H}_{n,i}$. If $h \in \mathcal{H}_{n,i}$ is any eigenvector for T_k with eigenvalue $a_{n,i}$, then $g \cdot h$ is an eigenvector for the same eigenvalue as $g \cdot T_k(h) = T_k(g \cdot h)$. Since $\mathcal{H}_{n,i}$ is G-irreducible, $Span\{g \cdot h\} = \mathcal{H}_{n,i}$. This proves 3(i).

For 3(ii) and 3(iii), just observe that $k_{\mathcal{H}_{n,i}}(\mathbf{x},\mathbf{p}) \in \mathcal{H}_{n,i}$, so is an eigenfunction for T_k with eigenvalue $a_{n,i}$ by 3.i. Thus

$$f_{n,i}(\mathbf{x}) = T_{k_{\mathcal{H}_{n,i}}}(k_\mathbf{p})(\mathbf{x}) = T_k(k_{\mathcal{H}_{n,i},\mathbf{p}})(\mathbf{x}) = a_{n,i}k_{\mathcal{H}_{n,i},\mathbf{p}}(\mathbf{x}).$$

If we set $\mathbf{x} = \mathbf{p}$ in the last equality above and use 4) of Theorem 9, we get the formula for $a_{n,i}$. Moreover, the sum in 2) applied to $f = k_\mathbf{p}$ yields 3(iii) since $\mathbf{p}$ is arbitrary. ∎

§3. Approximation and Positive Definiteness Properties of Zonal Kernels

We are now in a position to exploit the Fourier-like decomposition to give alternate approaches to a number of questions about approximations and interpolation by translates of a G-invariant kernel k. The first question we turn to is how can we characterize those k for which

$$\mathcal{K} = Span\{k(\cdot,\mathbf{x}_i): \ \mathbf{x}_i \in M\}, \tag{2}$$

the span of arbitrary translates of k, is a useful space for approximations. The answer is immediate in view of the expansion of k given in Theorem 11 and provides a generalization of the main result in [5]. To simplify notation we shall write $k_{n,i}$ for the reproducing kernel $k_{\mathcal{H}_{n,i}}$.

Theorem 12. *Let the G-invariant kernel k have the Fourier expansion $k = \sum_n \sum_i a_{n,i}k_{n,i}$. Then $\mathcal{K}$, the span of translates of k, is dense (in $C(M)$ or $L^2(M;\mu)$) if and only if $a_{n,i} \neq 0$ for all n,i.*

Proof: The set of translates of the zonal function k_p is G-invariant since $g \cdot k_{\mathbf{x}_i} = k_{g \cdot \mathbf{x}_i}$. Hence the closure of $\mathcal{K}$ is G-invariant. From the orthogonal decomposition in Theorem 11 it follows that $\mathcal{K}$ is dense if and only if $k_{n,i,\mathbf{p}} \in \text{closure}(\mathcal{K})$ for all $\mathbf{p}$. Now if $a_{n,i} = 0$ for some n,i, then $T_k(k_{n,i,\mathbf{p}}) = T_{k_{n,i}}(k_\mathbf{p}) = 0$ from Theorem 11. Writing this as an integral we see $k_{n,i,\mathbf{p}}$ is orthogonal to $\mathcal{K}$. Conversely if $a_{n,i} \neq 0$ for all n,i, the eigenvalue formulae show

$$\int_M k_{n,i}(\mathbf{y},\mathbf{p})k(\mathbf{x},\mathbf{y})d\mu(\mathbf{y})/a_{n,i} = k_{n,i}(\mathbf{x},\mathbf{p}).$$

Riemann sums for the integral are in $\mathcal{K}$ and will converge to $k_{n,i}(\mathbf{x},\mathbf{p})$, uniformly in $\mathbf{x}$ since k is continuous. Thus $\mathcal{K}$ is dense. ∎

The Fourier expansion for G-invariant kernels can also be used to resolve questions relating to *positive definiteness*. Recall

Definition 13. *A kernel $k(\mathbf{x},\mathbf{y})$ is positive definite, resp. strictly positive definite of order n, if for every set of n distinct points $\{\mathbf{x}_i : 1 \leq i \leq n\}$ the matrix $[k(\mathbf{x}_i,\mathbf{x}_j)]$ is positive semi-definite, resp. positive definite. k is positive definite if it is positive definite of order n for all n.*

A key fact for analyzing this notion is

Lemma 14. *Suppose $k_\mathcal{V}$ is the reproducing kernel for some G-invariant subspace. Then for any $c_i \in \mathbb{R}$ and $\mathbf{x}_i \in M$,*

$$\sum_{i,j} c_i c_j k_\mathcal{V}(\mathbf{x}_i, \mathbf{x}_j) = \|\sum_j c_j k_\mathcal{V}(\cdot, \mathbf{x}_j)\|_2^2.$$

Proof: Use the reproducing kernel property and self-adjointness to get

$$k_\mathcal{V}(\mathbf{x}_i, \mathbf{x}_j) = \int_M k_\mathcal{V}(\mathbf{y}, \mathbf{x}_i) k_\mathcal{V}(\mathbf{y}, \mathbf{x}_j) d\mu(\mathbf{y}).$$

Substitute this into the left hand side and factor into the product of sums over i, j inside the integral. This gives the result. ∎

Proposition 15. *If a G-invariant kernel $k = \sum_{n,i} a_{n,i} k_{n,i}$ has all $a_{n,i} \geq 0$ then it is positive definite.*

Proof: Lemma 14 shows $\sum_{i,j} c_i c_j k(\mathbf{x}_i, \mathbf{x}_j) = \sum_{n,i} a_{n,i} \|\sum_m c_m k_{n,i}(\cdot, \mathbf{x}_m)\|_2^2$. Now if all $a_{n,i}$ are non-negative, then so is the sum. ∎

We can also give a direct proof of the generalizations of [6] given in [4, Theorem 4.1 and Corollary 4.2.a].

Theorem 16. *Suppose a G-invariant kernel $k = \sum_{n,i} a_{n,i} k_{n,i}$ is positive definite and has $a_{l,i} > 0$ for $l \leq m$. Then a set $\{\mathbf{x}_i : 1 \leq i \leq N\}$ of distinct points produces a strictly positive definite matrix $[k(\mathbf{x}_i, \mathbf{x}_j)]$ provided*

1) $N \leq 2m+1$ or,
2) $N = 2m+2$ and not all $\mathbf{x}_i$ lie on a single 2-plane.

Proof: If $\sum_{i,j} c_i c_j k(\mathbf{x}_i, \mathbf{x}_j) = 0$, then Lemma 14 shows

$$a_{n,i} \|\sum_l c_l k_{n,i}(\cdot, \mathbf{x}_l)\|_2^2 = 0.$$

In particular, for all $n \leq m$, and $1 \leq i \leq d_n$, $\sum_l c_l k_{n,i}(\cdot, \mathbf{x}_l) = 0$ as $a_{n,i} > 0$. But a sum of these over all $n \leq m$ and all $i \leq d_n$ yields a relation for the reproducing kernel of $\mathcal{P}_m$, $\sum_l c_l k_{\mathcal{P}_m}(\cdot, \mathbf{x}_l) = 0$. If we multiply by any $p \in \mathcal{P}_m$ and integrate, we get

$$\sum_{l=1}^N c_l p(\mathbf{x}_l) = 0, \text{ all } p \in \mathcal{P}_m. \tag{3}$$

In the next lemma we exploit some simple geometry to show that in each of the three cases considered, there exist polynomials $p_l \in \mathcal{P}_m$ with $p_l(\mathbf{x}_i) = \delta_{l,i}$. Plugging these into (3) we see $c_l = 0, 1 \leq l \leq N$, so the strict positive definiteness is proved. ∎

Lemma 17. *Given $\{\mathbf{x}_i\}$ distinct points as in Theorem 16, there exists $p_l \in \mathcal{P}_m$ with $p_l(\mathbf{x}_i) = \delta_{l,i}$.*

Proof: First consider case 1) with $2m + 1$ distinct points on M. First given any $\mathbf{x}_i$ and another pair $\mathbf{x}_j, \mathbf{x}_k$ they can not lie on a single line, since M lies in some sphere, and lines intersect a sphere in at most two points. Then if $\mathbf{p}_{i,j,k}$ is the projection of $\mathbf{x}_i$ on the line from $\mathbf{x}_j$ to $\mathbf{x}_k$, the first degree polynomial $p_{i,j,k}(\mathbf{x}) = \langle \mathbf{x} - \mathbf{p}_{i,j,k}, \mathbf{x}_i - \mathbf{p}_{i,j,k} \rangle / \|\mathbf{x}_i - \mathbf{p}_{i,j,k}\|^2$ is zero at $\mathbf{x}_j, \mathbf{x}_k$ and one at $\mathbf{x}_i$.

Now fix $i = l$ and partition the remaining $2m$ points into m pairs of j, k's. Take the product of the m first degree polynomials constructed above. This will be one at $\mathbf{x}_i$ and zero at the other $\mathbf{x}_j$ as needed.

For case 2) we need only observe that since not all points lie on a plane, and any three points determine a plane, given any $\mathbf{x}_i$ there are three other $\mathbf{x}_j$ whose affine span doesn't include $\mathbf{x}_i$. Then in the construction above replace $\mathbf{p}_{i,j,k}$ by the projection of $\mathbf{x}_i$ on the plane of the other three. Now this allows a partition of $2m + 1$ points into one set of 3 and $m - 1$ of 2 points, with associated first degree polynomials. The corresponding product is one at $\mathbf{x}_i$ and zero at the others. ∎

Acknowledgments. Partial support for the first author provided by a Visiting Fellowship under grant GR/K79710.

References

1. Levesley, J., W. Light, D. L. Ragozin, and X. Sun, Variational theory for interpolation on spheres, Preprint July 1996, 24 pages.

2. Ragozin, D. L., Uniform convergence of spherical harmonic expansions, Math. Ann. **195** (1972), 87–94.

3. Ragozin, D. L., Constructive polynomial approximation on spheres and projective spaces, Trans. Amer. Math. Soc. **162** (1971), 157–170.

4. Ron, A. and X. Sun, Strictly positive definite functions on spheres in Euclidean spaces, Math. Comp. **65** (1996), 1513–1530.

5. Sun, X., The fundamentality of translates of a continuous function on spheres, Numer. Algorithms **8** (1994), 131–134.

6. Xu, Y. and E. W. Cheney, Strictly positive definite functions on spheres, Proc. Amer. Math. Soc. **116** (1992), 977–981.

David L. Ragozin
Department of Mathematics
University of Washington
Seattle, WA 98195-4350
rag@math.washington.edu

Jeremy Levesley
Department of Mathematics
University of Leicester
Leicester LE1 7RH ENGLAND
jl1@mcs.le.ac.uk

Interpolation by Pieces of Euler's Elastica

Klaus-Dieter Reinsch

Abstract. Mechanical splines are plane curves representing the central line of thin rods under different loads with coplanar point forces. Frictionless rotating slides with prescribed location in a plane and coplanar forces acting at fixed points with respect to the arc length of the rod are considered for buckling the rod. In equilibrium, the potential energy of the rod assumes a stationary value. The resulting curve can be expressed piecewise in terms of elliptic functions. If stretching is neglected, the spline consists of pieces of Euler's elastica. In the case of given frictionless rotating slides a subproblem between two neighbouring nodes is formulated in terms of coordinates, tangent and curvature in the endpoints, length of the subarc and energy along the subarc. Piecing together these partial subarcs leads to a system of nonlinear equations. Examples of the solution will be shown for the case that no moments and no tension occur in the first and last slide.

§1. Buckling of an Ideal Elastic Rod

First the buckling of a rod is considered which passes through frictionless rotating slides located at points $(x_j, y_j), j \in \mathbf{J} \subseteq \mathbf{M} := \{1, \ldots, n\}$ of the Euclidean plane and which is loaded at fixed points $s_l, l \in \mathbf{L} \subset \mathbf{M} - \mathbf{J}$ of the central line with respect to its arc lenght in the undeformed case by given point forces $(K_l^{(x)}, K_l^{(y)})$. The following assumptions are made:

1) The material is linearly elastic and uniform.
2) The rod is originally straight and has uniform cross section.
3) The plane of loading coincides with the plane of bending.
4) A cross section (perpendicular to the central line) which is plane in unstressed state remains plane in stressed state.
5) There is no torsion in physical sense.

The arc length s of the undeformed rod, the tangent angle $\theta(s)$ and the stretch $\varepsilon(s)$ of the central line of the deformed rod are geometrically connected with the shape of the deflected central line called *mechanical spline* by

$$\begin{pmatrix} x(s) \\ y(s) \end{pmatrix} = \int^s (1 + \varepsilon(\hat{s})) \begin{pmatrix} \cos \theta(\hat{s}) \\ \sin \theta(\hat{s}) \end{pmatrix} d\hat{s}. \tag{1}$$

Curves and Surfaces with Applications in CAGD
A. Le Méhauté, C. Rabut, and L. L. Schumaker (eds.), pp. 379–386.
Copyright © 1997 by Vanderbilt University Press, Nashville, TN.
ISBN 0-8265-1293-3.
All rights of reproduction in any form reserved.

After buckling, unknown forces $(K_j^{(x)}, K_j^{(y)})$ are acting in the slides (x_j, y_j) passed by the rod at $(x(s_j), y(s_j))$ where s_j is also unknown. The unknown positions of the given point forces $(K_l^{(x)}, K_l^{(y)})$ in the Euclidean plane will be denoted by $(x(s_l), y(s_l)) = (x_l, y_l)$. With these notations, one obtains for the potential energy of the buckled rod with undeformed length L, constant flexural rigidity $\mathcal{E} \mathcal{I}$ and constant positive tensile rigidity $\mathcal{E} \mathcal{A}/\mathrm{L}$, the functional

$$
J(\theta, \varepsilon; \vec{v}) = \int_0^L \left(\frac{1}{2} \mathcal{E} \mathcal{A} \varepsilon^2(s) + \frac{1}{2} \mathcal{E} \mathcal{I} \dot{\theta}^2(s) \right) \mathrm{d}s
$$

$$
- \sum_{\nu=1}^n \left(K_\nu^{(x)}, K_\nu^{(y)} \right) \left[\int_0^{s_\nu} (1 + \varepsilon(s)) \begin{pmatrix} \cos\theta(s) \\ \sin\theta(s) \end{pmatrix} \mathrm{d}s - \begin{pmatrix} x_\nu - x_0 \\ y_\nu - y_0 \end{pmatrix} \right] . \quad (2)
$$

The dot denotes differentiation with respect to the independent variable s. The vector $\vec{v}$ contains the a priori unknown forces and positions. The first integral of the right hand side describes the strain energy of the rod consisting of stretching and bending energy, the sum describes the virtual work done by the external forces.

With the partition $0 = s_0 \le s_1 < s_2 < \cdots < s_n = L$ and the definition

$$
\begin{pmatrix} F_i^{(x)} \\ F_i^{(y)} \end{pmatrix} = \sum_{j=i+1}^n \begin{pmatrix} K_j^{(x)} \\ K_j^{(y)} \end{pmatrix},
$$

$$
J(\theta, \varepsilon; \vec{v}) = \sum_{i=0}^{n-1} \left\{ \int_{s_i}^{s_{i+1}} \left[\frac{1}{2} \frac{1}{\mathcal{E} \mathcal{A}} (\mathcal{E} \mathcal{A} \varepsilon)^2 + \frac{1}{2} \mathcal{E} \mathcal{I} \dot{\theta}^2 - F_i^{(x)} (1 + \varepsilon) \cos\theta \right. \right.
$$

$$
\left. \left. - F_i^{(y)} (1 + \varepsilon) \sin\theta \right] \mathrm{d}s + F_i^{(x)}(x_{i+1} - x_i) + F_i^{(y)}(y_{i+1} - y_i) \right\} . \quad (3)
$$

According to the principle of virtual displacements (see, e.g., [7]) the functional $J(\theta, \varepsilon; \vec{v})$ has a stationary value in the class of admissible variations of the displacements in the equilibrium state of the buckled rod. The necessary condition $\delta J = 0$ yields

- the Euler-Lagrange variational equations

$$
\mathcal{E} \mathcal{I} \ddot{\theta} - F_i^{(x)} (1 + \varepsilon) \sin\theta + F_i^{(y)} (1 + \varepsilon) \cos\theta = 0, \quad (4)
$$

$$
\mathcal{E} \mathcal{A} \varepsilon - F_i^{(x)} \cos\theta - F_i^{(y)} \sin\theta = 0 , \quad s \in]s_i, s_{i+1}[, \quad (5)
$$

- the continuity of $\dot{\theta}$ in $[s_0, s_n]$ due to the Weierstraß - Erdmann corner conditions and the natural boundary conditions $\dot{\theta}(s_1) = \dot{\theta}(s_n) = 0$ (and $\dot{\theta}(s_0) = 0$) – the conditions hold as all tangent directions $\theta(s_i)$ are not prescribed and the point $(x_0, y_0) = (x(s_0), y(s_0))$ describes the free end of the rod,
- the sum of all forces of attack is vanishing (a well known physical fact).

A conclusion of (4) is the equilibrium of moments. The natural boundary conditions are chosen only to simplify matters. The combination of (4) and (5) gives

$$\ddot{\theta} + \frac{f_i}{\mathcal{E}\,\mathcal{I}}\,\sin(\theta - \delta_i) - \frac{f_i^2}{2(\mathcal{E}\,\mathcal{I})\,(\mathcal{E}\,\mathcal{A})}\,\sin\left(2(\theta - \delta_i)\right) = 0, \tag{6}$$

$$f_i = \sqrt{{F_i^{(x)}}^2 + {F_i^{(y)}}^2}\,,\ \delta_i \text{ s.t. } F_i^{(x)} = -f_i\cos\delta_i\,,\ F_i^{(y)} = -f_i\sin\delta_i\,. \tag{7}$$

The general solution of (6) is solved by use of the Jacobian elliptic function tangent amplitude $\mathrm{tn}(u,k)$ and is given for $f_i \neq 0$ by

$$\theta(s) = 2\,\arcsin\frac{A_i\,\mathrm{tn}\left(\sqrt{B_i}(s - \sigma_i), k^{(i)}\right)}{\sqrt{1 + C_i\,\mathrm{tn}^2\left(\sqrt{B_i}(s - \sigma_i), k^{(i)}\right)}} + \delta_i\,,\ \sigma_i \text{ const. of integ.}$$

With $a_i = \frac{f_i}{\mathcal{E}\,\mathcal{I}}$, $b_i = \frac{f_i^2}{2(\mathcal{E}\,\mathcal{I})(\mathcal{E}\,\mathcal{A})}$ and the second const. of integ. c_i, one obtains

$$B_i = \sqrt{a_i^2 + 2b_i^2 + 4b_i c_i}\,,\qquad k_i^2 = \frac{2B_i + 3b_i + 2c_i}{4B_i} \quad\text{(modulus)},$$

$$A_i = \sqrt{\frac{2a_i - b_i + 2c_i}{4B_i}}\,,\qquad C_i = \frac{a_i - 2b_i + B_i}{2B_i}\,.$$

The Jacobian elliptic function tangent amplitude $w = \mathrm{tn}(u,k)$ is the analytic continuation of the inversion of the elliptic integral of the first kind

$$u = \mathrm{tn}^{-1}(w,k) = \int_0^w \frac{d\omega}{\sqrt{(1 + \omega^2)\,(1 + k_c^2\,\omega^2)}}\,,\quad k_c^2 = 1 - k^2\,.$$

Let $\mathrm{sn}(u,k) = \mathrm{tn}(u,k)[1 + \mathrm{tn}^2(u,k)]^{-1/2}$ and $\mathrm{cn}(u,k) = [1 + \mathrm{tn}^2(u,k)]^{-1/2}$.

§2. Euler's Elastica (piecewise)

Equation (5) shows that $\mathcal{E}\,\mathcal{A}\,\varepsilon(s)$ is bounded. If $\mathcal{E}\,\mathcal{A}$ tends to infinity, then the stretch $\varepsilon(s)$ is vanishing, and $J(\theta, \varepsilon; \vec{v})$ tends with the normalisation of the stiffness $\mathcal{E}\,\mathcal{I} = 1$ to

$$\hat{J}(\theta; \vec{v}) = \sum_{i=1}^{n-1} \int_{s_i}^{s_{i+1}} \left\{ \left[\frac{1}{2}\dot{\theta}^2 - F_i^{(x)}\cos\theta - F_i^{(y)}\sin\theta\right] ds \right.$$
$$\left. + F_i^{(x)}(x_{i+1} - x_i) + F_i^{(y)}(y_{i+1} - y_i) \right\}\,. \tag{8}$$

Then the arc length of the bent curve is equal to the arc length of the straight curve. The Euler-Lagrange equation is now the well-known equation of the mathematical pendulum

$$\ddot{\theta} + f_i\,\sin(\theta - \delta_i) = 0\,. \tag{9}$$

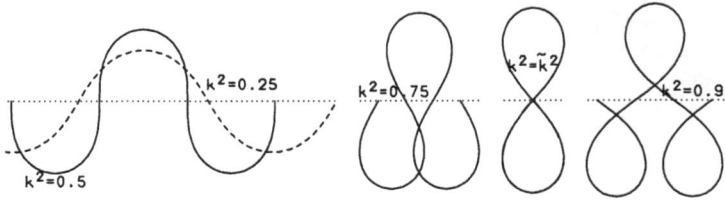

Fig. 1. Euler's elastica for $0 < k^2 < 1$ ($\tilde{k}^2 = 0.82593892\ldots$).

If $f_i \neq 0$, Eq. (9) has the analytical solution (c_i, σ_i constants of integration) :

$$\theta(s) = 2 \arcsin\left(k_i \operatorname{sn}\left(\sqrt{f_i}\,(s - \sigma_i), k_i\right)\right) + \delta_i \, , \ k_i^2 = \frac{c_i + f_i}{2 f_i} \, , \ s \in [s_i, s_{i+1}] \, .$$

The curvature κ of the curve is

$$\kappa(s) = \dot{\theta}(s) = 2\sqrt{f_i}\, k_i \operatorname{cn}\left(\sqrt{f_i}\,(s - \sigma_i), k_i\right) \, . \tag{10}$$

Integration of (9) leads with $\dot{x}(s) = \cos\theta(s)$, $\dot{y}(s) = \sin\theta(s)$ and Eq. (7) to

$$\kappa(s) - \kappa(s_i) = -F_i^{(y)}\big(x(s) - x_i\big) + F_i^{(x)}\big(y(s) - y_i\big) \, , \ s \in [s_i, s_{i+1}] \tag{11}$$

and with $\dot{x}\,\kappa = \cos\theta\,\dot{\theta}$, $\dot{y}\,\kappa = \sin\theta\,\dot{\theta}$ to

$$\frac{1}{2}\,\kappa^2(s) = -F_i^{(x)}\,\dot{x}(s) - F_i^{(y)}\,\dot{y}(s) + c_i \, , \ s \in [s_i, s_{i+1}] \, . \tag{12}$$

Integrating (12) gives

$$\frac{1}{2}\big(E(s) - E(s_i)\big) = \frac{1}{2}\int_{s_i}^{s}\kappa^2(\hat{s})\,\mathrm{d}\hat{s} =$$
$$- F_i^{(x)}\big(x(s) - x_i\big) - F_i^{(y)}\big(y(s) - y_i\big) + c_i(s - s_i) \, . \tag{13}$$

Equations (11) and (12) show with Eq. (10) that the shape between consecutive points s_i and s_{i+1} is a similarly transformed piece of Euler's elasticum with respect to the modulus k_i. For $f_i = 0$ one obtains a straight line section (if $c_i = 0$) or a circular arc (if $c_i > 0$). For $f_i > 0$ it follows

$$\begin{pmatrix} x(s) \\ y(s) \end{pmatrix} = \frac{1}{\sqrt{f_i}}\begin{pmatrix} -\dfrac{F_i^{(x)}}{f_i} & -\dfrac{F_i^{(y)}}{f_i} \\ -\dfrac{F_i^{(y)}}{f_i} & \dfrac{F_i^{(x)}}{f_i} \end{pmatrix}\begin{pmatrix} \xi\big(\sqrt{f_i}(s - \sigma_i), k_i\big) \\ \eta\big(\sqrt{f_i}(s - \sigma_i), k_i\big) \end{pmatrix} + \mathbf{t} \tag{14}$$

with $\mathbf{t} \in \mathbb{R}^2$. The similarity transform was already mentioned in [2].

Euler's elasticum $(\xi, \eta)(u, k)$ with respect to fixed modulus $k^2 \geq 0$ is defined by

$$\xi(u, k) = 2 k^2 \int_0^u \operatorname{cn}^2(\hat{u}, k)\,\mathrm{d}\hat{u} - (2 k^2 - 1)\,u \, , \ \eta(u, k) = 2\,k \operatorname{cn}(u, k) \, , u \in \mathbb{R}$$

and has inflection points for $0 < k < 1$, for $k \geq 1$ it does not. The arc length parameter of the curve $(\xi, \eta)(u, k)$ is u, $-\eta(u, k)$ is the curvature in $(\xi, \eta)(u, k)$.

Fig. 2. Euler's elastica for $k^2 \geq 1$.

For Euler's elastica the following periodicity property holds

$$\Big(\xi(u + p_u, k), \eta(u + p_u, k)\Big) = \Big(\xi(u, k) + \Delta\xi, \, v \cdot \eta(u, k)\Big) \,,$$

$$(p_u, \Delta\xi, v) = \begin{cases} \Big(2\,\mathbb{K}(k), \, 2\,(2\,\mathbb{E}(k) - \mathbb{K}(k)), \, -1\Big) & \text{if } 0 < k^2 < 1 \\[2mm] \Big(\tfrac{2}{k}\,\mathbb{K}(\tfrac{1}{k}), \, 2\,k\,(2\,\mathbb{E}(\tfrac{1}{k}) - (2 - \tfrac{1}{k^2})\,\mathbb{K}(\tfrac{1}{k})), \, 1\Big) & \text{if } k^2 > 1 \end{cases}$$

with the complete elliptic integral of first kind $\mathbb{K}(k)$ and second kind $\mathbb{E}(k)$.

Equations (11)–(14) show that the central line $\big(x(s), y(s)\big)$, $s \in [s_0, s_n]$ is uniquely determined by s_i, $Q_i = \big(x(s_i), \, y(s_i)\big)$, $\theta(s_i)$, $\kappa(s_i)$, $E(s_i)$ with $i = 0, \ldots, n$. If these values are known, each segment $\widetilde{Q_i Q_{i+1}}$ is a straight line or can be piecewisely parametrisized (between inflection points) by the tangent angle θ. Besides some start-up cost the calculation of each point in $\widetilde{Q_i Q_{i+1}}$ requires then one evaluation of a general elliptic integral of the second kind, see [5].

§3. Interpolation by Euler's Elastica

Now we treat the case of frictionless rotating slides located at given points $Q_i = (x_i, y_i)$, $i = 1, \ldots, n$. Equation (8) can be then interpreted as augmented functional of minimizing the bending energy of curves passing Q_i where the constraints are added using Lagrange multipliers (see [4]). In reverse interpretation the Lagrange multipliers are the components of the forces $(F_i^{(x)}, F_i^{(y)})$.

As $\Delta s_i = s_{i+1} - s_i$ is not prescribed, it follows as an additional condition for a stationary value of the potential energy from the Weierstraß - Erdmann corner conditions for the tension parameters c_i that

$$c_i = c\,(= \text{const.}), \qquad i = 0, \ldots, n - 1 \,.$$

By physical meaning the forces $\big(K_i^{(x)}, K_i^{(y)}\big)$ act on the rod at Q_i only in the normal direction.

Using (11) and (13), $F_i^{(x)}$ and $F_i^{(y)}$ can be expressed in terms of $\Delta x_i = x_{i+1} - x_i$, $\Delta y_i = y_{i+1} - y_i$, $S_i = \sqrt{\Delta x_i^2 + \Delta y_i^2}$, $\Delta\kappa_i = \kappa(s_{i+1}) - \kappa(s_i)$, $\Delta E_i = E(s_{i+1}) - E(s_i)$,

$$\begin{pmatrix} F_i^{(x)} \\ F_i^{(y)} \end{pmatrix} = \frac{\Delta\kappa_i}{S_i^2} \begin{pmatrix} \Delta y_i \\ -\Delta x_i \end{pmatrix} - \frac{\frac{1}{2}\,\Delta E_i - c\,\Delta s_i}{S_i^2} \begin{pmatrix} \Delta x_i \\ \Delta y_i \end{pmatrix} \,.$$

Equation (12) becomes with β_i s.t. $\Delta x_i = -S_i \sin \beta_i$, $\Delta y_i = S_i \cos \beta_i$,

$$\frac{S_i}{2} \kappa^2(s) = -\Delta\kappa_i \cos\left(\theta(s)-\beta_i\right) + \left(\frac{1}{2}\Delta E_i - c\,\Delta s_i\right) \sin\left(\theta(s)-\beta_i\right) + c\,S_i \,. \quad (15)$$

One obtains for $|c| < f_i$, i.e. $k_i^2 = \dfrac{f_i + c}{2\,f_i} < 1$ from the inversion of Eq. (2) :

$$s - s_i = \frac{1}{\sqrt{f_i}} \left[2\,l_s \; \mathrm{ell}(\infty, k_c^{(i)}) + \begin{cases} \mathrm{ell}(\infty, k_c^{(i)}) & , \kappa(s) = 0 \\ \mathrm{ell}(t(s), k_c^{(i)}) & , \kappa(s) \neq 0 \end{cases} \right.$$

$$\left. - \begin{cases} -\mathrm{ell}(\infty, k_c^{(i)}) & , \kappa(s_i) = 0 \\ \mathrm{ell}(t(s_i), k_c^{(i)}) & , \kappa(s_i) \neq 0 \end{cases} \right] \quad (16)$$

and from the first equality in (13)

$$E(s) - E(s_i) = 4\,k_i^2\,\sqrt{f_i} \left[2\,l_s \; \mathrm{el2}(\infty, k_c^{(i)}, 1, 0) + \begin{cases} \mathrm{el2}(\infty, k_c^{(i)}, 1, 0), & \kappa(s) = 0 \\ \mathrm{el2}(t(s), k_c^{(i)}, 1, 0), & \kappa(s) \neq 0 \end{cases} \right.$$

$$\left. - \begin{cases} -\mathrm{el2}(\infty, k_c^{(i)}, 1, 0), & \kappa(s_i) = 0 \\ \mathrm{el2}(t(s_i), k_c^{(i)}, 1, 0), & \kappa(s_i) \neq 0 \end{cases} \right] \quad (17)$$

with the complementary modulus $k_c^{(i)^2} = 1 - k_i^2$, the number $l_s \in \mathbb{N}_0$ of inflection points in $\,]\,s_i, s\,[\,$ and

$$t(s) = -\frac{2\left(\Delta\kappa_i \sin\left(\theta(s) - \beta_i\right) + \left(\frac{1}{2}\Delta E_i - c\,\Delta s_i\right) \cos\left(\theta(s) - \beta_i\right)\right)}{S_i\,\kappa(s)\,\sqrt{4\,f_i\,k_c^{(i)^2} + \kappa^2(s)}} \,.$$

For $k_i^2 \geq 1$ one obtains a similar expression by Jacobi's reciprocal modulus transformation. Here the general elliptic integral of the second kind is denoted by

$$\mathrm{el2}(x, k_c, a, b) = \int_0^x \frac{a + b\,\xi^2}{(1 + \xi^2)\,\sqrt{(1 + \xi^2)\,(1 + k_c^2\,\xi^2)}} \; d\xi \,,$$

and it holds for the elliptic integral of the first kind $\mathrm{ell}(x, k_c) = \mathrm{el2}(x, k_c, 1, 1)$. The elliptic integrals can be calculated very quickly by Landen - transformation, see [3].

Equation (15) with $s = s_i$ and $s = s_{i+1}$, Eqs. (16) and (17) with $s = s_{i+1}$ give together with the two natural boundary conditions and a condition for $c\,(s_n - s_1)$ $4n - 1$ nonlinear equations for the $4n - 1$ unknown Δs_1, $\theta(s_1)$, $\kappa(s_1)$, ΔE_1, ..., Δs_i, $\theta(s_i)$, $\kappa(s_i)$, ΔE_i, ..., $\theta(s_n)$, $\kappa(s_n)$ and c. Because of stability considerations (see [1]) it will be assumed for the numerical calculation that at most one inflection point exists in $[s_i, s_{i+1}]$. The system of nonlinear equations is solved by the modified Newton method. The

Jacobi matrix of the linearised system is computed together with the right hand side.

According to the given problem $s_n - s_1$ is not prescribed. The variation $\delta J = 0$ leads to $c = 0$ s.t. $k_i = k_c^{(i)} = \frac{1}{\sqrt{2}}$ in all intervalls $[s_i, s_{i+1}]$.

The vanishing of c stands for no tension by physical meaning. For $c = 0$ the quantities Δs_i are not needed. It remains a system of $3n - 1$ nonlinear equations for $3n - 1$ variables.

Since one can hardly find good initial data for solving the system of nonlinear equations if there are acute angles in the polygon $\left\{Q_i^{(0)}\right\}_i$, one starts with points $\left\{Q_i^{(0)}\right\}_i$ with the properties $\left\|\overrightarrow{Q_i^{(0)}, Q_{i+1}^{(0)}}\right\|_2 = \left\|\overline{Q_i, Q_{i+1}}\right\|_2$ and obtuse angle $\measuredangle\left(\overrightarrow{Q_i^{(0)}, Q_{i-1}^{(0)}}, \overrightarrow{Q_i^{(0)}, Q_{i+1}^{(0)}}\right)$ and draws a homotopy from $\left\{Q_i^{(0)}\right\}_i$ to $\{Q_i\}_i$ (if a solution for the given problem exists) by changing the relevant angles. The step size is controlled by the estimated change of $\Delta E_i^{(k-1)} - \Delta E_i^{(k)}$.

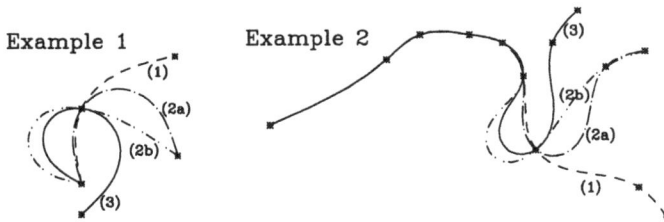

Example 1   Example 2

In Examples 1 and 2 the goal is the mechanical spline (3). The interpolation points are always marked with an asterisk. Each curve (1) – (3) is a mechanical spline. The homotopy starts with the point set due to curve (1) and is going to the point set due to curve (2a). Here the homotopy step size is below a given bound. Therefore a physically motivated "perturbation step" is carried out and leads to a second also physically true solution (2b) due to the same point set. From this solution, the homotopy tends to the point set due to curve (3).

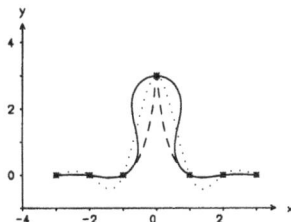

Example 3

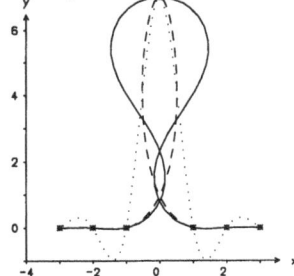

Example 4

In Examples 3 and 4, the solid line represents the mechanical spline passing the interpolation points marked with an asterisk, the dotted line represents the cubic spline, and the dashed line the curve $(\hat{\xi}, \hat{\eta})$ where $\hat{\xi}(t)$ resp. $\hat{\eta}(t)$ are cubic splines interpolating $(t_i, x_i)_{i=1,\ldots,n}$ resp. $(t_i, y_i)_{i=1,\ldots,n}$ with $t_1 = 0$ and $t_{i+1} = t_i + S_i$, $i = 1, \ldots, n-1$. All curves are satisfying natural boundary conditions. These two examples show that already in relatively simple cases of point sequences with $x_1 < x_2 < \cdots < x_n$ the interpolating mechanical spline *cannot* be defined as function $y(x)$. Especially the mechanical spline and the natural cubic spline have nothing in common here.

All examples demonstrate the difficulties which may occur in the computation of the mechanical spline interpolating a fixed sequence of points.

A fundamental difficulty is the uncertainty of the existence of a solution; e.g. for the sequence $(1,0), (2,0), (0,2), (0,1)$ no mechanical spline exists without additional constraints. This difficulty can be removed by an upper bound for $s_n - s_1$ which determines $c(s_n - s_1)$ and especially the tension parameter c. But a large upper bound for $s_n - s_1$ may not lead to a unique solution, see Examples 1 and 2.

Acknowledgments. This research was partly supported by the Deutsche Forschungsgemeinschaft through the Sonderforschungsbereich 255 Transatmosphärische Flugsysteme. The author is indebted to Prof. Dr. R. Bulirsch for his longstanding support and encouragement.

References

1. Born, M., Untersuchungen über die Stabilität der elastischen Linie in der Ebene und im Raum. *Inauguraldissertation und gekrönte Preisschrift,* Göttingen, 1906.

2. Birkhoff, G., H. Burchard, and D. Thomas, Nonlinear interpolation by splines, pseudosplines and elastica, Research Publication 468, General Motors Research Laboratories, Michigan, 1965.

3. Bulirsch, R., Numerical calculation of elliptic integrals and elliptic functions *Num. Math.* 7 (1965), 78–90.

4. Malcolm, M., On the computation of nonlinear spline functions, *SIAM J. Numer. Anal.* Vol 14 No 2 (1977), 254–282.

5. Reinsch, K.-D., Numerische Berechnung von Biegelinien in der Ebene, *Dissertation*, Technische Universität München, 1981.

6. Reinsch, K.-D., Fairing of ship lines with mechanical splines, *Proc. 4th Europ. Conf. on Math. in Industry*, St. Wolfgang, Austria, 1989, 379–383.

7. Sokolnikoff, I.S., *Mathematical Theory of Elasticity*, Mc Graw-Hill Book Comp. Inc., 1956.

Klaus-Dieter Reinsch
Mathematisches Institut, Technische Universität
D-80290 München, GERMANY
kladire@mathematik.tu-muenchen.de

A Knot Insertion Algorithm
for Weighted Cubic Splines

Mladen Rogina

Abstract. One of the main reasons why polynomial splines play an important role in computer–aided design as well as in diverse areas of approximation theory and numerical analysis is the fact that they can be represented as linear combination of B-splines. There are nice and stable algorithms for evaluation of such splines and their derivatives and integrals. The well-known tools of knot insertion and degree raising can be enhanced by introducing still more additional parameters, and relaxing the continuity conditions at the knots by prescribing jumps in their derivatives. The purpose of this paper is to derive recurrence formulae for some related B-splines, and to exploit the underlying connection with the theory of Chebyshev splines. The cubic version of the jump spline is then recognized as Foley's ν–spline, often used in minimizing functionals like
$$V(f) := \sum_{i=1}^{n}(w_i \int_{t_i}^{t_{i+1}}[D^2 f(t)]^2 dt + \nu_i \int_{t_i}^{t_{i+1}}[Df(t)]^2 dt), \; \nu_i \geq 0, \; w_i > 0.$$
The parametric version is often used as a polynomial alternative to the exponential spline in tension in computer–aided geometric design. It is shown how the associated B-splines can be calculated by a knot–insertion algorithm, and this in turn motivates a definition of certain generalized discrete splines.

§1. Jump Splines and Chebyshev Theory

Let us choose the knot sequence $\Delta = \{x_0, \ldots, x_{k+1}\}$ and for a given integer n and multiplicity vector $\boldsymbol{m} = (n_1, \ldots, n_k)^T$ let us define an extended partition $\{t_1 \ldots t_{2n+k}\}$ [10] in the usual way:

$$t_1 = \cdots = t_n = x_0$$

$$t_{n+k+1} = \cdots = t_{2n+k} = x_{k+1}$$

$$t_{n+1} \leq \cdots \leq t_{n+k} = \underbrace{x_1, \ldots, x_1}_{n_1}, \ldots, \underbrace{x_k, \ldots, x_k}_{n_k},$$

where n_i are integers satisfying $1 \leq n_i < n$. Let us assume that positive constants $\{p_j^i\}_{j=2,i=0}^{n-1,k}$ are given. We define the *jump* or *weighted spline* s of order

Curves and Surfaces with Applications in CAGD
A. Le Méhauté, C. Rabut, and L. L. Schumaker (eds.), pp. 387–394.

387

Copyright ⊚ 1997 by Vanderbilt University Press, Nashville, TN.
ISBN 0-8265-1293-3.
All rights of reproduction in any form reserved.

$n > 2$, associated with *weights* $\{p_j^i\}$, as a continuous piecewise polynomial of degree $n - 1$ having jumps in derivatives according to

$$(p_j^i D)(p_{j-1}^i D) \cdots (p_2^i D)s(x_i+) - (p_j^{i-1} D)(p_{j-1}^{i-1} D) \cdots (p_2^{i-1} D)s(x_i-) = 0,$$

$$j = 2, \ldots n - n_i, \quad i = 1, \ldots k; \qquad D := \frac{d}{dx}. \tag{1}$$

We could allow multiplicities of order $n_i = n$, but then we obtain discontinuous polynomial splines, in which weights play no role. The case $n_i = n - 1$ we treat as the continuity condition.

Let us define piecewise constant functions $p_j : [x_0, x_{k+1}] \rightarrow \mathbb{R}$, $j = 2, \ldots n - 1$, in such a way that

$$p_j |_{[x_i, x_{i+1})} := p_j^i, \qquad \text{for} \quad i = 0, \ldots, k - 1,$$
$$p_j |_{[x_k, x_{k+1}]} := p_j^k. \tag{2}$$

We may extend these functions to the whole of $\mathbb{R}$ by defining $p_j(t) = p_j^0$ for $t \leq x_0$, and $p_j(t) = p_j^k$ for $t > x_{k+1}$ and consider positive measures on intervals $\delta \subset \mathbb{R}$ generated by densities $1/p_j$:

$$d\sigma_j(\delta) := \int_\delta \frac{dt}{p_j(t)}, \qquad j = 2, \ldots n - 1.$$

Let $d\boldsymbol{\sigma} := (d\sigma_2, \ldots d\sigma_{n-1})$ be the *measure vector*. The positivity assumption enables us to think of jump splines as being in the special Chebyshev spline space [10] $\mathcal{S}(n, d\boldsymbol{\sigma}, \boldsymbol{m}, \Delta)$, associated with a CCT–system $\{1, u_2, \ldots u_n\}$:

$$u_2(x) = \int_0^x d\sigma_2(\eta_2)$$
$$u_i(x) = \int_0^x d\sigma_2(\eta_2) \int_0^{\eta_2} d\sigma_3(\eta_3) \ldots \int_0^{\eta_{n-2}} d\sigma_{n-1}(\eta_{n-1}) \int_0^{\eta_{n-1}} d\eta_n$$
$$i = 3, \ldots, n \qquad (n \geq 3).$$

It is easy to prove that the jump conditions (1) are equivalent to the continuity of *generalized derivatives* $L_{j,d\boldsymbol{\sigma}}$ [11] across the knots. For $p_j \equiv 1$, $j = 2, \ldots$, $n - 1$, we obtain the space of polynomial splines, which we shall hence refer to as $\mathcal{P}(n, \boldsymbol{m}, \Delta)$.

The existence of the local basis consisting of B-splines $\{T_{i,d\boldsymbol{\sigma}}^n\}$, $i = 1, \ldots, n + \sum_{i=1}^k n_i$, which make a partition of unity, thus follows from the theory of Chebyshev splines [11]. There are other benefits from Chebyshev approach, one of them being the derivative formula for Chebyshev splines [8,6]:

$$L_{1,d\boldsymbol{\sigma}} T_{i,d\boldsymbol{\sigma}}^n(x) = \frac{T_{i,d\tilde{\boldsymbol{\sigma}}}^{n-1}(x)}{C_{n-1}(i)} - \frac{T_{i+1,d\tilde{\boldsymbol{\sigma}}}^{n-1}(x)}{C_{n-1}(i+1)} \tag{3}$$

where

$$C_{n-1}(i) := \int_{t_i}^{t_{i+n-1}} T_{i,\tilde{\sigma}}^{n-1}(u)d\sigma_2,$$

and $d\tilde{\sigma} := (d\sigma_3, \ldots, d\sigma_{n-1})$ is the *reduced measure vector*. One is tempted to develop a recurrence for calculating Chebyshev B-splines by integration of the right side in the (3) with respect to the $d\sigma_2$, but unfortunately it yields a very unstable algorithm. We aim to show that for jump splines (at least of low orders), a derivative formula can be combined with knot insertion to produce numerically stable formulae for B-splines.

§2. Parabolic Jump Splines

For $n = 3$ there is only one piecewise constant function p_2, determining jumps in first derivatives. For notational purposes let us omit the index, and let $d\sigma_p$ be the measure with density $1/p$. The CCT–system consists of three functions:

$$u_1(x) = 1$$

$$u_2(x) = \int_0^x d\sigma_p(\eta_2)$$

$$u_3(x) = \int_0^x d\sigma_p(\eta_2) \int_0^{\eta_2} d\eta_3.$$

The B-splines in $\mathcal{S}(3, (d\sigma_p), m, \Delta)$ can be constructed by a de Boor–Cox type recurrence relation [8], which gives a parabolic jump B-spline $T_{i,p}^3$ as a combination of two linear "hat" functions, that is, B-splines in $\mathcal{P}(2, m, \Delta)$:

$$T_{i,p}^3(x) = \frac{d\sigma_p(t_i, x)}{d\sigma_p(t_i, t_{i+2})} B_i^2(x) + \frac{d\sigma_p(x, t_{i+3})}{d\sigma_p(t_{i+1}, t_{i+3})} B_{i+1}^2(x). \tag{4}$$

The recurrence also follows from the more general considerations [3,7], and, (for jumps in first derivatives only), can be generalized to arbitrary order. There is another way to compute T_i^3 by raising the multiplicity of the knots; the parabolic jump spline can thus be represented as a linear combination of three continuous parabolic B-splines which do not have to satisfy the jump condition. The same argument will be reapplied in Section 3 to construct the cubic version of the jump spline.

Lemma 1. *Let $T_{j,p}^3 \in \mathcal{S}(3, (d\sigma_p), m, \Delta)$ be a parabolic jump B-spline associated with the multiplicity vector $m = (1, \ldots 1)^T$, and let us assume that $\tilde{B}_j^3 \in \mathcal{P}(3, \tilde{m}, \Delta)$ are the parabolic B-splines associated with the multiplicity vector $\tilde{m} = (2, \ldots 2)^T$ on the same knot sequence. If $\{t_1, \ldots t_{k+6}\}$ and $\{\tilde{t}_1, \ldots \tilde{t}_{2k+6}\}$ are the associated extended partitions, and r an index such that $t_j = \tilde{t}_r < \tilde{t}_{r+1}$, then for $j = 1, \ldots k + 3$,*

$$T_{j,p}^3(x) = \frac{d\sigma_p(t_j, t_{j+1})}{d\sigma_p(t_j, t_{j+2})} \tilde{B}_r^3(x) + \tilde{B}_{r+1}^3(x) + \frac{d\sigma_p(t_{j+2}, t_{j+3})}{d\sigma_p(t_{j+1}, t_{j+3})} \tilde{B}_{r+2}^3(x).$$

Proof: Since $\mathcal{S}(3, (d\sigma_p), \boldsymbol{m}, \Delta) \subset \mathcal{P}(3, \tilde{\boldsymbol{m}}, \Delta)$, we conclude that $\delta(i) \in \mathbb{R}$ exist such that $T_{j,p}^3(x) = \sum_{i=r-2}^{r+4} \delta(i) \tilde{B}_i^3(x)$. For $x \in [t_j, t_{j+1}]$ we get

$$T_{j,p}^3(x) = \delta(r-2)\tilde{B}_{r-2}^3(x) + \delta(r-1)\tilde{B}_{r-1}^3(x) + \delta(r)\tilde{B}_r^3(x).$$

As $T_{j,p}^3(t_j) = 0$, $\tilde{B}_{r-2}^3(t_j) \neq 0$, we have $\delta(r-2) = 0$, and, by taking the first generalized derivative (3), $(pD)T_{j,p}^3(t_j+) = 0$, $(pD)\tilde{B}_{r-1}^3(t_j+) \neq 0$ imply that $\delta(r-1) = 0$. We conclude that

$$T_{j,w}^3(t_{j+1}) = \delta(r)\tilde{B}_r^3(t_{j+1}).$$

The spline $T_{j,p}^3$ can be evaluated by (4) at $x = t_{j+1}$, and, since $\tilde{B}_r^3(t_{j+1}) = 1$, it follows that

$$\delta(r) = \frac{d\sigma_p(t_j, t_{j+1})}{d\sigma_p(t_j, t_{j+2})}. \tag{5}$$

By taking $x \in (t_{j+1}, t_{j+2})$, we similarly obtain

$$\delta(r+4) = \delta(r+3) = 0, \quad \delta(r+2) = \frac{d\sigma_p(t_{j+2}, t_{j+3})}{d\sigma_p(t_{j+1}, t_{j+3})}.$$

Thus, for $x \in (t_{j+1}, t_{j+2})$

$$T_{j,p}^3(x) = \delta(r)\tilde{B}_r^3(x) + \delta(r+1)\tilde{B}_{r+1}^3(x) + \delta(r+2)\tilde{B}_{r+2}^3(x).$$

for $x \in (t_{j+1}, t_{j+2})$, and the remaining unknown coefficient $\delta(r+1)$ can be determined by comparing derivatives at $x = t_{j+1}$:

$$\frac{2}{d\sigma_p(t_j, t_{j+2})} = \frac{-2\delta(r)}{d\sigma_p(t_{j+1}, t_{j+2})} + \frac{2\delta(r+1)}{d\sigma_p(t_{j+1}, t_{j+2})}.$$

Upon substitution of $\delta(r)$ from (5), we have

$$\delta(r+1) = \frac{1}{d\sigma_p(t_j, t_{j+2})}[d\sigma_p(t_j, t_{j+1}) + d\sigma_p(t_{j+1}, t_{j+2})] = 1. \quad \blacksquare$$

For $p \equiv 1$ we obtain the well-known knot insertion formula [2] :

$$B_j^3(x) = \frac{t_{j+1} - t_j}{t_{j+2} - t_j}\tilde{B}_r^3(x) + \tilde{B}_{r+1}^3(x) + \frac{t_{j+3} - t_{j+2}}{t_{j+3} - t_{j+1}}\tilde{B}_{r+2}^3(x).$$

We may write $T_{j,p}^3(x) = \sum_i \delta_{j,p}^3(i)\tilde{B}_i^3(x)$, and call the mapping $i \mapsto \delta_{j,p}^3(i)$ a *discrete parabolic jump spline*. We assume that $\delta_{j,p}^3(i) = 0$ for $i \notin \{r, r+1, r+2\}$, where r is as in Lemma 1.

If we have two measures, $d\sigma_p$ and $d\sigma_w$ generated by the weights $\{p_i\}_0^k$ and $\{w_i\}_0^k$ as in (1), the following technical lemma shows how to calculate the respective integrals of jump splines.

Lemma 2. *Let $T_{i,w}^3 \in \mathcal{S}(3, (d\sigma_w), \boldsymbol{m}, \Delta)$ be the parabolic jump B-spline, and $C_3(i) := \int_{t_i}^{t_{i+3}} T_{i,w}^3 d\sigma_p$. Then for $i = 1, \ldots 3 + k$,*

$$C_3(i) = \frac{1}{3} \sum_j \delta_{i,w}^3(j) d\sigma_p(\tilde{t}_j, \tilde{t}_{j+3}), \qquad (6)$$

where $\delta_{i,w}^3$ is a discrete parabolic jump spline on the refined knot sequence associated with the multiplicity vector $\tilde{\boldsymbol{m}} = (n_i)^T$, $n_i \geq 2$.

Proof: We just integrate the representation formula for $T_{i,w}^3$ in Lemma 1 and use the fact that spline $\tilde{B}_i^3$ has a support of at most two intervals, its integral being of a very simple form: $\int_{t_j}^{\tilde{t}_{j+3}} \tilde{B}_i^3 d\sigma_p = d\sigma_p(\tilde{t}_j, \tilde{t}_{j+3})$. ∎

§3. Complete Cubic Jump Splines

We know how to calculate the parabolic jump splines, and let us further investigate the more interesting case of continuous piecewise cubics satisfying jump conditions for first and second derivatives:

$$(p^i D)s(x_i+) - (p^{i-1} D)s(x_i-) = 0 \quad i = 1, \ldots k. \qquad (7)$$

$$(w^i D)(p^i D)s(x_i+) - (w^{i-1} D)(p^{i-1} D)s(x_i-) = 0 \quad i = 1, \ldots k. \qquad (8)$$

Let $d\sigma_w$, $d\sigma_p$ be the measures with densities $1/p$, $1/w$ as in the general theory. The CCT–system $\{1, u_2, u_3, u_4\}$ associated with these measures is

$$u_2(x) = \int_0^x d\sigma_p(t_2)$$

$$u_3(x) = \int_0^x d\sigma_p(t_2) \int_0^{t_2} d\sigma_w(t_3)$$

$$u_4(x) = \int_0^x d\sigma_p(t_2) \int_0^{t_2} d\sigma_w(t_3) \int_0^{t_3} dt_4.$$

The cubic splines we wish to consider are Chebyshev in $\mathcal{S}(4, (d\sigma_p, d\sigma_w), \boldsymbol{m}, \Delta)$; the jump conditions (7), (8) are equivalent to the smoothnes of generalized derivatives (pD) and $(wD)(pD)$ across the knots.

Theorem 3. *Let $T_{j,p,w}^4$ and $\tilde{T}_{j,p}^4$ be the B-splines in $\mathcal{S}(4, (d\sigma_p, d\sigma_w), \boldsymbol{m}, \Delta)$, $\mathcal{S}(4, (d\sigma_p), \tilde{\boldsymbol{m}}, \Delta)$ respectively, where the multiplicity vectors are defined as in Lemma 1. There exist $\delta_{j,p,w}^4(i) > 0$ such that $T_{j,p,w}^4(x) = \sum_{i=r}^{r+3} \delta_{j,p,w}^4(i) \tilde{T}_{i,w}^4(x)$, where $r := r_j$ is such that $t_j = \tilde{t}_{r_j} < \tilde{t}_{r_j+1}$. If the associated extended partitions are $\{t_1, \ldots t_{k+8}\}$ and $\{\tilde{t}_1, \ldots \tilde{t}_{2k+8}\}$, then $\delta_{j,p,w}^4(i)$, $i = r, \ldots r + 4$*

can be expressed in terms of a discrete parabolic jump spline:

$$\delta^4_{j,p,w}(r) = \frac{\gamma^3_{j,p,w}(r)}{\|\gamma^3_{j,p,w}\|}$$

$$\delta^4_{j,p,w}(r+1) = \frac{\gamma^3_{j,p,w}(r) + \gamma^3_{j,p,w}(r+1)}{\|\gamma^3_{j,p,w}\|}$$

$$\delta^4_{j,p,w}(r+2) = \frac{\gamma^3_{j+1,p,w}(r+3) + \gamma^3_{j+1,p,w}(r+4)}{\|\gamma^3_{j+1,p,w}\|}$$

$$\delta^4_{j,p,w}(r+3) = \frac{\gamma^3_{j+1,p,w}(r+4)}{\|\gamma^3_{j+1,p,w}\|},$$

where $\gamma^3_{j,p,w}(i) = \delta^3_{j,w}(i) d\sigma_p(\tilde{t}_i, \tilde{t}_{i+3})$, and $\|\gamma^3_{l,p,w}\| = \sum_i \gamma^3_{l,p,w}(i)$ for $l = j$ and $l = j+1$.

Proof: Since $\mathcal{S}(4, (d\sigma_p, d\sigma_w), \boldsymbol{m}, \Delta) \subset \mathcal{S}(4, (d\sigma_p), \tilde{\boldsymbol{m}}, \Delta)$, we can find $\delta^4_{j,p,w}(i)$ such that $T^4_{j,p,w}(x) = \sum_{i=r-2}^{r+6} \delta^4_{j,p,w}(i) \tilde{T}^4_{i,p}(x)$. After applying the first generalized derivative $L_{1,d\sigma_p} = (pD)$ (3) to both sides, the change of the summation index leads to

$$\frac{T^3_{j,w}(x)}{C_3(j)} - \frac{T^3_{j+1,w}(x)}{C_3(j+1)} = \sum_i \frac{\delta^4_{j,p,w}(i) - \delta^4_{j,p,w}(i-1)}{\tilde{C}_3(i)} \tilde{T}^3_{i,p}(x),$$

where $C_3(i) = \int_{t_i}^{t_{i+3}} T^3_{i,w} d\sigma_p$ and $\tilde{C}_3(i) = \int_{\tilde{t}_i}^{\tilde{t}_{i+3}} \tilde{B}^3_i d\sigma_p$, since $\tilde{T}^3_{i,p} \equiv \tilde{B}^3_i$. We know that $\tilde{C}_3(i) = \frac{1}{3} d\sigma_p(\tilde{t}_i, \tilde{t}_{i+3})$, and, by Lemma 2, $C_3(i) = \sum_l \delta^3_{i,w}(l) \tilde{C}_3(l)$ for $i = j, j+1$. Splines $T^3_{j,w}$, $T^3_{j+1,w}$ can be expanded in $\mathcal{P}(3, \tilde{m}, \Delta)$ as in Lemma 1. By comparing the coefficients, we obtain a linear system for $\delta^4_{j,p,w}(i)$:

$$\frac{\delta^4_{j,p,w}(i) - \delta^4_{j,p,w}(i-1)}{d\sigma_p(\tilde{t}_i, \tilde{t}_{i+3})} = \frac{\delta^3_{j,w}(i)}{\sum_l \delta^3_{j,w}(l) d\sigma_p(\tilde{t}_l, \tilde{t}_{l+3})} - \frac{\delta^3_{j+1,w}(i)}{\sum_l \delta^3_{j+1,w}(l) d\sigma_p(\tilde{t}_l, \tilde{t}_{l+3})}.$$

On subtitution of $\gamma^3_{j,p,w}$, the common factor $d\sigma_p(\tilde{t}_i, \tilde{t}_{i+3})$ cancels and the ith-equation is

$$\delta^4_{j,p,w}(i) - \delta^4_{j,p,w}(i-1) = \frac{\gamma^3_{j,p,w}(i)}{\|\gamma^3_{j,p,w}\|} - \frac{\gamma^3_{j+1,p,w}(i)}{\|\gamma^3_{j+1,p,w}\|}.$$

The two–diagonal, 5th order matrix for $\delta^4_{j,p,w}(i)$ is easily inverted, and the proof follows. ∎

For $p \equiv 1$, $w \equiv 1$ we obtain the well-known formula expressing a C^2 cubic B-spline as a linear combination of C^1 cubic B-splines. The most interesting case, however, appears to be when only $p \equiv 1$. The remaining weights w_i can be determined in applications [4,5] such as monotone and/or convex data fitting.

The mapping $i \mapsto \delta^4_{j,p,w}(i)$ is a *discrete cubic jump spline*; it has other nice properties of ordinary discrete splines which we aim to prove.

Lemma 4. *Let Δ_j be Newton's forward difference with step 1:*

$$\Delta_j f(j) := f(j+1) - f(j),$$

$\{t_j\}$, $\{\tilde{t}_j\}$ *the knot partition and its refinement as in Theorem 3, and* $\delta^4_{i,p,w}(j)$ *a disrete cubic B–spline. Then for all* j, $t_4 \le \tilde{t}_j < t_{k+5}$,

$$\Delta_j \delta^4_{i,p,w}(j) = \tilde{C}_3(j+1)\left[\frac{\delta^3_{i,w}(j+1)}{C_3(i)} - \frac{\delta^3_{i+1,w}(j+1)}{C_3(i+1)}\right],$$

where $\tilde{C}_3$ i C_3 are defined as in Theorem 3.

Proof: By definition of discrete cubic spline $T^4_{i,p,w}(x) = \sum_j \delta^4_{i,p,w}(j)\tilde{T}^4_{j,p}(x)$. The derivative formula (3) yields

$$\frac{1}{C_3(i)}T^3_{i,w}(x) - \frac{1}{C_3(i+1)}T^3_{i+1,w}(x) = \sum_j \delta^4_{i,p,w}(j)\left[\frac{\tilde{B}^3_j(x)}{\tilde{C}_3(j)} - \frac{\tilde{B}^3_{j+1}(x)}{\tilde{C}_3(j+1)}\right].$$

The spline on the left side is in $\mathcal{S}(3, d\sigma_p, \boldsymbol{m}, \Delta)$ and can be expanded in terms of $\tilde{B}^3_j \in \mathcal{P}(3, \boldsymbol{m}, \Delta)$, the expansion coefficients following from Lemma 1. Therefore,

$$\sum_j \left[\frac{\delta^3_{i,w}(j)}{C_3(i)} - \frac{\delta^3_{i+1,w}(j)}{C_3(i+1)}\right]\tilde{B}^3_j(x) = \sum_j \frac{\delta^4_{i,p,w}(j) - \delta^4_{i,p,w}(j-1)}{\tilde{C}_3(j)}\tilde{B}^3_j(x),$$

and by comparing the coefficients we obtain

$$\frac{\delta^4_{i,p,w}(j) - \delta^4_{i,p,w}(j-1)}{\tilde{C}_3(j)} = \frac{\delta^3_{i,w}(j)}{C_3(i)} - \frac{\delta^3_{i+1,w}(j)}{C_3(i+1)}. \quad \blacksquare$$

Theorem 5. *The discrete cubic splines form a partition of unity, that is, for all j, $t_4 \le \tilde{t}_j < t_{k+5}$, we have*

$$\sum_i \delta^4_{i,p,w}(j) = 1.$$

Proof: Let $A(j; p, w) := \sum_i \delta^4_{i,p,w}(j)$. Then

$$\Delta_j A(j, p, w) = A(j, p, w) - A(j-1, p, w) = \sum_i \Delta_j \delta^4_{i,p,w}(j).$$

As a consequence of Lemma 4, this is equal to

$$\sum_i \tilde{C}_3(j+1)\left[\frac{\delta^3_{i,w}(j+1)}{C_3(i)} - \frac{\delta^3_{i+1,w}(j+1)}{C_3(i+1)}\right] = \tilde{C}_3(j+1)\left[-\sum_i \Delta_i \frac{\delta^3_{i,w}(j+1)}{C_3(i)}\right] = 0,$$

and $A(j, p, w) =: A(p, w)$ is independent of j. The proof follows if we substitute some special j, like $j = 1$. $\blacksquare$

It is easy to verify that parabolic discrete splines also satisfy a partition of unity, and that all mentioned discrete splines reduce to the ordinary polynomial ones for globally constant weights.

§3. Conclusions

We have found numerically stable algorithms for calculating with lower order jump splines, which are easy to implement as inner products of polynomial B–splines and certain discrete splines. They have already found its place in *CAGD* [9,4,5], where splines interpolating given data and minimizing a functional $\sum_{i=1}^{n}(w_i \int_{t_i}^{t_{i+1}}[D^2 f(t)]^2 dt$ are considered. Since in this case $p \equiv 1$, the formula for T–splines given in Theorem 3 is simplified even further. However, the shape preserving approximation requires the complete functional $V(f)$, with "point controls" added, and this leads us to splines in tension [1] as the solution of the associated minimization problem. The techniques discussed may well be applied in this setting, since the proofs rely on the existence of the derivative formula which we know for Chebyshev systems in general, and the recurrence for splines in the reduced system, which in this case is also known [6].

Acknowledgments. Supported in part by Grant 0-37-011 by the Ministry of Science and Technology of the Republic of Croatia.

References

1. Barsky, B. A., Exponential and polynomial methods for applying tension to an interpolating spline curve, Computer Vision, Graphics, and Image Processing **27** (1984), 1–18.
2. Cohen, E., T. Lyche, and R. Riesenfeld, Discrete *B*-splines and subdivision techniques in computer-aided geometric design and computer graphics, Comp. Graphics and Image Proc. **14** (1980), 87–111.
3. Dyn, N. and A. Ron, Recurrence relations for Tchebycheffian B-splines, J. Analyse Math. **51** (1988), 118–138.
4. Foley, T. A., Interpolation with interval and point tension controls using cubic weighted omega splines, ACM TOMS **13** (1987), 68–96.
5. Foley, T. A., A shape preserving interpolant with tension controls, Comput. Aided Geom. Design **5** (1988), 105–118.
6. Kvasov, B. I., Local bases for generalized cubic splines, Russ. J. Numer. Anal. Math. Modelling **10** (1995), 49–80.
7. Lyche, T., A recurrence relation for Chebyshevian B-splines, Constr. Approx. **1** (1985), 155–173.
8. Rogina, M., Basis of splines associated with some singular differential operators, BIT **32** (1992), 496–505.
9. Salkauskas, K., C^1 splines for interpolation of rapidly changing data, Rocky Mountain J. Math. **14** (1984), 239–250.
10. Schumaker, L. L., *Spline Functions: Basic Theory*, Wiley, New York, 1981.
11. Schumaker, L. L., On Tchebychevian spline functions, J. Approx. Theory **18** (1976), 278–303.

Rational Speed Pseudo-Quadratic B-Splines

Malcolm Sabin

Abstract. Much of the recent work on rational speed polynomial parametric curves has focussed on the quintic analogue of the parametric cubic. This paper fills in some results relating to the cubic analogues of the parametric quadratic and the quadratic B-spline curves. The new results support the view that these analogues have the same operational capabilities as the original quadratics, but are in some ways better. In particular the conditions for the curvature of a piece to be monotonic are shown to be a superset of those for the quadratic, and three variants of the B-spline are constructed, one of which has anomalously high continuity when the curvature is of constant sign.

§1. Background: Rational Speed Curves

In a series of papers, Farouki first showed that a subset of the parametric Bézier curves have the property of rational speed, then that these could be placed in correspondance with parametric curves of a lower order [3], and most recently, that these curves tend to have better aesthetic properties than the lower order curves satisfying the same interpolation conditions.

A rational speed curve of highest parametric power $2n - 1$ has the same number of freedoms as a 'normal' parametric polynomial curve of highest power n. One would therefore expect to be able to set up interpolation or approximation schemes using the rational speed curves which are close analogues of those using the lower order polynomial curves, and this has so far proved to be the case.

In particular, it has been shown that the quintic rational speed analogue of the cubic supports Hermite interpolation [4] and can also be used to form an interpolating spline [1].

Curves and Surfaces with Applications in CAGD
A. Le Méhauté, C. Rabut, and L. L. Schumaker (eds.), pp. 395–402.
Copyright ⓒ 1997 by Vanderbilt University Press, Nashville, TN.
ISBN 0-8265-1293-3.
All rights of reproduction in any form reserved.

§2. Background: The Parametric Quadratic Curve

The parametric quadratic is under-utilised in CAGD, probably because the cubic spline became established before it was widely realised that even order splines existed, but it has some significant advantages in contexts where continuity of curvature is not essential.

Using the Bernstein basis, the parametric quadratic segment has the equation

$$P(t) = (1-t)^2 A + 2t(1-t)B + t^2 C,$$

where A and C are the ends of the segment, and B is the point of intersection of the tangents at A and C.

The triangle ABC is often termed the *tangent triangle*, and it forms a convenient way of controlling the piece of curve. Notable properties are:

• that the piece of curve lies entirely within the triangle, and therefore simple culling tests can often detect that pieces do not intersect,

• that the piece is a piece of parabola, which has at most one maximum of curvature,

• that this maximum lies outside the piece, and so the piece has monotonic curvature, iff the tangent control point, B, lies inside either the circle whose diameter has endpoints A and the midpoint $[A + C]/2$ or the circle whose diameter has endpoints C and $[A + C]/2$ [6].

§3. Background: The Quadratic B-spline

The equal-interval quadratic B-spline curve may be constructed from a given control polygon by taking the midpoint of each edge except the first and the last, and constructing a vertex there. We call these new vertices *change-points*.

Then to each interior original vertex we can adjoin the adjacent change-points to form a tangent triangle. Fitting a quadratic into each tangent triangle gives a C^1 curve. The unequal-interval variant is given by taking some other positions than the midpoints. The ratio of distances from a change-point to the adjacent vertices is given by the ratio of successive parametric intervals. The notable properties are:

• that the complete curve lies within the convex hull of the original polygon vertices, and that individual pieces lie within the convex hull of the relevant groups of three successive vertices,

• that there are as many inflexions in the curve as in the polygon. Such inflexions happen at the junctions between one span and the next,

• that at a junction where there is no inflexion it is possible to achieve continuity of curvature by appropriate positioning of the change-point. This is equivalent to appropriate choice of the ratio of parametric intervals, see [7,9,8,2].

§4. The Pseudo-quadratic Rational Speed Curve

The pseudo-quadratic rational speed curve is a cubic satisfying the rational speed condition. It can be constructed by integrating the quadratic which is the square (in the complex plane) of a straight line segment [3].

For our purposes it is convenient to work with the conditions on the Bernstein cubic which result from this construction. Let a Bernstein cubic be

$$P(t) = (1-t)^3 A + 3t(1-t)^2 Q + 3t^2(1-t)R + t^3 C.$$

The rational speed conditions are:

(i) that the angle AQR should equal the angle QRC, and

(ii) that the length of QR should be the geometric mean of the lengths of AQ and RC.

If we choose to limit the total angle turned through to less than π, we can extend AQ and CR to meet at a point which we label B. ABC is now a tangent triangle.

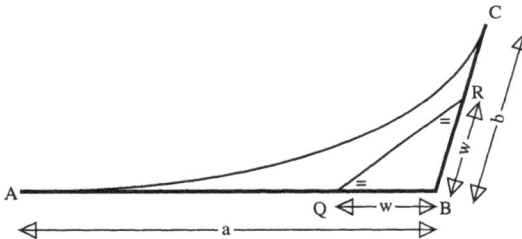

Fig. 1. PH cubic in a tangent triangle.

It is always possible to determine Q lying on a given AB and R lying on a given BC, so that the cubic satisfies the rational speed conditions.

The first condition is easily satisfied by merely ensuring that Q and R are equal distances from B, so that QBR is a isosceles triangle on base QR. Call the base angles QRB and RQB θ (marked by $=$ in Figure 1 above).

The second condition is met by choosing this distance appropriately. If we call this distance w, the lengths AB and BC a and b respectively, then the second condition becomes

$$(a-w)(b-w) = 4w^2 \cos^2 \theta,$$

and the closed form expression for w in terms of a, b and θ is just

$$w := \frac{2ab}{a+b+\sqrt{(a-b)^2 + 16ab\cos^2 \theta}}. \tag{1}$$

This always gives a value between 0 and $\min(a,b)$ by taking the positive square root. The negative square root always gives a w value outside the

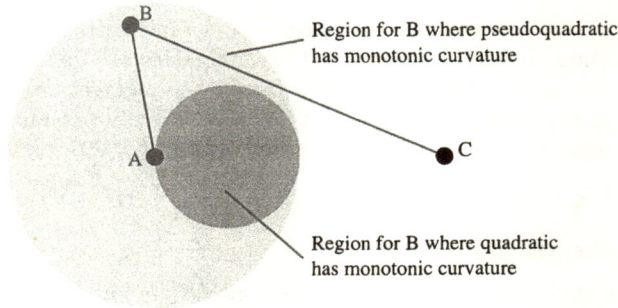

Fig. 2. Regions of monotonic curvature.

range between 0 and $\max(a, b)$. There is therefore a unique pseudo-quadratic fitting into a given tangent triangle.

We can then determine the Bézier points Q and R by stepping along the appropriate edges from B. Notable properties of this cubic are:

- that it always lies inside the tangent triangle. This follows from

$$0 < w < \min(a, b)$$

and from the convex hull property of the Bézier cubic.

- that provided B does not lie on AC, it always lies on that side of the quadratic which is further from B, so that if the quadratic does not cut a line, neither does the pseudo-quadratic. This follows from the facts (i) that the curvature at the ends of the piece is always tighter for the pseudo-quadratic than for the quadratic, so that the pseudo-quadratic is locally inside, (ii) that the two curves have tangency at A and C, (iii) that there must be two further intersections of the two curves in the region outside the pieces because of the asymptotic behaviour, and (iv) that a quadratic and a cubic can have at most six intersections.

- that it is part of a Tschirnhausen cubic [5], which has at most one maximum of curvature.

- that this maximum lies outside the piece, and so the piece has monotonic curvature, provided that the tangent control point, B lies inside either the circle centred on A of radius $|AC|/2$, or the circle of the same radius centred on C. This is no longer a strict condition. The locus where the maximum lies exactly at A is a quartic curve enclosing an area a few percent larger than the circle centred on A. The algebra deriving it works most conveniently by considering the locus of C meeting this condition given fixed A and B (an ellipse), and then inverting wrt A. This can be compared with the quadratic, for which the control point B must lie inside the heavily shaded circle in Figure 2.

§5. The Pseudo-quadratic Rational Speed Spline

Because the behaviour of individual pseudoquadratic pieces is so similar to that of quadratics, it is no surprise that such pieces can be assembled together into composite piecewise curves with similar properties to the quadratic B-spline curve. We call these *pseudo-quadratic B-spline curves*, even though the construction is not linear in the control points, and there is therefore no pseudo-quadratic B-spline basis function except in the differential sense as individual control points are moved infinitesimally.

We now have three alternative strategies for the positioning of the change-points, giving the three variants promised in the abstract.

1) the change-points can be at predefined ratios within the original polygon edges.

This is the simplest construction, since the entire curve is given in closed form. Suppose that the parametric knot intervals to left and right in Figure 3 are g and g' respectively. Then the condition for parametric C^1 continuity of the ordinary quadratic B-spline is just that

$$|BC|/g = |CB'|/g'. \tag{2}$$

Unfortunately this does not correspond to the condition of C^1 parametric continuity for the pseudo-quadratic, and so it cannot be used for grid curves of a tensor-product scheme.

2) the change-points can be chosen to match a predefined set of parametric intervals. The implementation of this is similar to the third case, described below, and the proofs are also similar in structure. The condition in Figure 3 above is that

$$|RC|/g = |CQ'|/g' \tag{3}$$

This means that C is dependent on R and Q', which are themselves dependent on C. It is therefore necessary to solve two sets of coupled equations, one set being non-linear.

3) the change-points can be chosen so that where there is no inflexion, the curvature will be continuous. It is convenient to use the condition that the magnitude of curvature should always be continuous, since this gives a good aesthetic result.

This is the condition that

$$|RC|^2/h1 = |CQ'|^2/h2 \tag{4}$$

where $h1$ and $h2$ are the perpendicular distances of Q and R' respectively from the line BB', as shown in Figure 3.

Again we have to solve two sets of coupled equations, now both non-linear. In fact we can linearise equation (4), giving it a unique closed-form solution, by noting that we seek the solution for C which lies between R and Q'. This is given by

$$|RC|/\sqrt{|h1|} = |CQ'|/\sqrt{|h2|},$$

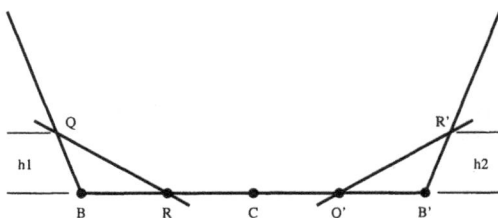

Fig. 3. Adjacent pieces.

and so

$$C := \frac{\sqrt{|h2|}R + \sqrt{|h1|}Q'}{\sqrt{|h2|} + \sqrt{|h1|}}.$$

This is sufficiently complicated that we need to show that a solution exists, that it is unique and that our algorithm for computing it converges.

The proof of existence is by induction on the number of vertices in the control polygon. Consider a control polygon of n vertices. If $n = 3$ the curve is just a single pseudoquadratic. If $n > 3$ then it is possible to divide it in two, each of fewer than n vertices, by choosing a change-point in an inner edge. If the change-point is chosen very close to one of the existing vertices, the curvature on that side will be very large. The discontinuity of curvature will therefore have one sense. If it is chosen very close to the other end of its edge, the curvature on the other side will be very large and the discontinuity will have the other sign. The curvatures, and therefore the discontinuity, vary continuously with the position of the change-point, and there must therefore be at least one point where the discontinuity is zero.

The proof of uniqueness depends in more detail on the actual equation of the pseudo-quadratic, but is enabled by showing that the variations of curvature on each side of the change-point are monotonic. This, in turn, requires induction on the number of vertices in the polygon.

§6. Construction

We are given a polygon, and require to construct the pseudo-quadratic B-spline. It is sufficient to construct the change-points and thence the Bézier points of the individual cubic pieces, since from them the curve is fully determined. The algorithms for the three variants (differing in the use of eqns (2), (3) or (4) respectively) are:

```
Begin
  set up initial change-points at midpoints of polygon edges
  compute w values and Bézier points from change-points (1)
  Until converged
  Do  compute change-points from Bézier points (2),(3) or (4)
      compute w values and Bézier points from change-points
  Enddo
End
```

The proof of convergence is dependent on the detailed algebra of the equations, but depends on noting that the vector of residuals from equation (4) at the start of the loop is multiplied at every step by a matrix dependent on the detailed geometry. However, each column of this matrix sums to a number strictly less than one, provided that no three consecutive control points are collinear, and so under this condition, the L^1 norm of the residuals must reduce at every step.

Further details of these proofs can be found in [10], which also includes examples and code for the computation of the curves from the control points.

Acknowledgments. This work was supported by the Royal Society of London under their Visiting Industrial Fellowships scheme.

References

1. G. Albrecht and R. T. Farouki, Construction of C^2 Pythagorean hodograph interpolating splines by the homotopy method, Advances in Comp Math, to appear.

2. W. L. F. Degen, High Accuracy Approximation of Parametric Curves, in *Mathematical Methods for Curves and Surfaces*, Morten Dæhlen, Tom Lyche, Larry L. Schumaker (eds.), Vanderbilt University Press, Nashville & London, 1995, 83–98.

3. R. T. Farouki The conformal map $z \mapsto z^2$ of the hodograph plane, Comput. Aided Geom. Design **10** (1995), 363–390.

4. R. T. Farouki and C. A. Neff, Hermite Interpolation by Pythagorean hodograph quintics, Math Comp, to appear.

5. R. T. Farouki and T. Sakkalis, Pythagorean Hodographs, IBM J Res Develop **34** (1990), 736–752.

6. D. Field and W. H. Frey, Constructing Bézier Conic Segments with Monotone Curvature, Proceedings of 5th IFIP Workshop on Solid Modeling in CAD, Chapman and Hall, 1997, to appear.

7. T. Pavlidis, Curve Fitting with Conic Splines, ACMToG **2**(1983), 1–31.

8. H. Pottmann, Locally Controllable Conic Splines with Curvature Continuity, ACMToG **10** (1991), 366–377.

9. V. Pratt, Techniques for Conic Splines, Proceedings of ACM SIGGRAPH 1985, 151–159.

10. M. A. Sabin,Pseudo-Quadratic Rational Speed Splines, Report NA17. Department of Applied Mathematics and Theoretical Physics, University of Cambridge.

Malcolm Sabin
Numerical Geometry Ltd.
26 Abbey Lane, Lode,
Cambridge CB5 9EP
ENGLAND
malcolm@geometry.demon.co.uk

Department of Applied Mathematics and Theoretical Physics
Silver Street
Cambridge
ENGLAND
mas33@damtp.cam.ac.uk

A Parameterization Technique
for the Control Point Form Method

Alessandra Sestini and Rossana Morandi

Abstract. The Control Point Form method for numerical grid generation [7] utilizes the transfinite interpolation techniques to carry out grids
which precisely conform to the boundaries of the domain. Through a set
of control points, it allows the manipulation of the coordinate curves of the
grid [4]. In the construction of computational grids, a criterion to control
the distribution of the coordinate curves is always desirable [2,6]. In [6]
the control of the grid-spacing through the parameter-spacing has been
obtained in the case of pseudo-rectangular domains, relating to the projection of the coordinate curves on a favourite a-priori assigned direction.
In this paper this approach has been extended to more general geometries,
relating to the arc-length of the coordinate curves.

§1. Introduction

The Control Point Form method is an efficient algebraic method [7] for generating computational grids that conform to the boundaries of a physical
domain. In this approach the grid is constructed by using the boundaries, as
well as a set of control points in order to control the shape and the position of
the coordinate curves [4,6]. In this environment, the capability of controlling
the grid-spacing by the parameter-spacing ("uniformity" property) is a fundamental tool for obtaining the desired behaviour of the computational grid. For
this purpose, in [6] a parameterization technique has been studied for "pseudo-
rectangular" domains. In practice, in [6] the control of the grid-spacing by
the parameter-spacing has been obtained with respect to the projection of the
coordinate curves on a favourite a-priori assigned direction. In this paper, the
geometric restrictions necessary in [6] for the parameterization technique are
removed, and the control of the grid-spacing is now obtained relating to the
arc-length of the coordinate curves. The approach here is probably the most
natural, and can be considered as an extension of the one proposed in [6],
particularly when almost orthogonal grids are sought.

Curves and Surfaces with Applications in CAGD   403
A. Le Méhauté, C. Rabut, and L. L. Schumaker (eds.), pp. 403–410.
Copyright © 1997 by Vanderbilt University Press, Nashville, TN.
ISBN 0-8265-1293-3.
All rights of reproduction in any form reserved.

§2. The Method

In this section the Control Point Form method is recalled to define a two-dimensional grid on a "quadrilateral" domain $\Omega \subset \mathbb{R}^2$, [4,6]. Let us assume that there exists a vector-valued function $\mathbf{F}(s,t) : I^2 \to \mathbb{R}^2$, $\mathbf{F} \in C^{p+1}(I^2)$, $p \geq 0$, $I^2 = [0,1]^2$, so that

$$Im(\mathbf{F}(s,t)|_{t=0}) = \partial\Omega_1, \quad Im(\mathbf{F}(s,t)|_{s=1}) = \partial\Omega_2,$$

$$Im(\mathbf{F}(s,t)|_{t=1}) = \partial\Omega_3, \quad Im(\mathbf{F}(s,t)|_{s=0}) = \partial\Omega_4,$$

with

$$\partial\Omega \equiv \cup_{i=1}^4 \partial\Omega_i,$$

where $\partial\Omega_1$, $\partial\Omega_3$, and $\partial\Omega_2$, $\partial\Omega_4$ are two pairs of opposite regular curves with

$$\partial\Omega_1 \cap \partial\Omega_3 = \emptyset, \quad \partial\Omega_2 \cap \partial\Omega_4 = \emptyset,$$

$$\partial\Omega_1 \cap \partial\Omega_2 = \mathbf{F}(1,0), \quad \partial\Omega_1 \cap \partial\Omega_4 = \mathbf{F}(0,0),$$

$$\partial\Omega_3 \cap \partial\Omega_2 = \mathbf{F}(1,1), \quad \partial\Omega_3 \cap \partial\Omega_4 = \mathbf{F}(0,1).$$

The Control Point Form method is an algebraic method [7] which, besides $\partial\Omega$, uses a set of control points in $\mathbb{R}^2$ to construct a transformation $\mathbf{B} : I^2 \to \Omega$ to define the grid. The transformation $\mathbf{B}(s,t)$ is defined as follows

$$\mathbf{B}(s,t) = \mathbf{B}_{\partial\Omega}(s,t) + \mathbf{Tp_0}(s,t), \tag{1}$$

where $\mathbf{B}_{\partial\Omega}(s,t) : I^2 \to \mathbb{R}^2$ is a function so that $\mathbf{B}_{\partial\Omega}|\partial I^2 = \partial\Omega$ and $\mathbf{Tp_0}(s,t) : I^2 \to \mathbb{R}^2$ is a control function vanishing on the boundaries of I^2. In particular, $\mathbf{B}_{\partial\Omega}(s,t)$ is defined as the Boolean sum $\mathbf{B}_{\partial\Omega}(s,t) = (P_1 \oplus P_2)[\mathbf{F}(s,t)]$, where

$$P_1[\mathbf{F}(s,t)] = \alpha_0(s)\mathbf{F}(0,t) + \alpha_1(s)\mathbf{F}(1,t),$$

$$P_2[\mathbf{F}(s,t)] = \beta_0(t)\mathbf{F}(s,0) + \beta_1(t)\mathbf{F}(s,1),$$

and the blending functions $\alpha_0(s), \alpha_1(s), \beta_0(t), \beta_1(t)$ are $C^{p+1}([0,1])$ cardinal functions with respect to the extremes of the interval $[0,1]$. Moreover, they have local support. The control function is defined as

$$\mathbf{Tp_0}(s,t) = \mathbf{Tp}(s,t) - (P_1 \oplus P_2)[\mathbf{Tp}(s,t)],$$

where

$$\mathbf{Tp}(s,t) = \sum_{i=1}^m \sum_{j=1}^n \mathbf{Q}_{i,j} G_j(s) H_i(t), \tag{2}$$

and $\{\mathbf{Q}_{i,j}, \ i = 1, \ldots, m, \ j = 1, \ldots, n\}$ is the assigned set of control points. The functions $G_j(s), j = 1, \ldots, n$ are defined as

$$G_1(s) = 1 - \frac{g_1(s)}{g_1(1)}, \quad G_n(s) = \frac{g_{n-1}(s)}{g_{n-1}(1)},$$

$$G_j(s) = \frac{g_{j-1}(s)}{g_{j-1}(1)} - \frac{g_j(s)}{g_j(1)}, \quad j = 2, \ldots, n-1,$$

where

$$g_j(s) = \int_0^s \phi_j(\sigma) \, d\sigma, \quad j = 1, \ldots, n-1. \tag{3}$$

The so called interpolation functions [4] $\phi_j(s), j = 1, \ldots, n-1$ used in (3) are $C^p([0,1])$ functions with the following properties:

$$
\begin{aligned}
&i \ ) \ \phi_j(s_k) = \delta_{j,k}, \quad k = 1, \ldots, n-1, \ j = 1, \ldots, n-1, \\
&ii \ ) \ \phi_j(s) = 0, \ s \notin [s_{max(1,j-u)}, s_{min(n-1,j+u)}] \\
&iii) \ \sum_{j=1}^{n-1} \phi_j(s) = 1, \quad \forall s \in [0,1],
\end{aligned}
\tag{4}
$$

where $\delta_{j,k}$ is the Kronecker's delta, $0 = s_1 < s_2 < \cdots < s_{n-1} = 1$ is an assigned s partition with $n \geq 2u+1$ and u is a positive integer related to the local support of the interpolation functions.

The functions $H_i(t), i = 1, \ldots, m$ can be defined analogously, and the coresponding interpolation functions $\psi_i(t), i = 1, \ldots, m-1$ are assumed to have the same properties as the ϕ_i.

It is easy to verify that hypothesis i) guarantees the interpolation of the directions defined by the control points, and hypothesis ii) allows to obtain a local influence of the control points on the grid. As will be clear after reading Section 3, hypothesis iii) and the analogous one related to the functions $\psi_i(t), i = 1, \ldots, m-1$, when combined with a suitable parameterization technique, guarantee the so called "uniformity" property [2], that is the capability of controlling the grid-spacing through the parameter-spacing.

For $p = 0$, the well-known C^0 piecewise linear hat functions are suggested in [2] as interpolation functions. In this case $u = 1$, and the interpolation functions are nonnegative. In [3] and in [6] two different sets of C^1 piecewise quadratic functions are proposed for $p = 1$. For both of them $u = 2$, and the negative oscillations can be suitably reduced. However, when $p = 1$ in the following, only the set of interpolation functions proposed in [6] is referred to, because it allows the preservation of some useful properties of the transformation $\mathbf{B}(s,t)$ (e.g. convex hull property) as well.

§3. The Parameterization Technique

In this section a strategy to choose the nodes $(s_j, t_i), \ j = 1, \ldots, n-1,$ $i = 1, \ldots, m-1$ is proposed . The presented approach is an extension of the one introduced in [6], where some restrictive hypotheses on the boundary of the domain Ω have been assumed. Here an attempt has been made to obtain the "uniformity" property for more general geometries. To this end, particularly when almost orthogonal grids are sought, it seems reasonable to require

$$\int_0^s |\frac{\partial \mathbf{B}(\sigma, t)}{\partial \sigma}| d\sigma = K_1(t)s, \ \forall s \in [0,1], \ \forall t \in [0,1], \tag{5}$$

$$\int_0^t |\frac{\partial \mathbf{B}(s,\tau)}{\partial \tau}| d\tau = K_2(s)t, \ \forall t \in [0,1], \ \forall s \in [0,1], \tag{6}$$

where $|\cdot|$ denotes the Euclidean norm and where

$$K_1(t) = \int_0^1 |\frac{\partial \mathbf{B}(\sigma,t)}{\partial \sigma}| d\sigma, \ \ K_2(s) = \int_0^1 |\frac{\partial \mathbf{B}(s,\tau)}{\partial \tau}| d\tau. \tag{7}$$

Since only (4) and a suitable parameterization technique to choose the nodes $(s_j, t_i), j = 1, \ldots, n-1, i = 1, \ldots, m-1$ can be used to obtain (5) and (6), the previous relations need to be relaxed. A first simplification can be obtained substituting $\mathbf{B}(s,t)$ by $\mathbf{Tp}(s,t)$ in (5), (6) and (7), thus considering the "uniformity" property only with respect to the control function. In regards to this, it should be noted that from a practical point of view, (5) and (6) can still be obtained if the possibility of a conforming boundary reparameterization is considered (see Section 4). However, in order to obtain two separate identities related to s and t respectively, further simplification is necessary. Thus (5) and (6) can be replaced by the following averaged identities

$$\sum_{i=1}^{m-1} |\frac{\partial \mathbf{Tp}}{\partial s}(s,t_i)| = K1, \ \sum_{j=1}^{n-1} |\frac{\partial \mathbf{Tp}}{\partial t}(s_j,t)| = K2, \tag{8}$$

where $K1 = \sum_{i=1}^{m-1} K_1(t_i)$ and $K2 = \sum_{j=1}^{n-1} K_2(s_j)$. Accordingly, only an averaged "uniformity" condition is looked for.

In the following we will refer only to the first identity of (8). Then, after some algebra, it is possible to obtain the following expression for $\frac{\partial \mathbf{Tp}}{\partial s}(s,t)$

$$\frac{\partial \mathbf{Tp}}{\partial s}(s,t) = \sum_{j=1}^{n-1} (\mathbf{P}_{j+1}(t) - \mathbf{P}_j(t)) \frac{\phi_j(s)}{g_j(1)}, \ \ \forall s \in [0,1], \ \forall t \in [0,1],$$

where $\mathbf{P}_j(t) = \sum_{i=1}^{m} \mathbf{Q}_{i,j} H_i(t), j = 1, \ldots, n$ are control curves. Thus, taking into account that the interpolation functions have local support and they are positive for $p = 0$ or quasi-positive for $p = 1$, [2], [6], for reasonable control point distributions the first identity in (8) can be replaced by

$$\sum_{j=1}^{n-1} sp_j(t_1, \ldots, t_{m-1}) \frac{\phi_j(s)}{g_j(1)} = K1, \tag{9}$$

where

$$sp_j(t_1, \ldots, t_{m-1}) = \sum_{i=1}^{m-1} |\mathbf{P}_{j+1}(t_i) - \mathbf{P}_j(t_i))|, \ \ j = 1, \ldots, n-1.$$

Then, through the partition of unity property of the interpolation functions it can be shown that (9) is equivalent to the system

$$g_j(1) = \frac{sp_j(t_1, \ldots, t_{m-1})}{K1}, \quad j = 1, \ldots, n-1. \tag{10}$$

If the sets of interpolation functions proposed in [2] when $p = 0$ and in [6] when $p = 1$ are used,

$$g_j(1) = \begin{cases} \frac{\delta_1}{2} & \text{if } j = 1 \\ \frac{\delta_{j-1}+\delta_j}{2} & \text{if } j = 2, \ldots, n-2 \\ \frac{\delta_{n-2}}{2} & \text{if } j = n-1 \end{cases}$$

where $\delta_j = s_{j+1} - s_j, j = 1, \ldots, n-2$.

Then, given $t_1, \ldots, t_{m-1}$, (10) becomes a linear system of $n-1$ equations in the $n-2$ constrained unknowns $\delta_1, \ldots, \delta_{n-2}$ which can be written as

$$\begin{cases} A_1\delta & = F_1 \\ \sum_{j=1}^{n-2} \delta_j & = 1 \\ 0 < \delta_j < 1 \end{cases} \tag{11}$$

where δ is the vector of the $n-2$ unknowns $\delta_1, \ldots, \delta_{n-2}$, and A_1 and F_1 are respectively the $(n-1) \times (n-2)$ matrix and the $(n-1)$ vector defined as

$$A_1 = \begin{pmatrix} 1 & 0 & 0 & \cdots & 0 & 0 \\ 1 & 1 & 0 & \cdots & 0 & 0 \\ 0 & 1 & 1 & \cdots & 0 & 0 \\ \cdot & \cdot & \cdot & \cdots & \cdot & \cdot \\ \cdot & \cdot & \cdot & \cdots & \cdot & \cdot \\ 0 & 0 & \cdot & \cdots & 1 & 1 \\ 0 & 0 & \cdot & \cdots & \cdot & 1 \end{pmatrix}, \quad F_1 = \frac{2}{K1} \begin{pmatrix} sp_1 \\ sp_2 \\ \cdot \\ \cdot \\ \cdot \\ sp_{n-2} \\ sp_{n-1} \end{pmatrix}.$$

Consequently, for general control point distributions, it is possible to find a solution of (11) in the sense of the constrained least-squares [5], that is to find δ^* so that

$$\begin{cases} |A_1\delta^* - F_1| = min_\delta|A_1\delta - F_1| \\ \sum_{j=1}^{n-2} \delta_j = 1, \\ 0 < \delta_j < 1. \end{cases} \tag{12}$$

The same procedure is used to deal with the second identity of (8). Thus the following iterative parameterization strategy is proposed. After a uniform initialization phase, the following two steps are executed at the typical k−th iteration

- a new set of values $s_1^{(k+1)}, \ldots, s_{n-1}^{(k+1)}$ for the parameter values $s_1, \ldots, s_{n-1}$ is computed from (12) where $t_i = t_i^{(k)}, i = 1, \ldots, m-1$;
- a new set of values $t_1^{(k+1)}, \ldots, t_{m-1}^{(k+1)}$ for the parameter values $t_1, \ldots, t_{m-1}$ is computed from the problem analogous to (12) related to the t parameter, where $s_j = s_j^{(k+1)}, j = 1, \ldots, n-1$.

§4. Numerical Tests

In this section some experiments to test the features of the proposed parameterization technique are illustrated. In particular, our approach ("criterion of the arc-length control") is compared with the uniform parameterization technique and with the approach presented in [6] ("criterion of the projections control").

The control point distributions are the same in all the figures related to the same geometry and, to make comparison easier, we always use a uniform tabulation with respect to both s and t. In all of the presented tests almost orthogonal grids are sought, as generally preferred to grids with distorted cells [7]. In Fig. 2 the grid obtained by the approach introduced in [6] has been constructed assuming as preferred directions $\eta = \mathbf{e}_1$ and $\theta = \mathbf{e}_2$, with $\mathbf{e}_1$ and $\mathbf{e}_2$ the canonical base of $\mathbb{R}^2$. It is also important to remark that the proposed parameterization strategy never needed more than four iterations for the convergence.

Relating to the reparameterization phase necessary for the boundaries, the following strategy has been used. Assuming the first and last rows and columns of control points distributed on the related boundary curves, a certain number of suitable points on each boundary curve $\partial\Omega_k, k = \ldots, 4$ has been selected and each of them has been associated to the parameter value of the nearest point on the corresponding control curve. Then, for each boundary curve, a reparameterization function belonging to $C^{p+1}[0,1]$ has been constructed using a suitable co-monotone interpolation scheme as in [1].

Figure 1 relates to a "pseudo-rectangular" domain [6], and it clearly shows the benefit of using the new approach in comparison with the uniform parameterization. The grid obtained by the "criterion of the projections control" is not reported here, since it is not very different from that obtained by the "criterion of the arc-length control". This result is not surprising, because the shape of the domain and the control point distribution cause the two criteria to be almost the same in this case (a slight difference between the two grids obtained by these two approaches was noticed near the two boundary curves $\partial\Omega_2$ and $\partial\Omega_4$).

Figures 2 and 3 (left) relate again to a "pseudo-rectangular" domain but in this case two opposite boundaries rapidly change their tangent vector in some areas, and the control point distribution is such that the grid is forced to preserve the shape of the boundaries. Accordingly, the right choice is the use of the "criterion of the arc-length control", as shown by the figures. In fact, this strategy guarantees a good correspondence between the uniform parameters tabulation and the almost uniform distribution of the related coordinate curves. Figure 3 (right) relates to a non "pseudo-rectangular" domain. Therefore, for this kind of geometry, the approach presented in [6] cannot be used. The grid obtained by the uniform parameterization technique is completely unsatisfactory, and is not reported here for reasons of brevity only. In Figure 3 (right) the grid obtained through the approach introduced in this paper is depicted.

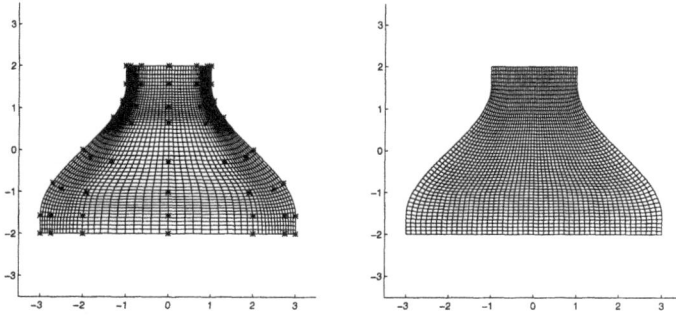

Fig. 1. (left) uniform parameterization, (right) parameterization by the criterion of the arc-length control.

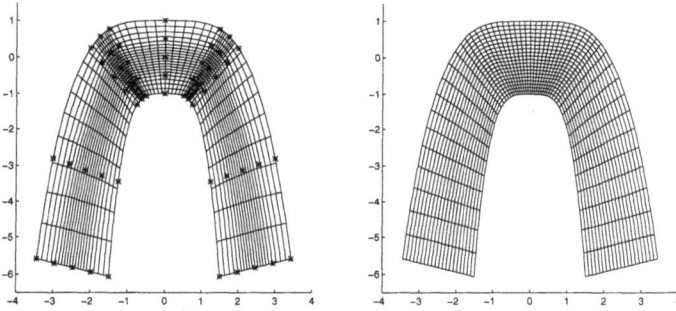

Fig. 2. (left) uniform parameterization, (right) parameterization by the criterion of the projections control.

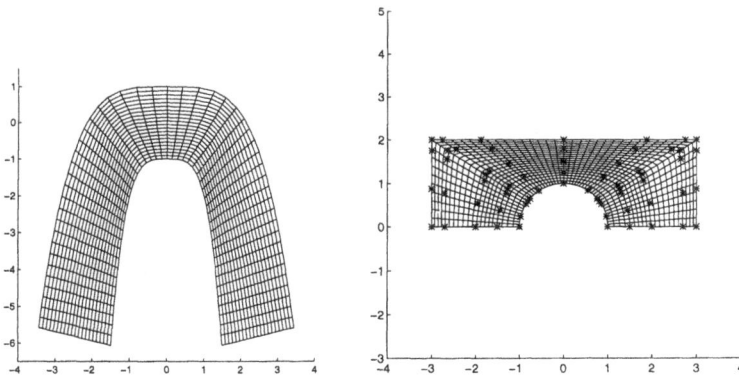

Fig. 3. Parameterization by the criterion of the arc-length control for two different geometries.

Acknowledgments. Work partially supported by M.U.R.S.T.

References

1. Costantini, P., Algorithms for Shape-Preserving Interpolation, in *Splines in Numerical Analysis*, Schmidt, J.W. and H. Spaeth (eds), Akademie Verlag, Berlin, 1989, 31–46.

2. Eiseman, P. R., Coordinate generation with precise controls over mesh properties, J. Comput. Phys. **47** (1981), 331-351.

3. Eiseman, P. R., High level continuity for coordinate generation with precise controls, J. Comput. Phys. **47** (1982), 352-374.

4. Eiseman, P. R., A control point form of algebraic grid generation, Int. J. for Numerical Methods in Fluids **8** (1988), 1165-1181.

5. Fletcher, R., *Practical Methods of Optimization 2*, Wiley and Sons Publishers, New York, 1981.

6. Morandi, R. and A. Sestini, Precise controls in numerical grid generation, in *Advanced Topics in Multivariate Approximation*, F. Fontanella, K. Jetter, and P.-J. Laurent (eds), World Scientific Publishing Co., Singapore, 1996, 211-226.

7. Thompson, J. F., Z.U.A. Warsi, and C.W. Mastin, *Numerical Grid Generation Foundations and Applications*, Elsevier Sciences Publishers, New York, 1985.

Alessandra Sestini
Dipartimento di Energetica "Sergio Stecco", Universitá di Firenze
Via Lombroso 6/17, 50134 Firenze, ITALY
sestini@ingfi1.ing.unifi.it

Rossana Morandi
Dipartimento di Matematica, Universitá di Perugia
Via Vanvitelli 1, 06123 Perugia, ITALY
morandi@gauss.dipmat.unipg.it

Planar Shape Enhancement and Exaggeration

Ami Steiner, Ron Kimmel, and Alfred M. Bruckstein

Abstract. A local smoothing operator applied in the reverse direction is used to obtain planar shape enhancement and exaggeration. Inversion of a smoothing operator is an inherently unstable operation. Therefore, a stable numerical scheme simulating the inverse smoothing effect is introduced. Enhancement is obtained for short time spans of evolution. Carrying the evolution further yields shape exaggeration or caricaturization effect. Introducing attraction forces between the evolving shape and the initial one yields an enhancement process that converges to a steady state. These forces depend on the distance of the evolving curve from the original one and on local properties. Results of applying the unrestrained and restrained evolution on planar shapes, based on a stabilized inverse Geometric Heat Equation, are presented showing enhancement and caricaturization effects.

§1. Introduction

In this paper we consider possible ways to design an automatic procedure for enhancing and caricaturizing planar shapes. There is a common trend in all caricatures: special, unusual or uncommon features in objects are detected and magnified. In [4] Brennan proposed a caricaturization algorithm based on exaggerating the differences between a given object and an 'average' one. Her algorithm requires a-priori knowledge of a set of items from the input class and the correspondence points between them.

We propose to evolve planar curves using reverse geometric smoothing operators in order to achieve exaggeration. Such an exaggeration scheme can also be used for deblurring and feature enhancement in 2D images.

Assume an image $I_0(x, y)$ is blurred in time by the differential equation $I_t(x, y; t) = \nabla^2 I(x, y; t)$, where $I(x, y; 0) = I_0(x, y)$ is the original image and $I(x, y; \Delta t)$ is the image distorted by the blurring process at time Δt. Deblurring the Gaussian blur (or inverting the heat equation) is a known problem in image processing. For short times, given the blurred image $I(x, y; \Delta t)$ one can approximate the original image $I(x, y; 0)$ using the Taylor expansion: $I(x, y; 0) = I(x, y; \Delta t) - \Delta t \nabla^2 I(x, y; \Delta t) + \mathcal{O}(\Delta t^2)$. We could try to use

Curves and Surfaces with Applications in CAGD
A. Le Méhauté, C. Rabut, and L. L. Schumaker (eds.), pp. 411–418.
Copyright © 1997 by Vanderbilt University Press, Nashville, TN.
ISBN 0-8265-1293-3.
All rights of reproduction in any form reserved.

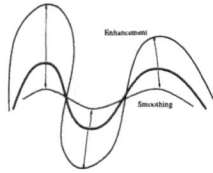

Fig. 1. Smoothing vs. Enhancement.

the same idea to restore and exaggerate curves using a reverse form of the curvature flow for planar curve evolution, also known as the geometric heat equation (GHE), which is a nonlinear smoothing operator described by the partial differential equation:

$$C_t = C_{ss}. \tag{1}$$

Here, $C(s,t) : [0, L] \times [0, T) \rightarrow \mathbb{R}^2$ is a planar curve deforming in time t according to its second derivative with respect to its arc-length.[1] By using the Taylor expansion, we could propose to reconstruct $C(s, 0)$ from $C(s, \Delta t)$ via: $C(s, 0) = C(s, \Delta t) - \Delta t C_{ss}(s, \Delta t) + \mathcal{O}(\Delta t^2)$. This suggests the reverse evolution

$$C_t = -C_{ss}, \tag{2}$$

where $C_{ss} = \kappa \hat{N}$ is the curvature vector. In [9] the inverse smoothing of curves was indeed identified as an unstable evolution. Malladi and Sethian [7] simulated the inverse geometric heat equation to exaggerate the borders of alpha-numeric input and thereby improve classification results of hand-written characters. Fig. 1 demonstrates the smoothing effects of the GHE and the desired enhancement/exaggeration effects to be obtained from the inverse flow.

However, (2) is not stable and cannot be implemented numerically. In the sequel we introduce two new approaches for simulating the effects of the inverse GHE on outlines of shapes given as polygons or via gray-level images on a grid of pixels. The proposed evolution enhances the curve for short time, but it does not reach a steady-state. In Section 3 we formulate "restrained" evolutions by introducing restraining forces that are added to the original evolution. Results of applying the unrestrained and restrained evolutions based on the Inverse GHE of some planar shapes are demonstrated on gray level images as well as on polygonal curves.

§2. Unrestrained Evolution

In this section we present two approaches to simulating the inverse smoothing operation for curves and numerical schemes for approximating the process.

---

1. See the detailed treatment of the GHE in [6], which is the steepest decent flow for minimizing the arclength of a curve, and is therefore the natural geometric analog to the heat equation. It was found to be an accurate model for many phenomena in nature.

For continuous curves,[2] the level-set Eulerian formulation [8] is used and a generalized model that controls the stability and the rate of the exaggeration is suggested. For polygonal approximation of curves, two discrete evolutions are applied: the discrete analogue to the reverse GHE evolution, and an evolution based on inverting the evolution equations analyzed in [1].

2.1. Simulating the Continuous Case

We follow the Osher-Sethian Eulerian formulation [8], creating a bivariate function $\phi : \mathbb{R}^2 \times [0, T) \to \mathbb{R}$, and evolve each of its level sets $\phi(x, y; t) = l$ (also denoted as $C(t) = \phi^{-1}(l)$) according to the following equation, geometrically identical to (2):

$$C_t = -\kappa(s, t)\hat{N}(s, t). \tag{3}$$

Using the results in [8] for this case, the ϕ-surface evolution equivalent to (3) is:

$$\phi_t = -\kappa(x, y; t)|\nabla\phi|. \tag{4}$$

where $C(t) = \phi^{-1}(0)$ for all t. One possible way of choosing the initial $\phi(x, y; 0)$ is the distance from the curve $C(s, 0)$, with negative signs in the interior and positive signs in the exterior of the curve. The curvature of each level set curve $\phi^{-1}(l)$ is given by

$$\kappa(x, y; t) = \kappa(\phi) = \nabla \cdot \left(\frac{\nabla\phi}{|\nabla\phi|} \right) = \frac{\phi_{xx}\phi_y^2 - 2\phi_x\phi_y\phi_{xy} + \phi_{yy}\phi_x^2}{(\phi_x^2 + \phi_y^2)^{3/2}}.$$

In order to stabilize this inherently instable evolution we modify (4) so that each level set evolves in lockstep with the zero level set, thus preserving the zero level set along the propagation (see [10] for details). We replace $\kappa(x, y; t)$ in (4) by $K(x, y; t) = \frac{\kappa(x,y,t)}{1-\kappa(x,y,t)\phi(x,y,t)}$, and fix the flow field to its initial value $K(x, y; 0)$ throughout the evolution:

$$\phi_t = -K(x, y; 0)|\nabla\phi|. \tag{5}$$

We also smooth the flow field $K(x, y; 0)$ by convolving it with a Gaussian kernel, to suppresses the effects of small perturbations and discontinuities on the propagating curves, and limit K so that $|K| < 2$ (setting higher values to 2) to increase stability under the reasonable assumption that curvature values of the outline contour of shapes given on a pixel grid do not exceed the value of 2, i.e., a curvature radius of half a pixel. This limit on $|K|$ also allows us to maintain the C.F.L. condition without forcing very short time steps in the numerical approximation.[3] Finally, since ϕ does not remain a distance map

---

2. In this case, the curve is given as the boundary of a shape in a gray level image, and the gray level image serves as the natural presentation of the curve (an implicit representation of the curve which is a given level set of the gray level image).

3. The Courant, Friedrichs, Lewy (CFL) condition is a necessary stability condition for any numerical scheme: The domain of dependence of each point in the domain of the numerical scheme should include the domain of dependence of the PDE itself.

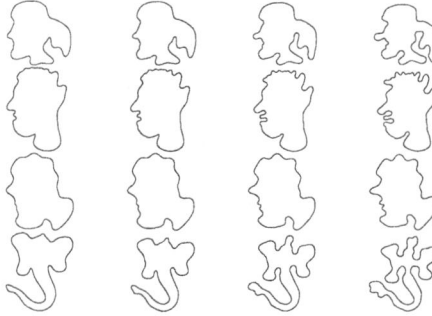

Fig. 2. Exaggeration of several curves.

while evolving, we adjust it to be the distance from its zero-level-set, after every few iterations.

Fig. 2 demonstrates some curve exaggerations using this procedure with the numerical implementation of (5). The original curves are at the left, and the evolving curves are sampled at $\lfloor n\Delta t \rfloor$=10, 30 and 50. Finally, we suggest an evolution law that controls the rate of enhancement. Substituting $K(x,y;0)$ in (5) by: $K_G = K(x,y;0)(1+\alpha|\phi(x,y;0)|^\beta)$; α, $\beta \geq 0$ increases the enhancement effect, since $\frac{K_G(x,y)}{K(x,y;0)} \geq 1$ for all (x,y) and it increases as the point (x,y) departs from the initial curve.

2.2. Simulating the Polygonal Case

For curves given as polygons, we can use a discrete non-linear evolution rule analogue to the continuous case. Alternatively, a linear affine invariant evolution similar to the reverse GHE can be introduced by reversing the direction of the discrete smoothing transformation introduced in [5] and discussed in [1,2].

Direct approximation: Let a polygonal contour be defined by its vertices, $\{P_i\}_{i=1}^M$. The discrete evolution analogue to the GHE shifts each vertex $P_i = (x_i, y_i)$, according to

$$P_i^{(n+1)} = P_i^{(n)} + \kappa_i^{(n)} \cdot \hat{N}_i^{(n)}. \tag{6}$$

Exaggeration is achieved by inverting the direction of movement. That is:

$$P_i^{(n+1)} = P_i^{(n)} - \kappa_i^{(n)} \cdot \hat{N}_i^{(n)} \tag{7}$$

Here $P_i^{(n)} = (x_i^{(n)}, y_i^{(n)})$ indicates the location of vertex i after n iterations, $\hat{N}_i^{(n)}$ is a unit normal to the curve at vertex i, defined as a unit vector in the direction of the bisector of that vertex, and $\kappa_i^{(n)}$ is the curvature at vertex i defined in [2] as: $\kappa_i^{(n)} \equiv c \cdot \theta_i^{ext}$, where θ_i^{ext} is the external angle between the two edges, for which $P_i^{(n)}$ is a common vertex, and c is a normalization factor.

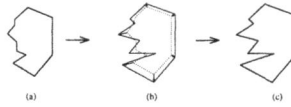

Fig. 3. Polygon exaggeration.

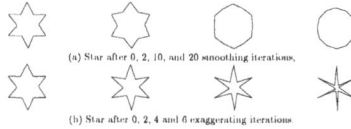

(a) Star after 0, 2, 10, and 20 smoothing iterations,

(b) Star after 0, 2, 4 and 6 exaggerating iterations

Fig. 4. Smoothing vs. exaggerating.

Fig. 3 demonstrates polygon exaggeration using (7). For each vertex of the initial polygon (a), the normal $\hat{N}_i^{(n)}$ and curvature $\kappa_i^{(n)}$ are calculated. Then, (b) each vertex is moved in the direction of $\hat{N}_i^{(n)}$ by a step proportional to $\kappa_i^{(n)}$ thus creating a caricature effect (c).

Other Affine and Euclidean approximations to the GHE: In [1] the following general smoothing operator is proposed:

$$P_i^{(n+1)} = (1 - \alpha)P_i^{(n)} + \alpha\Gamma_- P_{i-1}^{(n)} + \alpha\Gamma_+ P_{i+1}^{(n)}$$

or, in matrix form:

$$P^{(n+1)} = MP^{(n)} \tag{8}$$

where $\{P_i^{(n)}\}_{i=1}^N$ are the N vertices polygons after n iterations, M is an N by N matrix. We achieve shape enhancement by inverting (8). We calculate the shift from the original polygon to the smoothed one, $(M - I)P^{(n)}$, and move the vertex in the opposite direction:

$$P^{(n+1)} = (I - (M - I))P^{(n)} = (2I - M)P^{(n)}. \tag{9}$$

The evolution given by (6)–(9) are closely related for the case $\Gamma_- = \frac{d_+}{d_+ + d_-}$, $\Gamma_+ = \frac{d_-}{d_+ + d_-}$. Figs. 4, 5 show results obtained by applying the two evolution laws ((8)–(9)) in their linear form (i.e. $\Gamma_- = \Gamma_+ = \frac{1}{2}$). Finally, we compare our automated results with the work of Yaacov (Zeev) Farkash, a leading Israeli caricaturist. Fig. 6 shows some curves, their exaggeration using our curve evolution algorithm, and Zeev's caricatures based on the same initial curves.

§3. Restrained Polygon Evolution

So far we have defined several stable, yet non-converging, shape enhancing evolutions. Applying the above evolution laws for infinite time spans expands

Fig. 5. Polygon exaggeration examples.

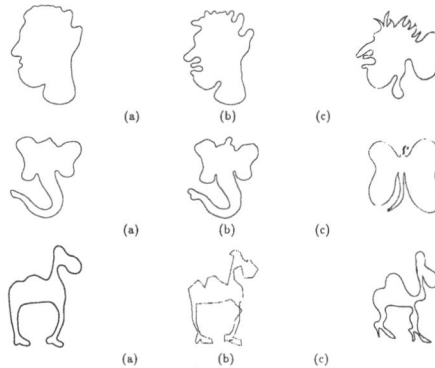

Fig. 6. Automated vs. hand-made caricatures: (a) Original (b) Automated (c)Artist .

the initial curve to infinity. Let us define evolution processes that converge to steady-states. We introduce "imaginary strings" that connect the original curve with its evolving "image" so that each point on the evolving curve is attracted back to its initial position. For polygonal approximations of curves, a set of strings binding each vertex of the original polygon with the corresponding vertex in the evolving polygon will create the desired attraction. The condition for the existence of steady-states, and an explicit formula for the steady-state in the polygonal case are given.

The attraction forces are assumed to act at each vertex. As before, let $\{P_i^0\}_{i=1}^N$ define the initial polygon, and let $\{P_i^n\}_{i=1}^N$ be the evolved versions at discrete time steps $n = 1, 2, \ldots$, the evolution being governed by (8). We introduce N attracting strings so that string (i) is attached on one side to the evolving vertex P_i^n and on the other side to the initial vertex P_i^0, and has an elasticity constant of β_i. The attraction force will be proportional to the distance of the evolving vertex from its original position. Adding these restraining forces to the smoothing evolution (8), we arrive at an evolution of

(a) Elephant after 0, 10, 30 and 100 unrestrained iterations.

(b) Elephant after 0, 10, 30 and 100 restrained iterations.

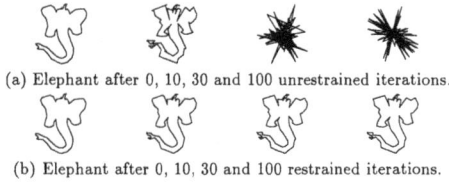

Fig. 7. Restrained versus unrestrained evolutions - the polygonal case.

the form

$$P^{(n+1)} = MP^{(n)} + B \cdot (P^{(0)} - P^{(n)}). \tag{11}$$

In a similar way, the reversed (exaggerating) evolution is given by

$$P^{(n+1)} = (I - (M - I))P^{(n)} + B \cdot (P^{(0)} - P^{(n)}). \tag{12}$$

The last term in (11)–(12) is the restraining force. B is an $N \times N$ diagonal matrix: $B = diag(\beta_0, ..., \beta_N)$. We are particularly interested in diagonal matrices with elements related to the curvature at each vertex. Fig. 7 shows an example of restrained evolution using (12) compared with unrestrained evolution using (9).

The following theorems proved in [10] state the conditions for the convergence of the evolving polygon to a steady-state, and explicitly express the steady-state in terms of the initial polygon and the restraining matrix B for both the linear smoothing and exaggeration cases.

Theorem 1. *Given the polygon smoothing equation:* $P^{(n+1)} = MP^{(n)} + B(P^{(0)} - P^{(n)})$, *where* $P^{(n)}$ *is an N-element vector of the polygon's coordinates after n iterations (in complex notation), M is an $N \times N$ circulant matrix with first row defined as* $M_{1,(*)} = \{1 - \alpha, \alpha/2, 0, ..., 0, \alpha/2\}$, *and* $B = diag(\beta_0, ..., \beta_N)$ *where β_i are the restraining coefficients, so that* $0 < \beta_i < 2(1 - \alpha)$, $\forall\, 0 \leq i \leq N - 1$, *there exists a steady-state polygon defined by*

$$P^{(\infty)} = \lim_{n \to \infty} P^{(n)} = (I - M + B)^{-1}BP^{(0)}. \tag{13}$$

Theorem 2. *Given the polygon exaggeration evolution:* $P^{(n+1)} = (2I - M)P^{(n)} + B(P^{(0)} - P^{(n)})$, *so that* $2\alpha < \beta_i < 2$, $\forall\, 0 \leq i \leq N - 1$, *there exists a steady-state polygon given by*

$$P^{(\infty)} = \lim_{n \to \infty} P^{(n)} = (M + B - I)^{-1}BP^{(0)}. \tag{14}$$

§4. Conclusions

The reverse GHE can be used to enhance features in planar curves. For a given initial curve, known to have been distorted by a smoothing operation (such as blurring), evolution using the reverse GHE for short times can approximately restore it. Longer time evolution will further enhance the curve yielding an exaggeration effect. For continuous curves, the level-set Eulerian formulation [8] was utilized and a generalization of the reverse GHE, which enables control over the intensity of exaggeration, was introduced, leading to suppressed or enhanced exaggeration. For polygonal shapes, two different evolution laws were explored. One derived directly from the continuous GHE, and the second is a discrete approximation of the GHE given by [1]. The relation between the two was shown and, in the linear case, conditions for convergence to a steady-state polygon were explicitly derived, as well as a closed form formula for the steady-state polygon itself.

Using our approach, planar curves are exaggerated using only their intrinsic features, without a-priori knowledge on their classification and with no need for further information. This is an advantage over previously stated exaggeration methods which require such knowledge. We introduced tools to control the parameters of the exaggeration, and in some cases allow the evolution to converge to a well defined steady-state curve.

Acknowledgments. The work of RK is supported in part by the OER under DE-AC03-76SFOOO98, and ONR grant under NOOO14-96-1-0381.

References

1. A. M. Bruckstein, G. Sapiro, D. Shaked, Evolutions of Planar Polygons, IJPRAI, 1996.
2. A. M. Bruckstein and D. Shaked, On Projective Invariant Smoothing and Evolutions of Planar Curves and Polygons, Technion - CIS Report #9328, Nov., 1993.
3. A. M. Bruckstein, Analyzing and Synthesizing Images by Evolving Curves, Proc. of ICIP'94, Nov., 1994.
4. S. E. Brennan, Caricature Generator: The Dynamic Exaggeration of Faces by Computer, Leonardo **18(3)** (1985), 170–178.
5. M. G. Darboux, Sur un problème de géométrie élémentaire, Bulletin Sci. Math **2** (1878), 298–304.
6. M. Grayson, The Heat Equation Shrinks Embedded Plane Curves to Round Points, J. Diff. Geom. **26** (1987), 285–314.
7. R. Malladi and J. A. Sethian, A unified Framework for Shape Segmentation, Representation, and Recognition, LBL-36039, UC-Berkely, Aug. 1994.
8. S. J. Osher and J. A. Sethian, Fronts Propagating with Curvature-Dependent Speed: Algorithms Based on Hamilton-Jacobi Formulations, J. Comp. Phys. **79** (1988), 12–49.
9. J. A. Sethian, Curvature and the Evolution of Fronts, Comm. in Math. Physics **101** (1985), 487–499.
10. A. Steiner, R. Kimmel and A. M. Bruckstein, Planar Shape Enhancement and Exaggeration, Technion - EE Pub. #977, July, 1995.

On Geometric Continuity of Isophotes

Holger Theisel

Abstract. It is a well-known fact that we can deduce G^n continuous isophotes from a G^{n+1} continuous surface. This paper gives an answer to the reverse problem: we deduce a G^{n+1} continuous surface from G^n continuous isophotes on the surface. We show how many families of isophotes we have to consider and what constraints apply. Furthermore, we apply the geodesic curvature and the "thickness" of isophotes as a surface interrogation tool.

§1. Introduction

Isophotes are a widely used interrogation tool in the design of various surfaces. First introduced in [4], they provide both an impression of global shape features and information about the continuity of the surface.

A family of isophotes on a surface $\boldsymbol{x}(u,v)$ is defined by a light direction vector $\boldsymbol{r}$ ($\|\boldsymbol{r}\| = 1$). Then the isophotes are the equipotential lines of the scalar field

$$s(u,v) = \boldsymbol{r} \cdot \boldsymbol{n}(u,v), \tag{1}$$

where $\boldsymbol{n} = \frac{\boldsymbol{x}_u \times \boldsymbol{x}_v}{\|\boldsymbol{x}_u \times \boldsymbol{x}_v\|}$ denotes the normalized normals of $\boldsymbol{x}$. This means that an isophote on the surface contains all surface points which have the same angle between the light direction and the surface normal. Silhouette lines are a special case of isophotes.

In this paper we use the following usual definition of geometric continuity: Two curves are G^n continuous at a common point $\mathbf{x}$ iff there exists a regular parametrization with respect to which they are C^n at $\mathbf{x}$. Two surfaces are G^n along a common line l iff there exists a regular parametrization with respect to which they are C^n along l.

It is a well known fact that a G^{n+1} continuous surface implies G^n continuous isophotes (see [4] and [2]) . Section 3 of this paper gives answers to the reverse questions:

1) Is it possible to deduce G^{n+1} continuity of the surface from the G^n continuity of isophotes ?

A. Le Méhauté, C. Rabut, and L. L. Schumaker (eds.), pp. 419–426.
Copyright © 1997 by Vanderbilt University Press, Nashville, TN.
ISBN 0-8265-1293-3.
All rights of reproduction in any form reserved.

2) If so, how many families of isophotes do we have to consider, and what constraints apply ?

The answers to the questions 1) and 2) are quite important for using isophotes to analyze the continuity of surfaces. It shows how many families of isophotes have to be considered in order to get reliable statements about the continuity of the surface.

Isophotes cannot generally be computed in a closed form but only as the numerical solution of partial differential equations. Nevertheless we want to compute local properties of isophotes, such as geodesic curvature and a new property called "thickness" in a closed form. In Section 4, these properties are applied as surface interrogation methods.

Notation and abbreviations: $x^{[i]}(t)$ denotes the i-th derivative vector of a parametrized curve $x(t)$. $x^{[i,j]}(u,v)$ denotes the partial derivative (i times in u-direction, j times in v-direction) of the parametrized surface $x(u,v)$. For instance, $x^{[2,1]}$ denotes x_{uuv}. The partials $n^{[i,j]}$ of the surface normals can be obtained by applying basic differentiation rules to n. Furthermore, we use the classical abbreviations $E = x_u \cdot x_u$, $F = x_u \cdot x_v$, $G = x_v \cdot x_v$, $L = n \cdot x_{uu}$, $M = n \cdot x_{uv}$, $N = n \cdot x_{vv}$. From these scalar fields we can also compute the partial derivatives.

In this paper we only consider regularly parametrized curves and surfaces. This means for curves that $\|\dot{x}(t)\| \neq 0$ for every t of the domain. For surfaces we assume that $\|x_u \times x_v\| = \sqrt{E \cdot G - F^2} \neq 0$.

§2. Theoretical Background

We will be analyzing isophotes on a parametric surface by interpreting them as tangent curves of vector fields. Before we discuss the surface case, we briefly describe the case of 2D vector fields.

Given is a 2D vector field $V : \mathbb{R}^2 \to \mathbb{R}^2$. V assigns a vector $V(u,v) = (vx(u,v), vy(u,v))^T$ to any point (u,v) of the domain. A curve $t \subseteq \mathbb{R}^2$ is called a *tangent curve* (stream line, flow line, characteristic curve) of the vector field V if the following condition is satisfied: For all points $(u,v) \in t$, the tangent vector of the curve in the point (u,v) has the same direction as the vector $V(u,v)$.

Tangent curves do not depend on the magnitudes of the vectors in V but only on their directions. A point $(u,v) \in \mathbb{R}^2$ is called a *critical point of V* if $V(u,v) = 0$ is the zero vector.

We consider a non-critical point (u_0, v_0) in the domain of V. Then we know that one and only one tangent curve $t(t) = (u(t), v(t))$ passes through (u_0, v_0). We assume $t(t_0) = (u_0, v_0)$. From the definition of tangent curves we know about the tangent vector of t in (u_0, v_0):

$$\dot{t}(t_0) = \begin{pmatrix} \dot{u}(t_0) \\ \dot{v}(t_0) \end{pmatrix} = V(t(t_0)) = \begin{pmatrix} vx(u_0, v_0) \\ vy(u_0, v_0) \end{pmatrix}. \tag{2}$$

Applying the chain rule to (2), we can compute the second derivative vector of t in (u_0, v_0):

$$\ddot{t}(t_0) = (vx \cdot V_u + vy \cdot V_v)(u_0, v_0). \tag{3}$$

If we consider the domain of the vector field V as the domain of a surface x as well, the tangent curves of V are curves in the domain of x, and therefore are mapped onto surface curves on x. Let $y(t) = x(t(t))$ be the map of the tangent curve $t(t)$ onto x. Applying the chain rule to $x(t(t))$, we obtain for the tangent vectors of y:

$$\dot{y}(t_0) = y^{[1]}(t_0) = x_u(t(t_0)) \cdot \dot{u}(t_0) + x_v(t(t_0)) \cdot \dot{v}(t_0)$$
$$= (vx \cdot x_u + vy \cdot x_v)(u_0, v_0). \tag{4}$$

Defining

$$x_0 = x$$
$$x_{r+1} = vx \cdot (x_r)_u + vy \cdot (x_r)_v \quad \text{for} \quad r = 0, 1, 2, ... \tag{5}$$

we obtain for higher order derivatives of y in a similar way to (4):

$$y^{[r]}(t_0) = x_r(t(t_0)) = x_r(u_0, v_0) \quad \text{for} \quad r = 1, 2, 3, ... \quad . \tag{6}$$

A vector field defining the isophote directions in the domain of x is the perpendicular vector field to the gradient vector field of s defined in (1):

$$V(u, v) = \begin{pmatrix} vx(u, v) \\ vy(u, v) \end{pmatrix} = \begin{pmatrix} -r \cdot n_v(u, v) \\ r \cdot n_u(u, v) \end{pmatrix}. \tag{7}$$

The tangent curves of V are the isophotes in the domain, their maps onto x are the actual isophotes on the surface. Since

$$n_u = \frac{F \cdot M - G \cdot L}{\|x_u \times x_v\|^2} \cdot x_u + \frac{F \cdot L - E \cdot M}{\|x_u \times x_v\|^2} \cdot x_v$$
$$n_v = \frac{F \cdot N - G \cdot M}{\|x_u \times x_v\|^2} \cdot x_u + \frac{F \cdot M - E \cdot N}{\|x_u \times x_v\|^2} \cdot x_v$$

we can write the isophotes vector field V in the domain as

$$V = \begin{pmatrix} vx \\ vy \end{pmatrix} = \begin{pmatrix} -r \cdot (c \cdot x_u + d \cdot x_v) \\ r \cdot (a \cdot x_u + b \cdot x_v) \end{pmatrix}, \tag{8}$$

where

$$a = F \cdot M - G \cdot L , \quad b = F \cdot L - E \cdot M$$
$$c = F \cdot N - G \cdot M , \quad d = F \cdot M - E \cdot N. \tag{9}$$

Critical points occur where the isophotes vector field has a zero vector, i.e. $vx = 0$ and $vy = 0$. We obtain a critical point in $x(u, v)$ iff at least one of the following conditions is satisfied:

- r is parallel to $n(u, v)$,
- $x(u, v)$ has a zero Gaussian curvature and r is in the plane defined by the normal and the principal direction with the zero normal curvature,
- $x(u, v)$ is a flat point.

A proof of this can be found in [7].

§3. The Continuity of Isophotes

In this section we show how to infer a G^{n+1} surface from G^n isophotes. The result is formulated in Theorem 2. To prove this we need the following

Lemma 1. *Given are two regularly parametrized curves $x(t)$ and $\tilde{x}(t)$ which join C^n $(n > 0)$ in the point $x_0 = x(0) = \tilde{x}(0)$. Then the following statement is valid: x and $\tilde{x}$ are G^{n+1} in x_0 iff $(\tilde{x}^{[n+1]}(0) - x^{[n+1]}(0))$ is parallel to $x^{[1]}(0)$.*

Proof: see [6]. ∎

Now we can formulate the following

Theorem 2. *Given are two regularly parametrized surfaces x and $\tilde{x}$ which join along a common line l. Then x and $\tilde{x}$ are G^{n+1} continuous $(n \geq 1)$ along l if there is one family of isophotes on x and $\tilde{x}$ (defined by the direction vector r) with the following properties:*

1) *In no point of l do the isophotes on x and $\tilde{x}$ have critical points.*

2) *In no point of l are the isophotes on x and $\tilde{x}$ tangent to l.*

3) *In no point of l is the projection of r into the tangent plane of x and $\tilde{x}$ tangent to l.*

4) *All isophotes of the family are G^n continuous across l.*

Proof: The direction vector r defines vx and vy with the values a, b, c, d on x (see (8) and (9)). In a similar way, r defines $\tilde{v}x$ and $\tilde{v}y$ with the values $\tilde{a}, \tilde{b}, \tilde{c}, \tilde{d}$ on $\tilde{x}$. We assume that the junction line l is $(0, v), 0 \leq v \leq 1$. This can be done by a linear reparametrization of x and $\tilde{x}$ without loss of generality. Assumption 2) of the theorem can then be written in the form $vx(0, v) \neq 0$. We express r as $r = q_1 \cdot x_u + q_2 \cdot x_v + q_3 \cdot n$ where q_1, q_2 and q_3 are bivariate scalar functions over the domain of x. Then assumption 3) of the theorem holds $q_1(0, v) \neq 0$. Since $q_1 \cdot (F^2 - E \cdot G) = (-G \cdot x_u + F \cdot x_v) \cdot (q_1 \cdot x_u + q_2 \cdot x_v + q_3 \cdot n)$ we obtain

$$(-G \cdot x_u + F \cdot x_v)(0, v) \cdot r \neq 0. \tag{10}$$

The G^n continuity of the family of isophotes gives the G^n continuity of x and $\tilde{x}$ along l. To show this, we can imagine a reparametrization of x and $\tilde{x}$ in such a way that the isophotes defined by r are the isoparametric lines $v = const$ on x and $\tilde{x}$. We thus can assume that x and $\tilde{x}$ are parametrized in such a way that they are C^n along l. Since l is the isoparametric line $u = 0$, we can deduce $x^{[i,j+1]}(0, v) = \tilde{x}^{[i,j+1]}(0, v)$ from $x^{[i,j]}(0, v) = \tilde{x}^{[i,j]}(0, v)$. We obtain

$$x^{[i,j]}(0, v) = \tilde{x}^{[i,j]}(0, v) \quad \text{for} \quad i, j \in \mathbb{N}, \ i + j \leq n + 1, \ i \neq n + 1. \tag{11}$$

(9) and (11) yield along l:

$$
\begin{aligned}
c^{[i,j]} &= \tilde{c}^{[i,j]}, \quad d^{[i,j]} = \tilde{d}^{[i,j]} \quad \text{for} \quad i + j < n \\
a^{[i,j]} &= \tilde{a}^{[i,j]}, \quad b^{[i,j]} = \tilde{b}^{[i,j]} \quad \text{for} \quad i + j < n, \ i \neq n - 1 \\
a^{[n-1,0]} &- \tilde{a}^{[n-1,0]} = (-G \cdot n \cdot (x^{[n+1,0]} - \tilde{x}^{[n+1,0]})) \\
b^{[n-1,0]} &- \tilde{b}^{[n-1,0]} = (F \cdot n \cdot (x^{[n+1,0]} - \tilde{x}^{[n+1,0]})).
\end{aligned}
\tag{12}
$$

From (8) and (12) we obtain along l

$$vx^{[i,j]} = \tilde{vx}^{[i,j]} \quad \text{for} \quad i+j < n$$

$$vy^{[i,j]} = \tilde{vy}^{[i,j]} \quad \text{for} \quad i+j < n\,, \ i \neq n-1$$

$$vy^{[n-1,0]} - \tilde{vy}^{[n-1,0]} = \boldsymbol{r} \cdot [(a^{[n-1,0]} - \tilde{a}^{[n-1,0]}) \cdot \boldsymbol{x}_u + (b^{[n-1,0]} - \tilde{b}^{[n-1,0]}) \cdot \boldsymbol{x}_v]$$

$$= (\boldsymbol{n} \cdot (\boldsymbol{x}^{[n+1,0]} - \tilde{\boldsymbol{x}}^{[n+1,0]})) \cdot (\boldsymbol{r} \cdot (-G \cdot \boldsymbol{x}_u + F \cdot \boldsymbol{x}_v)).$$

$$(13)$$

Let $\boldsymbol{y}^{[1]}(u,v)$ and $\tilde{\boldsymbol{y}}^{[1]}(u,v)$ be the tangent vectors of the isophotes on $\boldsymbol{x}$ and $\tilde{\boldsymbol{x}}$. From (5), (6) and (13) we obtain

$$\boldsymbol{y}^{[i]}(0,v) = \tilde{\boldsymbol{y}}^{[i]}(0,v) \quad \text{for} \quad i \leq n-1$$

$$(\boldsymbol{y}^{[n]} - \tilde{\boldsymbol{y}}^{[n]})(0,v) = (vx^{n-1} \cdot (vy^{[n-1,0]} - \tilde{vy}^{[n-1,0]}) \cdot \boldsymbol{x}_v)(0,v).$$

$$(14)$$

(14) yields that the family of isophotes is C^{n-1} across l. To achieve G^n of the isophotes we must have (see Lemma 1):

$$(\boldsymbol{y}^{[n]} - \tilde{\boldsymbol{y}}^{[n]})(0,v) \quad \text{parallel to} \quad (vx \cdot \boldsymbol{x}_u + vy \cdot \boldsymbol{x}_v)(0,v). \tag{15}$$

(14), $vx(0,v) \neq 0$ and the assumption that $\boldsymbol{x}$ and $\tilde{\boldsymbol{x}}$ are regularly parametrized yield the necessary condition for G^n of the isophotes across l, i.e. for (15):

$$(vy^{[n-1,0]} - \tilde{vy}^{[n-1,0]})(0,v) = 0. \tag{16}$$

Inserting (13) into (16) and keeping (10) in mind yields

$$\boldsymbol{n}(0,v) \cdot (\boldsymbol{x}^{[n+1,0]} - \tilde{\boldsymbol{x}}^{[n+1,0]})(0,v) = 0. \tag{17}$$

Because of (17), there exist two scalar functions $p_1(v)$ and $p_2(v)$ so that

$$\tilde{\boldsymbol{x}}^{[n+1,0]}(0,v) = \boldsymbol{x}^{[n+1,0]}(0,v) + p_1(v) \cdot \boldsymbol{x}_u(0,v) + p_2(v) \cdot \boldsymbol{x}_v(0,v). \tag{18}$$

We consider the reparametrization $\hat{\boldsymbol{x}}$ of $\boldsymbol{x}$ which is defined as

$$\hat{\boldsymbol{x}}(u,v) = \boldsymbol{x}(\hat{u}(u,v), \hat{v}(u,v))$$

$$\hat{u}(u,v) = u + \frac{u^{n+1}}{(n+1)!} \cdot p_1(v) \quad, \quad \hat{v}(u,v) = v + \frac{u^{n+1}}{(n+1)!} \cdot p_2(v). \tag{19}$$

Computing the u-partials of $\hat{\boldsymbol{x}}$ by applying the chain rule to (19) yields for $u = 0$:

$$\hat{\boldsymbol{x}}^{[i,0]}(0,v) = \boldsymbol{x}^{[i,0]}(0,v) = \tilde{\boldsymbol{x}}^{[i,0]}(0,v) \quad \text{for} \quad 0 \leq i \leq n$$

$$\hat{\boldsymbol{x}}^{[n+1,0]}(0,v) = \boldsymbol{x}^{[n+1,0]}(0,v) + p_1(v) \cdot \boldsymbol{x}_u(0,v) + p_2(v) \cdot \boldsymbol{x}_v(0,v). \tag{20}$$

From (18) and (20) we see that $\hat{\boldsymbol{x}}$ and $\tilde{\boldsymbol{x}}$ are C^{n+1} along l. Since $\hat{\boldsymbol{x}}$ is obtained from $\boldsymbol{x}$ by reparametrization, we have shown that $\boldsymbol{x}$ and $\tilde{\boldsymbol{x}}$ are G^{n+1} along l. ∎

Remark: The special case $n = 1$ of Theorem 2 is already shown in [5]. The constraints there are formulated in a slightly different way but coincide with the constraints of Theorem 2.

§4. Local Properties of Isophotes and Surface Interrogation

Since we were able to compute the first and second derivative vector of the isophote through a given surface point $x(u_0, v_0)$ (see (6)), we can compute the geodesic curvature of the isophote in this point:

$$\ddot{y}_p(t_0) = \ddot{y}(t_0) - (n(u_0, v_0) \cdot \ddot{y}(t_0)) \cdot n(u_0, v_0)$$

$$\kappa(u_0, v_0) = \text{sign}(\det[\dot{y}(t_0), \ddot{y}_p(t_0), n(u_0, v_0)]) \cdot \frac{\|\dot{y}(t_0) \times \ddot{y}_p(t_0)\|}{\|\dot{y}(t_0)\|^3}. \qquad (21)$$

$\ddot{y}_p$ denotes the projection of $\ddot{y}$ into the tangent plane. Since the geodesic curvature of a surface curve can be considered as the curvature of a 2D curve, it can be equipped with a sign.

The "thickness of isophotes" (or "distance between adjacent isophotes") is a measure of how strong the value of $s(u, v)$ changes locally. A strong change in s implies that "many isophotes are close together", one isophote is "thin". For the isophotes in the domain of x the measure of the "thickness" is $th = \frac{1}{\|\text{grad}(s)\|} = \frac{1}{\|V\|}$. Mapping this onto the surface, we obtain for the "thickness" of the isophotes through $x(u_0, v_0)$:

$$th(u_0, v_0) = \frac{\|x_u(u_0, v_0) \times x_v(u_0, v_0)\|}{\|\dot{y}(t_0)\|}. \qquad (22)$$

Note that neither the geodesic curvature nor the "thickness" of the isophote through $x(u_0, v_0)$ depends on the parametrization of x. Also note that we were able to compute geodesic curvature and "thickness" of the isophote in $x(u_0, v_0)$ in a closed form even if a closed form of the isophote itself does not exist.

Except for critical points of isophotes we can compute geodesic curvature and "thickness"of the isophotes for every surface point. Around critical points, geodesic curvature and "thickness" of isophotes diverge to infinity.

For using geodesic curvature and "thickness" as a surface interrogation method we compute and color code these measures for every surface point. For doing this we use a continuous color coding map with the following properties: a negative value gets a green color, a positive value gets a red color, the higher the magnitude of the value the lighter the color gets. In fact, a zero value gives black; if the value diverges to plus (minus) infinity the red (green) color tends to white.

The upper left picture of Figure 1 shows the ray traced image of the shoe-shaped test surface. This surface consists of 29×10 piecewise bicubic patches and is G^2 continuous along the patch boundaries. The surface looks smooth, and imperfections are hardly detectable. The color version of Figure 1 can be found on the author's home page.

The middle left picture shows the usual way of visualizing isophotes on the surface. The isophotes here are computed in the following way: choose a (small) interval and mark all points on the surface where the values of $s(u, v)$

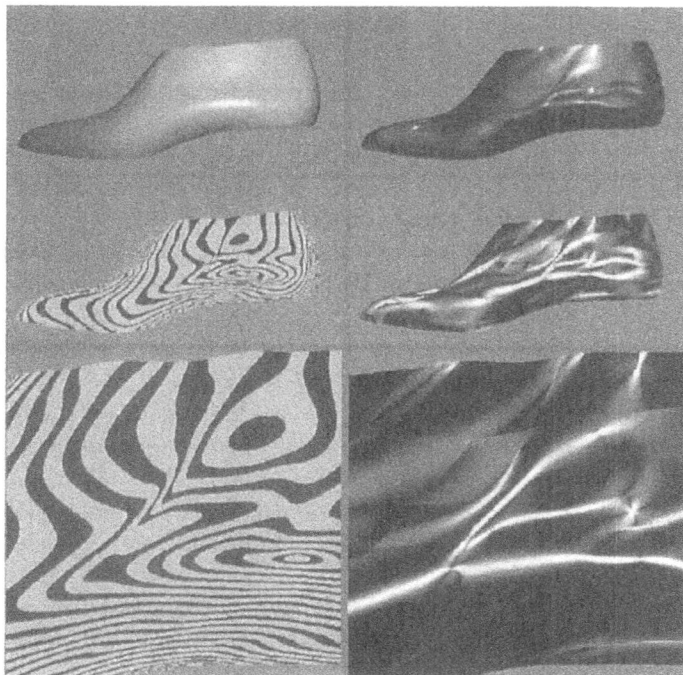

Fig. 1. Isophotes and their local properties on a test surface.

are in the interval. The result are not the isophotes themselves but point sets on the surface which give an impression of the behavior of the isophotes. In particular we can see that the point sets have a varying "thickness".

The upper right picture of Figure 1 shows the visualization of the "thickness" of the isophotes. Here we clearly detect areas of the surface where a redesign is necessary. The critical points of isophotes appear as highlights in the visualization.

The middle right picture of Figure 1 shows the visualization of the geodesic curvature of the isophotes. Again, the critical points of isophotes appear as highlights. We can clearly detect that the isophotes are not curvature (i.e. G^2) continuous at the patch boundaries. Therefore the surface is not G^3 continuous.

The lower left and the lower right pictures of Figure 1 are magnifications of the middle left and the middle right picture.

Acknowledgments. The author would like to thank Prof. Heidrun Schumann from the University of Rostock and Prof. Gerald Farin from Arizona State University for their constant support and encouragement for this work.

References

1. Farin, G., *Curves and Surfaces for Computer Aided Geometric Design*, Academic Press, Boston, 1992, third edition.

2. Hagen, H., S. Hahmann, T. Schreiber, Y. Nakajima, B. Wordenweber and P. Hollemann–Grundstedt, Surface interrogation algorithms, Comp. Graphics and Applics. **12(5)** (1992), 53–60.

3. Herron, G., Techniques for visual continuity, in *Geometric Modeling: Algorithms and New Trends*, G. Farin, editor, SIAM, Philadelphia, 1987, 163–174.

4. Poeschl, T., Detecting surface irregularities using isophotes, Comput. Aided Geom. Design **1(2)** (1984), 163–168.

5. Pottmann, H., Eine Verfeinerung der Isophotenmethode zur Qualitäts-analyse von Freiformflächen, *CAD und Computergraphik*, **4(4)** (1988), 99–109.

6. Theisel, H., Isophotes and geometric continuity of surfaces, *Rostocker Informatik–Berichte*, **19** (1996), ISSN 0233–0784, also available on my home page.

7. Theisel, H., Vector field curvature and applications, dissertation, Dept. of Computer Science, University of Rostock, Germany, 1996, also available on my home page.

Holger Theisel
University of Rostock, Computer Science Department
PostBox 999, 18051 Rostock
GERMANY
theisel@informatik.uni-rostock.de
http://www.icg.informatik.uni-rostock.de/~theisel/

A Sequence of Bézier Curves Generated by Successive Pedal-Point Constructions

Kenji Ueda

Abstract. The outer n legs of contiguous similar right-angled triangles constructs a Bézier n-gon. The Bézier polygon becomes a discrete logarithmic spiral. The Bézier curve of degree n is the negative pedal curve of the Bézier curve of degree $n-1$, and has its own pole, i.e., the pedal point. Each curve is traced by a point which winds around the pole. These Bézier curves are polynomial sinusoidal spirals which include: a straight line ($n = 1$), parabola ($n = 2$) and Tschirnhausen cubic ($n = 3$). All of the curves and their evolutes are offset-rational.

§1. Introduction

There are many operations for constructing one plane curve from another. The pedal curve of a given curve with respect to a fixed point is the locus of the feet of the perpendiculars drawn from the point to the tangents to the given curve. The given curve is called the *negative pedal* (or *antipedal*) curve of the pedal curve. If $\mathbf{Q}$ is a point on the curve $\mathbf{P}(t)$, and $\mathbf{O}$ is a fixed point, then the envelope of a line through $\mathbf{Q}$ perpendicular to $\mathbf{OQ}$ is the negative pedal curve of the curve $\mathbf{P}(t)$ with respect to $\mathbf{O}$ (see Figure 1). The construction of a negative pedal $\mathbf{C}(t)$ of the curve $\mathbf{P}(t)$ is called a *pedal-point construction* on $\mathbf{P}(t)$ with respect to $\mathbf{O}$ [4]. The parabola is the negative pedal of a straight line. The Tschirnhausen cubic is the negative pedal of a parabola with respect to its focus.

Tschirnhausen cubics are also known as *Pythagorean-hodograph (PH) cubics* [2] in CAGD. PH curves are curves with a polynomial speed. Thus the offset curve of a PH curve is rational. To investigate PH curves, it is convenient that plane curves are recognized as complex-valued functions of a real parameter [3]. Offset-rational curves are characterized by a class of complex functions [6].

In this paper, a sequence of Bézier curves constructed by successive pedal-point constructions with respect to the origin are investigated in the complex

A. Le Méhauté, C. Rabut, and L. L. Schumaker (eds.), pp. 427–434.
Copyright © 1997 by Vanderbilt University Press, Nashville, TN.
ISBN 0-8265-1293-3.
All rights of reproduction in any form reserved.

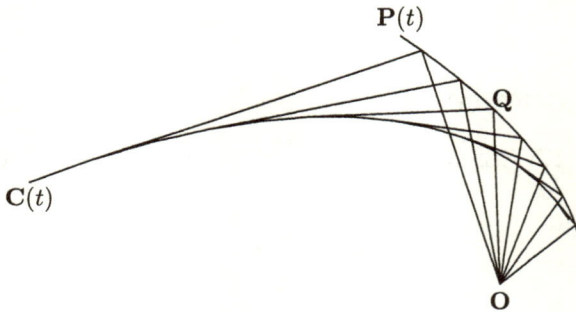

Fig. 1. Pedal-point construction.

plane. The sequence starts with a straight line, followed by a parabola, then the Tschirnhausen cubic, and so on. The curves in the sequence are sinusoidal spirals [5]. The properties of the Bézier polygons of the curves and geometric operation for constructing the Bézier polygons are shown in the following sections.

§2. A Sequence of Negative Pedal Curves

We begin with a point p and a straight line l. The distance from p to l is a. We consider the linear Bézier curve of length b that coincides with the line l. If one of the Bézier points coincides with the foot of the perpendicular dropped from p to l, the point p and the Bézier points form a right-angled triangle. Now we continue by constructing another similar right-angled triangle so that the cathetus adjacent to p has the same length of the hypotenuse as the adjacent right-angled triangle (see Figure 2).

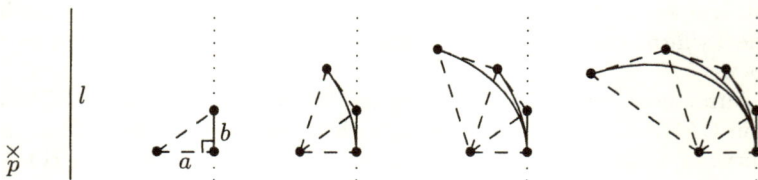

Fig. 2. Pedal-point constructions using Bézier points.

The constructed points lie on the logarithmic spiral $r = \exp \frac{\log \sqrt{2}}{\pi/4} \theta$, with the center at p. The constructed $n+1$ points form a Bézier polygon of degree n and the Bézier curve $\mathbf{Z}_n(t)$ is expressed as

$$\mathbf{Z}_n(t) = a \sum_{j=0}^{n} \left(1 + \mathrm{i}\frac{b}{a}\right)^j \mathrm{B}_j^n(t) = a\left(1 + \mathrm{i}\frac{b}{a}t\right)^n, \qquad t \in [0,1], \qquad (1)$$

where $\mathrm{B}_j^n(t)$ is the Bernstein basis function.

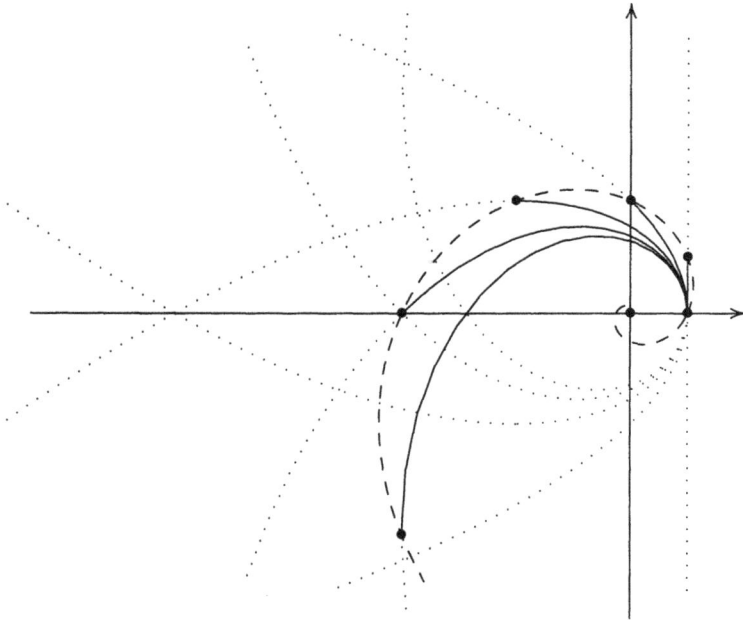

Fig. 3. The function $(1 + \mathrm{i}\,t)^k$ and the logarithmic spiral.

Since the foot of a perpendicular dropped from p to the tangent to $\mathbf{Z}_n(t)$ lies on $\mathbf{Z}_{n-1}(t)$, the pedal curve of $\mathbf{Z}_{n+1}(t)$ is $\mathbf{Z}_n(t)$, i.e., the negative pedal curve of $\mathbf{Z}_n(t)$ is $\mathbf{Z}_{n+1}(t)$. In other words, the constructions of a Bézier polygon are pedal-point constructions.

In this paper, we consider the simplest case of $a = b = 1$, namely

$$\mathbf{Z}_n(t) = \sum_{j=0}^{n} (1 + \mathrm{i})^j \mathrm{B}_j^n(t) = (1 + \mathrm{i}\,t)^n, \qquad 0 < n. \tag{2}$$

The right-angle triangles are isosceles triangles in this case. In Figure 3 the curves $\mathbf{Z}_n(t)$ and the logarithmic spiral are illustrated.

The conjugate of $\mathbf{Z}_n(t)$ is equal to $\mathbf{Z}_n(-t)$, that is, $\overline{\mathbf{Z}}_n(t) = \mathbf{Z}_n(-t)$. Therefore, the curve $\mathbf{Z}_n(t)$ is symmetric with respect to the real axis.

By setting $a = 1$, $b = 1/n$ and $t = \hat{t}/n$, the sequence of curves $\mathbf{Z}_n(t)$ tends to the unit circle centered at the origin, namely

$$\lim_{n \to \infty} \mathbf{Z}_n(t) = \lim_{n \to \infty} \left(1 + \frac{\mathrm{i}}{n}\hat{t}\right)^n = e^{\mathrm{i}\,\hat{t}} = \cos\hat{t} + \mathrm{i}\sin\hat{t}. \tag{3}$$

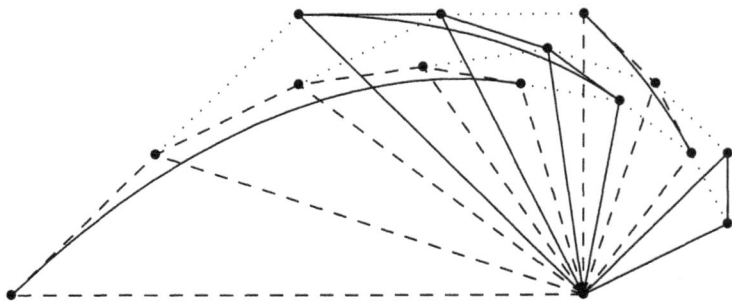

Fig. 4. Segments of $(1 + \mathrm{i}\,t)^k$ for parameter $t \in [\frac{1}{2}, 1]$.

§3. Properties of Each Curve in the Sequence

A segment of the curve $\mathbf{Z}_n(t)$ for the parameter value $t \in [t_0, t_1]$ can be obtained by the parameter transformation $t = (\hat{t} - t_0)/(t_1 - t_0)$. The segment curve $\mathbf{Z}_n(\hat{t})$ is given by

$$\mathbf{Z}_n(\hat{t}) = \left(1 - \mathrm{i}\,\frac{t_0}{t_1 - t_0}\right)^n \sum_{j=0}^{n} \left(1 + \frac{-t_0 + \mathrm{i}\,(t_1 - t_0)}{t_0^2 + (t_1 - t_0)^2}\right)^j B_j^n(\hat{t}), \qquad \hat{t} \in [0, 1]. \tag{4}$$

We can obtain these Bézier points by stepping along a logarithmic spiral in steps of a constant angle, as illustrated in Figure 4. Conversely, the Bézier curve has its own pole, the center of the logarithmic spiral. The pole is also the pedal point of the pedal-point constructions, and can be recognized as the focus of the parabola of which the Bézier points are three successive Bézier points of $\mathbf{Z}_n(t)$.

The derivative of the curve $\mathbf{Z}_n(t)$ is given by

$$\mathbf{Z}'_n(t) = \mathrm{i}\,n(1 + \mathrm{i}\,t)^{n-1} = \mathrm{i}\,n\mathbf{Z}_{n-1}(t). \tag{5}$$

Hence, the speed of the curve becomes

$$|\mathbf{Z}'_n(t)| = |\mathrm{i}\,n\mathbf{Z}_{n-1}(t)| = n\sqrt{1 + t^2}^{\,n-1}. \tag{6}$$

The curve $\mathbf{Z}_n(t)$ has polynomial speed for odd n, that is, the curve $\mathbf{Z}_n(t)$ is a PH curve.

Via the parameter transformation $t = (1 - s^2)/(2s)$, the term $1 + t^2$ is rewritten to

$$1 + t^2 = \left(\frac{1 + s^2}{2s}\right)^2. \tag{7}$$

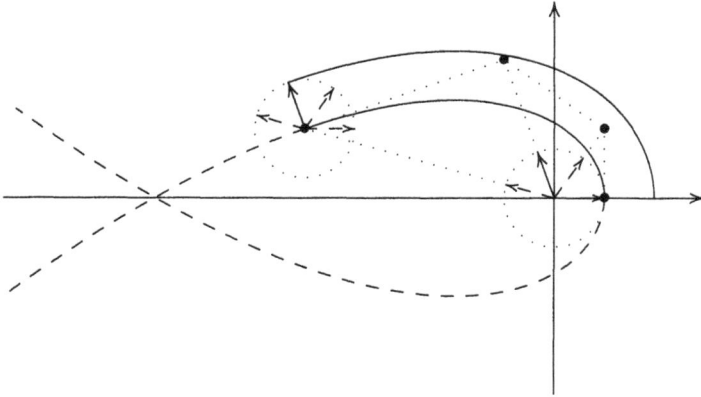

Fig. 5. Offset curve of Tschirnhausen cubic.

Therefore, we can remove the radical for even n from Equation (6). All the curves in the sequence have rational speed.

In the complex plane, the curvature of $\mathbf{Z}_n(t)$ is defined as

$$\kappa_n(t) = \mathrm{i}\,\frac{\mathbf{Z}_n'(t)\overline{\mathbf{Z}}_n''(t) - \overline{\mathbf{Z}}_n'(t)\mathbf{Z}_n''(t)}{2|\mathbf{Z}_n'(t)|^3} = \frac{n-1}{n}\,\frac{1}{(1+t^2)^{\frac{n+1}{2}}}. \tag{8}$$

The curve $\mathbf{Z}_n(t)$ has a vertex with the parameter value $t = 0$, because of $\kappa_n'(0) = 0$.

By using the unit normal $-\mathrm{i}\,\mathbf{z}'(t)/|\mathbf{z}'(t)|$ to a curve $\mathbf{z}(t)$, the offset at distance d from $\mathbf{z}(t)$ is given by

$$\mathbf{O}_d(\mathbf{z}(t)) = \mathbf{z}(t) - \mathrm{i}\,d\frac{\mathbf{z}'(t)}{|\mathbf{z}'(t)|}. \tag{9}$$

From Equation (9), the offset curve of $\mathbf{Z}_n(t)$ becomes

$$\mathbf{O}_d(\mathbf{Z}_n(t)) = \mathbf{Z}_n(t) + d\frac{\mathbf{Z}_{n-1}(t)}{|\mathbf{Z}_{n-1}(t)|} = \mathbf{Z}_n(t) + d\,\exp\left(\mathrm{i}\,\frac{n-1}{n}\,\arg(\mathbf{Z}_n(t))\right). \tag{10}$$

For $n = 3$, this situation is illustrated in Figure 5.

The evolute of a curve is the locus of the center of curvature of the curve. In the complex plane, the evolute of a curve $\mathbf{z}(t)$ is defined as

$$\mathbf{E}(\mathbf{z}(t)) = \mathbf{z}(t) + \frac{2\mathbf{z}'(t)\overline{\mathbf{z}}'(t)}{\overline{\mathbf{z}}''(t)\mathbf{z}'(t) - \mathbf{z}''(t)\overline{\mathbf{z}}'(t)}\mathbf{z}'(t). \tag{11}$$

The evolute of the curve $\mathbf{Z}_n(t)$ and its derivative becomes

$$\mathbf{E}(\mathbf{Z}_n(t)) = \frac{-1 + \mathrm{i}\,nt}{n-1}\mathbf{Z}_n(t), \quad \text{and} \quad (\mathbf{E}(\mathbf{Z}_n(t)))' = \mathrm{i}\,\frac{n+1}{n-1}t\mathbf{Z}_n'(t). \tag{12}$$

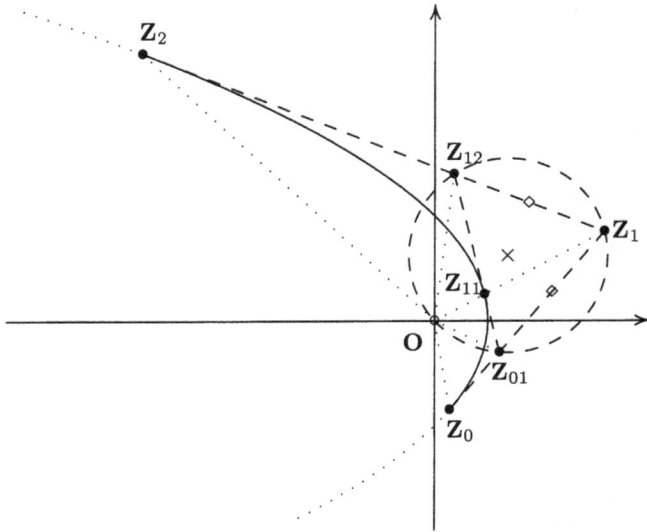

Fig. 6. Focus of parabola.

Since the curve $\mathbf{Z}_n(t)$ has its vertex with the parameter value $t = 0$, the evolute is a cuspidal curve. The square of the speed for the evolute is given by

$$|(\mathbf{E}(\mathbf{Z}_n(t)))'|^2 = \left(\frac{n+1}{n-1}t\right)^2 |\mathbf{Z}_n'(t)|^2. \tag{13}$$

The evolute can also have rational speed.

§4. Geometric Construction of the Negative Pedal Bézier Curves

The relationship between the focus of a conic and its Bézier polygon has been shown [7]. The pole of $\mathbf{Z}_n(t)$ can be recognized as the focus of the parabola, of which the Bézier points are three successive Bézier points of $\mathbf{Z}_n(t)$, i.e. $\mathbf{z}_{j-1}$, $\mathbf{z}_j$ and $\mathbf{z}_{j+1}$. The focus $\mathbf{z}$ is obtained as the point in the complex plane such that

$$(\mathbf{z}_j - \mathbf{z})^2 = (\mathbf{z}_{j-1} - \mathbf{z})(\mathbf{z}_{j+1} - \mathbf{z}). \tag{14}$$

The focus of a parabola can also be obtained by a geometric operation. Let $\mathbf{Z}(t)$ be the quadratic Bézier curve

$$\mathbf{Z}(t) = \mathbf{Z}_0(1-t)^2 + \mathbf{Z}_1 2(1-t)t + \mathbf{Z}_2 t^2. \tag{15}$$

Suppose $\mathbf{O}$ is the focus of the parabola $\mathbf{Z}(t)$, the triangles of points $\mathbf{OZ}_0\mathbf{Z}_1$ and $\mathbf{OZ}_1\mathbf{Z}_2$ are similar, so that

$$\frac{|\mathbf{Z}_1 - \mathbf{O}|}{|\mathbf{Z}_0 - \mathbf{O}|} = \frac{|\mathbf{Z}_2 - \mathbf{O}|}{|\mathbf{Z}_1 - \mathbf{O}|} = \frac{|\mathbf{Z}_2 - \mathbf{Z}_1|}{|\mathbf{Z}_1 - \mathbf{Z}_0|}. \tag{16}$$

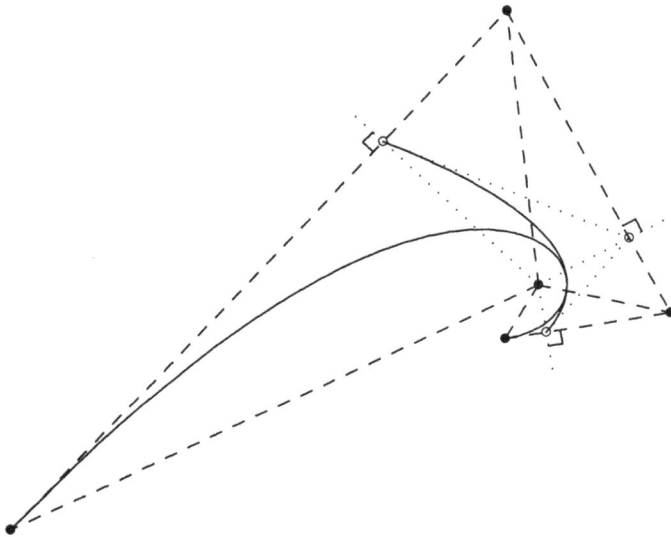

Fig. 7. Pedal-point construction of Bézier polygons.

Calculating the point with the parameter value

$$t^* = \frac{|\mathbf{Z}_1 - \mathbf{Z}_0|}{|\mathbf{Z}_1 - \mathbf{Z}_0| + |\mathbf{Z}_2 - \mathbf{Z}_1|} \tag{17}$$

on the Bézier curve by de Casteljau algorithm, we obtain the following de Casteljau points:

$$\mathbf{Z}_{01} = (1 - t^*)\mathbf{Z}_0 + t^*\mathbf{Z}_1, \qquad \mathbf{Z}_{12} = (1 - t^*)\mathbf{Z}_1 + t^*\mathbf{Z}_2, \tag{18}$$

$$\mathbf{Z}_{11} = (1 - t^*)\mathbf{Z}_{01} + t^*\mathbf{Z}_{12} = \mathbf{Z}(t^*). \tag{19}$$

Note that $\triangle \mathbf{Z}_{01}\mathbf{Z}_1\mathbf{Z}_{12}$ becomes an isosceles triangle.

If AD is an angle bisector of triangle ABC which meets BC at D, then $BD/DC = AB/AC$. Therefore $\overline{\mathbf{O}\mathbf{Z}_{01}}$ and $\overline{\mathbf{O}\mathbf{Z}_{12}}$ are angle bisectors of the similar triangles $\triangle \mathbf{O}\mathbf{Z}_0\mathbf{Z}_1$ and $\triangle \mathbf{O}\mathbf{Z}_1\mathbf{Z}_2$ respectively, and

$$\frac{|\mathbf{Z}_{12} - \mathbf{O}|}{|\mathbf{Z}_{01} - \mathbf{O}|} = \frac{|\mathbf{Z}_2 - \mathbf{Z}_1|}{|\mathbf{Z}_1 - \mathbf{Z}_0|}. \tag{20}$$

As $\overline{\mathbf{O}\mathbf{Z}_{11}}$ is also the angle bisector of triangle $\mathbf{O}\mathbf{Z}_{01}\mathbf{Z}_{12}$, the points $\mathbf{Z}_{11}$ $\mathbf{Z}_1$ and $\mathbf{O}$ are collinear. Moreover, the focus of a parabola belongs to the circle circumscribed around the triangle, if the three sides of a triangle are tangent to the parabola [1]. Thus the focus of the parabola $\mathbf{Z}(t)$ is the intersection of the line $\overline{\mathbf{Z}_1\mathbf{Z}_{11}}$ and the circumscribed circle of the triangle $\mathbf{Z}_{01}\mathbf{Z}_1\mathbf{Z}_{12}$ (see Figure 6).

Once the pole of the Bézier polygon is found, the Bézier polygon of its negative pedal curve can be constructed geometrically. Figure 4 suggests an operation for constructing the Bézier polygon of the negative pedal curve from the original Bézier polygon. Figure 7 shows the operation which generates the perpendiculars of the radials of the original control polygon at the Bézier points (hollow circles) and make the intersections of the perpendiculars the Bézier points (solid circles) of the negative pedal curve. Two end Bézier points are obtained by extrapolating polygon legs to construct similar triangles.

Conversely, the Bézier points of the pedal curve are the feet of the perpendiculars dropped from the pole to the legs of the original Bézier polygon.

§5. Conclusion

A sequence of Bézier curves generated by successive pedal-point constructions starting with a straight line have been shown. Geometrically, each curve in the sequence has its own pole and the Bézier polygon is divided into similar triangles of which one corner coincides with the pole. It has been shown algebraically that all of the curves and their evolutes can have rational speed. Therefore their offset curves are expressed by rational curves exactly.

The curves in the sequence are sinusoidal spirals which are polynomial. This paper investigated the properties of the curves from the Bézier polygon point of view. Sinusoidal spirals, which may be either polynomial or rational, have many interesting properties.

References

1. Berger, M., *Geometry*, Springer-Verlag, 1987.

2. Farouki R. T. and T. Sakkalis, Pythagorean hodographs, IBM J. Res. Develop. **34** (1990), 736–752.

3. Farouki, R. T., The conformal map $z \rightarrow z^2$ of the hodograph plane, Computer Aided Geometric Design **11** (1994), 363–390.

4. Hilbert D. and S. Cohn-Vossen, *Geometry and the Imagination*, Chelsea Publishing Co., 1952.

5. Lawrence, J. D., *A Catalog of Special Plane Curves*, Dover, 1972.

6. Lü, W., Offset-rational parametric plane curves, Computer Aided Geometric Design **12** (1995), 601–616.

7. Sànchez-Reyes, J., Single-valued curve in polar coordinates, Computer-aided Design **22** (1990), 19–26.

Kenji Ueda
Ricoh Company, Ltd.
1-1-17 Koishikawa, Bunkyo-ku, Tokyo 112, JAPAN
ueda@src.ricoh.co.jp

From Degenerate Patches to Triangular and Trimmed Patches

Marc Vigo, Núria Pla, and Pere Brunet

Abstract. CAD systems are usually based on a tensor product representation of free form surfaces. In this case, trimmed patches are used for modeling nonrectangular zones. However, several commercial CAD systems represent certain nonrectangular surface regions through degenerate rectangular patches. Degenerate patches produce rendering artifacts and can lead to malfunctions in the subsequent geometric operations. In the present paper two algorithms for converting degenerate tensor-product patches into triangular and trimmed rectangular patches are presented. In both algorithms, the final surface approximates the initial one in a quadratic sense while inheriting its boundary curves. Moreover, in the second one, almost G^1 continuity is achieved.

§1. Introduction

Although extensive research has been performed on nonrectangular patches for free-form surface design and representation, the standard polynomial tensor-product patch still remains as the classical way of representing surfaces in most commercial CAD systems. This representation is not suitable for general surface topologies. Nonrectangular patch shapes must be introduced at specific surfaces features, like corners or the intersections with the axis in revolution surfaces [4]. The basic tool for modeling nonrectangular zones has been the use of trimmed surfaces [2]. Usual data interchange formats as IGES and VDA [6,8] use this representation.

Several commercial CAD systems, however, represent some nonrectangular surface regions through degenerate patches. A degenerate rectangular patch is a tensor-product patch having several coincidental control points in the Euclidean space. Such patches can lead to significant problems and to malfunctions in the algorithms using the representations. As an example, the surface of revolution in Figure 2 (from a commercial CAD system) has been modeled by degenerate rectangular patches at the triangular regions that surround its top. Control points in the upper boundaries of the patches that

Curves and Surfaces with Applications in CAGD
A. Le Méhauté, C. Rabut, and L. L. Schumaker (eds.), pp. 435–444.
Copyright © 1997 by Vanderbilt University Press, Nashville, TN.
ISBN 0-8265-1293-3.
All rights of reproduction in any form reserved.

collapse at the surface top center coincide at this top central point. With this configuration, the tangent plane at this point is in general not defined [4], producing artifacts in the rendering of this zone.

In the present paper, two algorithms for converting degenerate patches to triangular and trimmed polynomial patches are presented and discussed. The first algorithm is based on the Watkins-Worsey degree reduction algorithm for curves [10]. The second algorithm also preserves the tangent plane along boundary curves, with a boundary ε-G^1 continuity in the sense of [3]. Both algorithms work by first obtaining a suitable triangular patch approximation to the initial degenerate patch. In a second step, the triangular patch is converted onto a standard polynomial rectangular patch trimmed by its diagonal in the parametric plane using the algorithm from [1].

The structure of the paper is as follows. Section 2 introduces the main definitions and terminology, while Section 3 presents the proposed algorithms. Approximation errors are analyzed in Section 4, and some examples are presented in Section 5.

§2. Problem Statement

Given an $m \times n$ (with $m < n$) tensor product patch, it can always be converted into a $n \times n$ rectangular patch by degree elevation; thus, it is not restrictive to consider tensor product patches of degree $n \times n$. Then, the problem to be solved can be stated as follows. Given

$$s(u,v) = \sum_{i=0}^{n}\sum_{j=0}^{n} \mathbf{P}_{ij} B_i^n(u) B_j^n(v) \quad \text{with} \quad \mathbf{P}_{0n} = \mathbf{P}_{1n} = \cdots = \mathbf{P}_{nn},$$

a set $\{\mathbf{Q}_{ijk} \mid i,j,k \geq 0, \; i+j+k = n\}$ of control points of the triangular patch

$$g(u',v',w') = \sum_{i+j+k=n} \mathbf{Q}_{ijk} B_{ijk}^n(u',v',w')$$

must be found such that the following conditions are fulfilled:

- $g(u',v',w')$ is a good approximation of $s(u,v)$, i.e., $\|g(u',v',w') - s(u,v)\|$ can be bounded by a small number using a suitable parameter mapping $(u,v) \longrightarrow (u',v',w')$.
- $g(u',v',w')$ has the same boundary Bézier curves as $s(u,v)$.
- $g(u',v',w')$ is ε-G^1 with respect to $s(u,v)$, see the definition below.

Definition 2.1. *It is said that a surface is ε-G^1 if the discontinuity in the surface normal is bounded by ε, (see [3]). The discontinuity of the surface normal is the maximum angle between two surface normals at any point. A surface g is ε-G^1 with respect to another surface s if the maximum angle between surface normals of g and s at homologous points along g and s boundaries (according to a certain parameter mapping) is bounded by ε.*

Note that the goal is to obtain a good approximation of $s(u,v)$. In general, it is not possible to convert a rectangular degenerate patch s into a

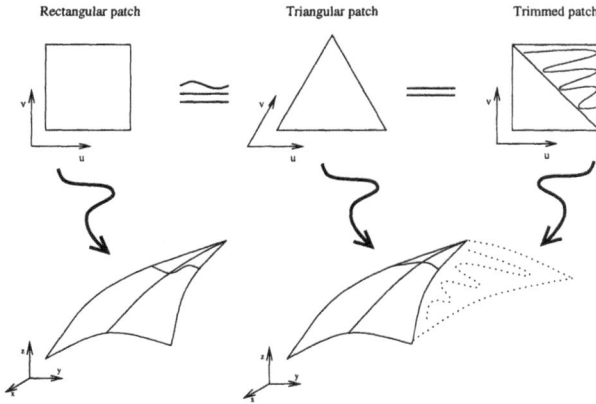

Fig. 1. Conversion scheme.

triangular one g without approximation error, as it is not guaranteed that the control points $\mathbf{P}_{0n-1}, ..., \mathbf{P}_{nn-1}$ define a unique tangent plane in the degenerate point $\mathbf{P}_{nn}$ of s. For the same reason, only approximate continuity, ε-G^1 continuity, can be expected instead of G^1.

§3. Conversion Algorithms

Two different algorithms are proposed in this paper to approximate a degenerate Bézier tensor product patch using a triangular Bézier patch. This triangular Bézier patch is converted into a trimmed rectangular patch using Brueckner's algorithm [1]. Figure 1 represents the steps of the complete algorithm.

Both algorithms are based on Lemma 3.1, which proves that any triangular Bézier patch can be represented as a degenerate rectangular Bézier patch.

Lemma 3.1. *Let* $g(u', v', w') = \sum \mathbf{Q}_{ijk} B_{ijk}^n(u', v', w')$ *be an arbitrary triangular Bézier patch. The patch can be always transformed into a degenerate rectangular Bézier patch through a parameter mapping.*

Proof: The mapping from the unit square $[0,1] \times [0,1]$ to the triangle with vertices $(1,0,0)$, $(0,1,0)$, $(0,0,1)$, in barycentric coordinates, defined by $(u,v) \rightarrow (u(1-v), v, (1-u)(1-v))$ gives the following transformation:

$$s(u,v) = g(u(1-v), v, (1-u)(1-v)) = \sum_{j=0}^{n} \left(\sum_{i=0}^{n-j} \mathbf{Q}_{ijn-i-j} B_i^{n-j}(u) \right) B_j^n(v)$$

and by degree elevation of the curves to degree n, the lemma is proved. ■

Corollary 3.2. *Given a tensor product Bézier patch* $\sum \mathbf{P}_{ij} B_i^n(u) B_j^m(v)$, *if the u-control curves* $\sum \mathbf{P}_{ij} B_i^n(u)$ *come from curves* $\sum \mathbf{Q}_{ijn-i-j} B_i^{n-j}(u)$ *of* degree $n - j$ by degree elevation, then the patch $\sum \mathbf{P}_{ij} B_i^n(u) B_j^m(v)$ may be written as a triangular patch $g(u', v', w')$ with $u' = u(1 - v)$, $v' = v$, $w' = (1 - u)(1 - v)$.

The corollary will be used to construct a triangular Bézier patch from a degenerate rectangular Bézier patch.

Let $s(u, v) = \sum_{i,j} \mathbf{P}_{ij} B_i^n(u) B_j^n(v)$ be a degenerate rectangular Bézier patch such that $\mathbf{P}_{0n} = \mathbf{P}_{1n} = \cdots = \mathbf{P}_{nn}$. A triangular Bézier patch has to be defined to approximate the degenerate patch. Using the previous transformation, the triangular patch can be represented as

$$g(u, v) = \sum_{j=0}^{n} \left(\sum_{i=0}^{n-j} \mathbf{Q}_{ijn-i-j} B_i^{n-j}(u) \right) B_j^n(v).$$

The control points $\{\mathbf{Q}_{ijk} \mid i, j, k \geq 0, \ i + j + k = n\}$ will be defined in order to minimize $\max \|g(u, v) - s(u, v)\|$, which can be represented as

$$\max_{\substack{0 \leq u \leq 1 \\ 0 \leq v \leq 1}} \left\| \sum_{j=0}^{n} \left(\sum_{i=0}^{n} \mathbf{P}_{ij} B_i^n(u) - \sum_{i=0}^{n-j} \mathbf{Q}_{ijn-i-j} B_i^{n-j}(u) \right) B_j^n(v) \right\|.$$

Due to the convex hull property of the Bernstein polynomials, control points $\{\mathbf{Q}_{ijk} \mid i, j, k \geq 0, \ i + j + k = n\}$ are sought which minimize

$$\max_{0 \leq u \leq 1} \left\| \sum_{i=0}^{n} \mathbf{P}_{ij} B_i^n(u) - \sum_{i=0}^{n-j} \mathbf{Q}_{ijk} B_i^{n-j}(u) \right\|, \qquad \forall j = 0, \ldots, n.$$

The problem is therefore reduced to the problem of degree reduction of Bézier curves.

3.1 C^0 Algorithm

The first method uses a degree reduction algorithm that transforms a nth degree Bézier curve into a $(n - 1)$th degree curve with the same endpoints. The method used here is due to Watkins and Worsey, which is in turn based on the minimax approximation, [10].

Given a degenerate patch $\sum \mathbf{P}_{ij} B_i^n(u) B_j^n(v)$ such that $\mathbf{P}_{0n} = \cdots = \mathbf{P}_{nn}$, the C^0-algorithms applies the Watkins-Worsey degree reduction algorithm to the control curves

$$\sum_{i=0}^{n} \mathbf{P}_{ij} B_i^n(u), \qquad j = 0, \ldots, n,$$

to convert them into a $(n - j)$th degree curves with the same endpoints.

As result, a new set of $n + 1$ Bézier curves is obtained. The jth curve is a $(n - j)$th degree curve, having the same endpoints as the jth curve of the original set. The new curves define a triangular patch with the same boundary curves as the original patch.

3.2 ε-G^1 **Algorithm**

The Watkins-Worsey curve degree reduction method assures C^0 continuity, but any kind of continuity cannot be expected of the tangent at the boundary of the endpoints. The ε-G^1 algorithm also controls the variation of the tangent plane along the boundary curves of the patch.

Lemma 3.2. *Let* $\sum \mathbf{P}_{ij} B_i^n(u) B_j^n(v)$ *be a degenerate tensor product Bézier patch with* $\mathbf{P}_{0n} = \cdots = \mathbf{P}_{nn}$, *and let* $\sum \mathbf{Q}_{ijk} B_{ijk}^n(u', v', w')$ *be a triangular Bézier patch. Assume that for each* j, *by degree elevation of the Bézier curve* $\sum_i \mathbf{Q}_{ijn-i-j} B_i^{n-j}(u)$, *it is obtained* $\sum_i \mathbf{Q}'_{ij} B_i^n(u)$ *with the properties:*

$$\mathbf{Q}'_{0j} = \mathbf{P}_{0j}, \qquad and \qquad \mathbf{Q}'_{nj} = \mathbf{P}_{nj}, \qquad \forall j. \qquad (1)$$

$$\mathbf{Q}'_{1j} = \mathbf{P}_{1j} \qquad and \qquad \mathbf{Q}'_{n-1j} = \mathbf{P}_{n-1j} \qquad \forall j \qquad (2)$$

Then, the boundary curves converging at $\mathbf{P}_{0n}$ *in both patches coincide and have the same tangent plane along them.*

Proof: The conditions (1) imply that both patches define the same boundary curves. On the other hand, the tangent planes along the curves given by $u = 0$ and by $u = 1$ are defined by the control points $\{\mathbf{P}_{0j}, \mathbf{P}_{1j} \mid \forall j\}$ and $\{\mathbf{P}_{nj}, \mathbf{P}_{n-1j} \mid \forall j\}$, respectively. Thus, from (1) and (2), both patches define boundary curves with the same tangent planes. ∎

The previous lemma leads to the following degree reduction algorithm that is the basis of the ε-G^1 approximation method: Given an Bézier curve $\sum \mathbf{P}_i B_i^n(u)$ with $n > 3$, an $(n-1)-th$ degree Bézier curve $\sum \mathbf{Q}_i B_i^{n-1}(u)$ can be defined such that

– The endpoints coincide, that is

$$\mathbf{Q}_0 = \mathbf{P}_0 \quad and \quad \mathbf{Q}_{n-1} = \mathbf{P}_n \qquad (3)$$

– By degree elevation of the computed curve $\sum \mathbf{Q}_i B_i^{n-1}(u)$, an $n-th$ degree curve $\sum \mathbf{P}'_i B_i^n(u)$ is obtained such that $\mathbf{P}'_1 = \mathbf{P}_1$ and $\mathbf{P}'_{n-1} = \mathbf{P}_{n-1}$, i.e., Q_1 and Q_{n-2} must be computed as

$$\mathbf{Q}_1 = \frac{n}{n-1}\mathbf{P}_1 - \frac{1}{n-1}\mathbf{P}_0 \quad and \quad \mathbf{Q}_{n-2} = \frac{n}{n-1}\mathbf{P}_{n-1} - \frac{1}{n-1}\mathbf{P}_n. \quad (4)$$

The rest of the control points $\mathbf{Q}_2, \ldots, \mathbf{Q}_{n-3}$ are computed by imposing

$$\mathbf{P}_2 - \left(\frac{2}{n}\mathbf{Q}_1 + \frac{n-2}{n}\mathbf{Q}_2\right) = 0,$$

$$\vdots$$

$$\mathbf{P}_{n-3} - \left(\frac{n-3}{n}\mathbf{Q}_{n-4} + \frac{3}{n}\mathbf{Q}_{n-3}\right) = 0$$

$$\mathbf{P}_{n-2} - \left(\frac{n-2}{n} \mathbf{Q}_{n-3} + \frac{2}{n} \mathbf{Q}_{n-2} \right) = 0,$$

together with the constraints (4). In fact, this system can be seen as three different overdetermined linear systems, corresponding to the x, y and z coordinates, with the same matrix A_n. Each system has $n-3$ equations and $n-4$ unknowns, and they are solved by the least square method. The matrix A_n depends only on the degree n. From now on, this system will be denoted by $A_n Q = P'$.

The degree reduction applied to a Bézier curve of degree $n = 3$ uses (4), and computes Q_2 as the middle point of the two points that have been computed. In case of a degree $n = 2$ curve, the endpoints are taken.

Now, the ε-G^1 algorithm can be described. Let $\sum_{ij} \mathbf{P}_{ij} B_i^n(u) B_j^n(v)$ be a degenerate patch with $\mathbf{P}_{0n} = \cdots = \mathbf{P}_{nn}$. For any j value, the previous degree reduction algorithm is sequentially applied j times in order to obtain a $n - j$ degree curve from the $j - th$ control curve $\sum_{i=0}^{n} \mathbf{P}_{ij} B_i^n(u)$, where $j = 0, \ldots, n$. The resulting set of Bézier curves describe the approximated triangular Bézier patch.

§4. Error Analysis

The goal of this section is to bound the error due to the approximation of a degenerate rectangular patch by a triangular Bézier patch, in each one of the proposed algorithms.

Let $\{P_{ij}\}_{i,j=0,\ldots,n}$ be the control points of the degenerate patch $s(u,v)$, and let $\{Q_{ijk}\}_{i+j+k=n}$ be the control points of the triangular patch, $g(u,v)$, obtained using one of the algorithms. The approximation error bounds the difference $\|g(u,v) - s(u,v)\|$. Using the properties of Bernstein polynomials, this difference is bounded by the maximum error produced by the degree reduction process applied to the control curves of the original patch.

The C^0-algorithm uses the degree reduction process proposed by Watkins and Worsey. The error incurred in the degree reduction of a nth degree curve is bounded by 2α, where α is the coeficient of the nth degree term of the curve represented in the Tchebycheff polynomial basis. This bound is used to compute the error produced when applying j times this algorithm. In this case, it can be proved that the approximation error can be reduced to any positive value using one of the following preprocesses: degree elevation of the original patch or subdividing the patch into vertical strips, see [9].

In case of the ϵ-G^1 algorithm, the degree reduction error can be bounded as follows. Let $c(u) = \sum \mathbf{P}_i B_i^n(u)$ be a nth degree Bézier curve with $n > 3$, and let $c'(u) = \sum \mathbf{Q}_i B_i^{n-1}(u)$ be the result of the degree reduction method. Then, using that the function $\sum_{i=2}^{n-2} B_i^n(u)$ attains its maximum value for $u = \frac{1}{2}$,

$$\|c(u) - c'(u)\| \le \max_{i=2,\ldots,n-2} \left\| \mathbf{P}_i - \left(\frac{i}{n} \mathbf{Q}_{i-1} + \frac{n-i}{n} \mathbf{Q}_i \right) \right\| \left(1 - \frac{n+1}{2^{n-1}} \right). \quad (5)$$

The differences $\left\|\mathbf{P}_i - \left(\frac{i}{n}\mathbf{Q}_{i-1} + \frac{n-i}{n}\mathbf{Q}_i\right)\right\|$ can be evaluated from the least square error, given by, see [5]:

$$\left\|\left(A_n \left(A_n^T A_n\right)^{-1} A_n^T - I\right) P'\right\| \leq \left\|A_n \left(A_n^T A_n\right)^{-1} A_n^T - I\right\| \cdot \|P'\|.$$

The norm of the matrix $A_n \left(A_n^T A_n\right)^{-1} A_n^T - I$ only depends on n and asymptotically tends to zero when n increases.

As in the case of the C^0 algorithm, a preprocess step can be applied in order to obtain an approximation error smaller than a prefixed value. The preprocess is based on degree elevation.

In the case of the ε-G^1 algorithm, the continuity error can also be studied. This error evaluates the difference between the normal directions along each one of the three boundary curves of the patch.

The three boundary curves do not receive the same treatment. The boundary curves that collapse to the degenerate vertex have a different treatment than the other boundary. The curve $u = 0$ is studied; an analogous reasoning is true in the case of the curve $u = 1$. In this case, as the boundary curves of both patches are identical and have the same parameterization, the difference between the normal directions will depend on the difference between the cross tangent directions along the curve. Let us call $D_1(v)$ the *cross tangent direction* of this boundary curve of the original patch and $D'_1(v)$ the corresponding tangent direction of the resulting patch. The difference between the tangent directions is

$$D_1(v) - D'_1(v) = \left(\mathbf{P}_{1n-2} - \mathbf{P}'_{1n-2}\right) B_{n-2}^n(v) + \left(\mathbf{P}_{1n-1} - \mathbf{P}'_{1n-1}\right) B_{n-1}^n(v) \quad (6)$$

where P_{ij} are the original control points and P'_{ij} are the control points of the new patch. From (6) and due to the properties of the Bernstein polynomials, it can be proved that the continuity error along this curve is only significant in a neighborhood of the degenerate vertex.

Theorem. *Let $\sum \mathbf{P}_{ij} B_i^n(u) B_j^n(v)$ be a degenerate Bézier patch, and let ϵ and δ be arbitrary numbers with $\epsilon > 0$ and $\delta > 0$. The given patch can be transformed by a degree elevation process into a patch $\sum \mathbf{R}_{ij} B_i^m(u) B_j^m(v)$ such that the patch obtained by the ε-G^1 algorithm presents a continuity error smaller than ϵ, for all v value with the condition that $1 - v > \delta$.*

Proof: A proof can be found in [9]. ∎

The continuity error along the boundary curve that does not converge to the degenerate vertex is given by the difference

$$D_1(u) - D_2(u) = \sum_{i=2}^{n-2} \left(\mathbf{P}_{i1} - \mathbf{P}'_{i1}\right) B_i^n(u). \quad (10)$$

This difference can be evaluated from the least square error. Moreover, notice that if one-step degree elevation of the original patch is performed, a zero continuity error along this boundary curve will result, because the two control curves that take part in the computation of the tangent plane along this curve have not changed.

§5. Results and Examples

Several tests have been made, transforming degenerate patches first into triangular patches and then into trimmed patches. Figures 2 and 3 (*upper left*) show surfaces made of degenerate patches where the produced error can be visually appreciated. The original surfaces (*upper left*) presents a dark zone close to the degenerate vertex. In the approximation resulting using the first algorithm (*upper right*) edges along common boundaries of the patches are produced, because only the C^0 continuity is maintained. When the object is transformed using the ε-G^1 algorithm (*lower left*), those edges are smoothed. Degree elevation is performed on the original surface to obtain a more accurate approximation of the object before applying the ε-G^1 algorithm (*lower right*).

Table 1 shows the upper bounds of the produced errors using our conversion methods on a 5×5 degenerate patch. The table shows the relative errors in relation to the diameter of the convex hull of the control points. In the case of the approximation errors, values are listed for each nonboundary control curve. As expected, the maximum error is produced near the degenerate vertex, in the region corresponding to the $(n-1)th$ curve.

	C^0 algorithm	ε-G^1 algorithm	
Degree elevation	Approximation error	Approximation error	Continuity error
0	0.00849 0.01172 0.10415	0.09059 0.02710 0.13936	0.11071
2	0.00278 0.00789 0.06943	0.01778 0.01042 0.07886	0.05123
4	0.00079 0.00684 0.05208	0.00430 0.00881 0.05565	0.02926

Table 1. Approximation error of the algorithms.

§6. Conclusions and Future Work

In the present paper, two algorithms for converting degenerate patches into triangular and trimmed tensor-product patches have been proposed and discussed. It has been shown that in general it is not possible to convert a rectangular degenerate patch into a triangular one without approximation error. Both algorithms minimize this approximation error in a quadratic sense while ensuring C^0 continuity between neighbor patches. In the second algorithm, ε-G^1 continuity is guaranteed. Bounds for the approximation errors have been derived. It has also been shown that the approximation error can be arbitrarily decreased through the degree elevation of the degenerate patches.

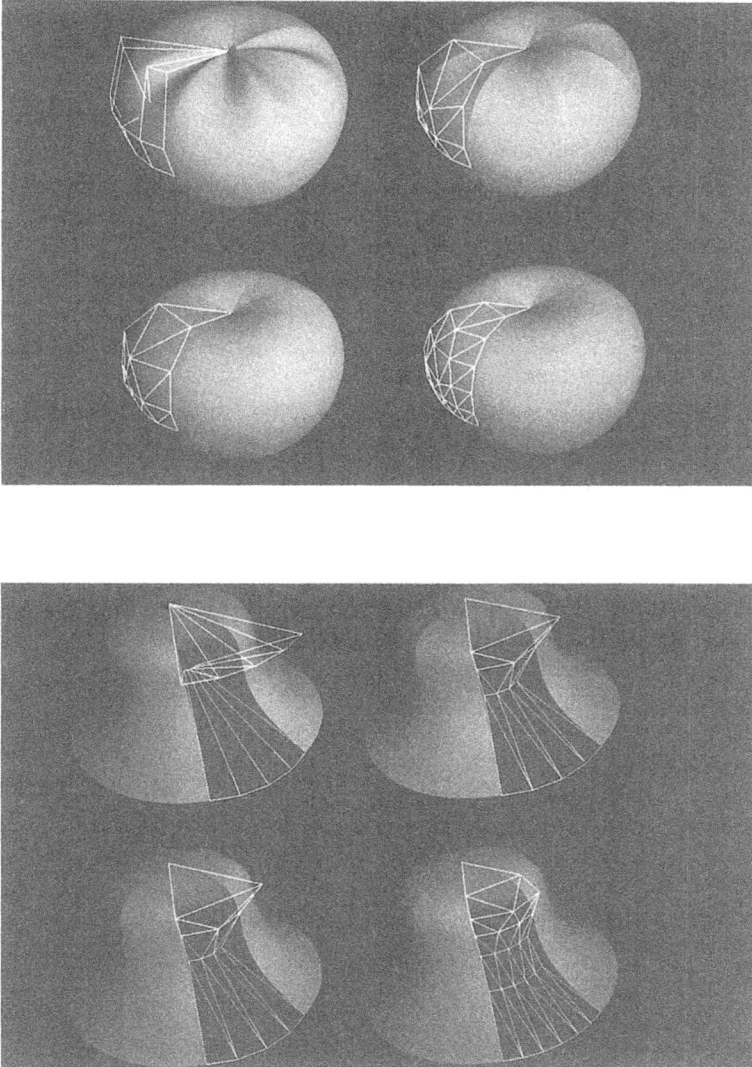

Fig. 3. Example II.

Acknowledgments. This work has been partially supported by the CEC, under Brite-Euram project BRE2-CT92-0228 and by the CICYT under the project TIC 95-0630-C05.

References

1. I. Brueckner, Construction of Bézier points quadrilaterals from those of triangles, Computer-Aided Design **12(1)** (1980), 21–24.

2. M. Casale, Freeform solid modeling with trimmed patches, IEEE Comp. Graph. Appl. **7** (1987), 33–43.

3. T. DeRose, S. Mann, An approximately G^1 cubic surface interpolant, in *Mathematical Methods in Computer Aided Geometric Design II*, T. Lyche and L. L. Schumaker (eds.), Academic Press, 1992, 185–196.

4. G. Farin, *Curves and Surfaces for Computer Aided Geometric Design*, Academic Press, 1990.

5. G.E. Forsythe, M.A. Malcolm, C.B.Moler, *Computer Methods for Mathematical Computations*, Prentice-Hall, 1977.

6. IGES/PDES Organization, *The Initial Graphic Exchange Specification (IGES) Version 5.0*, U.S. Dept. of Commerce, 1990.

7. J. Peters, Smooth interpolation of a mesh of curves, Constructive Approximation **7** (1989), 221–247.

8. VDA Working Group CAD/CAM, *VDA Surface Data Interface (VDAFS), version 2.0*.

9. M. Vigo, N. Pla, P. Brunet, From degenerate to triangular and trimmed patches, Research report LSI-9520R, UPC, 1995.

10. M. Watkins, A. Worsey, Degree reduction of Bézier curves, Computer-Aided Design **20(7)** (1988), 398–405.

Marc Vigo, Núria Pla, and Pere Brunet
Universitat Politècnica de Catalunya
Computer Graphics section, Software Department, E.T.S.E.I.B.
Avda. Diagonal 647, Planta 8, 08028 Barcelona, SPAIN
marc@turing.upc.es
nuria@turing.upc.es
pere@turing.upc.es

Spline Orbifolds

Johannes Wallner and Helmut Pottmann

Abstract. In order to obtain a global principle for modeling closed surfaces of arbitrary genus, first hyperbolic geometry and then discrete groups of motions in planar geometries of constant curvature are studied. The representation of a closed surface as an orbifold leads to a natural parametrization of the surfaces as a subset of one of the classical geometries S^2, E^2 and H^2. This well known connection can be exploited to define spline function spaces on abstract closed surfaces and use them e. g. for approximation and interpolation problems.

§1. Geometries of Constant Curvature

We are going to define three geometries consisting of a set of *points*, a set of *lines*, and a group of *congruence transformations*: The geometry of the euclidean plane E^2, the geometry of the unit sphere S^2 of euclidean E^3, and the geometry of the hyperbolic plane H^2. The geometries of E^2 and S^2 are well known: the hyperbolic plane will be presented in the next subsections. For more details, see for instance (Alekseevskij et al., 1988).

It is possible to define hyperbolic geometry in a completely synthetic way. We could use a system of axioms for euclidean geometry and then negate the parallel postulate or one of its equivalents. Any structure satisfying the axioms would be called a *model* of hyperbolic geometry. We would have to verify that all models, including the classical ones, the *Poincaré* and the *Klein* model, are isomorphic. We start from a different point of view: We first define a set of points, lines and congruence transformations, as linear as possible, and then show some structures isomorphic to it. The reader then will see the difference to euclidean or spherical geometry.

Curves and Surfaces with Applications in CAGD
A. Le Méhauté, C. Rabut, and L. L. Schumaker (eds.), pp. 445–464.
Copyright © 1997 by Vanderbilt University Press, Nashville, TN.
ISBN 0-8265-1293-3.
All rights of reproduction in any form reserved.

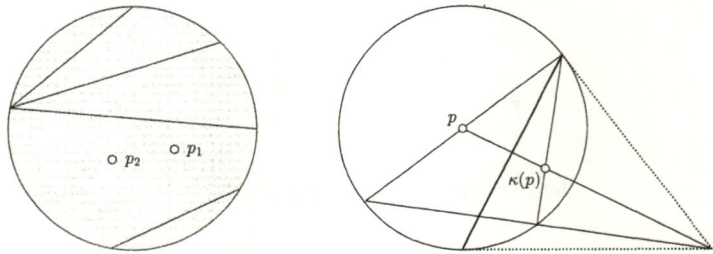

Fig. 1. Projective model of H^2: (a) points and non-intersecting lines, (b) hyperbolic reflection κ.

1.1 The Projective Model of Hyperbolic Geometry

Consider the real projective plane P^2 equipped with a homogeneous coordinate system, where a point with homogeneous coordinates $(x_0 : x_1 : x_2)$ has affine coordinates $(x_1/x_0, x_2/x_0)$. We will not distinguish between the point and its homogeneous coordinate vectors. Every time when a coordinate vector of a point appears in a formula, it is tacitly understood that any scalar multiple of this coordinate vector could be there as well.

We define an *orthogonality relation* between points: Let β be a symmetric bilinear form defined in $\mathbb{R}^3$, and let β have two negative squares, for instance

$$\beta(x,y) = x_0 y_0 - x_1 y_1 - x_2 y_2.$$

An equivalent formulation is $\beta(x,x) = x^T J x$, J being the diagonal matrix with entries 1, -1 and -1. We call x and y *orthogonal*, if $\beta(x,y) = 0$. Points with $\beta(x,x) = 0$ are called *ideal* points. The set of all ideal points is a conic and will be called the *ideal circle*. If we choose β as above, the ideal circle is nothing but the euclidean unit circle.

Now a point $x \in P^2$ shall belong to the *hyperbolic plane* H^2 if it is contained in the interior of the ideal circle,

$$x \in H^2 \iff \beta(x,x) > 0.$$

The *lines* of the hyperbolic plane are the intersections of projective lines with H^2. We define two lines to be parallel if they have no point in common. It is now obvious that for all lines l and all non-incident points p, there are a lot of lines parallel to l and containing p. A picture of the projective model can be found in Figure 1.

So far we have dealt with the incidence structure of the hyperbolic plane. We now come to metric properties. We define the *hyperbolic distance* $d(x,y)$ between points x and y of H^2 by

$$\cosh d(x,y) = \frac{|\beta(x,y)|}{\sqrt{\beta(x,x)\beta(y,y)}}.$$

We leave the verification of the fact that always $\beta(x,x)\beta(y,y) \le \beta(x,y)^2$ to the reader. This metric satisfies the triangle inequality and is compatible with the definition of lines, in the sense that they are precisely the geodesic curves with respect to this metric.

Hyperbolic congruence transformations will be those projective transformations, which map H^2 onto H^2 and preserve hyperbolic distances. For this reason and also because it is shorter, we will call them *isometries* or *motions*. We express the isometric property in matrix form: for each projective transformation κ there is a matrix such that in homogeneous coordinates

$$\kappa(x) = A \cdot x.$$

It is easy to see that the condition $d(x,y) = d(\kappa(x), \kappa(y))$ for all $x \in H^2$ is equivalent to

$$A^T J A = \lambda J \text{ with } \lambda > 0,$$

and that there are the following types of hyperbolic isometries:

1) the identity transformation;

2) *hyperbolic reflections*, which leave the points on a hyperbolic line fixed and reverse orientation (see Figure 1b);

4) *hyperbolic translations*, which preserve orientation and leave no point of H^2 fixed, but a hyperbolic line is mapped onto itself;

5) *hyperbolic rotations*, which leave one point of H^2 fixed and preserve orientation (for a picture in a different model, see Figure 3);

6) *ideal hyperbolic transformations* which leave no point of H^2 fixed, and no line is mapped to itself, but orientation is preserved;

7) the remaining hyperbolic isometries reverse orientation and are the product of a hyperbolic reflection by one of the above.

The model of the hyperbolic plane just described is called the *projective* or *Klein* model. In this model hyperbolic geometry appears as a subset of projective geometry: the point set is a subset of the projective point set, the lines are the appropriate subsets of projective lines, and hyperbolic isometries can be expressed in matrix form.

What remains to be defined is the *hyperbolic angle*. We will do this in a different model, which will also explain the name "hyperbolic".

1.2 The Hyperboloid Model of Hyperbolic Geometry

In $\mathbb{R}^3$, $\beta(x,x) = 0$ is the equation of a quadratic cone with apex at the origin, and $\beta(x,x) = 1$ is the equation of a two-sheeted hyperboloid, which can be seen as the unit sphere with respect to the *pseudo-euclidean* scalar product β. We call the "upper sheet" of this unit sphere the *hyperbolic plane*:

$$x \in H^2 \iff \beta(x,x) = 1 \text{ and } x_0 > 0.$$

There is an obvious one-to-one correspondence between the hyperbolic plane defined in Section 1.1 and the hyperbolic plane defined in this subsection.

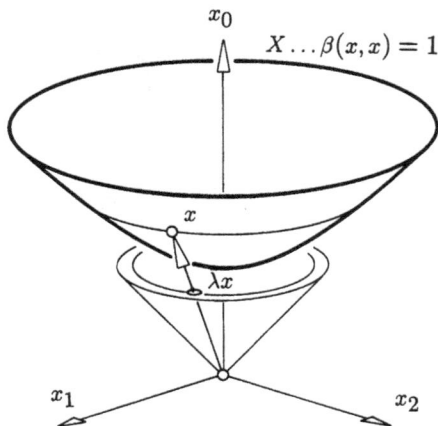

Fig. 2. The hyperboloid model $X \subset \mathbb{R}^3$ of H^2 and the correspondence between hyperboloid and projective model, which appears as a unit disk tangent to X.

Given a projective point, its uniquely defined coordinate vector x with $\beta(x, x) = 1$ and $x_0 > 0$ defines the corresponding point of the hyperboloid model. It is easy to transfer lines and hyperbolic isometries to the hyperboloid model: Hyperbolic lines have linear equations and therefore are intersections of H^2 with two-dimensional linear subspaces of $\mathbb{R}^3$. A picture of the hyperboloid model is given in Figure 2.

In the projective model, a hyperbolic isometry given by its matrix A is equivalently described by any scalar multiple of A. Now scale A such that

$$A^T J A = J.$$

Then the unit hyperboloid $\beta(x, x) = 1$ is invariant under multiplication by A. Conversely, as scalar products can be expressed in terms of distances, the invariance of the unit hyperboloid implies $A^T J A = J$. If A interchanges the two sheets of the unit hyperboloid, multiply A by -1. Thus, without loss of generality, we call all linear automorphisms of $\mathbb{R}^3$ which map H^2 onto itself *hyperbolic isometries* and this definition is compatible with the definition given in Section 1.1.

A scalar product β always defines an angle between vectors x and y: In $\mathbb{R}^3$ the *pseudo-euclidean angle* $\angle(x, y)$ is partially defined by

$$\lambda = \frac{\beta(x, y)}{\sqrt{\beta(x, x)\beta(y, y)}} \quad \text{if } \beta(x, x)\beta(y, y) > 0$$

$$\cos \angle(x, y) = \lambda \quad \text{if } |\lambda| \leq 1$$

$$\cosh \angle(x, y) = |\lambda| \quad \text{if } |\lambda| \geq 1$$

In every case where we will calculate an angle it has to be verified that $\beta(x, x) \cdot \beta(y, y) > 0$. In most cases we leave this verification to the reader. It is

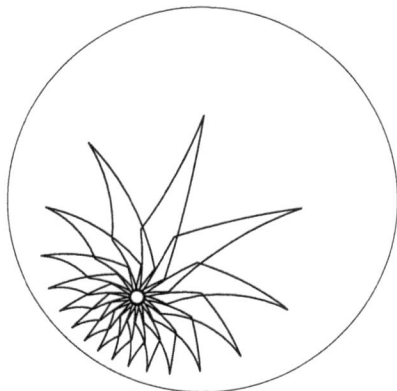

Fig. 3. Conformal model: hyperbolic rotation.

clear that the pseudo-euclidean angle of vectors x and y corresponds to the hyperbolic distance defined in Section 1.1. The *hyperbolic angle* between lines meeting in x is defined as the pseudo-euclidean angle of tangent vectors at x. Because all vectors v tangent to H^2 in a point x satisfy $\beta(v, v) < 0$, the angle between them is defined.

We define the *geodesic distance* between points x and y on a smooth surface X in $\mathbb{R}^n$ as the infimum of the arc-lengths of smooth curves c joining x and y in X. Arc-lengths are measured by means of the scalar product β: We can define the *norm* of a vector by $\|x\|^2 = |\beta(x, x)|$ and measure the arc length by $\int \|\dot{c}(t)\| dt$. It is easy to see that for H^2 we can explicitly find the curves for which the infimum, actually then the minimum, is attained: The geodesic distance is the arc-length of the unique hyperbolic line joining x and y and equals the hyperbolic distance $d(x, y)$.

1.3 The Conformal Model of Hyperbolic Geometry

Distorting the projective model leads to a new model of hyperbolic geometry with some other special metric properties: Let H^2 be the interior of the unit circle and define $\sigma : H^2 \to H^2$ in affine coordinates by

$$(x, y) \mapsto \frac{1}{1 - \sqrt{1 - x^2 - y^2}}(x, y).$$

Thus points will be moved a bit towards the origin. Hyperbolic lines will be σ-images of hyperbolic lines defined in Section 1.1. If κ is a hyperbolic isometry as defined in Section 1.1, then $\sigma \kappa \sigma^{-1}$ shall be a congruence transformation. This geometry which is obviously isomorphic to the projective and the hyperboloid model is called the *conformal* or *Poincaré* model of the hyperbolic plane. It has the following interesting properties:

Fig. 4. M. C. Escher's "Circle Limit IV", (c) 1997 Cordon Art – Baarn – Holland. All rights reserved.

1) Hyperbolic lines appear as euclidean circular arcs or straight line segments which intersect the ideal circle orthogonally.

2) The hyperbolic angle appears as the euclidean angle between circular arcs or straight line segments. This is why the model is called conformal (see Figure 3).

3) Hyperbolic reflections appear as *inversions*. The group of hyperbolic isometries is generated by the hyperbolic reflections, so in the conformal model it appears as the subgroup of Möbius transformations which map H^2 onto itself.

Because the euclidean radius of hyperbolic distance circles with center in the origin is smaller in the conformal model than it is in the projective model, usually the conformal model is used for illustrations. In Figure 3b you can see an iterated hyperbolic rotation in the conformal model.

The conformal properties of this model have also been exploited by the Dutch artist *M. C. Escher* in some of his famous drawings. One of them is depicted in Figure 4.

1.4 An Overview

We can assume that the reader is familiar with the geometry of the euclidean plane E^2 and the unit sphere S^2. In this section we will present these two together with hyperbolic geometry from a unified point of view. S^2 and H^2 will in some places be dual to each other, whereas euclidean geometry does sometimes not fit so nicely into the description. Also the generalizations of S^2, E^2 and H^2 to higher dimensions are obvious: E^n is euclidean n-space, S^n and H^n are the unit spheres with respect to a scalar product in $\mathbb{R}^{n+1}$ with zero or n negative squares, respectively. It may be stated that almost everything in this paper, except, of course, the classification of surfaces in Section 2.5, holds for any dimension with only slight notational changes.

- *Linear incidence structure:* For each of the three geometries there is a model as a subset X of $\mathbb{R}^3$ such that lines in the geometry are intersections of two-dimensional linear subspaces with X. For X we can choose the unit sphere, the plane with coordinate $x_0 = 1$, and the upper sheet of the hyperboloid described in Section 1.2.

- *Linear model and metric:* Given a scalar product β in $\mathbb{R}^3$, then dependent on the number of negative squares, the unit sphere will be an ellipsoid, a one-sheeted hyperboloid, a two-sheeted hyperboloid, or empty. If β is positive definite, the unit sphere carries the structure of a spherical geometry. If β has two negative squares, then each of the two connected components (sheets) of the unit sphere carries the structure of a hyperbolic geometry. Distances of points are given in terms of angles between the corresponding vectors, as are angles between tangent vectors. The geodesic distance in X equals the distance previously defined.

- *Congruence transformations:* In the linear models $X \subset \mathbb{R}^3$ of S^2 and H^2, the group of motions or isometries consists of the restrictions $L|_X$ of those linear automorphisms L of $\mathbb{R}^3$ which map X onto itself.

- *Curvature:* The sphere, the euclidean plane and the hyperbolic plane are Riemannian manifolds of constant Gaussian curvature, the value of which is 1, 0 and -1, respectively. From the Gauss-Bonnet theorem it then follows that the angle sum in a triangle is greater than, equal to, or less than π, respectively. Moreover, the absolute value of the difference is the *area* of the triangle as of a Riemannian manifold.

§2. Discrete Motion Groups and Orbifolds

In this section we define the factor orbifold X/H where X is one of E^2, S^2 or H^2, and H is a discrete transformation group acting on X. X will always denote one of the three geometries, and its motion group will be denoted by G. We will not be able to present a complete theory, and we simplify some notions in some places.

For a detailed presentation, see for instance (Ratcliffe, 1994), (Vinberg and Shvartsman, 1988) or (Zieschang et al., 1980). For a well illustrated book which is easy to read, see for instance (Week, 1985).

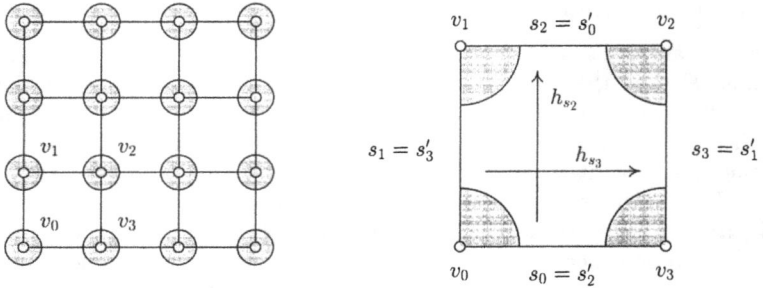

Fig. 5. The torus as an orbifold.

2.1 Discrete Transformation Groups

We will consider groups H of motions *acting* on X, which means that each $h \in H$ is an isometry $h : X \to X$ and $h_1(h_2(x)) = (h_1 \cdot h_2)(x)$. The identity transformation will always be denoted by e. We write $h(x)$ for the h-image of an $x \in X$ and $h(K)$ for the h-image of a subset $K \subset X$. We call a group H acting on X *discrete* if for every compact set K the intersection $K \cap h(K)$ is nonempty only for finitely many $h \in H$. This implies that the *orbit* $\{h(x), h \in H\}$ of a point x is discrete, i. e., it has no accumulation point. An example for this is the group $H = \mathbb{Z}^2$ acting as a group of translations on the euclidean plane: The pair (i, j) of integers acts on $X = E^2$ by $(x, y) \mapsto (x + i, y + j)$. It is only a change in notation if we consider H as a subgroup of the euclidean motion group. A picture can be seen in Figure 5.

For a group H acting on X, the *stabilizer* H_x of x is the subgroup of all those $h \in H$ with $h(x) = x$. If H is discrete, obviously H_x is finite. The *order* of x is the cardinality of its stabilizer. In the example given above all stabilizers are trivial. We call such actions *free*.

If x and y are not antipodal points of the sphere, the unique shortest segment joining them is called their convex hull, and a set C is called *convex*, if for all $x, y \in C$ the convex hull of x and y is in C. Then a *convex polygon* is defined as the convex hull of a finite non-collinear set of points. *Edges* and *vertices* are defined in the obvious way. Note that a convex polygon is always the closure of its interior.

2.2 Fundamental Domains

A *fundamental domain* F of a discrete motion group H is a set which is the closure of its interior and fulfills the following conditions: 1) the sets $h(F)$, $h \in H$ cover X, and 2) if $h_1(F)$ and $h_2(F)$ have an interior point in common, then $h_1 = h_2$. There are discrete groups of motions which have no convex polygons as fundamental domains, for instance the discrete group of translations along integer multiples of one fixed vector in E^2. We will not try to generalize the notion of polygon such that it covers all discrete motion groups (which is possible), but we restrict ourselves to groups which possess convex fundamental polygons.

We denote the edges of the fundamental polygon F by $s_0, \ldots, s_{n-1}, s_n = s_0$. The intersection of edges $s_i \cap s_{i+1}$ is a vertex v_i. By subdividing finitely many edges and introducing new vertices it is possible to achieve that the intersection of F with any $h(F)$ is either empty or an edge. We call the uniquely defined motion $h \in H$ for which $F \cap h(F) = s_i$ the *adjacency transformation* of the edge s_j. We call a sequence $h_1(F), \ldots, h_n(F)$ a chain of polygons, if the intersection $h_i(F) \cap h_{i+1}(F)$ is an edge. Because any two $h \in H$ can be connected by a chain, the group H is entirely generated by the finitely many adjacency transformations of one fundamental polygon.

If an adjacency transformation maps s_i to s_j, then we write $s_i = s_j'$. Obviously then the inverse adjacency transformation maps s_j to s_i, so $s_i' = s_j$. For the example given above, the adjacency transformations are indicated in Figure 5.

2.3 Defining Relations

We write h_s for the adjacency transformation with $F \cap h_s(F) = s$. A sequence $h_{s_1}, \ldots, h_{s_n}$ of adjacency transformations with $h_{s_1} \cdot \ldots \cdot h_{s_n} = e$ corresponds to a chain $F_0 = F, F_1 = h_{s_1}(F), F_2 = h_{s_1}(h_{s_2}(F)), \ldots, F_n = F_0$ of polygons. Such a chain is called a *cycle*.

Let s, s' be edges with $h_s(s') = s$ and $h_{s'}(s) = s'$. Then obviously $h_s h_{s'} = e$ and $F, h_s(F), h_s h_{s'}(F) = F$ is a cycle. Formally, we write

$$ss' = e.$$

Also for all vertices v there is a cycle of polygons consisting of all polygons containing v in the order in which they are encountered when cycling v. The corresponding sequence $h_{s_1}, \ldots, h_{s_n}$ of adjacency transformations gives the formal relation

$$s_1 s_2 \ldots s_n = e,$$

which is called a *Poincaré relation*. The importance of the Poincaré relations is shown by the following

Theorem. *Let H be a group with a convex fundamental polygon. Denote its set of edges by S and the set of relations $ss' = e$ together with all Poincaré relations with R. Then the abstract group with generator set S and relations R is isomorphic to H.*

In the example given above, all adjacency transformations are translations. They correspond to the edges s_0, s_1, s_2, s_3 and $s_0' = s_2$, $s_1' = s_3$ (see Figure 5). The four Poincaré relations are $s_0 s_1 s_2 s_3 = e$, $s_1 s_2 s_3 s_0 = e$, $s_2 s_3 s_0 s_1 = e$ and $s_3 s_0 s_1 s_2 = e$. Obviously $s_2 s_0 = e$ and $s_1 s_3 = e$. So we can eliminate s_2 and s_3. Each Poincaré relation implies the other three. It follows that H as an abstract group is isomorphic to the group with generators s_0, s_1 and the single relation $s_0 s_1 s_0^{-1} s_1^{-1} = e$, or, equivalently, $s_0 s_1 = s_1 s_0$. This means that H is a free abelian group with free generators s_0 and s_1.

A natural question to ask now is: Given a convex polygon F and for each edge s an adjacency transformation h_s, such that (a) $F \cap h_s(F) = s$,

(b) $h_s(s') = s$ implies $h_{s'}(s) = s'$, and (c) $h_s h_{s'} = e$. Suppose further that (d) for each vertex v of F there are adjacency transformations $h_{s_1}, \ldots h_{s_n}$ such that their product equals e and the polygons $h_{s_1} h_{s_2} \ldots h_{s_i}(F)$ form a "circuit" around v. Does there exist a discrete group of motions having F as fundamental polygon and h_s as adjacency transformations? The answer, due to Poincaré, is yes.

2.4 Orbifolds

Let H be a discrete group of motions in one of the three geometries E^2, S^2 or H^2. By identifying all points $h(x)$, $h \in H$, we get the points of the *orbifold* X/H. This definition, however, gives only the orbifold as a set, without additional structures. They are to be defined by means of the *canonical projection* $p : X \to X/H$ which maps an $x \in X$ to its orbit. The topology on X/H is defined as the final topology of p: U is open if and only if $p^{-1}(U)$ is open. The incidence structure is directly mapped by p: A line segment in X/H is the p-image of a line segment of X. The distance between x, y in X/H is the minimum of distances of points in $p^{-1}(x)$ and $p^{-1}(y)$ measured in X.

An example for an orbifold which is very well known, but, in some sense is not typical, is the torus. It appears as the orbifold X/H if $X = E^2$ and H is the discrete group of translations along integer multiples of two basis vectors e_1 and e_2 (see Figure 5). The order of all points x equals 1, and so for all y in $p^{-1}(x)$ there is a neighborhood of y which is mapped isometrically (and, of course, homeomorphically) to X/H. This need not be the case, and happens if and only if some $h \in H$ has a fixed point. These orbifolds have metric singularities and could also be used for modeling surfaces, but we will omit them in order to keep the presentation simple.

2.5 Surfaces

Our aim is to find discrete groups H in a geometry X of constant curvature such that the corresponding orbifold X/H is a compact surface, orientable or nonorientable, of arbitrary genus g. It is well known that the compact surfaces without boundary are precisely the spheres with g handles and the spheres with g crosscaps. For the classification of surfaces from the topological, differentiable, or combinatorial viewpoint, see textbooks of algebraic topology, differential topology or combinatorial topology, for instance (Hirsch, 1976) or (Kinsey, 1993).

It is well known that the following discrete transformation groups H in various geometries X lead to all compact surfaces:

- *The sphere:* S^2 itself as the orientable surface of genus 0 is one of the primitive geometries. Formally, let $X = S^2$ and $H = \{e\}$.
- *The projective plane:* P^2 is obtained by identifying antipodal points in S^2. If s denotes the antipodal map, then $P^2 = S^2/H$ with $H = \{e, s\}$. A fundamental polygon is the upper hemisphere.
- *The torus:* Letting $X = E^2$ and H equal the group generated by the translations along two linearly independent vectors gives the torus, which

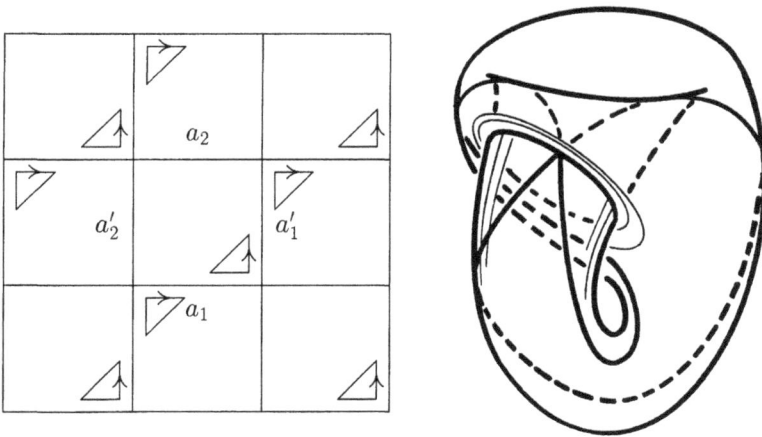

Fig. 8. Klein bottle.

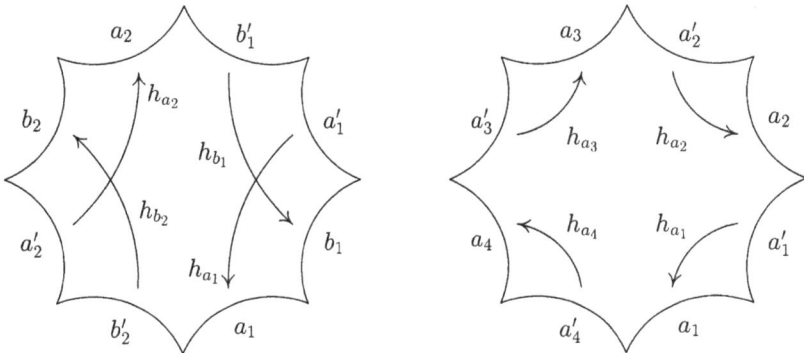

Fig. 9. Regular octogon as fundamental domain (a) of a group whose orbifold is an orientable surface of genus 2 (b) of a group whose orbifold is a non-orientable surface of genus 4.

is the orientable surface of genus 1. A fundamental polygon is the parallelogram spanned by a lattice basis.

• *The Klein bottle:* Letting $X = E^2$ and H equal the group generated by the adjacency transformations depicted in Figure 8 gives the nonorientable surface of genus 2, which is called the Klein bottle.

• *Orientable surfaces of higher genus:* In the conformal model of the hyperbolic plane, consider the points $(r\cos(2k\pi/l), r\sin(2k\pi/l))$ with $l = 4g$, $g \geq 2$ and $k = 0, \ldots, l-1$. The convex hull F of these points is a regular $4g$-gon, with interior hyperbolic angles α depending on the value of r (see

Figure 9). It is easily seen that α tends to 0 as r tends to 1, and α tends
to $\pi - 2\pi/l$ as r tends to 0. By continuity, for all l there is a value of r
such that the interior angle α equals $2\pi/l$. Now denote the edges of F by
$a_1, b_1, a_1', b_1', \ldots, a_g, b_g, a_g', b_g'$ and define orientation-preserving adjacency
transformations which map a_k to a_k', b_k to b_k' and vice versa, for all k.
Then the Poincaré relations will be equivalent to the relation

$$a_1 b_1 a_1' b_1' a_2 b_2 a_2' b_2' \ldots a_g b_g a_g' b_g' = e.$$

This shows that the group H generated by the adjacency transformations
defined above is, as an abstract group, isomorphic to the group with
generators $a_1, b_1, \ldots, a_g, b_g$ and the single relation

$$a_1 b_1 a_1^{-1} b_1^{-1} \ldots a_g b_g a_g^{-1} b_g^{-1} = e.$$

The order of all vertices is 1, and therefore X/H is a manifold. Gluing
those edges of the fundamental polygon together which are mapped onto
each other by adjacency transformations, gives precisely X/H. From the
gluing construction it is clear that X/H is a sphere with g handles. A
picture of the gluing for $g = 2$ can be seen in Figure 11a. This shows
that the orientable surface of genus g with $g > 1$ is an orbifold, even a
manifold, of the form X/H where X is the hyperbolic plane and H is the
group generated by the adjacency transformations defined above. The
torus fits into this description, if we set $g = 1$ but instead of a polygon
in H^2 use a euclidean square.

- *Nonorientable surfaces of higher genus:* In analogy to the previous construction, construct a regular $2g$-gon ($g \geq 3$) in the hyperbolic plane with
angles π/g. Denote the edges by $a_1, a_1', \ldots, a_g, a_g'$. To the edge a_k corresponds the uniquely determined adjacency transformation which reverses
orientation and maps a_k to a_k', and for a_k' vice versa (see Figure 9). Then
all Poincaré relations are equivalent to the relation $a_1^2 a_2^2 \ldots a_g^2 = e$. Thus
the discrete motion group H generated by these adjacency transformations is, as an abstract group, isomorphic to the group with generators
$a_1, \ldots, a_g$ and the single relation

$$a_1^2 \ldots a_g^2 = e.$$

The order of all vertices equals 1, and therefore X/H is a manifold. It
is nonorientable because H contains orientation reversing motions. From
the gluing construction it is clear that X/H is a sphere with g crosscaps.

The Klein bottle ($g = 2$) and the projective plane ($g = 1$) fit into this
formalism, if we choose a euclidean square or a spherical 2-gon (such as
the northern hemisphere) instead.

§3. Functions on Surfaces

3.1 Group-Invariant Functions

We call a function $\widetilde{f} : X \to R$ *invariant* with respect to the group H, if

$$\widetilde{f}(h(x)) = \widetilde{f}(x) \quad \text{for all } x \in X, h \in H.$$

If $p : X \to X/H$ denotes the canonical projection, an H-invariant function directly leads to a function f whose domain is the factor orbifold:

$$f : X/H \to R, \ f(p(x)) = \widetilde{f}(x)$$

and vice versa: a function f defined on X/H gives rise to an H-invariant function

$$\widetilde{f} : X \to R, \ \widetilde{f}(x) = f \circ p(x).$$

If the range R is the real number field $\mathbb{R}$ and $\widetilde{f}$ is a function defined on X, then we can build an H-invariant function $\widetilde{g}$ from $\widetilde{f}$ by letting

$$\widetilde{g}(x) = \sum_{h \in H} \widetilde{f}(h(x)).$$

Of course it has to be verified that this sum makes sense. If X is the sphere, every discrete motion group is finite, and the sum above is finite. So every property of f which is invariant with respect to finite sums is preserved, so for instance continuity or differentiability.

If f has compact support, then for all x there is a neighborhood U of x such that the sum defined above is finite in U, by discreteness of H. So all local properties which are invariant with respect to finite sums are preserved, for instance continuity or differentiability.

If X/H is a manifold, it is clear that $f : X/H \to \mathbb{R}$ is continuous (differentiable, of class C^r, of class C^∞) if and only if the corresponding $\widetilde{f} : X \to \mathbb{R}$ has this property. If X/H is an orbifold with metric singularities, we avoid difficulties by *defining* that an f defined on X/H is differentiable (of class C^r, of class C^∞) if the corresponding $\widetilde{f}$ has this property.

The above sum can make sense even if f does not have compact support. It is sufficient that f decreases fast enough. An example for a summable function whose sum is of class C^∞ is the Gaussian $\widetilde{f}(x) = \exp(-d(x, m)^2)$ in E^2 and H^2 (note that in S^2 $\widetilde{f}$ is not differentiable everywhere).

3.2 Polynomial and Rational Functions

For each of the three geometries S^2, E^2 and H^2 we have found a model as a subset X of $\mathbb{R}^3$. This enables us to define polynomial or rational functions on X as the restriction of polynomial or rational functions defined in $\mathbb{R}^2$ to X. It is well known that both S^2 and H^2 possess rational parametrizations which can be given by stereographic projections: The mapping σ defined by

$$\sigma : \mathbb{R}^2 \to S^2 \setminus \{(-1, 0, 0)\}, \ (p, q) \mapsto \frac{1}{1 + p^2 + q^2}(1 - p^2 - q^2, 2p, 2q)$$

is one-to-one. Also, the mapping σ defined by

$$\sigma : D \to H^2, \ (p,q) \mapsto \frac{1}{1 - p^2 - q^2}(1 + p^2 + q^2, 2p, 2q)$$

with D being the interior of the unit circle, is one-to-one. If f is a polynomial defined in $\mathbb{R}^3$, then $f \circ \sigma$ is a rational function defined in the domain of σ.

We want to indicate how modeling of closed surfaces with the aid of piecewise rational functions is possible. First we give an easy example which shows how to proceed in the not so trivial cases: The B-spline basis functions on the real line are well known, and so are tensor product B-splines in E^2. We define a knot sequence on the x_1-axis which is periodic and has period 1. This means that if t is a knot, then $t + k$ is a knot for every integer k. The same we do for the x_2-axis, and then we consider the B-spline basis functions $B_i(x_1)$ and $B_j(x_2)$ which correspond to this knot sequences. Their products $B_{ij}(x_1, x_2)$ defined in the plane form a partition of unity. There are finitely many functions $B_{ij}(x_1, x_2)$ such that all others can be expressed in the form $B(x_1, x_2) = B_{ij}(h(x_1, x_2))$ where h is an element of the translation group H generated by translations along the unit vectors in x_1- and x_2-direction. All B_{ij}'s are compactly supported, so the functions

$$\tilde{C}_{ij}(x) = \sum_{h \in H} B_{ij}(h(x))$$

are well defined, are group-invariant, and form a partition of unity. Thus there are finitely many functions C_{ij} defined on the torus E^2/H such that

$$C_{ij}(p(x)) = \tilde{C}_{ij}(x) \text{ and } \sum C_{ij}(p(x)) = 1 \text{ for all } x \in E^2,$$

where p is the canonical projection which maps a point $x = (x_1, x_2)$ to its orbit.

The preceding paragraph contained nothing new. It could be said that it is a complicated formulation of the simple fact that "closing" B-spline curves is also possible in the plane, and analogously to closed curves which can be viewed as defined on the circle, this closing operation yields a closed surface defined on the torus. On other surfaces the process of making a function group-invariant may be more complicated, but the principle is the same and has been shown in Section 3.1.

3.3 Simplex Splines and a DMS-Spline Space

It is well known that the restriction of homogeneous B-splines to the sphere leads to spline spaces of functions whose domain are subsets of the surface of the sphere, see e. g. (Alfeld et al., 1996). We want to show that the concept of simplex spline is not restricted to the sphere and that there is a natural generalization to abstract surfaces of higher genus.

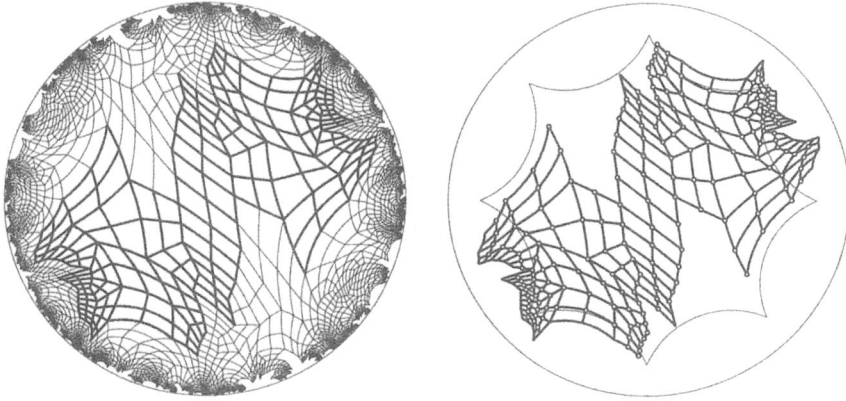

Fig. 10. Group-invariant tesselation.

Choose a basis $b_1, \ldots, b_n \in \mathbb{R}^n$. Then for all $v \in \mathbb{R}^n$ there is a unique linear combination $v_1 b_1 + \cdots + v_n b_n$ equal to v. For all n-tuples $k = (k_1, \ldots, k_n)$ of integers we define the function

$$B_k : \mathbb{R}^n \to \mathbb{R}, \quad v \mapsto \frac{(k_1 + \ldots + k_n)!}{k_1! \ldots k_n!} v_1^{k_1} \ldots v_n^{k_n},$$

which is called a *homogeneous Bernstein basis polynomial* of *degree* $|k| = k_1 + \cdots + k_n$. Any linear combination of homogeneous Bernstein basis polynomials of the same degree is called a *homogeneous Bernstein polynomial*. For such a polynomial $p = \sum_{|k|=d} c_k B_k$ the equation $p(\lambda v) = \lambda^d p(v)$ holds. If X is the linear model of one of the three geometries S^2, E^2 or H^2, the restrictions $p|X$ are called *spherical, planar* or *pseudo-spherical* Bernstein polynomials. Note that the planar Bernstein polynomials are just the well-known triangular Bernstein polynomials in the plane.

Also the notion of *simplex spline* has a natural meaning in the linear model of S^2, E^2 and H^2. We recall that the homogeneous simplex spline $M_B : \mathbb{R}^n \to \mathbb{R}$ is well defined for a set $B = \{b_1, \ldots, b_m\}$ of vectors as follows:

$$M_B(v) = \chi_B(v)/\det(b_1, \ldots, b_n) \quad \text{if } |B| = n$$
$$M_B(v) = \sum_{b \in T} \lambda_b M_{B \backslash b}(v) \quad \text{if } |B| > n \text{ and } v = \sum_{b \in T} \lambda_b b.$$

Here $\chi_B(v)$ equals 1 if all coordinates λ_i with respect to the basis B are positive and zero if at least one is negative. T denotes an arbitrary n-element subset of B which is a basis of $\mathbb{R}^n$. This defines the simplex spline in $\mathbb{R}^n$ except in some subspaces. Now extend the simplex spline continuously. This gives a C^{m-n} function. It is natural to define spherical, planar or pseudo-spherical simplex splines as the restriction $M_B|X$ of simplex splines M_B.

This allows the definition of a spline space analogous to the DMS-spline spaces introduced in (Dahmen et al., 1992). This is of theoretical interest, because it shows the existence of a spline space consisting of piecewise rational functions of arbitrary finite differentiability defined on a surface of genus g over an arbitrary triangulation. The planar and the spherical variant of the DMS-spline space have already been defined, for instance in (Pfeifle and Seidel, 1994).

Simplex splines are most easily made group-invariant if they are defined over a group-invariant triangulation. Here group-invariant means that every motion $h \in H$ maps the triangulation onto itself. One possibility to construct such a triangulation is the following: Choose a set V of vertices in a fundamental domain of the group H and consider the set $\widetilde{V} = \{h(v), h \in H, v \in V\}$. Then apply an algorithm which finds the edges of a triangulation with vertex set $\widetilde{V}$ and is designed such that it uses only information which can be expressed in terms of the geometry, for instance distance. Then a congruence transformation κ applied to $\widetilde{V}$ must result in edges which are just the κ-images of the previous ones. Now it is clear that if $\widetilde{V}$ is group-invariant, so is the whole triangulation. An example of a triangulation which is invariant with respect to the group corresponding to the orientable surface of genus 2 is shown in Figure 10.

A function defined by means of the triangulation and the geometry alone then is group-invariant. This is especially true for all functions which are defined by means of one of the linear models $X \subset \mathbb{R}^3$ and are linearly dependent on the coordinate vectors of the vertices. One example of this is given by the simplex splines defined above.

3.4 Approximation, Interpolation, Visualization

These tools can be used for approximation and interpolation of functions defined on a compact surface and also for visualizing such surfaces. This has been pointed out by Ferguson and Rockwood (1993). The spherical, euclidean or hyperbolic area $d\widetilde{\mu}$ defines a measure in X. If we assume that the boundary of the fundamental domain F has measure zero, $d\widetilde{\mu}$ naturally defines a measure $d\mu$ on X/H and we can define the space $L^2(X/H)$ with the scalar product

$$(f, g) = \int_{X/H} fg\,d\mu = \int_F \widetilde{f}\widetilde{g}\,d\widetilde{\mu}$$

in the well known way. The resulting norm will be denoted by $\|f\|_2$. One typical problem now is the following: Given a finite set $B = \{b_1, \ldots, b_n\}$ of basis functions and a function f on X/H, we seek a linear combination of the b_i such that

$$\|f - \sum \lambda_i b_i\|_2 \to \min.$$

This is a classical least squares problem and can easily be solved: If the b_i are orthonormal, $\lambda_i = (f, b_i)$ is the solution. If not, apply the Gram-Schmidt orthogonalization process. For *interpolation* we for instance introduce the space $L^2_\Delta(X/H)$ with the scalar product

$$(f, g)_\Delta = \int_{X/H} \Delta f \Delta g\,d\mu,$$

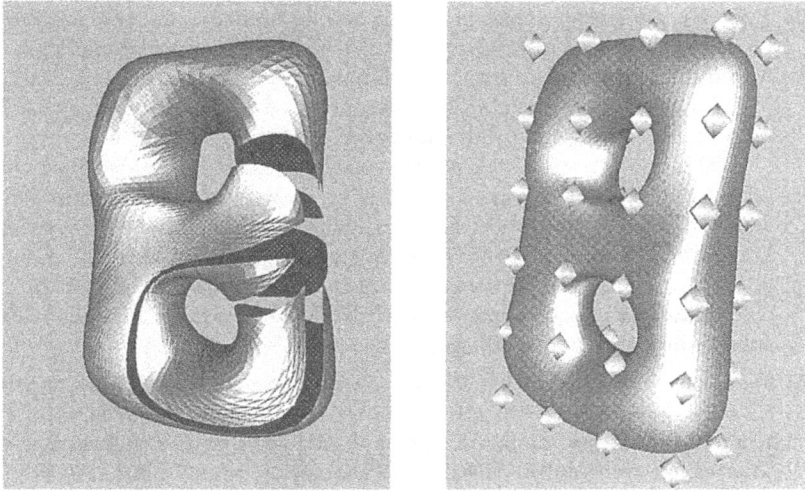

Fig. 11. (a) Gluing the boundary of an octogon together yields a surface of genus 2, (b) C^∞-approximation of a polyhedron.

where Δ is the Laplace-Beltrami operator on X/H which is inherited from X. This allows us to find in the linear solution space of the interpolation problem $\sum \lambda_i b_i(x_j) = c_j$ a solution of *minimal energy*. It is also possible to extend interpolation schemes which have been successfully employed for sphere-like surfaces to the linear models of the S^2, E^2 and H^2, for instant the hybrid patch of (Liu and Schumaker).

Modeling and *visualization* of compact surfaces is possible in the following way: A closed surface in $\mathbb{R}^3$ can be seen as an embedding (or, at least, an immersion) of an abstract closed surface X/H into $\mathbb{R}^3$. For each abstract point $p \in X/H$ three coordinate values $x_1(p)$, $x_2(p)$ and $x_3(p)$ are given. This means that we have three real functions x_1, x_2 and x_3 whose domain is X/H. Equivalently, we have three H-invariant functions $\tilde{x}_i$ whose domain is X. They have the property that the (x_1, x_2, x_3)-image of X/H is homeomorphic to X/H. Thus approximation, interpolation and modeling of surfaces is nothing but approximation, interpolation and modeling of three separate coordinate functions.

Suppose we are given a polyhedron with vertices $p_1, \ldots, p_n$ and we seek a C^∞ approximation to it. We describe our solution to this problem, which is typical for the sort of problems arising in this context. The algorithm is the following:

1) Cut the polyhedron along four closed curves passing through one fixed base point, such that the resulting surface becomes simply connected. An example of such a cutting is shown in Figure 11a. The cuts are in correspondence to the fundamental polygon which is shown in Figure 9a.

2) Find finitely many points $q_1, \ldots, q_n$ in the fundamental octogon corresponding to the appropriate group H with the following property: If we construct a group-invariant triangulation with vertices $h(q_i)$, $i = 1, \ldots, n$, $h \in H$, then this triangulation, when factored to the orbifold, is combinatorically equivalent to the triangulation of the polyhedron.

    This triangulation need not necessarily consist of triangles, it can also be a tesselation with n-gons of different shape and different number of vertices. The vertices do not necessarily have to lie inside the fundamental polygon. The cuts and the 8-gon are merely a guide where to put the q_i. For instance the tesselation shown in Figure 10 after factoring is combinatorically equivalent to the polyhedron shown in Figure 11 after subdividing each of the squares in four parts.

3) Optimize the triangulation/tesselation with respect to appropriate criteria. For instance we can try to optimize the shape of the faces of the triangulation/tesselation. In our case, the faces of the polyhedron are squares, so we want the faces of the tesselation be as square-like as possible. Because not for all vertices the number of faces containing this vertex equals four, we have to compromise.

4) Find a one-to-one correspondence between X/H and the polyhedron, or, equivalently, find a covering map from X to the polyhedron which is compatible with the triangulation. In our case this is done easily by mapping the 4-gons of the tesselation to the squares of the polyhedron in the obvious way.

5) To each point of X we assign the three coordinate values x_1, x_2, x_3 of its corresponding point on the polyhedron. This defines three continuous H-invariant functions on X.

6) Approximate the x_i by functions y_i which are linear combinations of C^∞ basis functions, for instance Gaussians.

7) Use the three functions y_1, y_2 and y_3 as coordinate functions of a surface in $\mathbb{R}^3$ whose parameter domain is X/H or X depending on the level of abstraction. This is how Figure 11b was made.

If the correspondence between the points $q_1, \ldots, q_n$ of X/H and the vertices $p_1, \ldots, p_n$ is established, interactive modeling of polyhedra of similar shape is easy. We can construct the correspondence between X/H a further polyhedron P of the same shape by finding a correspondence between the model polyhedron and P. This is especially trivial if P combinatorically is just a *refinement* of the model polyhedron. The approximation problem for P is then just the problem of approximation of three new coordinate functions.

If the basis functions b_i are already orthonormal, approximation can be done very quickly. Moving the vertices of the polyhedron, which can now be seen as *control points* of the surface, influences the approximant surface. Depending on the type of basis function, the influence will be local or global. As Gaussians decrease rapidly, and in addition can be multiplied by compactly supported bump functions to become compactly supported without essentially changing their global shape, we have local control. For implementation purposes, the basis functions with compact support are very convenient, because the handling of infinite sums can be avoided completely.

Modeling of C^r surfaces with polynomial coordinate functions is possible if we choose the b_i as simplex splines or homogeneous DMS-splines. This gives an algorithm which makes it possible to model surfaces of arbitrary differentiability, of arbitrary genus, over an arbitrary triangulation, without any boundary and gluing conditions. A more detailed theory and further examples of this can be found in (Wallner, 1996).

References

1. Alekseevskij D. V., E. B. Vinberg and A. S. Solodovnikov, Geometry of spaces of constant curvature, in: Itogi Nauki i Tekhniki, Sovremennye Problemy Matematiki, Fundamentalnye Napravleniya, Vol. **29**, VINITI, Moscow 1988; English translation in: Encyclopedia of Mathematical Sciences **29**, Springer Verlag 1993.

2. Alfeld P., M. Neamtu and L. L. Schumaker, Bernstein-Bézier polynomials on spheres and sphere-like surfaces, Comput. Aided Geom. Design **13** (1996), 333–349.

3. Dahmen W., C. A. Micchelli and H.-P. Seidel, Blossoming begets B-spline bases built better by B-patches, Math. Comp. **59** (1992), 97–115.

4. Ferguson H. and A. Rockwood, Multiperiodic functions for surface design, Comput. Aided Geom. Design **10** (1993), 315-328.

5. Hirsch M. W., *Differential Topology*, Graduate Texts in Mathematics **33**, Springer Verlag 1976.

6. Kinsey L. C., *Topology of Surfaces*, Undergraduate Texts in Mathematics, Springer Verlag, 1993.

7. Liu X. and L. L. Schumaker, Hybrid Bézier patches on sphere-like surfaces, J. Comp. Appl. Math. 73 (1996), 157–172.

8. Pfeifle R. and H.-P. Seidel, Spherical triangular B-splines with application to data fitting, Technical report 7/94, University of Erlangen.

9. Ratcliffe J. G., *Foundations of Hyperbolic Manifolds*, Graduate Texts in Mathematics **149**, Springer Verlag 1994.

10. Vinberg E. B. and O. V. Shvartsman, Discrete groups of motions of spaces of constant curvature, in: Itogi Nauki i Tekhniki, Sovremennye problemy matematiki, Fundamentalnye napravleniya, Vol. **29**, VINITI, Moscow 1988; English translation in: Encyclopedia of Mathematical Sciences **29**, Springer Verlag 1993.

11. Wallner J., Geometric Contributions to Surface Modeling, Ph. D. Thesis, Vienna University of Technology, 1996.

12. Week J. R., *The Shape of Space*, Pure & Applied Mathematics, Vol. **96**, Marcel Dekker, 1985.

13. Zieschang H., E. Vogt and H.-D. Coldewey, *Surfaces and Planar Discontinuous Groups*, Lecture Notes in Mathematics **835**, Springer Verlag 1980.

Johannes Wallner/Helmut Pottmann
Institut für Geometrie,
Technische Universität Wien,
Wiedner Hauptstraße 8–10/113,
A-1040 Vienna, AUSTRIA
hannes@geometrie.tuwien.ac.at
pottmann@geometrie.tuwien.ac.at

Numerically Stable Conversion Between the Bézier and B-Spline Forms of a Curve

Joab R. Winkler

Abstract. A numerically stable method for the conversion between the Bézier and B-spline forms of a curve is presented. It is shown that truncated singular value decomposition (TSVD) can be used to stabilise this potentially ill-conditioned problem if it is known that the theoretically exact solution satisfies the discrete Picard condition. An example of the conversion of a Bézier curve into its B-spline form using TSVD is given, and it is shown that the improvement in the quality of the computed answer is significant.

§1. Introduction

High integrity data exchange between different computer aided design (CAD) systems is an important industrial requirement and necessitates the accurate and stable transformation between different mathematical representations (Bézier, B-spline, etc.) of curves and surfaces. Even when the transformation between the representations is theoretically exact, numerical difficulties may occur if the transformation is ill-conditioned, in which case special methods must be used in order to guarantee that the solution is computationally reliable. The adverse effects of ill-conditioning are most likely to be observed when derived data, such as the computed curve of intersection of two surfaces, is transformed.

This paper describes the use of truncated singular value decomposition (TSVD) for the regularisation (stabilisation) of the transformation between the Bernstein and B-spline polynomial bases. It is assumed that the polynomials in both bases are of the same degree. The conversion between the power and Bernstein polynomial bases has been considered in [3], and it is noted that regularisation may also be required for this conversion, especially for polynomials of high degree.

The conversion between the Bézier and B-spline forms of a curve can be interpreted geometrically [12, pp. 239–240]. The transformation of a B-spline

Curves and Surfaces with Applications in CAGD
A. Le Méhauté, C. Rabut, and L. L. Schumaker (eds.), pp. 465–472.
Copyright © 1997 by Vanderbilt University Press, Nashville, TN.
ISBN 0-8265-1293-3.
All rights of reproduction in any form reserved.

curve segment into its Bézier form can be accomplished by (i) an algorithm due to Boehm [1], and (ii) the Oslo algorithm [2]. It is noted on page 152 of [8] that *'the generalized Oslo algorithm is stable in some cases ... . To determine whether the algorithm is unstable for given knot vectors seems like a more difficult task'.* The advantage of performing the basis transformation numerically by TSVD is that potential numerical difficulties are revealed and remedial action follows immediately if the discrete Picard condition is satisfied [4].

It is shown in [9] that it is necessary to consider the transformation between the Bernstein and B-spline forms of a curve in $\mathbb{R}^2$ as two separate transformations, one for the x-coordinates of the vertices of the control polygons of the Bézier and B-spline forms of the curve, and one for the y-coordinates. This yields two linear algebraic equations that have the same coefficient matrix, and thus it is desirable to consider the general linear algebraic equation

$$\mathbf{Ax} = \mathbf{b}, \tag{1}$$

where $x, b \in \mathbb{R}^m$ and the transformation matrix $\mathbf{A} \in \mathbb{R}^{m \times m}$ between the polynomial bases is non-singular and ill-conditioned. The degree of the polynomial is n, where $m = n + 1$.

§ 2. Regularisation and Truncated Singular Value Decomposition

It is shown in [9] that the condition number $\kappa(\mathbf{A})$ of $\mathbf{A}$ with respect to the spectral norm increases with the polynomial degree. Figure 1 shows the variation of the logarithm of the singular values s_i of $\mathbf{A}$ with the index i for polynomials of degree 12 ($m = 13$), and it is seen that $\kappa(\mathbf{A}) \approx 10^9$ for polynomials of this degree. It is assumed that the knots in the B-spline basis are uniformly spaced.

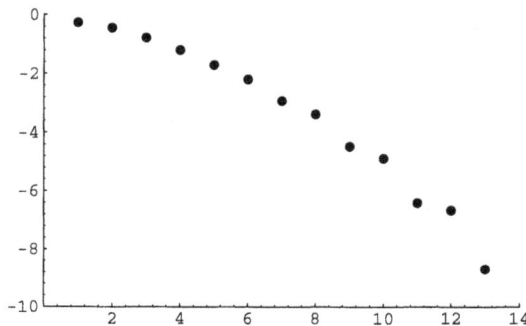

Fig. 1. The variation of $\log_{10} s_i$ with i.

It is important to note that even if $\kappa(\mathbf{A})$ is large, it does not follow that for the given right hand side vector $\mathbf{b}$, equation (1) is ill-conditioned [10].

This follows because $\kappa(\mathbf{A})$ is a worst case error measure, taken over all $\mathbf{b}$ and $\delta\mathbf{b}$. Moreover, TSVD can only regularise a particular class of ill-conditioned linear algebraic equation, specifically, the class for which the discrete Picard condition is satisfied; a large value of $\kappa(\mathbf{A})$ is a necessary but not sufficient condition for the correct use of TSVD. Conventional methods such as Gaussian elimination cannot be used to solve equation (1) when it is ill-conditioned, and special methods, called regularisation procedures, must be used. The basis transformation may, for example, be numerically stable for the x-components of the vertices of the control polygon, but unstable for the y-components. In this circumstance, equation (1) may be solved directly for the transformed x-components (or any stable algorithm may be used for this transformation), but regularisation must be applied to the equation that defines the transformation of the y-components.

The solution that is obtained from a regularisation procedure must satisfy two conditions: (i) It must be less sensitive to noise than is the simple (direct) solution of the equation that is being regularised, and (ii) it must be a good approximation to the exact solution. It is shown in [9] that the first condition is satisfied if the exact solution satisfies the discrete Picard condition. Attention is therefore restricted to the second condition.

The singular value decomposition of $\mathbf{A}$ is $\mathbf{USV}^T$ where $\mathbf{U}, \mathbf{S}, \mathbf{V} \in \mathbb{R}^{m \times m}$, $\mathbf{U}$ and $\mathbf{V}$ are orthogonal matrices and $\mathbf{S}$ is the diagonal matrix whose diagonal elements $s_{ii} = s_i$, $i = 1..m$, are the singular values of $\mathbf{A}$, arranged in non-increasing order. In TSVD the small singular values of $\mathbf{A}$ are deleted, and the minimum norm solution that minimises the residual $\|\mathbf{Ax} - \mathbf{b}\|$, where $\|\ \|$ denotes the 2-norm, is

$$\mathbf{x}(k) = \mathbf{V}\mathbf{S}_k^\dagger\mathbf{U}^T\mathbf{b}, \qquad (2)$$

where

$$\mathbf{S}_k^\dagger = \mathrm{diag}\left(\frac{1}{s_1}\ \frac{1}{s_2}\ \ldots\ \frac{1}{s_{k-1}}\ \frac{1}{s_k}\ 0\ 0 \ldots 0\ 0\right),$$

and the integer k is the regularisation parameter. The notation $\mathbf{x}(k)$ denotes that the solution is obtained when rank $\mathbf{A} = k$. The exact solution of eq. (1) is obtained when $k = m$, and in this case $\mathbf{S}_m^\dagger = \mathbf{S}^{-1}$.

Consider the error between the solution $\mathbf{x}(k)$ and the exact solution $\mathbf{x}(m)$. Equation (2) can be written in the form

$$\mathbf{x}(k) = \sum_{i=1}^{m} f_i \frac{c_i}{s_i}\mathbf{v}_i,$$

where $\mathbf{v}_i$ is the i'th column of $\mathbf{V}$, $\mathbf{c} = \{c_i\} = \mathbf{U}^T\mathbf{b}$ and the filter factor f_i satisfies

$$f_i = \begin{cases} 1 & \text{for} \quad i \leq k \\ 0 & \text{for} \quad i > k. \end{cases} \qquad (3)$$

The square of the relative error Δ_k is

$$\Delta_k^2 = \frac{||\mathbf{x}(k) - \mathbf{x}(m)||^2}{||\mathbf{x}(m)||^2} = \frac{\sum\limits_{i=1}^{m}(1 - f_i)\left(\dfrac{c_i}{s_i}\right)^2}{\sum\limits_{i=1}^{m}\left(\dfrac{c_i}{s_i}\right)^2}.$$

This equation shows that Δ_k^2 is the weighted average of the function $1 - f_i$. The requirement that Δ_k be small implies that large values of $1 - f_i$ be associated with small values of $|c_i|/s_i$, and vice-versa. It follows from eq. (3) that Δ_k is small if $|c_i|/s_i$ decreases, on average, as the index i increases. In particular, it is necessary that as $i \to m, |c_i|/s_i \to 0$. Since the singular values are arranged in non-increasing order, the components $|c_i|$ must decrease towards zero more rapidly than do the singular values s_i. This is the discrete Picard condition and its satisfaction is crucial to the success of TSVD.

When the right hand side of equation (1) is corrupted by noise, the ratio $|c_i + \delta c_i|/s_i$ must be examined. Assuming that the discrete Picard condition is satisfied and $|\delta c_i| \approx \epsilon, i = 1..m$, the components $|c_i + \delta c_i|/s_i$ typically decrease to a minimum level that is determined by the noise, and then increase. If the minimum occurs at the index $i = p + 1$, then

$$|c_i| >> |\delta c_i| \qquad \text{for} \quad i = 1..p, \tag{4a}$$

$$|c_i| \approx |\delta c_i| \qquad \text{for} \quad i = p + 1, \tag{4b}$$

$$|c_i| << |\delta c_i| \qquad \text{for} \quad i = p + 2 .. m. \tag{4c}$$

If the regularisation parameter is chosen such that

$$f_i = \begin{cases} 1 & \text{for} \quad i \leq p \\ 0 & \text{for} \quad i > p, \end{cases} \tag{5}$$

the noise corrupted terms are removed and the regularised solution of eq. (1) when $\mathbf{b}$ is perturbed to $\mathbf{b} + \delta\mathbf{b}$ is

$$\mathbf{x}(p) + \delta\mathbf{x}(p) = \sum_{i=1}^{p}\frac{(c_i + \delta c_i)}{s_i}\mathbf{v}_i \approx \sum_{i=1}^{p}\frac{c_i}{s_i}\mathbf{v}_i,$$

since equation (4a) is satisfied.

The square of the relative error in the presence of noise is

$$\Delta_k^2 = \frac{||(\mathbf{x}(k) + \delta\mathbf{x}(k)) - \mathbf{x}(m)||^2}{||\mathbf{x}(m)||^2} = \frac{\sum\limits_{i=1}^{m}\left[f_i\left(\dfrac{\delta c_i}{s_i}\right)^2 + (1 - f_i)\left(\dfrac{c_i}{s_i}\right)^2\right]}{\sum\limits_{i=1}^{m}\left(\dfrac{c_i}{s_i}\right)^2}. \tag{6}$$

On substituting equations (4a,b,c, 5) into equation (6) and noting that the discrete Picard condition is satisfied

$$\sum_{i=1}^{m} \frac{c_i}{s_i} \mathbf{v}_i \approx \sum_{i=1}^{p} \frac{c_i}{s_i} \mathbf{v}_i, \quad \sum_{i=p+1}^{m} \frac{c_i}{s_i} \mathbf{v}_i \approx \mathbf{0},$$

it follows that

$$\Delta_p = \left(\frac{\sum_{i=1}^{p} \left(\frac{\delta c_i}{s_i} \right)^2}{\sum_{i=1}^{m} \left(\frac{c_i}{s_i} \right)^2} \right)^{1/2} \approx \epsilon \left(\frac{\sum_{i=1}^{p} \frac{1}{s_i{}^2}}{\sum_{i=1}^{m} \left(\frac{c_i}{s_i} \right)^2} \right)^{1/2} \approx \frac{\epsilon}{|c_1|} \ll 1.$$

It is seen that if the discrete Picard condition is satisfied, there exists a value p of k such that the error between the solution from TSVD and the exact solution is small.

The success of TSVD is dependent upon the correct choice of k, even if the discrete Picard condition is satisfied. If the chosen value of k is too large, $\mathbf{x}(k)$ may be corrupted by noise, but if it is too small, a significant portion of the desired solution may be removed from $\mathbf{x}(k)$. It is shown in [5,6] that the L-curve is a good method for the calculation of the optimum value of k. The L-curve is a parametric plot of $\log \|\mathbf{x}(k)\|$ against $\log \|\mathbf{A}\mathbf{x}(k) - \mathbf{b}\|$. If the discrete points are joined up to form a continuous curve and three assumptions are made [5,6], the curve assumes the shape of an L. The optimum value of k is its value at the corner of the L because this is the best value that minimises the residual and the norm of the solution. If the discrete Picard condition is satisfied in both coordinate directions, an L-curve must be plotted for each coordinate because different degrees of smoothing may be required.

§3. Example

This section contains an example of the conversion of a curve segment of degree 12 from the Bernstein basis to the B-spline basis. A Bézier curve was generated and its B-spline form calculated. More details are in [9,11]. It must be noted that polynomials of degree 12 and even higher are used in commercial CAD systems [7]. The B-spline curve and the convex hull of its control polygon are shown in Figure 2. The vertices of the control polygon in this figure and Figures 3, 5 are denoted by •.

Zero mean Gaussian noise was added to $\mathbf{b}$ and its variance was adjusted so that $\|\mathbf{b}\|/\|\delta\mathbf{b}\| \approx 10^3$ for both the x- and y-coordinates of the vertices of the control polygon of the Bézier curve. Figure 3 shows the computed convex hull of the control polygon of the B-spline curve in this circumstance, and Figure 4 shows the associated B-spline curve. Although the convex hull of the control polygon has suffered gross distortion, the integrity of the curve is retained. It is seen that although the transformation of the convex hull between the two representations is unstable, the transformation of the curve is stable.

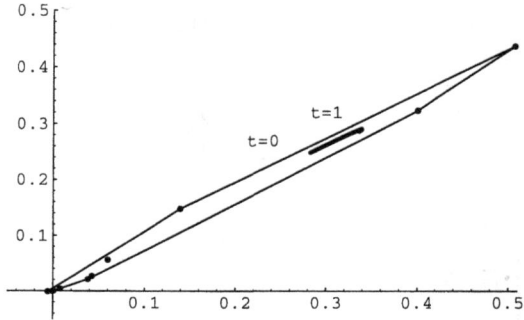

Fig. 2. The B-spline curve and convex hull of its control polygon.

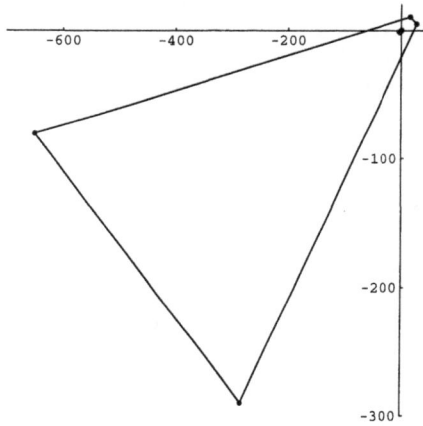

Fig. 3. The convex hull of the control polygon of the B-spline curve in the presence of noise.

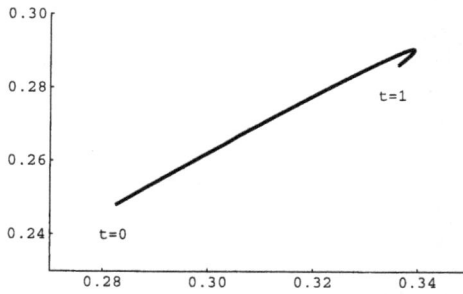

Fig. 4. The B-spline curve in the presence of noise when regularisation is not used.

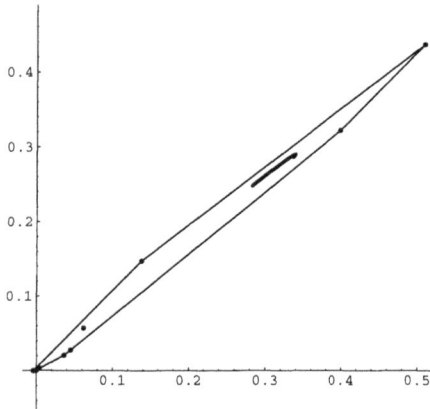

Fig. 5. The B-spline curve and convex hull of its control polygon after regularisation.

Figure 5 shows the solution that is obtained after regularisation by TSVD using the regularisation parameters that are determined by the L-curves. It is seen that Figures 2 and 5 are very similar, and in particular, the integrity of the transformed convex hull is restored.

Acknowledgments. The support of the Engineering and Physical Sciences Research Council (EPSRC) by an Advanced Research Fellowship, reference number B/93/AF/1703, is acknowledged.

References

1. Boehm, W., Generating the Bézier points of B-spline curves and surfaces, Computer-Aided Design **13** (1981), 365–366.

2. Cohen, E., T. Lyche, and R. F. Riesenfeld, Discrete B-splines and subdivision techniques in computer-aided geometric design and computer graphics, Comp. Graphics and Image Proc. **14** (1980), 87–111.

3. Farouki, R. T., On the stability of transformations between power and Bernstein polynomial forms, Comput. Aided Geom. Design **8** (1991), 29–36.

4. Hansen, P. C., The discrete Picard condition for discrete ill-posed problems, BIT **30** (1990), 658–672.

5. Hansen, P. C., Analysis of discrete ill-posed problems by means of the L-curve, SIAM Review **34** (1992), 561–580.

6. Hansen, P. C. and D. P. O'Leary, The use of the L-curve in the regularization of discrete ill-posed problems, SIAM J. Sci. Comput. **14** (1993), 1487–1503.

7. Lachance, M. A., Piecewise polynomial approximation of polynomial curves, Second Int. Conf. on Algorithms for Approximation, Royal Military College of Science, Shrivenham, UK, J. C. Mason, and M. G. Cox (eds.), (1988), 125–133.

8. Lyche, T., K. Mørken, and K. Strøm, Conversion between B-spline bases using the generalized Oslo algorithm, in *Knot Insertion and Deletion Algorithms for B-spline Curves and Surfaces*, R. N. Goldman, and T. Lyche (eds.), SIAM, 1993, 135–153.

9. Winkler, J. R., Polynomial basis conversion made stable by truncated singular value decomposition, submitted to Applied Mathematical Modelling, 1996.

10. Winkler, J. R., The condition number of a matrix and the stability of linear algebraic equations, submitted to Applied Mathematical Modelling, 1996.

11. Winkler, J. R., Tikhonov regularisation in standard form for polynomial basis conversion, submitted to Applied Mathematical Modelling, 1996.

12. Yamaguchi, F., *Curves and Surfaces in Computer Aided Geometric Design*, Springer-Verlag, 1988.

Joab R Winkler
Department of Computer Science
The University of Sheffield
Regent Court
211 Portobello Street
Sheffield S1 4DP, ENGLAND
j.winkler@dcs.shef.ac.uk

G² Continuous G-Splines: An Interpolation Property

Rainer Zeifang

Abstract. Based on the theory of G-splines developed by Höllig and Mögerle [8], a method for representing G^2 surfaces over nearly arbitrary rectangular meshes is presented. The only assumption made is that singular vertices, i.e. vertices where $n \neq 4$ edges meet, are separated by at least two edges. Our goal is to obtain surfaces of low degree extending the theory of parametric continuity (C^2), which will be used wherever possible in the mesh. We use Mögerle's diffeomorphisms [9] to define geometric continuity for singular edges, i.e. edges of the mesh ending in a singular vertex, which improve Hahn's approach [6,7] in the sense of lower degrees. To illustrate the flexibility of these spaces, we discuss the interpolation problem of extending a single arbitrary Bézier patch of at least degree 2 around a singular vertex. The results show that polynomial degrees $d = 6, 7$ are suitable depending on the type of singular vertices in the mesh. As we are looking for low degree solutions and using optimal diffeomorphisms, it is hard to separate the smoothness conditions to reduce the complexity of the linear systems. The systems arising from this problem are quite large. Computer algebra tools to investigate the solvability of these systems are discussed and evaluated.

§1. Introduction

Surface patches with rectangular domains are most widely used in CAD due to manufacturing reasons [6,12]. Using rectangular patches lead to problems, e.g. when oriented closed surfaces have to be modeled. Due to the Euler characteristic, only surfaces of genus 1 can be modeled using a regular rectangular mesh, where 4 edges meet at all vertices [14]. To overcome this problem for arbitrary closed surfaces, several approaches are possible, either using trimmed rectangular patches, or degenerate patches, where e.g. one isoparametric boundary curve vanishes [1], or untrimmed rectangular patches and allowing *singular vertices*, i.e. vertices where $n \neq 4$ edges meet, and singular parametrizations [10,11].

Curves and Surfaces with Applications in CAGD
A. Le Méhauté, C. Rabut, and L. L. Schumaker (eds.), pp. 473–480.

Copyright © 1997 by Vanderbilt University Press, Nashville, TN.
ISBN 0-8265-1293-3.
All rights of reproduction in any form reserved.

Geometric continuity (G^r) is a method to overcome the problems of singular parametrizations and trimming and is used in this paper.

Based on the concept of geometric continuity, G-spline spaces are introduced in Section 2 as an appropriate tool to model closed surfaces following [8] and [9]. This approach generalizes a method by Goodman [4], and improves the diffeomorphisms used in the G^2 case by Hahn [6]. Analyzing the systems of smoothness constraints requires computer algebra, therefore a Gaussian elimination algorithm suited for those systems is discussed in Section 3. In Section 4 we focus on G^2 G-spline spaces and the interpolation property, which is essential for practical use.

§2. G-Splines

Looking at tensor product splines, a topological connection between surface patches is given by double indices (i, j). A similar method for identifying neighboring patches in arbitrary meshes has to be employed. To this end we label the patches by $F := \{0, 1, \ldots, \sharp F - 1\}$, the boundary curves by $E := \{0, 1, \ldots \sharp E - 1\}$, and the vertices by $C := \{0, 1, \ldots, \sharp C - 1\}$. Then the topological structure of the network can be described by mappings

$$g : F \times \{1, 2, 3, 4\} \mapsto E, \qquad h : F \times \{1, 2, 3, 4\} \mapsto C.$$

For each $f \in F$, $g(f, \nu)$ are the labels of the boundary curves and $h(f, \nu)$ are the labels of the vertices of the patch f, ordered counterclockwise with respect to the outward surface normals. Thus, two patches f and $\tilde{f}$ have a common boundary curve e if

$$e = g(f, \nu) = g(\tilde{f}, \tilde{\nu}) \qquad (1)$$

for some indices $1 \leq \nu, \tilde{\nu} \leq 4$.

Considering two neighboring polynomial patches $f, \tilde{f}$ in Bézier representation

$$(u, v) \mapsto p(u, v, f) := \sum_{i=0}^{d} \sum_{j=0}^{d} b_{i,j}^{f} B_i^d(u) B_j^d(v), \qquad (u, v) \in [0, 1]^2 \qquad (2)$$

of common degree d and setting $q := p(\cdot, f), \tilde{q} := p(\cdot, \tilde{f})$, we can assume that they have a common boundary curve

$$q([0, 1], 0) = \tilde{q}(0, [0, 1]), \qquad (3)$$

which can be achieved by a rotation of the parameter square. As is well known, these two patches are *geometric continuous of order* r if there exists a diffeomorphism

$$\phi(\cdot, e) : \mathbb{R}^2 \to \mathbb{R}^2, \qquad \phi(u, 0, e) = (u, 0),$$

such that the transversal (with respect to the common boundary curve) derivatives up to order r of the parametrizations $q \circ \phi$ and $\tilde{q}$ are continuous along the common boundary curve. This means that

$$\left(D_v^l q(\phi(u, v, e))\right)_{|(u,v)=(t,0)} = (-1)^l \left(D_u^l \tilde{q}(u, v)\right)_{|(u,v)=(0,t)}, \tag{4}$$

$0 \le t \le 1, 0 \le l \le r$. The equations (4) are linear in q and $\tilde{q}$ for any choice of $\phi(\cdot, e)$ and hold simultaneously in each component. We call these equations G^r *smoothness constraints*, and for later reference we abbreviate them by

$$L(q, \tilde{q}, \phi(\cdot, e), l) = 0 \qquad 0 \le l \le r. \tag{5}$$

Collecting Bézier coefficients $b_{i,j}^f$ of f into a vector

$$\mathbf{b}^\nu := (b_{0,0}^\nu, b_{1,0}^\nu, \dots, b_{d,0}^\nu, b_{0,1}^\nu, \dots, b_{d,1}^\nu, \dots, b_{d,d}^\nu)',$$

G^r constraints are given as linear constraints in matrix form

$$A_0(r, n)\mathbf{b}^f + A_1(r, n)\mathbf{b}^{\tilde{f}} = 0,$$

where n is the order of the common edge.

With the above definitions we define the G-spline space as in [8].

Definition 1. *Given diffeormorphisms $\phi(\cdot, e)$ for each edge of a mesh defined by a mapping g a G-spline is a function*

$$s : \Omega := [0, 1]^2 \times F \to \mathbb{R}$$

which satisfies the G^r smoothness constraints (5) for any interior boundary curve e related to f and $\tilde{f}$ by (1) and (3). The linear space of all such functions s is denoted by $\mathcal{S}(d, r, g, \phi)$. A G-spline surface has a parametrization of the form (2) with coordinate functions in $\mathcal{S}$.

We use the same terminology as in [14], so that the *order* of a vertex is the number of edges meeting at these vertex. A vertex with order $n \ne 4$ is called *singular*. An edge (patch) containing a singular vertex is called *singular edge (patch)*. A nonsingular edge (patch) is called *regular*. A vertex of order 4 on a singular edge is called a *G-vertex* and we only consider meshes, where every G-vertex is contained in exactly one singular edge, so called *separated meshes or meshes with isolated singular vertices*. Throughout this paper g always describes a mesh with isolated singular vertices.

To keep everything as simple as possible for regular patches of the mesh, we want to have C^r constraints there, meaning $\phi = id$. To enable regular parametrizations along singular edges, ϕ has to fulfill some *consistency conditions* [7,9]. A possible choice of ϕ for $\mathcal{S}(d, 1, g, \phi)$ is given in [8]. For $\mathcal{S}(d, 2, g, \phi)$ we choose Mögerle's diffeomorphisms [9]

$$\phi(u, v, e) := \phi_n(u, v), \qquad \text{if } e \text{ singular of order } n,$$

where

$$\phi_n(u,v) := \begin{cases} (u - 3(1-v)^3 u^2, \\ v + (1-v)^3 u - 3/2 \left((1-v)^6 + (1-v)^5\right) u^2\right), & n = 3, \\ (u,v), & n = 4, \\ (u, v + \lambda_n(1-v)^3 u - 3\lambda_n^2/2(1-v)^3 u^2), & n \geq 5, \end{cases} \quad (6)$$

and $\lambda_n = -2\cos(2\pi/n) = -\omega - \omega^{n-1}$. The choice of λ_n is motivated by symmetry considerations and $\omega = \exp(2\pi i/n)$. These diffeomorphisms are different from Hahn's diffeomorphisms [7]

$$\phi_{H,n}(u,v) := \begin{cases} (u, v + \lambda_n(1-v)^3(1 + 3v + 6v^2 + 10v^3)u) & n \neq 4, \\ (u,v), & n = 4, \end{cases} \quad (7)$$

and allow to work with patches of lower degree.

G^1 constraints are calculated manually in [14] for arbitrary d. G^2 constraints defined by ϕ_n are calculated in [15] by computer algebra tools. We observe that the boundary conditions

$$\sum_{j=0}^{d}(-1)^j \binom{d}{j} b_{j,0}^0 = 0, \qquad \sum_{j=1}^{d}(-1)^j \binom{d-1}{j} b_{j,0}^0 = 0$$

for singular edges lead to a degree reduction to $d - 2$ along the boundary. For Hahn's diffeomorphisms (7) the situation is much worse, as they lead to a degree reduction to $d - 5$.

§3. Gaussian Elimination for Computer Algebra

As Gaussian elimination is our basic tool for analyzing G-spline spaces, we look at it in more detail especially in connection with computer algebra.

Gaussian elimination is well known as a stable method for the numerical solution of systems of linear equations, when *pivoting* is applied. Following [3], Gaussian elimination is used to factor a $m \times n$ matrix A in the form $PAQ = LU$, where P is a $m \times m$ and Q a $n \times n$ permutation matrix, L is an unit lower triangular matrix and U is an upper triangular matrix. The basic idea of this factorization is the *Gauss transformation*. For a vector $x = (x_1, \ldots, x_m)'$ the components $x_{k+1}, \ldots, x_m$ are annihalated by the Gauss transformation $M_k = I - \tau e'_k$, if $x_k \neq 0$, where

$$\tau' = (\underbrace{0, \ldots, 0}_{k}, \tau_{k+1}, \ldots, \tau_m), \qquad \tau_i = \frac{x_i}{x_k}, \quad i = k+1, \ldots, m, \quad (8)$$

so that $M_k x_i = x_i$, $i = 1, \ldots, k$ and $M_k x_i = 0$, $i = k+1, \ldots, m$.

Starting with $A^{(0)} = A$, a sequence $A^{(k)}$, $k = 0, \ldots, m - 1$, is generated, where the Gauss transformation M_k is applied to $A^{(k-1)}$. To get a numerically stable algorithm, rows E_k and columns F_k are exchanged, so that

$$|(E_k A^{(k-1)} F_k)|_{kk} = \max_{k \leq i \leq n, k \leq j \leq m} |A_{ij}^{(k-1)}|. \quad (9)$$

This method is called *pivoting*. Altogether we obtain

$$U = A^{(m-1)} = M_{m-1}E_{m-1}\cdots M_1 E_1 A F_1 \cdots F_{m-1} = L^{-1}PAQ.$$

In computer algebra one wants to avoid division, i.e. working in the field $\mathbb{Q}$. As we will see in Section 4, we need a version of Gaussian elimination which works in $R := \mathbb{Z}[\omega]/p(\omega)$, where p is the n-th cyclotomic polynomial, i.e. $\omega = \exp(2\pi i/n)$.

In this case we use the Gauss transformation without division

$$M_k^R = \text{diag}(\sigma) - \tau e_k',\tag{10}$$

where

$$\tau' = (\underbrace{0,\ldots,0}_{k}, x_{k+1},\ldots,x_m), \qquad \sigma' = (\underbrace{1,\ldots,1}_{k}, \underbrace{x_k,\ldots,x_k}_{m-k})$$

and $\text{diag}(\sigma)$ is a $m \times m$ diagonal matrix with σ as its diagonal. In this case pivoting is used to keep the system as simple as possible. Setting

$$d_{\min} := \min_{k\leq i\leq m, k\leq j\leq n} \{\deg(A_{ij}^{(k-1)}) : A_{ij}^{(k-1)} \neq 0\},$$

where deg is the degree function for polynomials , we achieve optimal pivoting in $\mathbb{Z}[\omega]/p(\omega)$ by

$$(E_k^R A^{(k-1)} F_k^R)_{kk} = \arg\min_{k\leq i\leq m, k\leq j\leq n} \{\text{lc} \,|A_{ij}^{(k-1)}| : \deg(A_{ij}^{(k-1)}) = d_{\min}\},\tag{11}$$

where lc is the leading coefficient function for polynomials.

§4. Extension Property for G^2 G-Spline Spaces

Analyzing properties of G^2 G-spline spaces $\mathcal{S}(d,2,g,\phi_n)$ over separated meshes locally is possible as long as a decomposition in linear independent subspaces exists. For $d \geq 5$

$$\mathcal{S}(d,2,g,\phi_n) = \bigoplus_{c=1}^{\sharp C'} \mathcal{S}_c(d,2,g,\phi_n),\tag{12}$$

where $C' \subset C$ is the set of all regular and singular vertices in the mesh, i.e. all vertices with the exception of G-vertices. Using $d = 5$ for regular patches the decomposition is very simple. If c is regular the support of $\mathcal{S}$ are those patches $f_0, \ldots, f_3$ surrounding the vertex. For singular vertices $\mathcal{S}$ consists of n patches $f_0, \ldots, f_{n-1}$ containing a vertex of order n as its center and all patches sharing an edge or a vertex with them, i.e. in total $4n$ patches.

There is no overlap in nonzero Bézier coefficients as in both cases the 3 outermost rows or columns of control points are zero. Moreover, it is obvious that they span the whole space $\mathcal{S}(d,2,g,\phi_n)$

So it is sufficient to analyze the subspaces consisting of n patches with a common center. We call the mapping defining these meshes g_n, so that the G-spline space is referred to as $\mathcal{S}(d,2,g_n,\phi_n)$. Beside dimensions and interpolation problems as discussed in [9,14] an indicator for the flexibility of G-spline spaces is defined by

Problem 2. *Given*
- *an arbitrary Bézier patch p of degree $d' \leq d$,*
- *a G-spline space $\mathcal{S}(d, r, g_n, \phi)$ consisting of n patches, where*
  - *$d(0) = d'$ and $d(f) = d$ for $f = 1, \ldots, n - 1$,*
  - *$r = 2$,*
  - *ϕ as in (6),*

we are looking for the minimal degree d, for which a G-spline function $q \in \mathcal{S}(d, r, g_n, \phi)$ exists where $q(\cdot, 0) = p$, meaning p is extendable.

The minimal degree d' to consider is $d' = 2$. The difficulties in these cases arise as the system related to the problem is not cyclic. Starting with a patch f_0 of degree $d' = 2$ with Bézier coefficients

$$\mathbf{a} := (a_{0,0}, a_{1,0}, a_{2,0}, a_{0,1}, a_{1,1}, a_{2,2}, a_{0,2}, a_{1,2}, a_{2,2})'$$

one has to apply degree elevation up to degree d [2] which is a linear process and can be written in matrix form as

$$\mathbf{b}^0 = G_{d'}^d \mathbf{a}. \tag{13}$$

Therefore the system is given by

$$
\begin{pmatrix}
A_1 & 0 & 0 & \cdots & & 0 \\
A_0 & A_1 & 0 & & & \\
0 & A_0 & A_1 & & & \\
\vdots & \ddots & \ddots & \ddots & & \vdots \\
0 & \cdots & 0 & A_0 & A_1 & \\
0 & 0 & \cdots & 0 & A_0 &
\end{pmatrix}
\begin{pmatrix}
\mathbf{b}^1 \\
\mathbf{b}^2 \\
\mathbf{b}^3 \\
\vdots \\
\mathbf{b}^{n-2} \\
\mathbf{b}^{n-1}
\end{pmatrix}
=
\begin{pmatrix}
A_0 G_{d'}^d \\
0 \\
0 \\
\vdots \\
0 \\
A_1 G_{d'}^d
\end{pmatrix}
\mathbf{a}, \tag{14}
$$

which is not a cyclic system, and cannot be reduced by Fourier transforms, but contains entries in $\mathbb{Z}[\omega]/p(\omega)$. This makes it difficult to solve it for arbitrary n. But of course a solution for all practical cases is possible. A system like (14) in the form $\mathbf{A}x = \mathbf{B}\mathbf{a}$ is solvable for arbirary $\mathbf{a}$, if $\mathrm{rank}(\mathbf{A}) = \mathrm{rank}(\mathbf{A}, \mathbf{B})$. Using this fact we are able to prove

Theorem 3. *The solution of Problem 2 for $d' = 2$, $r = 2$ and*
- *$n = 3$ is $d = 7$,*
- *$5 \leq n \leq 10$ is $d = 7$.*
- *For $d' = 2$, $n = 3$ a quadratic patch is extendable for $d = 5$, if*

$$a_{0,0} - 2a_{1,0} + a_{2,0} = 0, \tag{15a}$$

$$25a_{0,0} - 47a_{1,0} + 22a_{2,0} = 0, \tag{15b}$$

$$a_{0,0} - 2a_{0,1} + a_{0,2} = 0, \tag{15c}$$

$$25a_{0,0} - 47a_{0,1} + 22a_{0,2} = 0, \tag{15d}$$

$$-203a_{1,0} + 78a_{2,0} + 203a_{0,1}$$
$$- 18a_{2,1} - 78a_{0,2} + 18a_{1,2} = 0. \tag{15e}$$

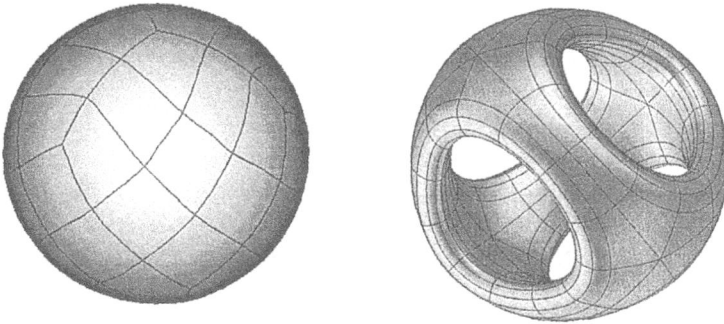

Fig. 1. Interpolation of data on a sphere and a closed surface of genus 3.

- For $r = 2, n = 3$ a quadratic patch is extendable for $d = 6$, if (15a) and (15c) hold.
- For $r = 2, 5 \leq n \leq 10$ a quadratic patch is extendable for $d = 5$, if (15a) and (15c) hold.

Proof: Just apply Gaussian elimination (11) to the system (14). Equations (15) belong to rows, which are zero in $\mathbf{U}$, where $\mathbf{PAQ} = \mathbf{LU}$, but are not zero in $\tilde{\mathbf{U}}$ in the extended system, $\mathbf{P(AB)Q} = \mathbf{L\tilde{U}}$. Gaussian elimination is implemented using MATHEMATICA [13] and can be found in [15]. ∎

§5. Examples

The examples in this section were constructed by solving interpolation problems using local bases [15] and minimizing the distance of control points to a given initial representation. This initial representation of the surface is not G^2, but C^0. $d = 7$ is used for singular patches and $d = 5$ for regular patches. For algebraic surfaces, we calculate this initial representation by using standard least squares approximation techniques of the given algebraic equation. For the sphere in Figure 1 we achieve mean curvature values $0.8 \leq H \leq 1.5$ and Gaussian curvature values $0.5 \leq K \leq 2.0$. For the other example in Figure 1 where only point data are given, first a C^2 interpolation of the regular part of the mesh is calculated using degree $d = 3$. For singular patches tangent data are estimated using standard techniques and patches of degree $d = 3$ are built up. This initial representation is degree elevated to get the initial representation.

Acknowledgments. I would like to thank Professor Klaus Höllig for many helpful hints and comments on this paper. Furthermore I would like to thank Matthias Leber who supplied the grids for the surfaces in Figure 1.

References

1. Bézier, P., *The Mathematical Basis of the UNISURF CAD System*, Butterworth & Co Ltd, London, 1986.

2. Farin, G., *Curves and Surfaces for Computer Aided Geometric Design*, Academic Press, London, 1st edition, 1988.

3. Golub, G. H. and C. F. van Loan, *Matrix Computations*, The John Hopkins University Press, 2nd edition, 1989.

4. Goodman, T. N. T., Closed biquadratic surfaces, Constr. Approx. **7** (1991), 149 – 160.

5. Grandine, T. A., An iterative method for computing multivariate C^1 piecewise polynomial interpolants, Comput. Aided Geom. Design **4** (1987), 307–319.

6. Hahn, J. M., Filling polygonal holes with rectangular patches, in *Theory and Practice of Geometric Modeling*, W. Straßer and H.-P. Seidel (eds.), Springer, 1989, 81–91.

7. Hahn, J. M., Geometric continuous patch complexes, Comput. Aided Geom. Design **6** (1989), 55–67.

8. Höllig, K and H. Mögerle, G-Splines, Comput. Aided Geom. Design **7** (1990), 197–207.

9. Mögerle, H., G-Splines höherer Ordnung, PhD thesis, University of Stuttgart, 1992.

10. Reif, U., Neue Aspekte in der Theorie der Freiformflächen beliebiger Topologie, PhD thesis, University of Stuttgart 1993.

11. Reif, U., A refinable space of spline surfaces of arbitrary topological genus, to appear in J. Approx. Theory.

12. Sarraga, R. F., G^1 interpolation of generally unrestricted cubic Bézier curves, Comput. Aided Geom. Design **4** (1987), 23–39.

13. Wolfram, S., *Mathematica. A System for Doing Mathematics by Computer*, Addison-Wesley Publishing Company, Inc., 2nd edition, 1991.

14. Zeifang, R., Interpolation with g-splines in *Mathematical Methods in Computer Aided Geometric Design II*, T. Lyche and L. Schumaker (eds.), Academic Press, New York, 1992, 615–626.

15. Zeifang, R., Interpolationsmethoden mit g-Splines, PhD thesis, University of Stuttgart, 1994.

Rainer Zeifang
CoCreate Software GmbH
Posener Straße 1
71065 Sindelfingen, GERMANY
rainer_zeifang@hp.com

www.ingramcontent.com/pod-product-compliance
Lightning Source LLC
Chambersburg PA
CBHW021428180326
41458CB00001B/171